Distillation and Hydrocarbon Processing Practices

DISTILLATION & HYDROCARBON PROCESSING PRACTICES

Ashis Nag

PennWell®

Disclaimer. The recommendations, advice, descriptions, and the methods in this book are presented solely for educational purposes. The author and publisher assume no liability whatsoever for any loss or damage that results from the use of any of the material in this book. Use of the material in this book is solely at the risk of the user.

PennWell Corporation
1421 South Sheridan Road
Tulsa, Oklahoma 74112-6600 USA

800.752.9764
+1.918.831.9421
sales@pennwell.com
www.pennwellbooks.com
www.pennwell.com

Marketing Manager: Sarah De Vos
National Account Executive: Barbara McGee Coons

Director: Mary McGee
Managing Editor: Stephen Hill
Production Manager: Sheila Brock
Production Editor: Tony Quinn
Book Designer: Susan E. Ormston
Cover Designer: Karla Womack

Library of Congress Cataloging-in-Publication Data

Nag, Ashis.
Distillation and hydrocarbon processing practices / Ashis Nag.
pages cm
Includes bibliographical references and index.
ISBN 978-1-59370-343-1
1. Distillation, Fractional. 2. Extractive distillation. 3. Hydrocarbons. 4. Petroleum refineries. I. Title.
QD526.N34 2015
665.5'32--dc23

2014044494

Printed in the United States of America

1 2 3 4 5 20 19 18 17 16

Contents

Preface

This book is intended primarily for practicing process engineers and students of chemical engineering.

I have noticed that some process principles, their details, and design considerations are often not well-documented for use by the engineers engaged in process design and troubleshooting. My role in both process engineering as well as plant operation made me realize that such pieces of knowledge lie scattered over different textbooks, research papers, design codes, safety codes, and individual notes of experienced engineers and operators.

This work is my humble attempt to compile and elucidate the practices in the hydrocarbon industry and their fundamental principles in a concise form. The advanced practices in distillation are explained and illustrated with the help of examples, using process simulations along with the basic principles. The practical examples and design guidelines are based on all the knowledge and experience I have gathered in my professional career and from various other resources, including personal interaction with masters in the field.

I sincerely hope that this work shall add value to your academic and professional pursuits.

Acknowledgments

I'd like to express my sincere thanks to Indian Oil Corporation Limited for providing the opportunity to work in different process plants of its several refineries.

I am indebted to my friend Mr. Subhabrata Ray, a faculty member in the Department of Chemical Engineering, Indian Institute of Technology, Kharagpur, India, who inspired me to write this book based on practical experiences supplemented by the underlying process principles and case study examples.

I also wish to thank my colleagues Rakesh Gagat, Prabhat Ranjan Choudhary, Mukesh Kumar Sharma, and S. G. Venkatesh for their support through numerous useful discussions and suggestions. The contribution of Mr. M. M. Khan, my colleague, in Chapters 7 and 10 is also gratefully acknowledged.

I wish to thank PennWell's Professional Education Products for agreeing to publish the work, and Mr. Stephen Hill, acquisitions editor, and Mr. Tony Quinn, production editor, for their continued association and support.

I cannot afford to forget my wife Indrani and my daughter Shreya for their unconditional support and continuous encouragement.

I shall be pleased to hear any comments or responses of any reader at my personal email nag.ashis@gmail.com.

1 Distillation Practices

1.1 Distillation Fundamentals

Distillation, a well-known and much practiced process, separates products from a homogeneous mixture based on their boiling point difference or relative volatility. In a typical distillation tower, which contains a reboiler and a condenser, heat is applied in the reboiler to vaporize the bottom liquid and condense it to produce overhead product and generate reflux and high-boiling product from the bottom. At a specified reflux ratio, a minimum number of trays is required for desired separation. A minimum reflux is always required to attain the specified separation. The requirements for minimum stages and minimum reflux are discussed in the section Short-Cut (FUG or FUEM) Procedure.

Distillation is the main unit operation in a refinery. Distillation alone consumes around 40% of a refinery's energy requirement. Thus, the high energy needed to perform distillation processes has become a major concern. As a result, process designers have begun putting more emphasis on energy-efficient design and retrofitting in addition to the recovery of products of a specified purity and yield.

Basic definitions

For a two-component system with components A and B:

Relative volatility $\alpha = Y(1 - X)/X(1 - Y)$ or $Y = \alpha X/(1 + (\alpha - 1)X)$, where X and Y are the liquid and vapor phase mole fractions respectively for the more volatile component.

$\alpha = K_A/K_B = P_A/P_B$ when ideal in vapor and liquid phase, where K_A, K_B are equilibrium constants and P_A, P_B are the vapor pressures of components A and B.

The stages and reflux ratio can be estimated by the Fenske–Underwood–Gilliland (FUG) or Fenske–Underwood–Erbar-Maddox (FUEM) correlations. The correlations can estimate the number of stages and reflux ratio required for a specified separation. A rigorous calculation using a steady state simulator can also be used to determine the number of stages and reflux ratio.

Short-cut (FUG or FUEM) Procedure

The Fenske equation calculates the minimum number of stages required at total reflux. Thus, the Fenske equation is used for estimating N_m (minimum number of stages) as

$N_m = \log[X_{LK}/X_{HK}]_D[X_{HK}/X_{LK}]_B/\log \alpha_{avg.\,LK-HK}$, where $\alpha_{avg.\,LK-HK} = (\alpha_{top} \times \alpha_{bottom})^{1/2}$ or $\alpha_{avg.\,LK-HK} = (\alpha_{top} \times \alpha_{feed} \times \alpha_{bottom})^{1/3}$, where $[X_{LK}/X_{HK}]_D$ and $[X_{HK}/X_{LK}]_B$ are the ratio of concentrations (mole fraction) of light key to heavy key in the overhead and heavy key to light key in the bottom product, respectively. $\alpha_{avg.\,LK-HK}$ is the relative volatility of the light key with respect to the heavy key. $\alpha_{avg.\,LK-HK}$ can also be taken as the arithmetic mean of α_{top} and α_{bottom} or can simply be evaluated at feed tray temperature.

The Underwood equation calculates the minimum reflux ratio required for a given separation as

$$\sum_{i=1}^{i=n} \frac{\alpha_i X_{iF}}{\alpha_i - \theta} = 1 - q \qquad \text{(Equation 1.1.1)}$$

where q depends on the condition of the feed (defined in section 1.1.2) and θ root of the Underwood equation and derived from the above equation.

$$\sum_{i=1}^{i=n} \frac{\alpha_i X_{iD}}{\alpha_i - \theta} = R_m + 1 \qquad \text{(Equation 1.1.2)}$$

X_{iF} is the mole fraction of the ith component in the feed, X_{iD} is the mole fraction of the ith component in the overhead distillate, R_m is the minimum reflux, and n is the total number of components in the feed.

Feed tray location using the Kirkbride equation:

$$\log(N/M) = 0.206\log\left[\left(\frac{B.XF_{HK}}{D.XF_{LK}}\right)\left(\frac{XB_{LK}}{XD_{HK}}\right)^2\right] \qquad \text{(Equation 1.1.3)}$$

In this equation, N and M are the number of stages above feed and below feed, respectively. D and B are molar flows of the top and bottom products, respectively. XF_{HK}, XF_{LK} are concentrations (mole fractions) of heavy and light key in the feed. XB_{LK} and XD_{HK} are concentrations (mole fractions) of light and heavy key in the bottom product and top product. A Gilliland chart, an Erbar-Maddox correlation (chart), or an Eduljee correlation calculates the number of stages required at actual reflux based on the data of minimum reflux and minimum number of stages calculated using the above equations of the FUEM correlation.

The method outlined above is popularly known as the Short-cut Method or FUG/FUEM method.

1.1.1 Reflux ratio, heat and material balance

Reflux ratio

Reflux ratio is the ratio of reflux flow to distillate product, or *R*:*D*. Actual reflux rate is derived from the minimum reflux rate required for a given separation. Distillation columns are typically designed to operate at 1.2–1.5 times the minimum reflux ratio (R_m) based on economic considerations. A higher reflux ratio will call for a lower number of stages to achieve the desired separation but would need higher reboiler/condenser duties. As a result, the tower's height will be lower, resulting in lower capital cost. However, the operating cost will increase due to higher reboiler/condenser duties. Additionally, with an increase in reflux ratio, the diameter of a column may increase, adding to capital cost.

Heat and material balance

Fig. 1.1.1 demonstrates the heat and material balance of a distillation column.

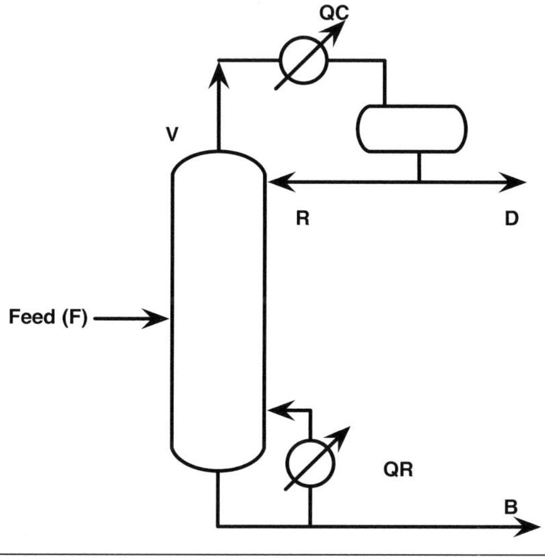

Fig. 1.1.1 Heat and material balance of a column

From Fig. 1.1.1, overall material balance can be written as $F = D + B$, and the component balance equation can be written as $FX_F = DX_D + BX_B$, where X_F is the liquid phase mole fraction of one component in the feed, X_D is the liquid phase mole fraction of the same component in the overhead product (D), X_B is the liquid phase mole fraction of the same component in the bottom product (B), and F is molar flow of feed.

Overall heat balance is written as $Fh_F = Dh_D + Bh_B + Q_C - Q_R$, where h_F is the molar enthalpy of feed, h_D is the molar enthalpy of overhead liquid product, h_B is the molar enthalpy of bottom liquid product, Q_C is the condenser duty, and Q_R the reboiler duty.

Q_C is a function of reflux rate and overhead product rate, column top temperature, and reflux drum temperature. The vapor leaving the column top is $V = R + D$: Q_C, which can be estimated as $Q_C = (R + D)(H_{\text{top}} - h_{drum})$, where H_{top} is the enthalpy of top vapor and h_{drum} is the enthalpy of drum liquid. Once Q_C is calculated, Q_R can be calculated from the overall energy balance equation: $Q_F = Q_D + Q_B + Q_C - Q_R$, where Q_F is equal to Fh_F, Q_D is equal to Dh_D, and Q_B is equal to Bh_B. Thus, all the parameters, including condenser duty and reboiler duty, can be estimated once the reflux rate (reflux ratio) is estimated for a given number of stages.

Reboiler duty can be also lowered by introducing heat at any location in the column below the feed tray. If reboiler duty is inadequate then additional heat can be injected above the reboiler (but below the feed tray) of the column. The heat supplemented below the feed tray is often called a staged reboiler (inter-reboiler).

No heat should be injected above the feed tray in a distillation column (except for side strippers in some cases) so as not to raise the condenser duty. Fig. 1.1.2 demonstrates heat material balance with an inter-reboiler.

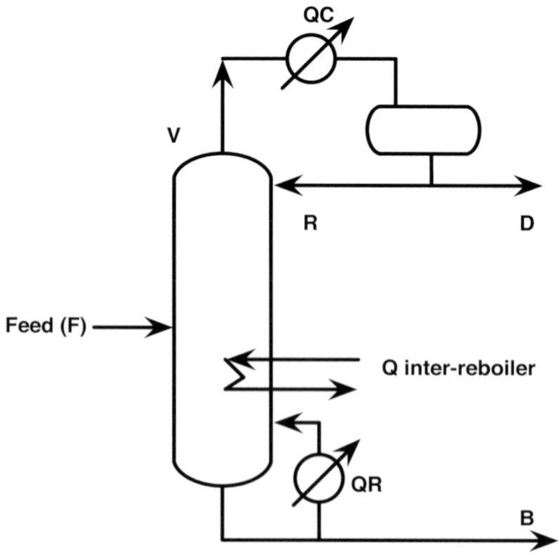

Fig. 1.1.2 Heat material balance with an inter-reboiler

Heat balance with an inter-reboiler (Fig. 1.1.2) can be written as $Q_R + Q_F + Q$ inter-reboiler $= Q_D + Q_B + Q_C$. Thus, reboiler duty with an inter-reboiler will be less than reboiler duty without an inter-reboiler. Similarly, when an intercooler is added above the feed tray, the heat balance can be written as $Q_R + Q_F = Q_D + Q_B + Q_C + Q$ intercooler. Thus the condenser duty with an intercooler will be less than the condenser duty without an intercooler.

1.1.2 McCabe–Thiele method

The McCabe–Thiele method provides insight for the analysis of many distillation problems. The basic assumption of the McCabe–Thiele is equal molar heat of vaporization and condensation of components, and it reduces to equal molar liquid–vapor rates in all stages in the sections above the feed stage and equal molar liquid and vapor flow rates in all the stages below the feed stage. Fig. 1.1.3 shows the enrichment section and stripping section envelopes used to derive the enrichment section equation and stripping section equation.

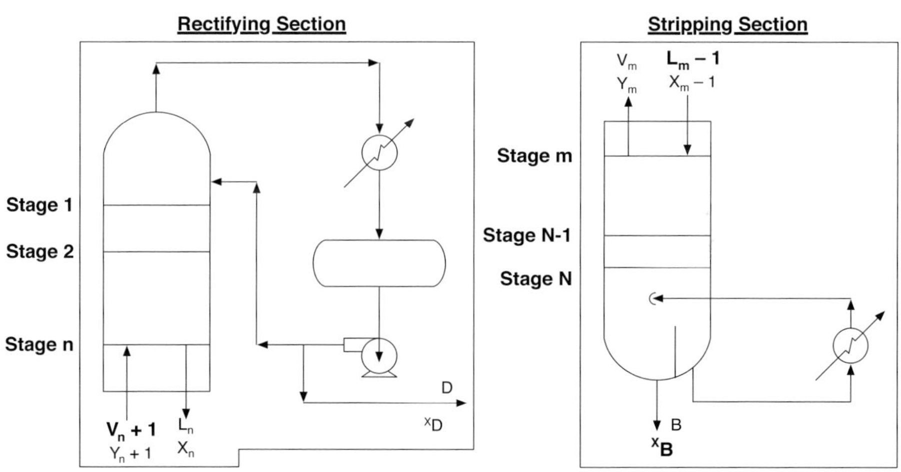

Fig. 1.1.3 Envelopes of the rectification and stripping section

Enrichment line equation

The material balance equation on the enrichment section is $V = L + D$.

$Y_{n+1} = (L/(L + D))X_n + (D/(L + D))X_D$ is the equation of the enrichment line for n^{th} stage.

The enrichment line equation can be rewritten as $Y_{n+1} = (R/(R + 1))X_n + (1/(R+1))X_D$ with substitution of $R = L/D$ in the equation. Thus, $(R/(R+1))$ is the slope and $(1/(R+1))X_D$ is the intercept of the enrichment line and can be plotted on an X-Y chart.

Stripping line equation

$$Ym = (L'/V')X_{m-1} - (B/V')X_B$$

q line equation

$$Y = (q/(q-1))X + (1/(q-1))X_F$$

$q/(q-1)$ is the slope of the line, where q = (heat required to vaporize one mole of feed/molar latent heat of vaporization of the feed). Thus, $L' = L + qF$, $V = V' + (1-q)F$, where F is the molar flow rate of the feed and X_F is the mole fraction of the more volatile component in the feed. V, L are the molar flow rates of vapor and liquid in the rectification section, and V', L' are the molar flow rates of vapor and liquid respectively in the stripping section. Table 1.1.1 gives the values of q with the thermal conditions of the feed.

Table 1.1.1 Values of *q* with thermal condition of feed

Condition	Value of *q*
BP liquid	1.0
DP vapor	0
Subcooled	>1.0
Superheated vapor	<1.0
Part vapor	Mole fraction of liquid

BP liquid is liquid at boiling point, which is also known as saturated liquid. *DP* is vapor at dew point, or saturated vapor. Fig. 1.1.4 shows q lines with different slopes in benzene–toluene system.

To provide an example, Example Problem 1.1.1 is solved first with the short-cut method (FUG/FUEM) and then with the McCabe–Thiele method.

Example Problem 1.1.1

Feed = 200 kmoles/hr (50:50 mole:mole, benzene:toluene), benzene purity at top product = 95 mole %. Toluene purity in bottom product is 90 mole %. Feed is 25% vaporized ($q = 0.75$).

The overall material (mole) balance equation is written as

$$F = D + B \text{ or } 200 = D + B \quad (1)$$

Component (benzene) balance: $200 \times 0.5 = D \times 0.95 + B \times 0.1$ (2)

From the equations, $D = 94.1$ kmoles/hr and $B = 200 - 94.1 = 105.9$ kmoles/hr.

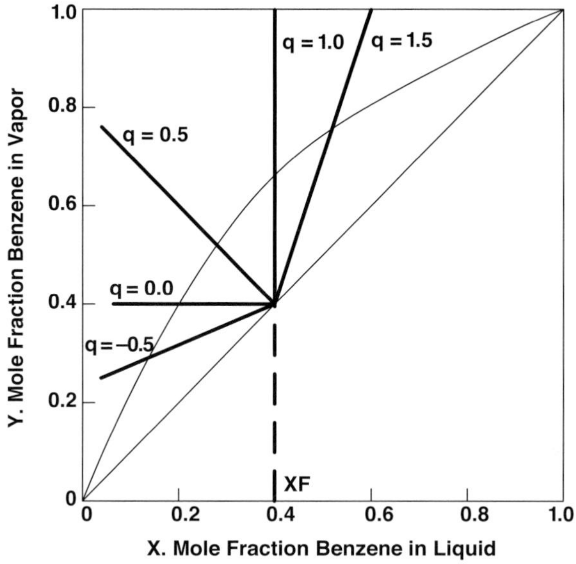

Fig. 1.1.4 q lines with different slopes in a benzene–toluene system

Solution using short-cut (FEUM/FUG) method

Short-cut method (FEUM) calculations are performed through a simulator to avoid tedious calculations required to solve problems using the Underwood equation.

Summary of results of Example Problem 1.1.1 (at minimum operating pressure of 1.3 kg/cm^2 (a))

- Minimum reflux ratio (R_{min}) required = 1.476
- Minimum reflux rate as per Underwood equation = 1.476 × 94.1 = 138.9 kmoles/hr
- FENSKE minimum stages = 6.2.

Table 1.1.2 provides the results for a few alternative solutions derived through simulation using the short-cut method for Example Problem 1.1.1. The results given in the table can be used to find the optimum number of trays, reflux ratio, and feed tray based on the capital cost and operating cost of the selected option. In Table 1.1.2, the actual reflux ratio for each value R/R_{min} is evaluated by multiplication of R/R_{min} with the value of R_{min} (1.476).

Table 1.1.2 Results of the simulation of a short-cut column

Total stages	Feed tray	R/R_{min}	N/Nm	Reflux ratio	Duty condenser, Gcal/hr	Duty reboiler, Gcal/hr
14	8	1.205	2.293	1.779	−1.62E−03	1.32E−03
13	8	1.307	2.096	1.93	−1.71E−03	1.41E−03
12	7	1.41	1.943	2.081	−0.80E−03	1.49E−03
11	7	1.512	1.821	2.233	−1.89E−03	1.58E−03
11	6	1.615	1.741	2.384	−1.98E−03	1.67E−03

Thus, in solving Example Problem 1.1.1, all the parameters of the example column were evaluated.

While computing manually, the Gilliland or other correlations can be used to evaluate the actual number of stages at actual reflux. Alternatively, the Eduljee correlation, as explained below, can be used to calculate the actual number of stages (N) at actual reflux (R) with the following equation: $Y = 0.75(1 - X0.5668)$, where $Y = (N - Nm)/(N + 1)$ and $X = (R - Rm)/(R + 1)$.

Calculation of the number of stages, minimum reflux ratio, and feed location (with McCabe–Thiele and simulator)

Example Problem 1.1.1 is solved with the McCabe–Thiele method.

Calculation using the McCabe–Thiele method

The purity of the top product benzene X_D (0.95) and bottom composition X_B (0.1) along with feed composition X_F (0.50) are marked on the x-axis. The point X_F is marked on $Y = X$ line (45° line in the X-Y plot) (Fig. 1.1.5).

Step 1. Calculation of minimum reflux

For the problem, feed is 25% vapor and the corresponding q is 0.75. The q line is drawn with slope of $q/(q - 1)$ to cut the equilibrium line. Then, the enrichment line is drawn through the point of intersection of the q line and equilibrium line. The intercept is read from the chart as 0.40 (Fig. 1.1.5).

From equations (1) and (2), $D = 94.1$ kmoles/hr and $B = 200 - 94.1 = 105.9$ kmoles/hr.

From Fig. 1.1.5, the intercept of the rectification line at minimum reflux is read as 0.4, so $0.4 = (D/(L + D))X_D$.

Minimum reflux: $L = 129$ kmoles/hr.

A reflux higher than minimum reflux is assumed, one where $L = 183$ kmoles/hr.

$L/L_{min} = 183/129 = 1.41$.

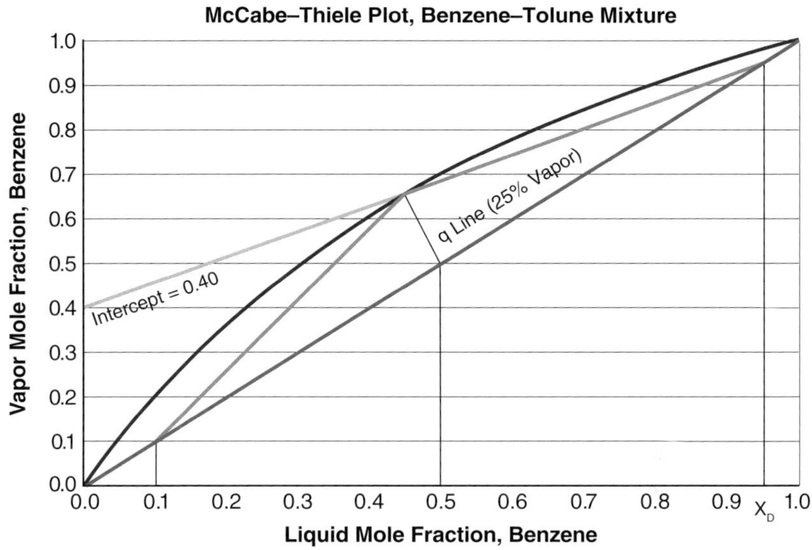

Fig. 1.1.5 Determination of intercept for minimum reflux evaluation using McCabe-Thiele

Step 2. Calculation of stages at actual reflux

Minimum reflux: L = 129 kmoles/hr.

Actual reflux quantity: L = 183 kmoles/hr, $V = L + D = 183 + 94.1 = 277.1$ kmoles/hr.

Intercept calculated as $(D/(L + D))X_D = (94.1/(183 + 94.1)) \times 0.95 = 0.324$.

The value of the intercept is plotted on the X-Y chart on the Y-axis (Fig. 1.1.6), the enrichment line at actual reflux is drawn through X_D, and the point marked on the Y-axis as the intercept. The stripping line is drawn through the intersection of the q line and enrichment line through X_B. The steps are drawn to show the number of total stages required in addition to the feed stage. Fig. 1.1.5 gives the determination of intercept for minimum reflux evaluation using the McCabe–Thiele method.

Fig. 1.1.6 shows the calculation of stages at actual reflux using the McCabe–Thiele method.

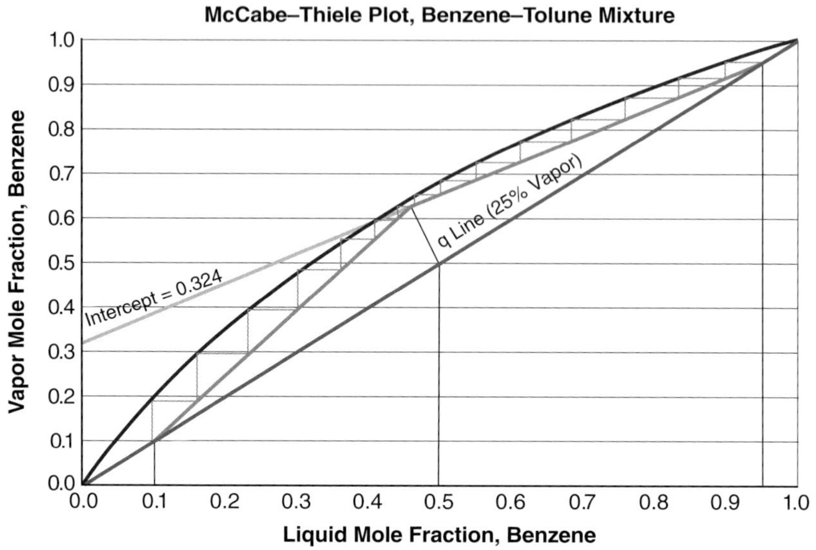

Fig. 1.1.6 Calculation of stages at actual reflux using McCabe-Thiele

From Fig. 1.1.6, the total number of stages required is approximately 15 stages. The feed stage is 8 from the top.

Rigorous calculation using simulator

A rigorous simulation is carried out (Example Problem 1.1.1) with 15 stages. The same purity percentage of overhead was used (95% benzene and bottom toluene of 90%). Fig. 1.1.7 provides the simulation results of the example problem with feed condition (25% vaporized).

The results of the rigorous simulation in Fig. 1.1.7 show that the reflux ratio closely matches that found using the McCabe–Thiele method for the same separation.

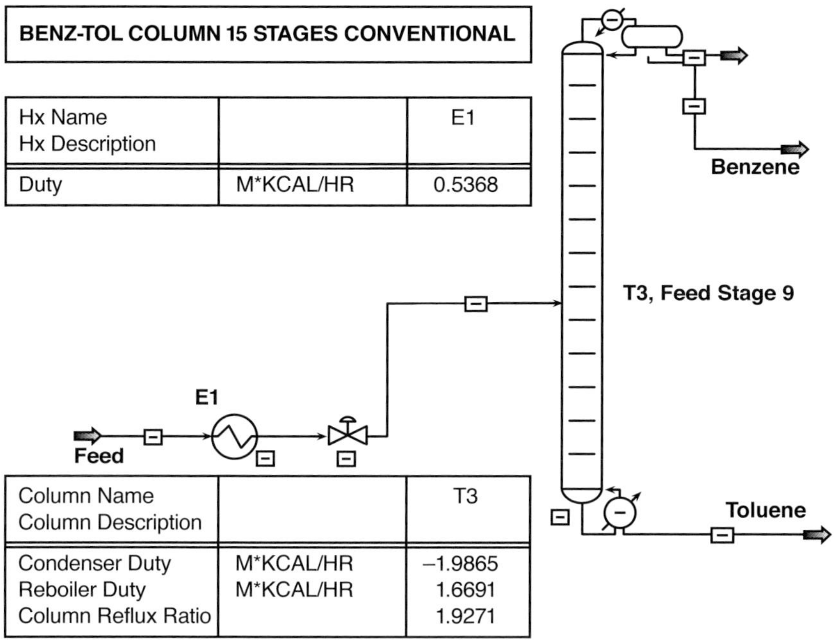

Hx Name Hx Description		E1
Duty	M*KCAL/HR	0.5368

Column Name Column Description		T3
Condenser Duty	M*KCAL/HR	–1.9865
Reboiler Duty	M*KCAL/HR	1.6691
Column Reflux Ratio		1.9271

Fig. 1.1.7 Feed stage 9, 25% vaporized

1.1.3 Concept of Exergy

A distillation column can be viewed as a heat engine in which a reboiler supplies heat at a constant temperature. Part of the heat gets rejected in the condenser, and part is extracted as work (work of separation). The energy balance and entropy balance work as follows:

The equation (Sum($nh + Q + Ws$)) outlet – (Sum($nh + Q + Ws$)) inlet = 0 describes the energy balance.

The equation (Sum($ns + Q/Ts$)) outlet – (Sum($ns + Q/Ts$)) inlet = ΔS_{irr} describes the entropy balance.

Lost work, *LW*, which is also called the loss of availability (or exergy), can be found using the equation $LW = T_0 \Delta S_{irr}$. Alternatively, lost work can be estimated from an availability balance.

Availability (or exergy) balance

(Sum($nb + Q(1 - T_0/T_s) + Ws$) inlet – (Sum($nb + Q(1 - T_0/T_s) + Ws$) outlet = *LW*, where n = molar flow rate, h = molar enthalpy, s = molar entropy, b = molar availability, T_0 = reference temperature, LW = lost work, Q = heat flow in or out of the system, and W_s is shaft work in/out of the system boundary.

- Molar availability function is defined as $b = h - T_o S$.
- Minimum work of separation in a distillation process is represented by W_{min} = Sum(nb(inlet)) – (Sum nb(outlet)).
- Net work available from heat energy in distillation process is written as W_{heat}, and is found using this equation:
$$W_{heat} = Q_R(1 - (T_0/T_R)) - Qc(1 - (T_0/T_C)).$$
- $LW = n_F b_F - (n_D b_D + n_B b_B) - W_{heat}$, where n_F, n_D, and n_B are moles of feed, overhead and bottoms respectively, b_F, b_D, and b_B are molar availability functions of feed, overhead and bottoms respectively and Q_C and Q_R are condenser and reboiler duties. T_0 is surrounding temperature, T_C is the average temperature of the coolant in the condenser, and T_R is the average temperature of the heating agent in the reboiler.
- Efficiency: $E = W_{min}/(LW + W_{min})$.

1.2 Feed Tray Location

Feed tray location plays a vital role in optimizing reboiler and condenser duty in a distillation column. Example Problem 1.2.1 illustrates this process.

Example Problem 1.2.1

A BZ–TOL–PX system with 30 total stages is considered for investigation. The feed of 100 MT/hr is fed with feed compositions of 50%, 40%, and 10% wt of benzene, toluene, and *para*-xylene respectively. A benzene purity of 99.0 mole % in the top product and a maximum benzene slippage of 1 mole % in the bottom product are desired.

A separation factor, defined as Log(X_{LK}/X_{HK}), was plotted against stages generated with the help of a commercial simulator to investigate the effect of feed stage with a separation factor between benzene and toluene with a feed stage of 12 (Fig. 1.2.1).

Total stages: 30; feed temperature = 80°C; current feed stage: 12; Q_R = 16.4 Gcal/hr; Q_c = 15.2 Gcal/hr. The shape of the curve indicates that feed stage 12 is *not* an optimum feed location. The mole fraction—number of stages plot (Fig. 1.2.2)—shows practically no change of composition (no separation) in trays 12 to 15.

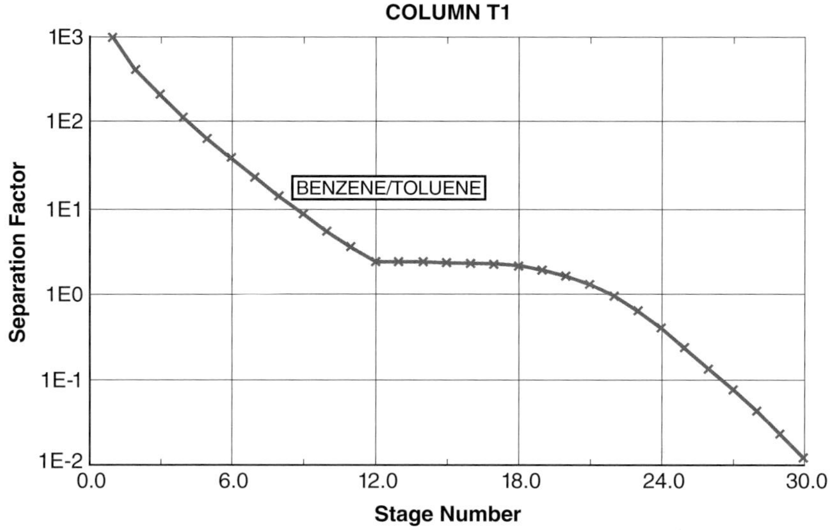

Fig. 1.2.1 Separation factor with stage number plot, feed stage 12

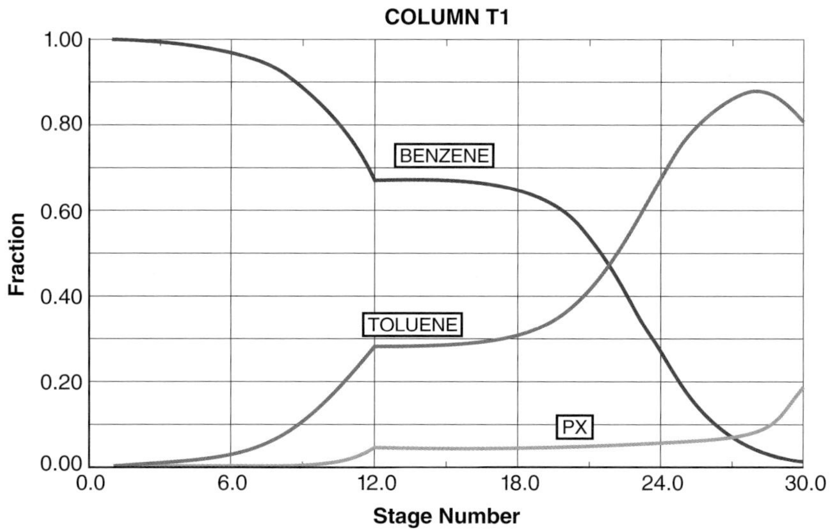

Fig. 1.2.2 BZ–TOL–PX stage-wise liquid composition, feed stage 12

Composition change from trays 12 to 15 is very low, suggesting that a relocation of feed should be considered.

Fig. 1.2.3 presents a temperature and stage number plot (feed stage 12).

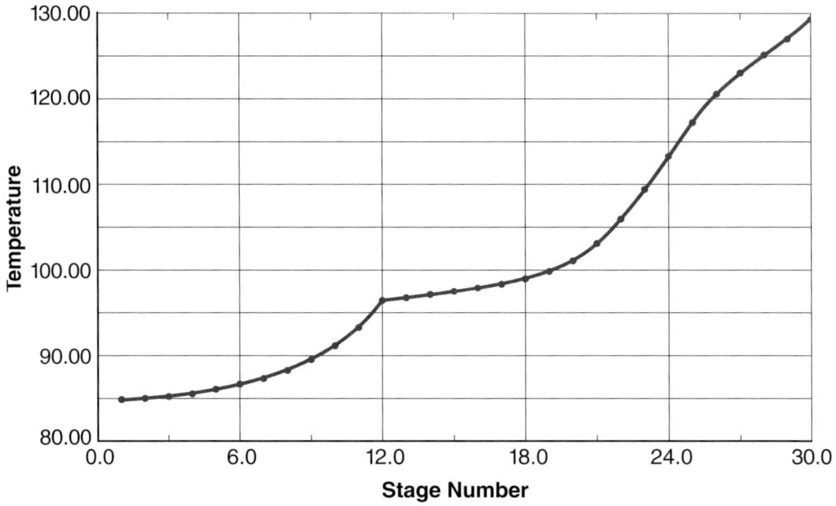

Fig. 1.2.3 Temperature and stage number plot, feed stage 12

Temperature change in trays from 12 to 15 is very small (Fig. 1.2.3). This also suggests that the feed tray location may not be optimal. The feed tray is subsequently relocated to tray 18 for the same example problem (Example Problem 1.2.1) and curves are generated for separation factor, liquid mole fraction on trays, and tray temperatures. Fig. 1.2.4 shows the separation factor with stage number plot for benzene–toluene separation for revised feed stage 18.

Total stages: 30; feed temperature = 80°C; current feed stage: 18; Q_R = 13.26, Q_c = 12.03;. By looking at the smooth curve of the separation factor with tray number (stage number) in Fig. 1.2.4, it can be concluded that this is almost the optimum feed location. The smooth curves of the concentration-stage diagram (Fig. 1.2.5) and the temperature-stage diagrams (Fig. 1.2.6.) corroborate the conclusion. The corresponding values of reboiler and condenser duties show that the reboiler and condenser duties have reduced due to the relocation of the feed stage from tray 12 to tray 18 with the same feed temperature. The reboiler duty of 16.4 Gcal/hr and a condenser duty of 15.2 Gcal/hr are reduced to 13.26 Gcal/hr and 12.03 Gcal/hr, respectively.

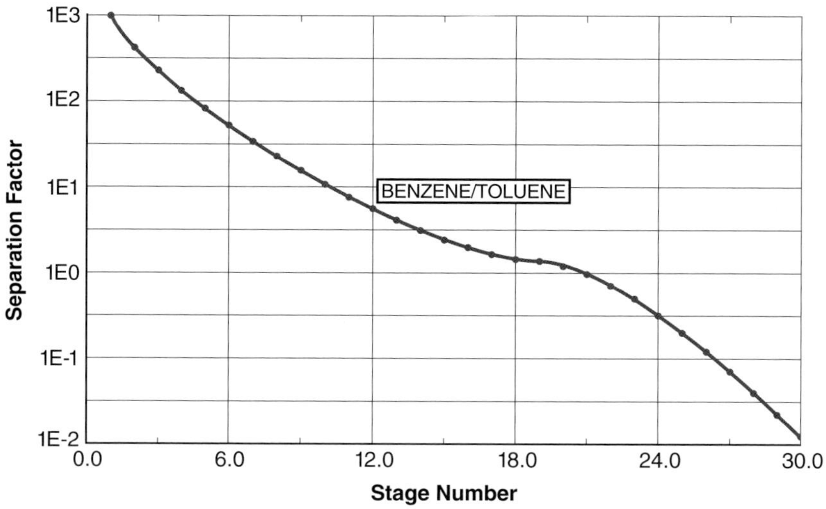

Fig. 1.2.4 Separation factor with stage number plot, feed stage 18

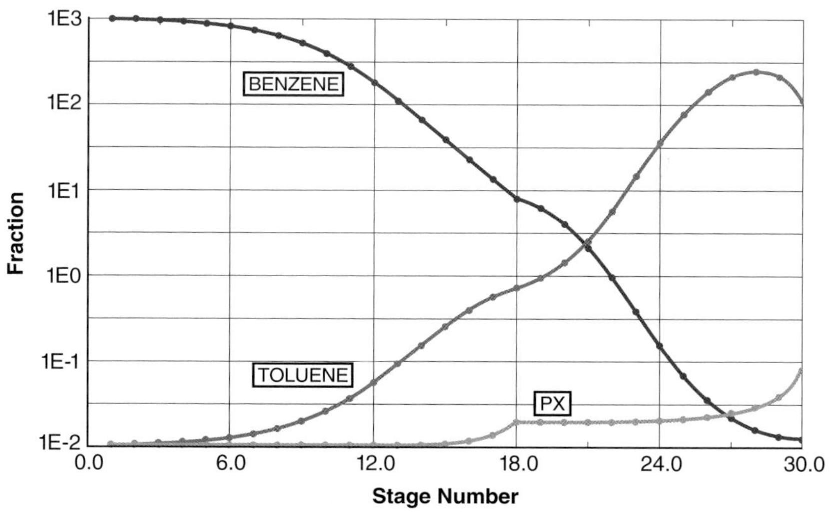

Fig. 1.2.5 BZ–TOL–PX stage-wise liquid composition, feed stage 18

The profile of liquid mole fractions of light key and heavy key gradually improved through stages, indicating that stage 18 is the better feed stage location. Fig. 1.2.6 gives the change in temperature of the stages (with feed stage 18).

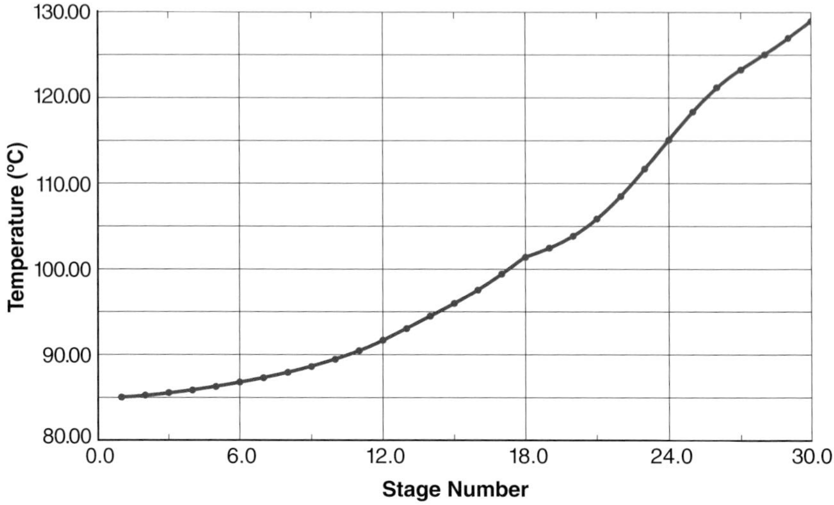

Fig. 1.2.6 Temperature of the stages, feed stage 18

In feed stage 18, temperature changes are more gradual and uniform, confirming that this is a better feed tray location. Commercial simulators can calculate optimum feed tray location. The location of the feed tray can be changed in the simulation and corresponding condenser and reboiler duties can be evaluated. Later, a plot of condenser and reboiler duties with different feed stages can be generated. The minimum values of condenser and reboiler duty can be read from the plot to identify the best feed point. Table 1.2.1 gives the reboiler and condenser duties generated, corresponding to different feed stages for Example Problem 1.2.1.

Optimum feed stage by simulator

Table 1.2.1 Reboiler/condenser duty variations with stage

No. of stages	Q_R	Q_c
11	17.94	16.71
12	16.39	15.15
13	15.32	14.09
14	14.57	13.34
15	14.04	12.80
16	13.65	12.41
17	13.39	12.16
18	13.26	12.03
19	13.27	12.03
20	13.45	12.22

Table 1.2.1 is derived using a simulator and also indicates that the minimum reboiler/condenser duty is at feed stage 18, confirming that feed stage 18 is the best location for the feed.

Fig. 1.2.7 plots the Q_R and Q_c data from Table 1.2.1.

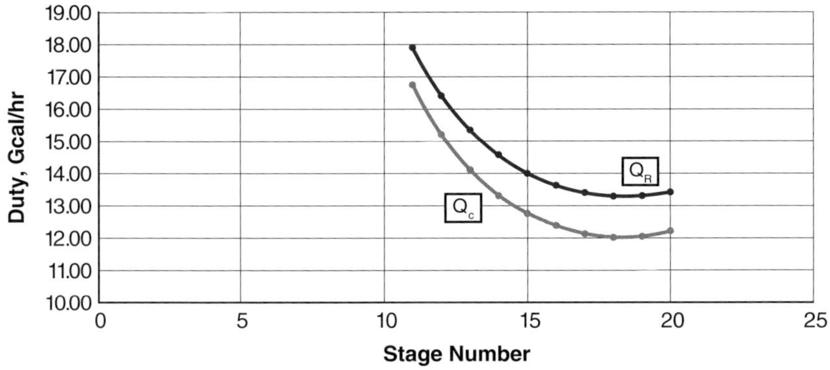

Fig. 1.2.7 Reboiler/condenser (Q_R/Q_c) duty plots with feed stage

As shown in Table 1.2.1 and Fig. 1.2.7, the optimum feed tray location is 18.

The minimum Q_R and Q_c values are to be read from the table or the chart as 13.26 and 12.06 Gcal/hr, respectively.

1.3 Effect of Feed Preheat

1.3.1 Effect of change in feed temperature or feed thermal condition

To explain the effect of feed preheat on the major parameters of a distillation column, several approaches have been used; for example, creating a model with a different feed condition using a simulator, the McCabe–Thiele method, or the computational method, etc.

Effect of feed condition on the reboiler/condenser duties and reflux ratio with a simulator

To capture the effect of change in feed temperature/feed condition and consequent effect on Q_R and Q_c and reflux ratio, the binary distillation of benzene and toluene is described in Example Problem 1.3.1.

Example Problem 1.3.1

Feed rate: 200 kmoles/hr; feed stage: 9.

Feed composition: 50:50 mole % benzene and toluene.

Column total stages: 15 stages.

Product purities: top 95% benzene in mole %, bottom 90% toluene in mole %.

The example problem is worked out with a simulator and shown in Fig. 1.3.1 with feed as saturated liquid.

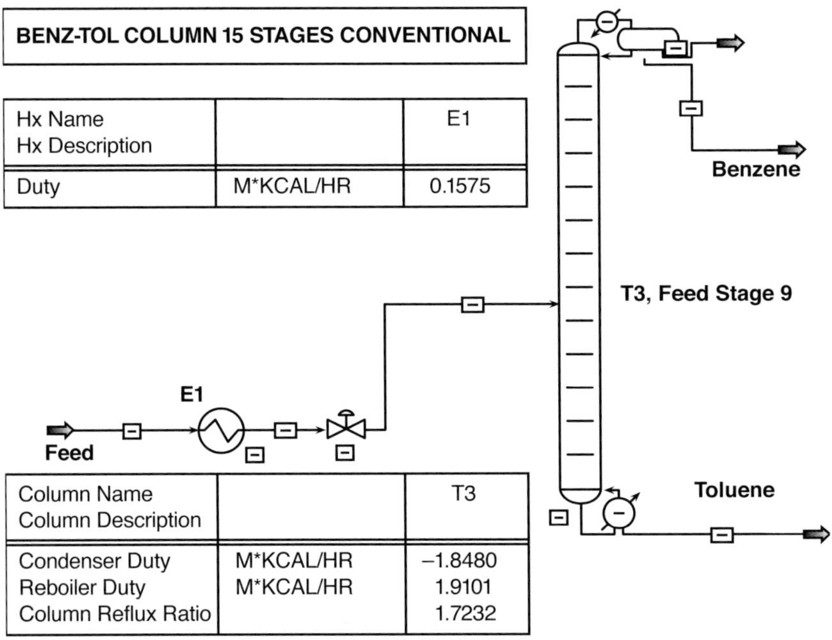

Hx Name Hx Description		E1
Duty	M*KCAL/HR	0.1575

Column Name Column Description		T3
Condenser Duty	M*KCAL/HR	–1.8480
Reboiler Duty	M*KCAL/HR	1.9101
Column Reflux Ratio		1.7232

Fig. 1.3.1 Feed as saturated liquid

Results with stage 9 feed that has been 25% vaporized are given in Fig. 1.1.7. Results for stage 9 feeds with higher vaporization (40%) are given in Fig. 1.3.2.

The results of simulation of feed as a saturated liquid (Fig. 1.3.1), feed 25% vaporized (Fig. 1.1.7), and feed 40% vaporized (Fig. 1.3.2) are tabulated in Table 1.3.1.

The results (Table 1.3.1) show that the reboiler duty reduces as the feed vaporization percentage increases from saturated liquid to 40% vaporization, while the condenser duty increases. It can be seen that as the percentage of feed vaporization increases, the reflux ratio also increases, which in turn results in a higher condenser duty.

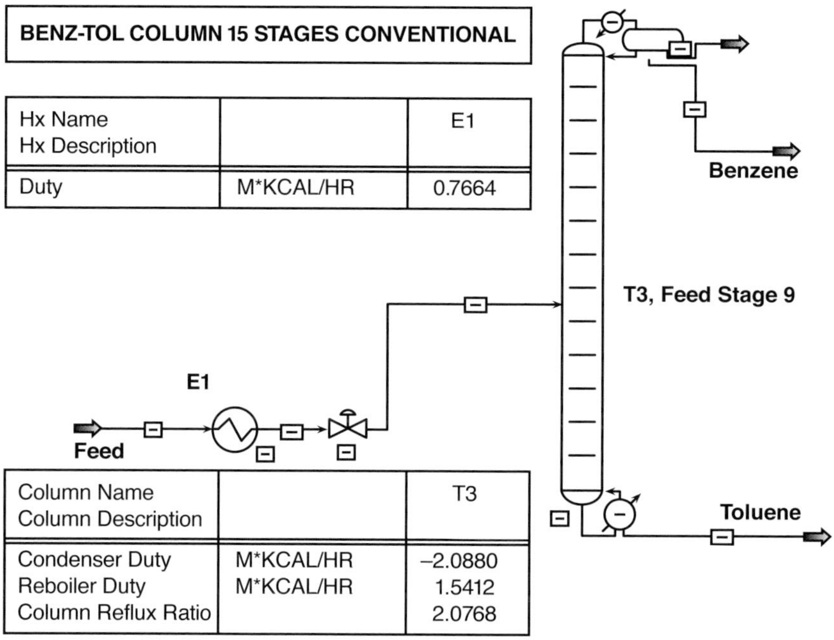

Fig. 1.3.2 Simulation with feed 40% vaporized

Table 1.3.1 Summary of results of cases with feed as a saturated liquid, 25% vaporized and 40% vaporized

Parameter	Feed saturated liquid	Feed 25% vaporized	Feed 40% vaporized
Condenser duty Gcal/hr	−1.8480	−1.9864	−2.0829
Reboiler duty Gcal/hr	1.9101	1.6692	1.5364
Reflux ratio	1.7232	1.9271	2.0700

Feed is normally preheated by hot products. Typically, the integration of the column's bottom product heat with the feed is the most rudimentary heat integration scheme used to conserve energy. Thus, feed temperature will be considered based on preheating the feed with hot column bottom product.

For revamp applications, the feed is often heated to a higher temperature to contain the reboiler heat requirement if augmentation of reboiler duty is a problem. While augmenting the capacity of a naphtha-splitting unit producing ISOM feed and reformer feed, the feed temperature was increased from 95°C to 139°C with feed reheaters using low-pressure steam and medium-pressure steam to avoid augmentation of the bottom reboiler furnace. The capacity increased by around 40%. The condenser duty increased and an additional condenser was installed.

The reason for the increase in reflux ratio is further examined using the McCabe–Thiele method.

Analysis of higher reflux requirement with increase of feed preheat with the McCabe–Thiele method

The phenomenon of the increase in reflux ratio with the increase in feed vaporization as seen above is further analyzed using the McCabe–Thiele method with the help of an example problem. The same problem as above (Problem 1.3.1) is worked out in a 15-stage column with feed stage 9.

Feed = 200 kmoles/hr (50:50 mole:mole benzene:toluene), benzene purity in top product is 95 mole %. Toluene purity in bottom product is 90 mole %.

Overall material (mole) balance (from earlier pages):

$$F = D + B \text{ or } 200 = D + B \tag{1}$$

Component (benzene) balance:

$$200 \times 0.5 = D \times 0.95 + B \times 0.1 \tag{2}$$

$D = 94.1$ kmoles/hr, $B = 200 - 94.1 = 105.9$ kmoles/hr

To further analyze the effect of feed condition on minimum reflux, the McCabe–Thiele method is used. The limiting reflux is calculated for three feed conditions: feed as a saturated liquid, feed 25% vaporized, and feed 40% vaporized.

Limiting reflux calculation using the McCabe–Thiele method for three conditions of feed

Feed as a saturated liquid

For saturated liquid where $q = 1$, the slope of the q line is ∞. The vertical q line is drawn and the intersection with the equilibrium line is marked. The enrichment line and stripping line are drawn through the point where the q line and the equilibrium line intersect.

Intercept from the chart (similar to Fig. 1.1.5) for enrichment line = 0.42 = $X_D \times D/(L + D)$.

Again, $V = L + D$; thus, $L = 118.7$ kmoles/hr and $V = 212.8$ kmoles/hr, and $V' = V - (1 - q)F$ or $V' = 212.8 - (0 \times 200) = 212.8$ kmoles/hr.

Feed 25% vaporized

Limiting reflux in the McCabe–Thiele method is calculated from the slope of the rectification line passing through the intersection of the q line and the equilibrium line. From the chart (Fig. 1.1.5), the limiting value of the intercept is 0.4 when the feed is 25% vaporized.

So, $0.4 = (94.1 \times 0.95)/V$. Therefore, $V = 223.1$ kmoles/hr.

Material balance at the top section provides the following equations:

$V = L + D$ or $229 = L + 94.1$ or $L_{min} = 129$ kmoles/hr or $V_{min} = 223.1$ kmoles/hr, and

$V' = V - (1 - q)F$ or $V' = 223.1 - (0.25 \times 200) = 173.1$ kmoles/hr.

Feed 40% vaporized

The minimum reflux found by the McCabe–Thiele method for 40% feed vaporization is 137.2 kmoles/hr, which is calculated in a similar way (drawing the rectification line through the point where the q line and the equilibrium lines intersect and noting the value of the intercept and calculating the value of reflux as minimum).

$V' = V - (1 - q)F$ or $V' = 231.1 - (0.4 \times 200) = 151.1$ kmoles/hr

The summary of the effects of feed preheat on minimum reflux and minimum vaporization as calculated with the McCabe–Thiele method is given in Table 1.3.2.

Table 1.3.2 Summary of the effects of feed condition on minimum reflux/vaporization (using the McCabe-Thiele method)

Feed condition	L_{min}	V'_{min}
Saturated liquid, $q = 1.0$	118.7 kmoles/hr	212.8 kmoles/hr
25% vaporized, $q = 0.75$	129 kmoles/hr	173.1 kmoles/hr
40% vaporized, $q = 0.6$	137.2 kmoles/hr	151.3 kmoles/hr

Table 1.3.2 shows that the minimum reflux requirement reduces with an increase in feed q and the minimum reflux requirement is lowest when the feed is saturated liquid. The lower minimum reflux requirement in turn reduces the actual reflux requirement, as can be seen in Table 1.3.3 with the assumption that L/L_{min} equals 1.42 in all the feed conditions considered.

Table 1.3.3 Effects of different feed conditions on actual reflux/minimum reflux and reflux ratios

Feed condition	q	$q/(q-1)$	Slope of q line	Intercept	L_{min}	L	$R = L/D$
Feed saturated	1.0	∞	90°	0.42	118.7	168.6	1.79
Feed 25% vaporized	0.75	−3.0	71°	0.324	129	183	1.94
Feed 40% vaporized	0.6	−1.5	56°	0.3094	137.2	194.8	2.07

Calculation of minimum vaporization and reflux using the computational method (for feed as saturated liquid)

Rectification section

The feed tray is the fth tray and $(f + 1)^{th}$ tray is located immediately above the feed tray. At minimum vaporization and reflux, the composition change between X_f and X_{f+1} is minimal, so the two can be read as equal. Thus: $X_f = X_{f+1}$.

$V_{min} \times Y_f = L_{min} \times X_f + DX_D$ describes the component balance of more volatile component.

$L_{min} = V_{min} - D$ is the material balance equation for the top section.

$$V_{min} = D(X_d - X_f)/(Y_f - X_f) \quad (1)$$

$$L_{min} = D(X_d - Y_f)/(Y_f - X_f) \quad (2)$$

For $X_f = 0.5$ and corresponding $Y_f = 0.7$ from the equilibrium curve of benzene–toluene shown in Fig. 1.1.5 or Fig. 1.1.6:

$V_{min} = 94.1 \times (0.95 - 0.5)/(0.7 - 0.5) = 211.725$ kmoles/hr

$L_{min} = 94.1 \times (0.95 - 0.7)/(0.7 - 0.5) = 117.625$ kmoles/hr

Stripping section

Similarly, in the stripping section:

$L' = V' + B$ and $L' \times X_f = V' \times Y_{f-1} + B \times X_B$, where $f - 1$ is the tray immediately below the feed tray.

At minimum vaporization, $Y_f = Y_{f\text{-}1}$ (No change in composition.)

$$V'_{min} = B \times (X_f - X_B)/(Y_f - X_f) \quad (3)$$

$$L'_{min} = B \times (Y_f - X_B)/(Y_f - X_f) \quad (4)$$

$V'_{min} = 105.9 \times (0.5 - 0.1)/(0.7 - 0.5) = 211.8$ kmoles/hr

$L'_{min} = 105.9 \times (0.7 - 0.1)/(0.7 - 0.5) = 317.5$ kmoles/hr

In Table 1.3.4, all the limiting values of liquid/vapor flow rates are tabulated and compared with values obtained through simulation and the McCabe–Thiele method.

Table 1.3.4 Limiting values of reflux and vaporization estimated using the McCabe-Thiele and computational methods

Limiting values of vapor/ liquid rates, kmoles/hr	McCabe-Thiele method (feed saturated liquid)	Computational method (feed saturated liquid)
L	118.7	117.63
V	212.8	211.73
L'	317.7	317.5
V'	211.8	211.8

It can be seen (Table 1.3.4) that the minimum vapor/liquid flow rates calculated by the McCabe–Thiele and computational methods are almost identical for the same feed condition.

1.3.2 Estimation of minimum reflux and minimum number of stages using a simulator

The problem (Example Problem 1.3.1) for the separation of a benzene–toluene mixture with two different feed conditions (as saturated liquid and 25% vaporized) is considered for the estimation of minimum reflux and minimum number of stages. Simulation is used and the reflux rate L is generated with a different number of total column stages for both the feed cases. The data generated is plotted in Fig. 1.3.3 and Fig. 1.3.4, and the minimum reflux and minimum stages can be seen in Fig. 1.3.3 and Fig. 1.3.4. The column is simulated at 1.3 kg/cm^2(a) for both the feed conditions.

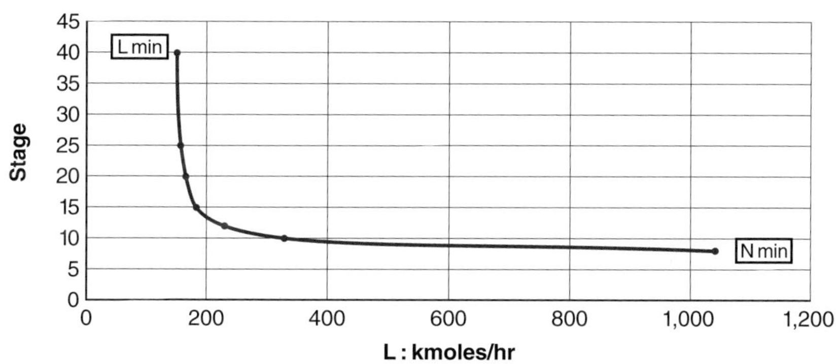

Fig. 1.3.3 Plot of L_{min} and N_{min} for 25% vaporized feed

A curve like the one in Fig. 1.3.3 can be developed by varying the number of stages and calculating the reflux requirement for the same separation. From Fig. 1.3.3, L_{min} is read as 180 kmoles/hr and N_{min} as 8. The same exercise is repeated with feed as a saturated liquid, and N_{min} and L_{min} are determined from the curve (Fig. 1.3.4).

In Fig. 1.3.4, where the feed is a saturated liquid, it can be seen that the L_{min} is reduced to 162 kmoles/hr and N_{min} remains equal to 8. The estimated minimum reflux requirements for feed 25% vaporized and feed as a saturated liquid are presented below in Table 1.3.5.

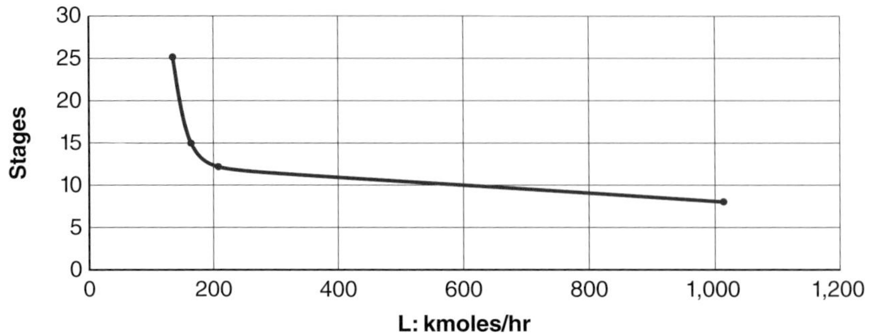

Fig. 1.3.4 Plot of L_{min} with stage for saturated liquid feed

Table 1.3.5 Minimum reflux with condition of feed (q)

Feed condition	Minimum reflux, kmoles/hr
q = 1 (feed saturated liquid)	162
q = 0.75 (feed 25% vapor)	180

Note: In the simulation column, pressure is higher (1.3 kg/cm^2 (a)) than considered in the McCabe–Thiele method. Thus, the minimum reflux calculated through a simulator is higher than that calculated using the McCabe–Thiele method at atmospheric pressure. The purpose of the simulation was solely to show the effect of feed preheat (feed condition) on minimum reflux and actual reflux and consequently on reboiler and condenser duties.

Summary of Findings:

- In the enrichment section, the minimum reflux requirement increases with an increase in feed temperature or feed vaporization.
- The minimum vaporization requirement decreases with an increase in feed temperature or feed vaporization.
- The minimum number of stages is not dependent on feed condition.
- The minimum reflux increases with an increase in pressure of the column as α approaches unity.

1.3.3 Efficiency of feed preheat

If feed preheat temperature (feed vaporization) is increased, reboiler duty reduces considerably up to a certain point. Beyond a certain feed temperature (vaporized conditions of feed), the reboiler duty does not decrease significantly, and the increase in condenser duty becomes significant. If the lighter fraction in the feed is large, a higher vaporization of the feed is optimum. If the heavier fraction in the feed is larger, then a lower vaporization is optimum. The typical optimum feed vaporization percentage is 40%. With lighter feed, preheat mainly vaporizes the lighter components, which are recovered as overhead distillate. With a higher proportion of heavy component in the feed, an increase in feed preheat vaporizes the heavier components, which are not recovered as top product, but rather are returned as bottom product after condensation. This increases condenser duty and does not reduce the reboiler duty considerably.

Optimum feed preheat from preheat efficiency

In Example Problem 1.3.1, data for reboiler duty (Q_r) as a function of preheat duty (H) is generated from a simulator for attaining different conditions of feed (values of q). The preheat efficiency is defined as the ratio dQ/dH, where dQ is the change in reboiler or condenser duty with respect to different feed conditions (feed q), and dH is the change in duty of the feed preheater with change in the corresponding feed condition. The change of reboiler duty (dQ_r) as a function of the change in heat duty of the feed preheater (dH) is given in Table 1.3.6. Heat duties are all in Gcal/hr.

Table 1.3.6 dQ_r as a function of feed heat dH

Qr	1.164	1.2699	1.3928	1.5364	1.6692	1.8103	1.9099
H	1.696	1.3847	1.0745	0.7664	0.5368	0.3087	0.1575
dQr	0.1059	0.1229	0.1436	0.1328	0.1411	0.0996	
dH	0.3113	0.3102	0.3081	0.2296	0.2281	0.1512	
E	0.340186	0.396196	0.466082	0.578397	0.618588	0.65873	
q	0	0.2	0.4	0.6	0.75	0.9	1

dH and dQ_r are generated as differential duties of the preheater and reboiler for corresponding values of feed q. The efficiencies (E) are calculated as dQ_r/dH and plotted against corresponding values of q. Similarly, the pattern of condenser duties Q_c can be also plotted.

The data of preheat efficiency from Table 1.3.6 is plotted against different q values in Fig. 1.3.5.

The heat duty in a reboiler is generally supplied by steam, furnace, or by hot streams, while condenser heat duty is removed by air coolers, circulating water, or by other colder streams. The preheat efficiency (dQ_r/dH) plot with feed q (Fig. 1.3.5) shows that up to $q = 1 - 0.8$, the preheat efficiency is good. For $q = 0.8$ and below, the efficiency drops substantially. This implies that with an increase in preheat, the reboiler duty does not decrease significantly while the condenser duty increases significantly. Thus, the feed preheat needs to be decided based on the requirement in each specific case.

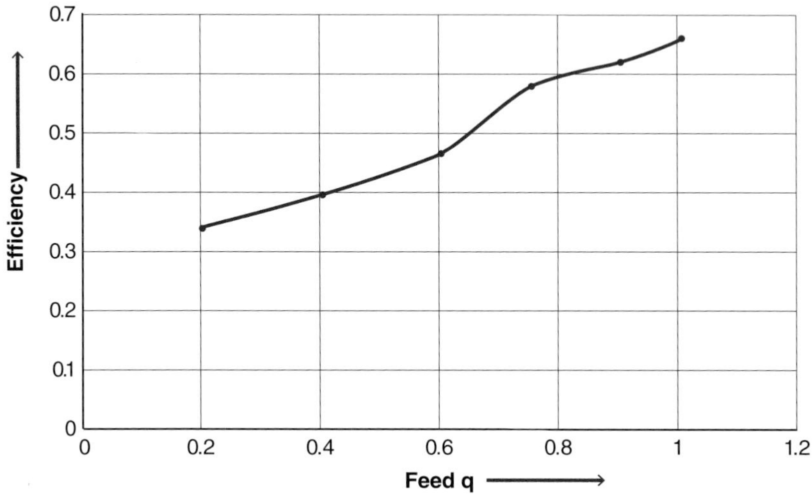

Fig. 1.3.5 Preheat efficiency (*dQ*/*dH*) with feed condition (*q*)

1.3.4 Analysis of the effects of feed preheat using the concept of exergy

There are methods that also allow the analysis of feed preheat using the concept of exergy. Using the same problem that calculated benzene–toluene separation (Example Problem 1.3.1), data has been generated for the loss of work (LW Mkcal/hr) for feeds with different q values using a commercial simulator. The results are tabulated below in Table 1.3.7, again with 15 stages.

Loss of work (*LW*) plot with feed conditions (q) is presented in Fig. 1.3.6.

The plot in Fig. 1.3.6 shows that with an increase in q, the exergy loss is reduced. The exergy loss is maximum when the feed is fully vaporized ($q = 0$). When the feed is almost entirely vaporized, most of the energy input to the feed will be bounced to the condenser instead of being used to lower the reboiler duty. A similar trend is shown by the preheat efficiency plot shown earlier in Fig. 1.3.5.

Table 1.3.7 *LW* variation with feed *q*

Feed *q* (fraction liquid)	*LW*
1	0.089
0.85	0.1123
0.8	0.1207
0.75	0.1292
0.7	0.1377
0.65	0.1464
0.6	0.1554
0.5	0.1736
0.2	0.2218
0	0.26

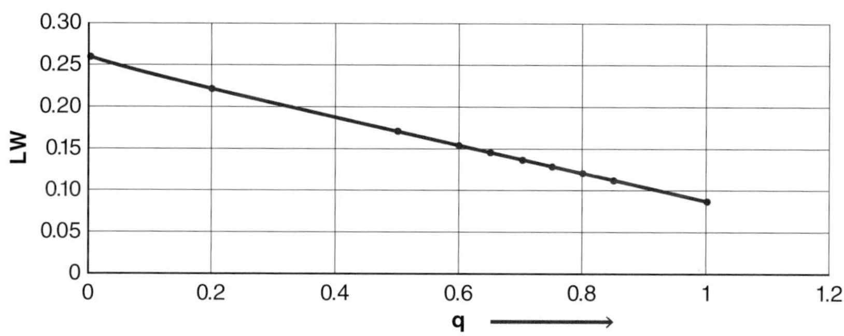

Fig. 1.3.6 Loss of work (*LW*) with feed conditions (*q*)

1.3.5 Best feed point location with feed *q* using the concept of exergy

The single feed of 200 kmoles/hr is considered in Example Problem 1.3.1, where it is assigned different *q* values and fed to different feed trays (15 total stages), and lost work (*LW*) is evaluated. *LW* in Mkcal/hr (Gcal/hr) at different feed stages is given in Table 1.3.8.

Table 1.3.8 *LW* variation with feed *q* and stage number

Feed stage	*LW* in Mkcal/hr @ $q = 0$	@ $q = 0.5$	@ $q = 1$
7			0.2487
8	0.26993	0.1679	0.2440
9	0.2597	0.1634	0.24416
10	0.2552	0.16355	0.2489
11	0.2551	0.16853	0.2601
12	0.25992	0.1806	0.2822

Table 1.3.8 shows that depending on the feed *q*, the best feed point changes as the location with minimum loss of exergy (*LW*) changes. The minimum loss is highlighted in Table 1.3.8.

This method can also be used for theoretical analysis of the best feed point location. However, feed point location based on minimum reboiler duty, condenser duty and reflux ratio may be a better approach for finding the best feed point location as it is practical and easier to compute.

1.4 Location of Multiple Feeds with Different Compositions

Example Problem 1.4.1

Two feed mixtures of benzene and toluene considered:

Feed F1 is 200 kmoles/hr (with 50:50, mole:mole benzene:toluene) at 102.48°C.

Feed F2 is 50 kmoles/hr (with 70:30, mole:mole benzene:toluene) at 80°C.

Column feed description: Feed F1 (leaner feed) is at stage 9 (lower location) and F2 (richer feed) is at stage 5 (upper location). Again F1 and F2 are fed together at stage 9, and the condenser/reboiler duties and reflux ratios are summarized in Table 1.4.1.

Table 1.4.1 Summary of the parameters of split feeds and feeding together

Parameters	Units	Feed 1 at stage 9 Feed 2 at stage 5	Both feeds at stage 9
Condenser duty	Mkcal/hr	–2.3077	–2.5173
Reboiler duty	Mkcal/hr	2.0371	2.2470
Column reflux ratio		1.4737	1.6988

The reboiler and condenser duties increase when two different compositions are fed into the same stage, as compared to two separate locations, the richer at the upper location and the leaner at the lower location. The reflux ratio also increases when the two are fed together.

Possible area of application:

When pre-topped naphtha and topper naphtha are fed to naphtha splitters in refineries for producing isomerization unit feed and catalytic reformer feed, often both the naphtha streams are mixed and fed together. However, it is more efficient to feed them at different locations in the splitter, as feeding at different points saves reboiler duty requirement and can help debottleneck a splitter. This is demonstrated in Example Problem 1.4.2.

Example Problem 1.4.2

An example problem of naphtha splitting is considered with pre-topped naphtha (80 MT/hr at 80°C) and topped naphtha (27 MT/hr at 110°C) as the feeds. The ASTM D-86, 95% of the overhead product is specified as 95°C and 5% of the bottom product as 110°C (ASTM gap of 15°C). They are fed together into stage 17 (the optimized stage), and the reboiler and condenser duties and reflux ratio are obtained. A total of 36 stages in the column are considered and shown in Fig. 1.4.1.

The feeds are then introduced at different locations, with pre-topped naphtha at stage 12 and topped naphtha at stage 25 with the same separation criterion as given in Fig. 1.4.2.

The figures (Fig. 1.4.1 and Fig. 1.4.2) show the effects of feeding the two different streams at two different locations. The percentage saving in reboiler duty is around 15% in this case, with a considerable reduction of the reflux ratio. An application of the same concept is practiced in aromatic plants. The xylene splitter of PX plants receives feeds of three different compositions, typically fed at three different locations in the xylene splitter to optimize the reboiler/condenser duties and reflux ratio.

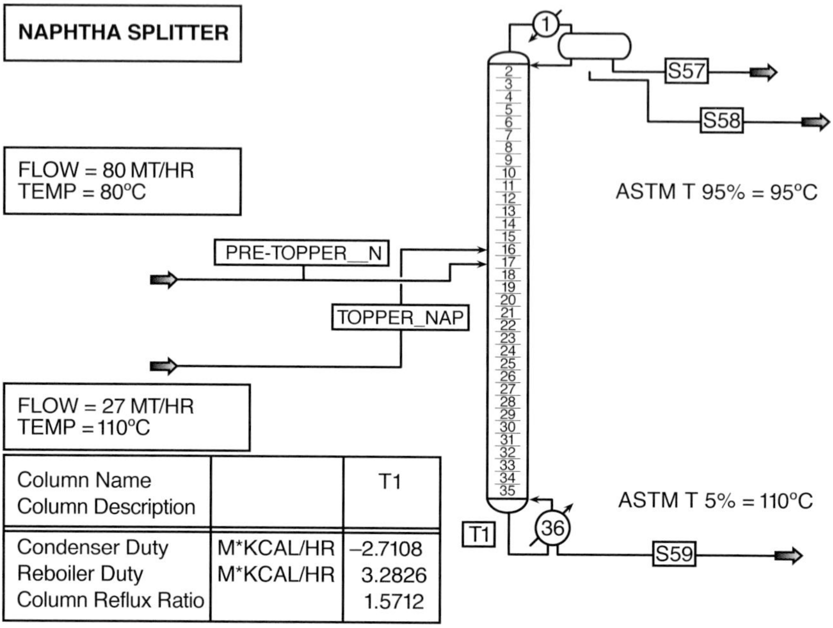

Column Name Column Description		T1
Condenser Duty	M*KCAL/HR	–2.7108
Reboiler Duty	M*KCAL/HR	3.2826
Column Reflux Ratio		1.5712

Fig. 1.4.1 Pre-topper and topper naphtha with a common feed point

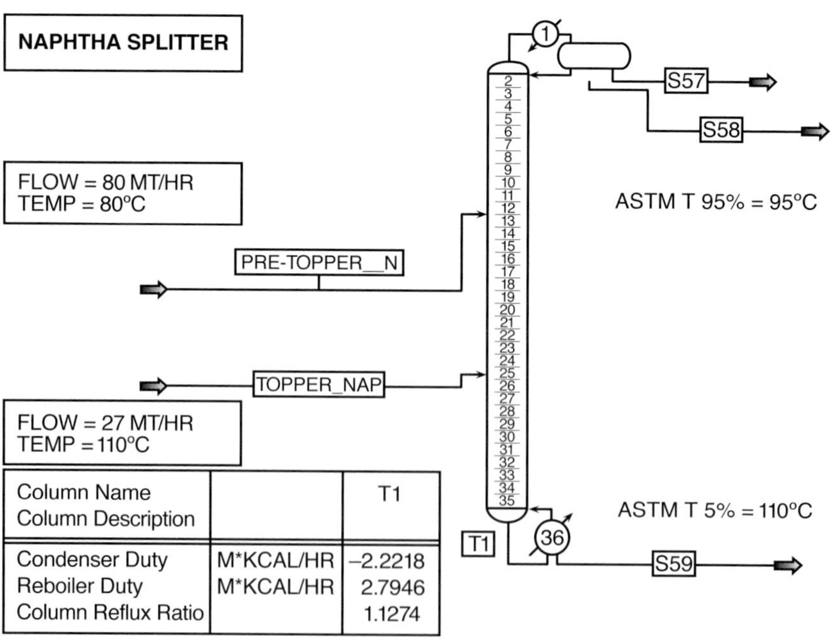

Column Name Column Description		T1
Condenser Duty	M*KCAL/HR	–2.2218
Reboiler Duty	M*KCAL/HR	2.7946
Column Reflux Ratio		1.1274

Fig. 1.4.2 Pre-topper and topper naphtha with separate feed points

1.5 Features of Two-Enthalpy Feeds with the Same Compositions

The feed with the composition 50:50 mole:mole benzene:toluene is fed at two locations with different thermal conditions. The base case is generated in Example Problem 1.5.1, with feed as saturated liquid.

Example Problem 1.5.1

Feed: 50:50 benzene: toluene (mole:mole)

Total feed: 200 kmoles/hr

Top product: 95 mole % benzene and 5 mole % benzene slip in the bottom product

Feed1 and Feed2 directly fed to same feed stage (stage 15)

Number of stages = 30; Preheater duty = 0.1521 Gcal/hr

A simulation with single point feed is shown in Fig. 1.5.1.

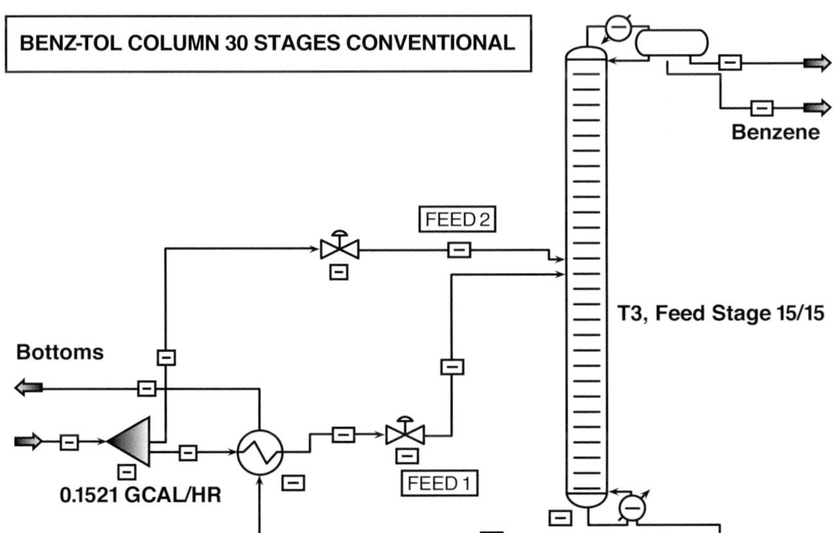

Column Name Column Description		T3
Condenser Duty	M*KCAL/HR	–1.7362
Reboiler Duty	M*KCAL/HR	1.8060
Column Reflux Ratio		1.4079

Fig. 1.5.1 B–T system with single point feed at stage 15

B–T system (two enthalpy, same composition) Example Problem 1.5.2

Feed1: 50:50 benzene:toluene (mole:mole).

Total feed = 200 kmoles/hr

Feed1 = 50% of total feed is fed to feed stage 20 with feed preheat achieved by the same preheater duty of 0.1521 Gcal/hr and the same separation as in Example Problem 1.5.1.

Feed2 = 50% of total directly fed to feed stage 12.

B–T system with two points of feed, cold: hot (50:50). Feed1 (hot) at stage 20 and Feed2 (cold) at stage 12 are shown in Fig. 1.5.2.

The case expressed in Fig. 1.5.2, where feed is split with the cooler feed at an upper location (stage 12) and the hotter feed at lower location (stage 20), resulted in a lower condenser duty, reboiler duty, and reflux ratio than the base case (Fig. 1.5.1).

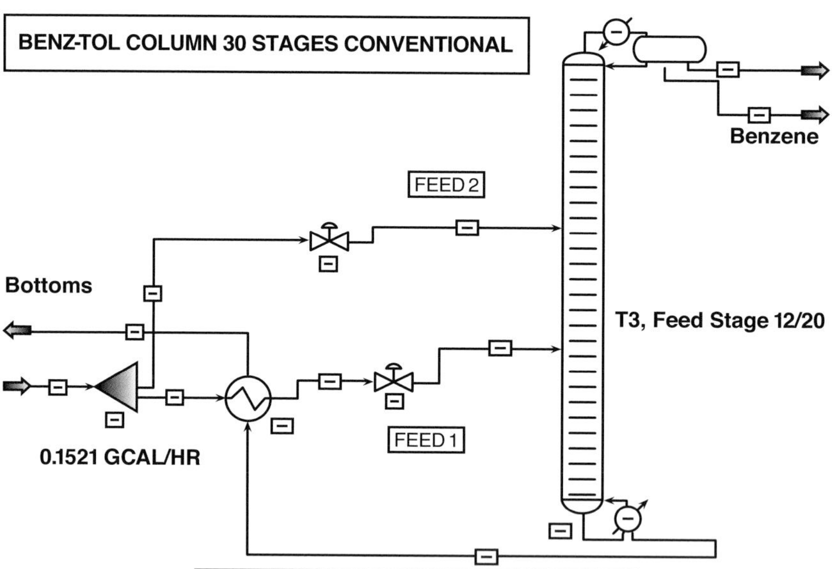

Column Name Column Description		T3
Condenser Duty	M*KCAL/HR	−1.7025
Reboiler Duty	M*KCAL/HR	1.7723
Column Reflux Ratio		1.3611

Fig. 1.5.2 B-T system, two points of feed: Cold:hot (50:50)

The example in Problem 1.5.1 was worked out with a different feed split of hot and cold streams.

Feed: 50:50 benzene:toluene (mole:mole).\

Total feed = 200 kmoles/hr

Feed1 = 75% of total feed is fed to feed stage 17 with feed preheat achieved with the same preheater duty of 0.1521 Gcal/hr

Feed2 = 25% of total directly fed to stage 14

Number of stages = 30. This two-split feed case is given below in Fig. 1.5.3.

A benzene–toluene system with a different feed split of hot:cold (25:75), where Feed1 (hot) is at stage 19 and Feed2 (cold) is at stage 13 is evaluated in Fig. 1.5.4.

It is found that this change in feed split requires a readjustment of the feed trays. The case in Fig. 1.5.4 shows the minimum reboiler duty, condenser duty and reflux ratio from the three cases considered, and the results are summarized in Table 1.5.1.

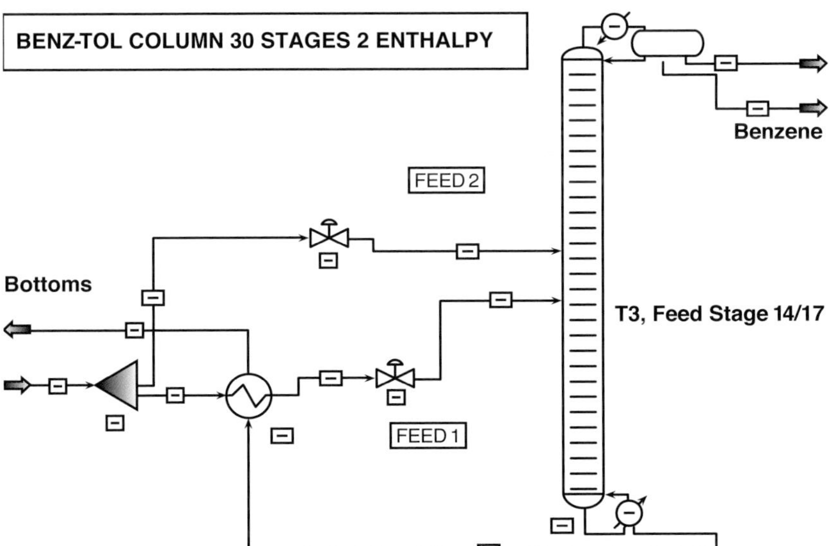

Column Name Column Description		T3
Condenser Duty	M*KCAL/HR	−1.7226
Reboiler Duty	M*KCAL/HR	1.7924
Column Reflux Ratio		1.3890

Fig. 1.5.3 B-T system with a different feed split of cold:hot (25:75)

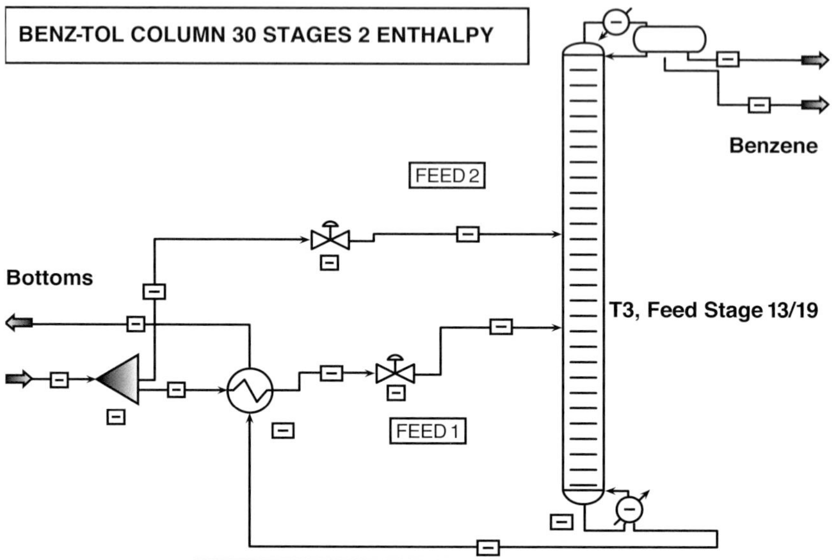

Column Name Column Description		T3
Condenser Duty	M*KCAL/HR	−1.6925
Reboiler Duty	M*KCAL/HR	1.7623
Column Reflux Ratio		1.3472

Fig. 1.5.4 Benzene–toluene system with a different feed split of hot:cold (25:75)

Table 1.5.1 Condenser and reboiler duties with single feed and two-feed splits

Different feed arrangements	Condenser duty, Gcal/hr	Reboiler duty, Gcal/hr	Reflux ratio
Single feed on stage 15	−1.7362	1.8060	1.4079
Split feed (50:50) on stage 20 and 12	−1.7025	1.7723	1.3612
Split feed (75:25) on stage 17 and 14	−1.7226	1.7924	1.3890
Split feed (25:75) on stage 19 and 13	−1.6925	1.7623	1.3472

In Table 1.5.1 the feed in the upper location is the cold feed. Table 1.5.1 shows that the condenser duty, reboiler duty, and reflux ratio are lowest when the cold feed in the upper location is higher than the hot feed

Analysis of two-enthalpy feed with the concept of exergy

For further analysis, loss of work (*LW*) is estimated for the two-enthalpy case from Example Problem 1.5.1, where a benzene–toluene system is again considered with 30 stages. The loss of work is calculated for the case with a:

- Single feed location
- Two-enthalpy feed introduction at two different locations

Lost work with a single feed location is given in Table 1.5.2.

Table 1.5.2 Lost work (LW) with stage number

LW @stage no. (Mkcal/hr)	@14	@15	@16
200 kmoles hot at single location	0.192857	0.192835	0.192995

Table 1.5.2 indicates that the loss of exergy is *minimum* at feed stage 15 for a single feed; therefore, feed stage 15 is the best feed point.

Lost work with two feed locations with different feed split

The feed is split between hot and cold and introduced at different locations, and shown in contrast with a preheated feed at a single location (stage 15).

In the same problem (Example Problem 1.5.1), a case is considered where feed is introduced at two points of the column with different enthalpies. The exergy loss is calculated for different locations of hot and cold feed introduction with different feed splits to identify minimum loss of exergy. Lost work (*LW*) with different feed split and stage number is presented in Table 1.5.3.

Table 1.5.3 Lost work (*LW*) with different feed split and stage number

LW @stage no. (Mkcal/hr)	Hot feed at stage 21 and cold at stage 11	Hot feed at stage 20 and cold at stage 12	Hot feed at stage 19 and cold at stage 13
Feed 50 (hot):50 (cold) split	0.190683	0.190411	0.190438
LW @stage no. (Mkcal/hr)	Hot feed at stage 20 and cold at stage 12	Hot feed at stage 19 and cold at stage 13	Hot feed at stage 21 and cold at stage 11
Feed 25 (hot):75 (cold) split	0.189732	0.189673	0.190167

Table 1.5.3 shows that the lost work (*LW*) is minimum with a 25:75 split of hot and cold feed split-fed to stages 19 and 13. This is corroborated by the data in Table 1.5.1, where it can be seen that the lowest reboiler duty (1.7623 Gcal/hr), condenser duty (−1.6932 Gcal/hr), and reflux ratio (1.3483) occur when feed is 25% hot at stage 19 and 75% cold at stage 13. This phenomenon is further examined with liquid composition through the stages with a benzene–toluene column of 30 stages in Fig. 1.5.4A for a single feed location.

Fig. 1.5.4A shows that separation was very poor from stages 12 to 19. This can be resolved by introducing feed at two locations with identical compositions but different enthalpies. A similar diagram is plotted with a two-enthalpy feed at two different locations in Fig. 1.5.4B.

Fig. 1.5.4B shows that once the feed is split, the separation improves, and the reboiler duties, condenser duties, and reflux ratio(s) come down. The above concept (two-enthalpy feed) is further applied to a three-component distillation system: a benzene–toluene–*para*-xylene system.

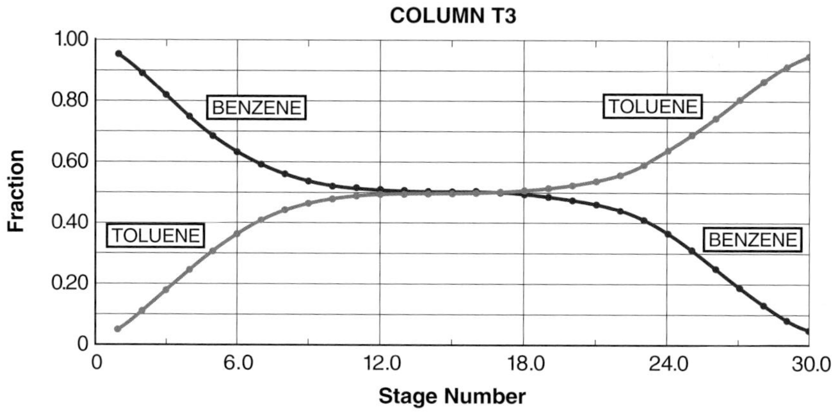

Fig. 1.5.4A Benzene–toluene liquid composition through the stages with a single feed

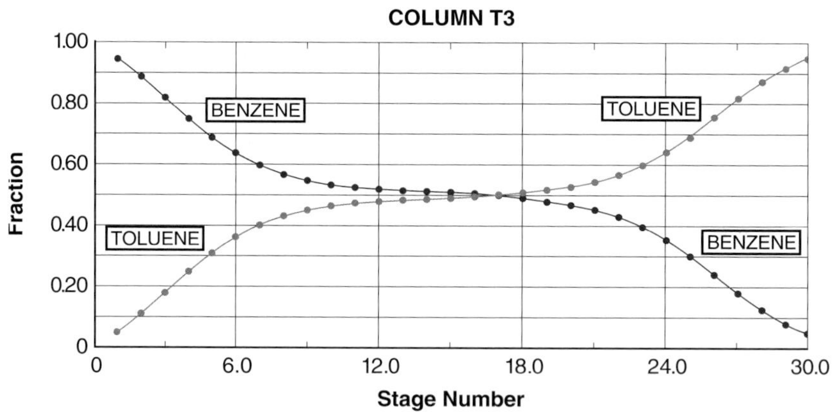

Fig. 1.5.4B Benzene–toluene liquid composition through the stages with a split feed

Example Problem 1.5.3

Feed: 60:20:20 benzene:toluene: p-xylene (mol:mol:mol)

Total feed = 200 kmoles/hr

Top product = 95% mole benzene with a 5 mole % benzene slip in bottom product

Feed1 = 50% of total feed is fed to feed stage 15 with feed preheat achieved by preheater duty of 0.1958 Gcal/hr

Feed2 = 50% of total directly fed to the same feed stage 15. Total number of stages = 30.

Cases of single-point feed and split feed can be compared using Example Problem 1.5.3.

Fig. 1.5.5 shows a B–T–PX column with the same feed location at stage 15.

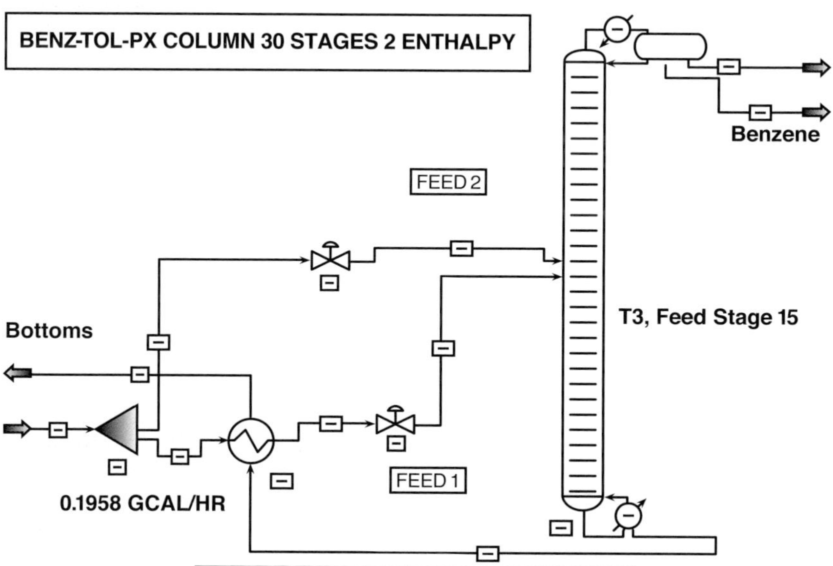

Column Name Column Description		T3
Condenser Duty	M*KCAL/HR	−1.5644
Reboiler Duty	M*KCAL/HR	1.6352
Column Reflux Ratio		1.1697

Fig. 1.5.5 B-T-PX column with the same feed location at stage 15

Feed1 = 50% of total feed with preheat fixed by the same preheater duty of 0.1958 Gcal/hr fed to feed stage 20

Feed2 = 50% of total directly fed to feed stage 12

Total number of stages = 30

A benzene–toluene–*para*-xylene split feed, where Feed1 is fed in at stage 20 and Feed2 at stage 12, is given in Fig. 1.5.6.

Summary of the results from Fig. 1.5.5 and Fig. 1.5.6 are given in Table 1.5.4.

The results of Fig. 1.5.6 show that in a three-component system, a split of the two-enthalpy feed is beneficial in terms of the reboiler duty, condenser duty, and reflux ratio required for separating products of the same purity compared to a single-point or conventional feed, which can be seen in Fig. 1.5.5 and Fig. 1.5.6.

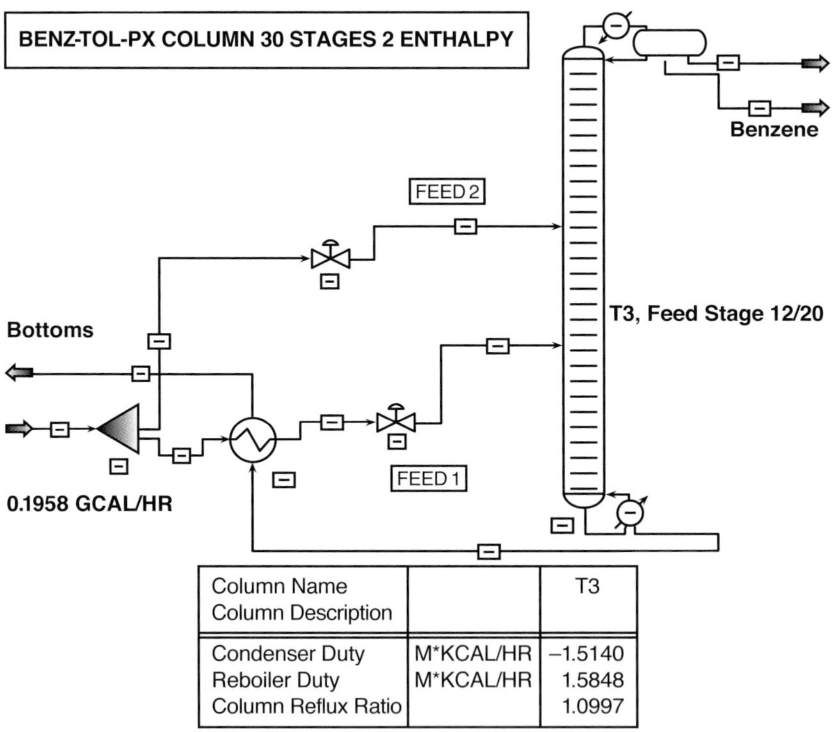

Fig. 1.5.6 B-T-PX split feed: Feed1 at stage 20, Feed2 at stage 12

Table 1.5.4 Summary of results from the comparison of B-T-PX columns with single and split feeds

Parameter	Units	Single feed location 15	Split feed at 12 & 20
Condenser duty	Mkcal/hr	−1.5619	−1.5140
Reboiler duty	Mkcal/hr	1.6326	1.5848
Column (R.R)		1.1661	1.0997

1.6 Industrial Applications of Two-Enthalpy Feeds

A debutanizer column (deployed for recovering butane and lighters from naphtha) in a refinery can be revamped for capacity augmentation through the introduction of a two-enthalpy feed concept.

Example Problem 1.6.1

A typical debutanizer is considered with a 30 MT/hr feed rate. The feed is split in two parts: 50% is routed through a heat exchanger and the other 50% fed cold directly to the column. The specification used for liquefied petroleum gas (LPG) is 1% wt pentane maximum, and the bottom naphtha specification for butane content is 5 mole %. The feed is introduced in the single feed location in a column with a total of 32 stages to generate the base case. Both preheated naphtha and cold naphtha are fed to stage 18. Preheated naphtha is at 115°C (partially vaporized with a preheater duty of 0.67 Gcal/hr). The reflux ratio, and reboiler and condenser duties are calculated. Data for a debutanizer with a single-point feed is given in Fig. 1.6.1.

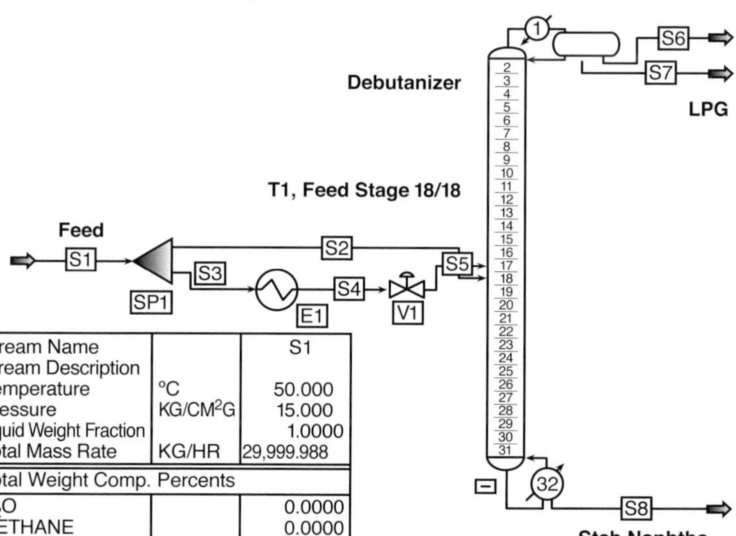

Stream Name		S1
Stream Description		
Temperature	°C	50.000
Pressure	KG/CM^2G	15.000
Liquid Weight Fraction		1.0000
Total Mass Rate	KG/HR	29,999.988
Total Weight Comp. Percents		
H_2O		0.0000
METHANE		0.0000
ETHANE		0.0000
PROPANE		12.1058
IBUTANE		11.9675
BUTANE		7.9783
PENTANE		19.8074
HEXANE		11.8291
HEPTANE		20.6319
OCTANE		15.6800
NONANE		0.0000

Column Name		T1
Column Description		
Condenser Duty	M*KCAL/HR	−1.3021
Reboiler Duty	M*KCAL/HR	2.0590
Column Reflux Ratio		0.8611

Fig. 1.6.1 Debutanizer with a single-point feed at stage 18

Now the feed is divided (50:50) and fed at two points. Cold feed is fed at stage 15 and hot at stage 25. All other parameters including preheater duty are kept the same. The data for a debutanizer with double-point feed (two-enthalpy, same composition) is given in Fig. 1.6.2.

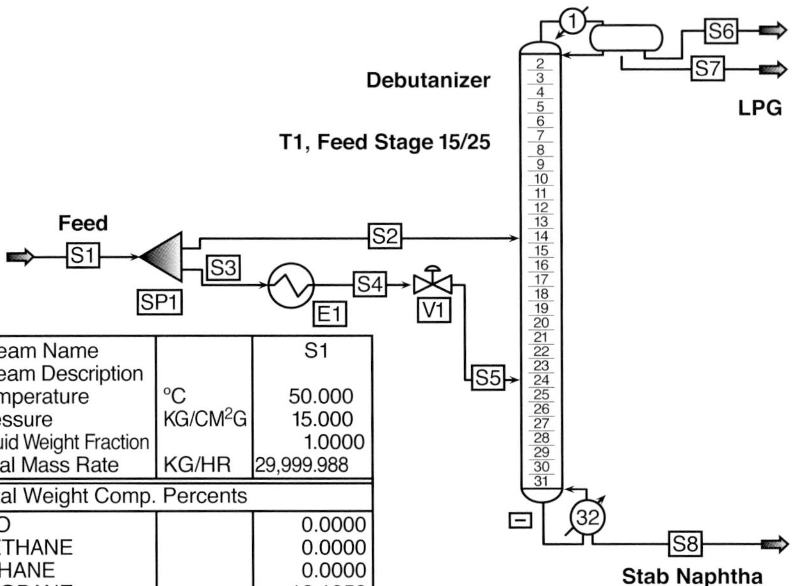

Stream Name		S1
Stream Description		
Temperature	°C	50.000
Pressure	KG/CM^2G	15.000
Liquid Weight Fraction		1.0000
Total Mass Rate	KG/HR	29,999.988
Total Weight Comp. Percents		
H_2O		0.0000
METHANE		0.0000
ETHANE		0.0000
PROPANE		12.1058
IBUTANE		11.9675
BUTANE		7.9783
PENTANE		19.8074
HEXANE		11.8291
HEPTANE		20.6319
OCTANE		15.6800
NONANE		0.0000

Column Name		T1
Column Description		
Condenser Duty	M*KCAL/HR	–1.1609
Reboiler Duty	M*KCAL/HR	1.9130
Column Reflux Ratio		0.6740

Fig. 1.6.2 Debutanizer with double-point feed

A summary of results from Fig. 1.6.1 (single feed point) and Fig. 1.6.2 (two feed points) is given in Table 1.6.1.

Table 1.6.1 Summary of results for a debutanizer with single vs. split feed

Parameter	Units	Single feed location 18	Split feed at 15 & 25
Condenser duty	Mkcal/hr	–1.3021	–1.1609
Reboiler duty	Mkcal/hr	2.0590	1.9130
Column (R.R)		0.8611	0.6740

Table 1.6.1 shows a substantial reduction in reboiler and condenser duties and reflux ratio when the two-enthalpy feed is introduced at two different locations. The principle of the two-enthalpy feed can be applied to a typical naphtha splitter used for splitting naphtha for the production of isomerate unit feed and feed for CCRU.

Example Problem 1.6.2

The example problem uses the example of a feed with 200 MT/hr of naphtha separating in isomerate feed and CCRU feed in the single (Fig. 1.6.3) and split feeds (Fig. 1.6.4) with the same specification of top (8% wt C7) and bottom product (benzene precursor of 3.5% wt) in a 22 stage column with the same preheater duty of 5.94 Gcal/hr. The simulation and data for an isomerate feed with CCRU separation and a single feed is given in Fig. 1.6.3.

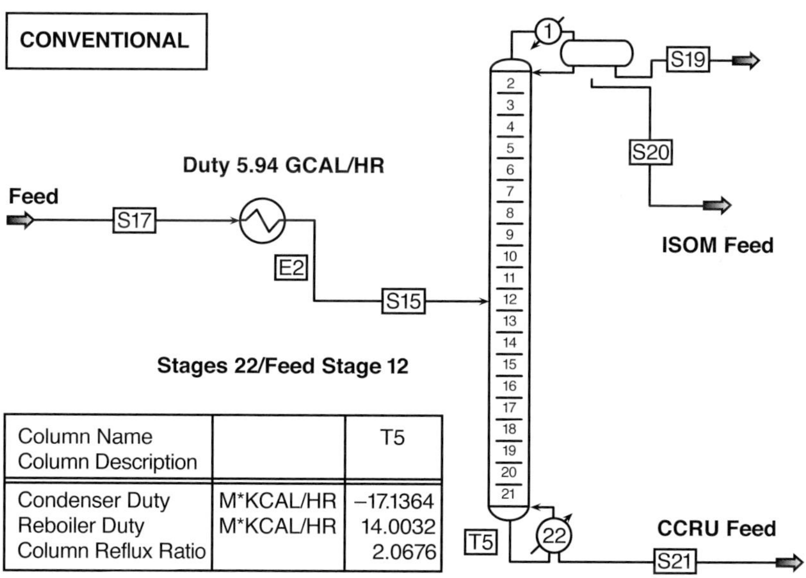

Fig. 1.6.3 Isomerate feed with CCRU separation and a single feed

Data for an isomerate feed with CCRU separation and a split feed is given in Fig. 1.6.4.

Fig. 1.6.4 shows a reduction of reboiler duty (around 9%–10%) and reflux ratio over the conventional (single feed) scheme shown in Fig. 1.6.3. The preheat duty is considered the same as 5.94 Gcal/hr.

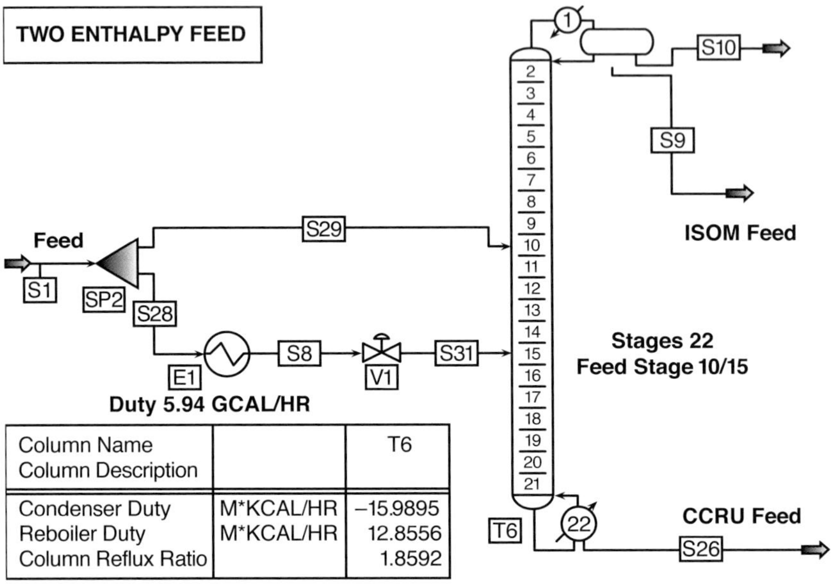

Fig. 1.6.4 Isomerate feed with CCRU separation and a split feed

1.7 Use of Intercoolers and Inter-Reboilers

1.7.1 Introduction

Intercoolers or inter-reboilers can be used to reduce condenser and rebioler duties. Some excess heat can be delivered to the inter-reboiler. Similarly, some excess heat can be removed through intercoolers. The inter-reboiler is normally placed below the feed tray and the intercooler is placed above it. The location and duties of inter-reboilers and intercoolers and their effects are explained in the example problem below.

Example Problem 1.7.1

Feed: 50:50 benzene: toluene, mole:mole

Total feed = 200 kmoles/hr

Top product is 95 mole % benzene, with 90 mole % toluene as bottom product.

The results of Example Problem 1.7.1 are listed below under Result 1.7.1 without the effects of an intercooler or inter-reboiler.

Results

Column reboiler/condenser duties without intercooler/inter-reboiler with saturated feed:

Column description: 15 stages

Condenser duty: –1.8462 Mkcal/hr

Reboiler duty: 1.9082 Mkcal/hr

Column reflux ratio: 1.7204

Now, in Example Problem 1.7.1, an intercooler duty of 0.6 Gcal/hr is used at stage 2 and stage 4 and results are listed below.

Example Problem 1.7.1A Column reboiler/condenser duties with intercooler and saturated feed (stage 2)

Intercooler duty = 0.6 Gcal/hr at stage 2

Results

Condenser duty: –1.2465 Mkcal/hr

Reboiler duty: 1.9087 Mkcal/hr

Column reflux ratio: 0.8367

Example Problem 1.7.1B Column reboiler/condenser duties with intercooler and saturated feed (stage 4)

Intercooler duty = 0.6 Gcal/hr at stage 4

Results

Condenser duty: –1.3044 Mkcal/hr

Reboiler duty: 1.9665 Mkcal/hr

Column reflux ratio: 0.9222

A summary of the results for parameters with and without intercoolers at different stages is presented in Table 1.7.1.

Table 1.7.1 Summary of parameters with and without intercoolers at different stages

Parameter	Units	Without intercooler	Intercooler at stage 2	Intercooler at stage 4
Condenser duty	Mkcal/hr	–1.8462	–1.2465	–1.3044
Reboiler duty	Mkcal/hr	1.9082	1.9087	1.9665
Reflux ratio		1.7204	0.8367	0.9222

Table 1.7.1 shows that reboiler duty and condenser duty both increase as the intercooler location is lowered. This suggests that the best point for installation of the intercooler is the topmost stage in the column; however, the temperature of the cooling agent has to be low, which means that the heat duty of the intercooler has to be delivered to a stream at a lower temperature or to a cold utility (air cooler or cooling water). Thus, the heat integration may not be very attractive from the point of view of heat recovery.

Use of an inter-reboiler

Wherever reboiler duty is a constraint, it can be supplemented by using an inter-reboiler with a lower-grade energy source. In Example Problem 1.7.1, an inter-reboiler is used at stage 10 (below the feed tray) and stage 13 and the results are listed.

Example Problem 1.7.1C Column reboiler/condenser duties with an inter-reboiler and saturated feed

Inter-reboiler duty = 0.6 Gcal/hr at stage 10 out of 15 total stages

Results

Condenser duty: –1.9514 Mkcal/hr

Reboiler duty: 1.4136 Mkcal/hr

Column reflux ratio: 1.8756

The reboiler duty has gone down but the condenser duty has increased. Thus, there is a penalty on the condenser duty.

Example Problem 1.7.1D Column reboiler/condenser duties with an inter-reboiler and saturated feed

Inter-reboiler duty = 0.6 Gcal/hr at stage 13 out of 15 total stages

Results

Condenser duty: –1.8628 Mkcal/hr

Reboiler duty: 1.3251 Mkcal/hr

Column reflux ratio: 1.7450

A summary of the results of parameters with and without an inter-reboiler at different stages is presented in Table 1.7.2.

Table 1.7.2 Summary of results of parameters with and without an inter-reboiler at different stages

Parameter	Units	Without inter-reboiler	Inter-reboiler at stage 10	Inter-reboiler at stage 13
Condenser duty	Mkcal/hr	−1.8462	−1.9514	−1.8628
Reboiler duty	Mkcal/hr	1.9082	1.4136	1.3251
Reflux ratio		1.7204	1.8756	1.7450

The results in Table 1.7.2 show that the reduction of reboiler duty is greater and the penalty on condenser duty decreased if the inter-reboiler is placed toward the bottom.

However, the temperature requirement for the heating fluid will increase as the inter-reboiler is lowered through the column.

An inter-reboiler gives the benefit of a feed preheater to reduce the reboiler duty more efficiently, but it is difficult to install in revamps. Therefore, feed preheaters are often preferred over inter-reboilers. Use of intercoolers and inter-reboilers has some penalty but is often helpful in reducing condenser/reboiler load as desired. When an intercooler is used, part of the intercooler duty can be imparted to the feed to preheat it. Often the same method is used in the refinery distillation illustrated in Chapter 3. With an inter-reboiler, some of the low-temperature excess heat can be meaningfully used.

1.8 Heat-Coupled Distillation/ Multi-Effect Distillation

1.8.1 Heat integration and coupling

The most preliminary heat integration is feed being heated by the hot bottom product of the column. Feed can also be heated by column over-head vapors and subsequently by bottom product if differential temperature permits. The bottom heat integration scheme is presented in Fig. 1.8.1.

Overhead heat integration with feed is shown in Fig. 1.8.2.

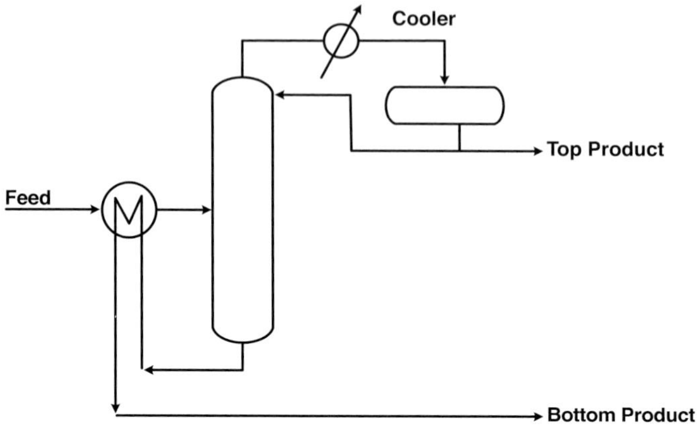

Fig. 1.8.1 Bottom heat integration with feed

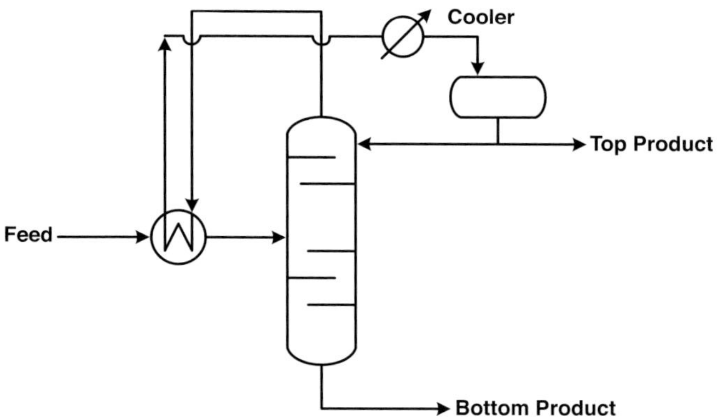

Fig. 1.8.2 Overhead heat integration with feed

1.8.2 Heat-coupled distillation and multi-effect distillation

Typically when successive columns are used, the overhead of the succeeding column is coupled with the reboiler of the preceding column; thus, the heat of the second column's overhead will be delivered to the reboiler of the first column. In order to achieve the heat transfer, the second column has to be operated at a higher pressure than the first column so that the dew point and the bubble point of the second column's overhead are at least 10°C higher than the dew point and the bubble point of the first column's bottom.

This will ensure heat transfer from the second column's overhead vapor to the bottom of the first column by maintaining a hot end and a cold end differential temperature (DT) of 10°C around the first column's reboiler. Often a trim condenser is provided in the second column to achieve total condensation of the overhead and to return the condensate to the reflux drum. Typically, with an increase in pressure, relative volatility tends to move toward unity, necessitating higher stages and reflux, and may need higher energy for separation in the second column. However, the net heat saved with an overhead heat coupling is appreciable. The first column is normally provided with a balancing reboiler over and above the vapor-coupled one for startup and balancing purposes. The heat input in the balancing reboiler gets reduced when vapor coupling becomes operational. The conventional processing of a two-column distillation with three components is shown in Fig. 1.8.3.

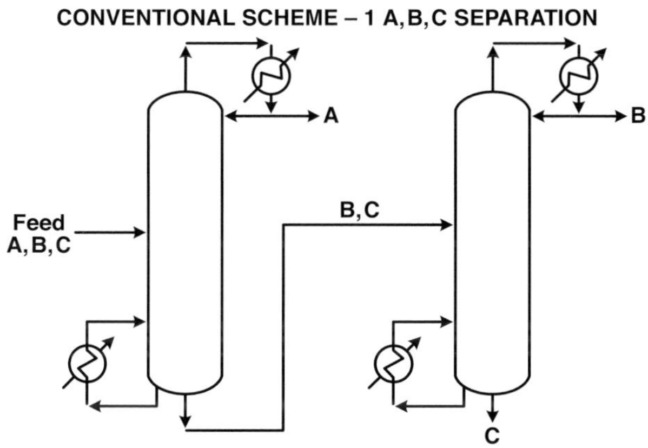

Fig. 1.8.3 Direct feed scheme with no heat coupling

Heat-coupled scheme

A scheme of heat coupling is shown below in Fig. 1.8.4. The second column operates at higher pressure to provide a temperature differential across the first column reboiler.

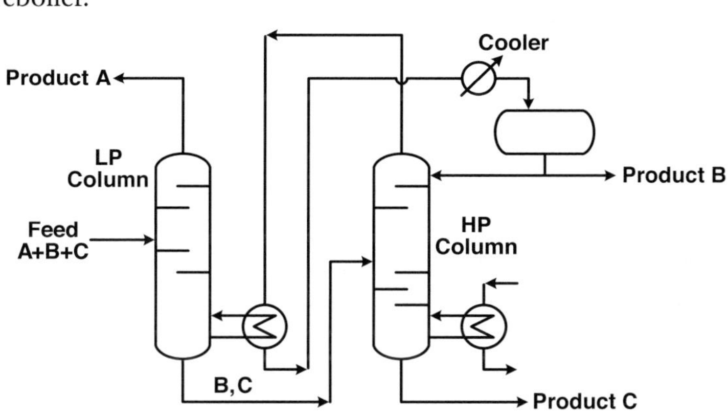

Fig. 1.8.4 Overhead of the second column coupled with the reboiler of the first

Some industrial applications of overhead heat coupling are shown below.

An overhead integration of a xylene column in a *para*-xylene complex is shown in Fig. 1.8.5.

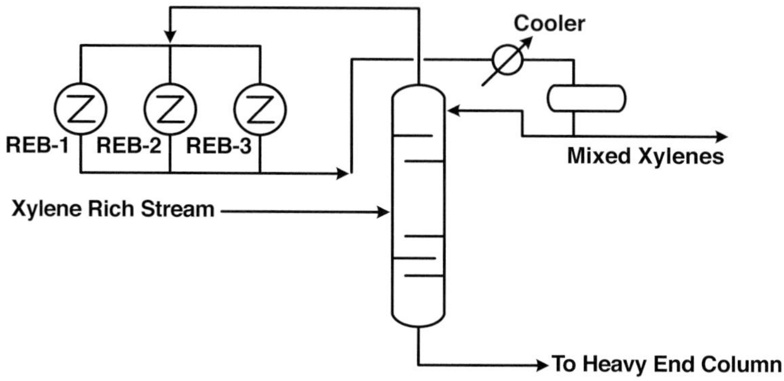

Fig. 1.8.5 Xylene column with overhead integration

In the scheme in Fig. 1.8.5, the overhead of the xylene column is supplying reboiler heat to various columns in the plant. The xylene fractionator is typically operated at about 7 kg/cm^2(g) to raise the overhead vapor dew point and bubble point temperatures so that overhead heat can be delivered to the reboilers of different columns.

A butene-1 plant with heat coupling with the reboiler of the first column is shown in Fig. 1.8.6.

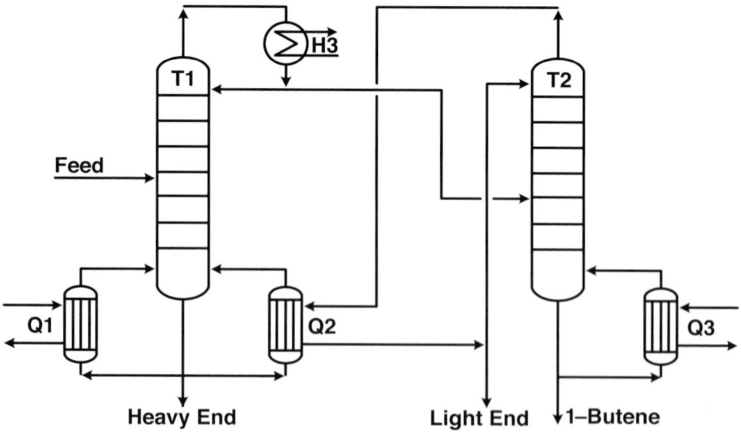

Fig. 1.8.6 Butene-1 plant with heat coupling with the reboiler of the first column

Some butene-1 plants also use the scheme of overhead coupling seen in Fig. 1.8.6. Q2 is shown as the overhead coupled reboiler and Q1 as the balancing reboiler of column T1. Q3 is shown as the reboiler of the second high-pressure column, T2.

1.8.3 Multi-effect distillation schemes

Following are the schemes used in multi-effect distillation:

1. Split feed
2. Forward feed
3. Backward feed
4. Column with a pre-splitter

A split feed arrangement with overhead heat coupling is shown in Fig. 1.8.7.

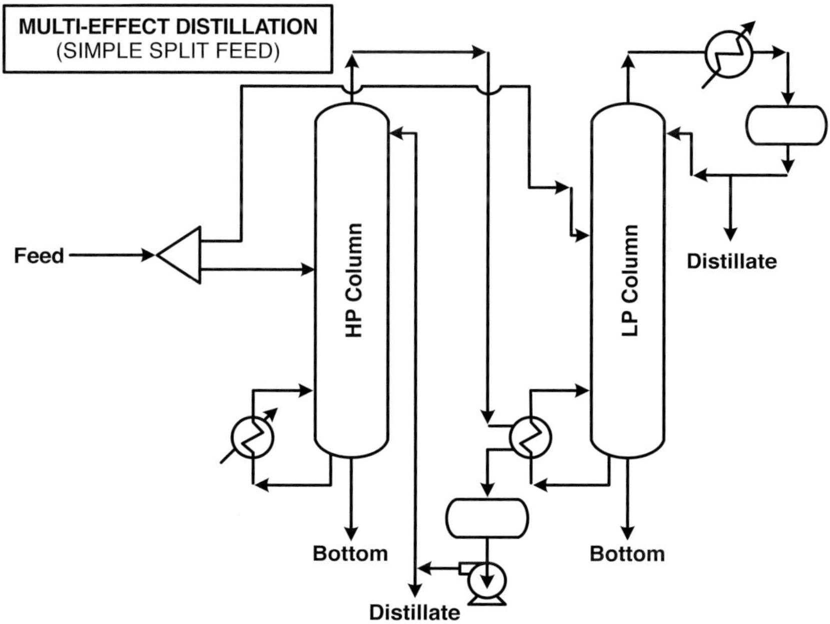

Fig. 1.8.7 Split feed arrangement with overhead heat coupling

Fig. 1.8.7 shows a split feed arrangement where the feed is split into two parts. One part is fed to the high-pressure column (HP column) and the rest to the low-pressure column (LP column).

A multi-effect forward/backward feeding scheme is given in Fig. 1.8.8.

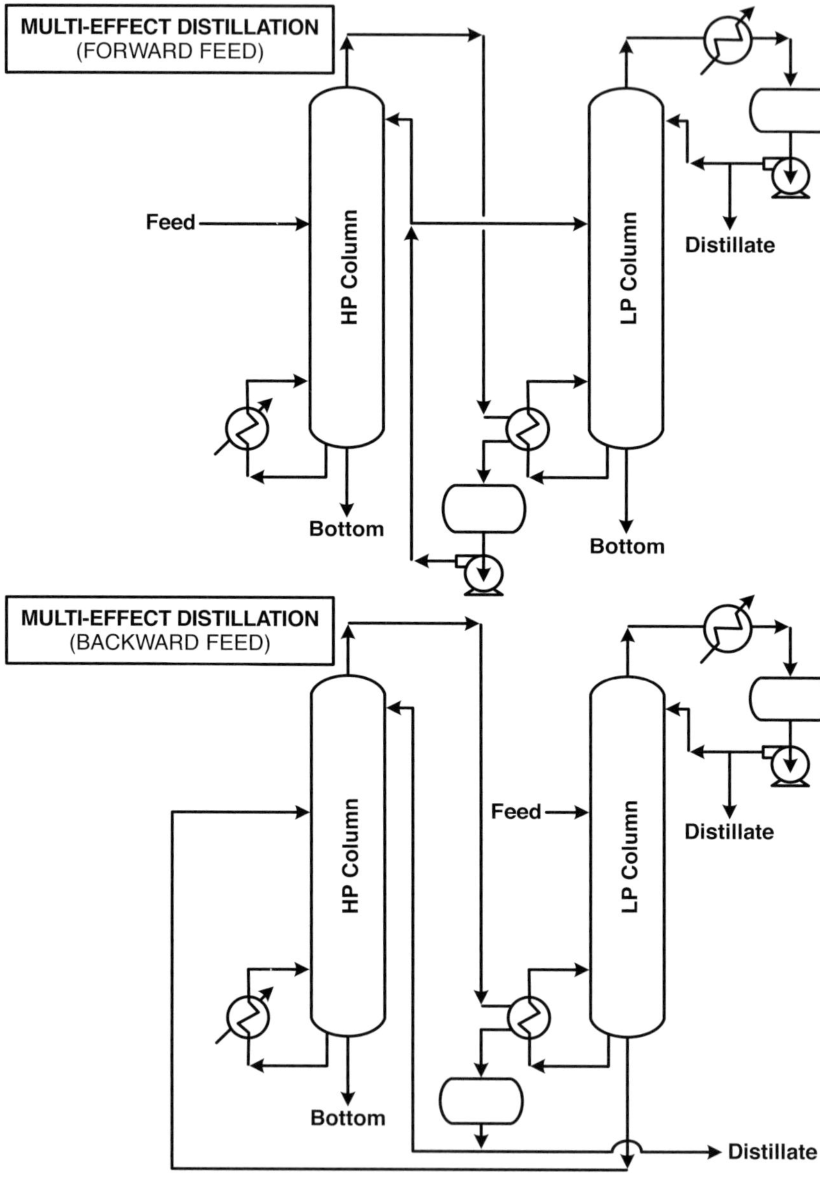

Fig. 1.8.8 Multi-effect forward/backward feeding scheme

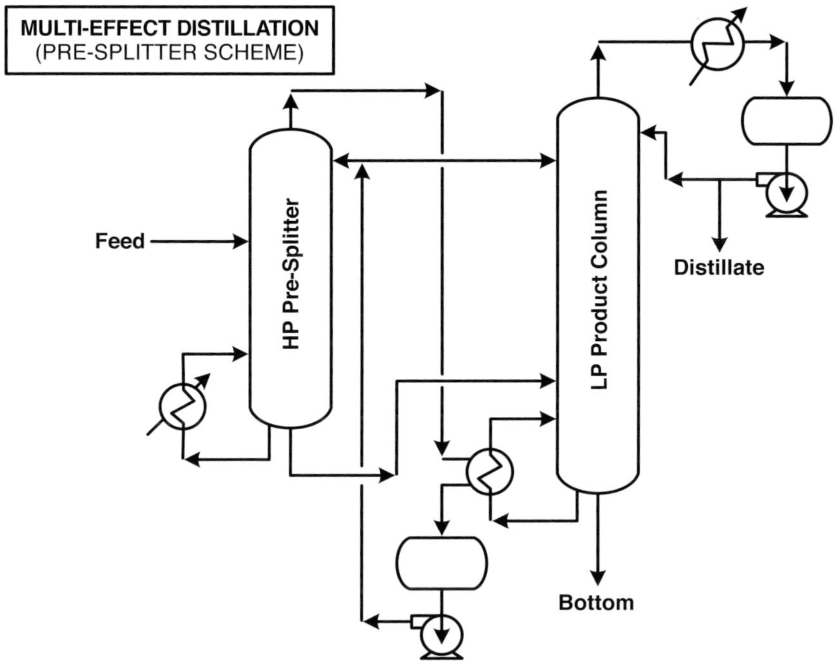

Fig. 1.8.9 Multi-effect with a high-pressure pre-splitter

Example of a benzene–toluene separation with multi-effect distillation

An example for comparing different multi-effect schemes along with their benefits is considered in Example Problem 1.8.1.

Example Problem 1.8.1

Feed rate: 200 kmole/hr

Feed composition: benzene:toluene (50:50 mole:mole)

The desired purity of benzene is 95 mole % and that of toluene is 90 mole %. The feed temperature considered is 80°C at 9 kg/cm^2(a).

Different multi-effect distillation schemes can be tried and compared with a conventional scheme with the same number of total stages (27). The conventional scheme of a B–T system is shown in Fig. 1.8.10.

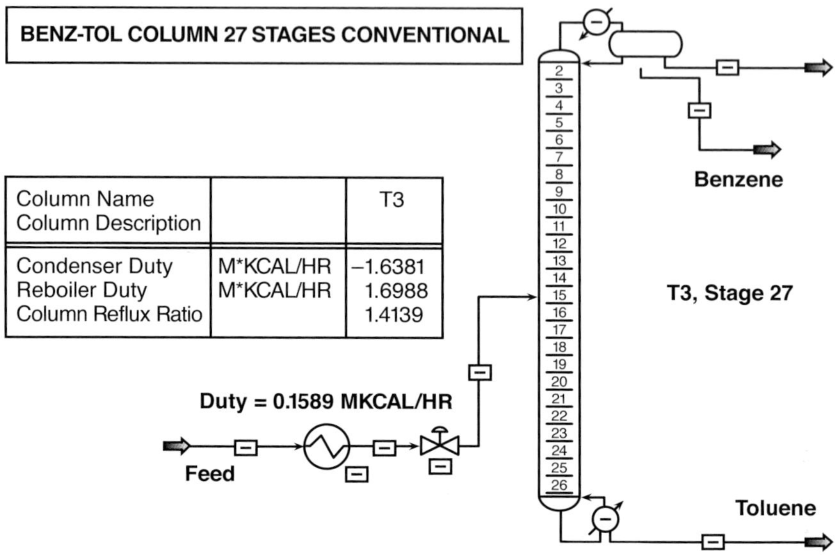

Fig. 1.8.10 Conventional B-T separation

In Fig. 1.8.10, the column is operated at 0.3 kg/cm^2(g) pressure. Multi-effect schemes for the example problem are configured with two columns: one (the HP column) at a higher pressure (4 kg/cm^2(g)) and the other (the LP column) at a lower pressure (0.3 kg/cm^2(g)). Each scheme deploys a total of 27 stages and the same feed rates, feed conditions, and product purities. Different multi-effect schemes are shown below.

1. **Split feed: First column with 12 stages, second column with 15 stages**

The feed can be split into two columns, with part going to the HP column and part to the LP column. The HP column's overhead supplies reboiler heat to the LP column. The reboiler in the HP column would be heated via external hot fluid. The total number of stages in the HP and LP columns is 27. The performance with respect to the reboiler and condenser duties of both columns is compared with the conventional 27 stages of a single distillation column and the same feed preheat of 80°C.

The simulation run in Fig. 1.8.11 using a heat-coupled reboiler in the LP column shows a conservation of approximately 25% in reboiler duty as compared to a conventional column.

Some additional benefit may be available, as preheat to the HP column can be increased with an increase in the temperature of the HP column's bottom fluid. This method shows substantial savings in reboiler duty. It may be a good practice to install a second column and couple the overhead heat to increase the capacity of a unit and save energy. A B–T system with this split feed arrangement is shown in Fig. 1.8.11.

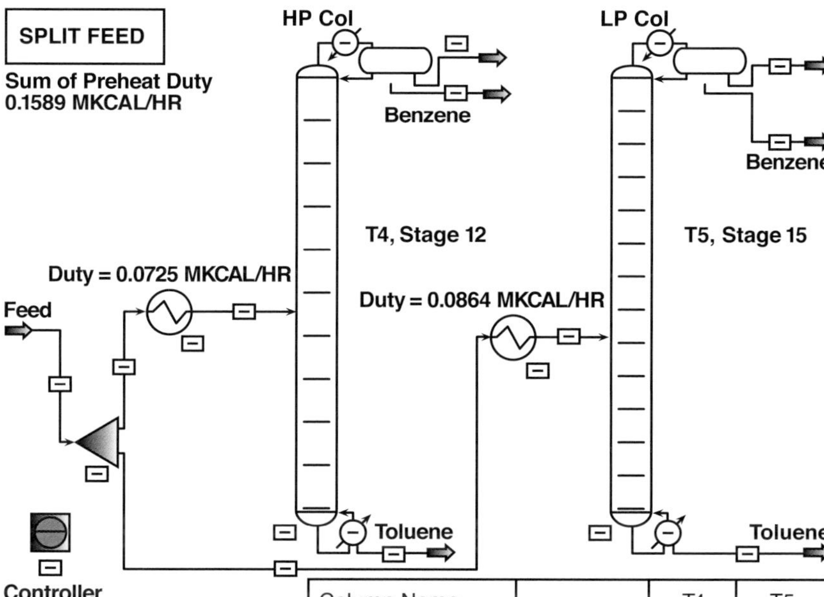

Column Name Column Description		T4	T5
Condenser Duty	M*KCAL/HR	−1.2044	−1.1668
Reboiler Duty	M*KCAL/HR	1.3968	1.2045
Column Reflux Ratio		0.0000	1.7243

Fig. 1.8.11 B–T system with split feed

Simulation features

In Fig. 1.8.11, a controller is used to make the HP column condenser duty equal to the LP column reboiler duty by varying the flow of feed (flow split) to the HP column and LP column. Thus, the HP column reboiler duty must be compared with the reboiler duty of a conventional column. HP column pressure is chosen to deliver the required differential temperature for the transfer of overhead duty from the HP column to the LP column reboiler.

The feed is split into two almost equal parts. One part (nearly 50%) is fed to the HP column, and the remainder is fed to the LP column.

2. **Forward feed with two stages of distillation: First column with 12 stages, second column with 15 stages**

The feed is introduced in the HP column. The overhead of the HP column supplies heat to the LP column reboiler and splits as feed to the LP column and reflux to the HP column. In the HP column, part of the separation will be done and the feed to the LP column will be an enriched feed. The heavier component will be recovered from both column bottoms. The lighter component will be recovered from the LP column top. Comparison with a conventional column shows an initial reboiler heat conservation of around 2%–3%, even without considering any additional heat savings from likely higher preheat of feed. The scheme is likely to provide higher conservation of reboiler duty when the feed is rich with components with low volatility (more toluene).

A benzene–toluene system with forward feed is shown in Fig. 1.8.12.

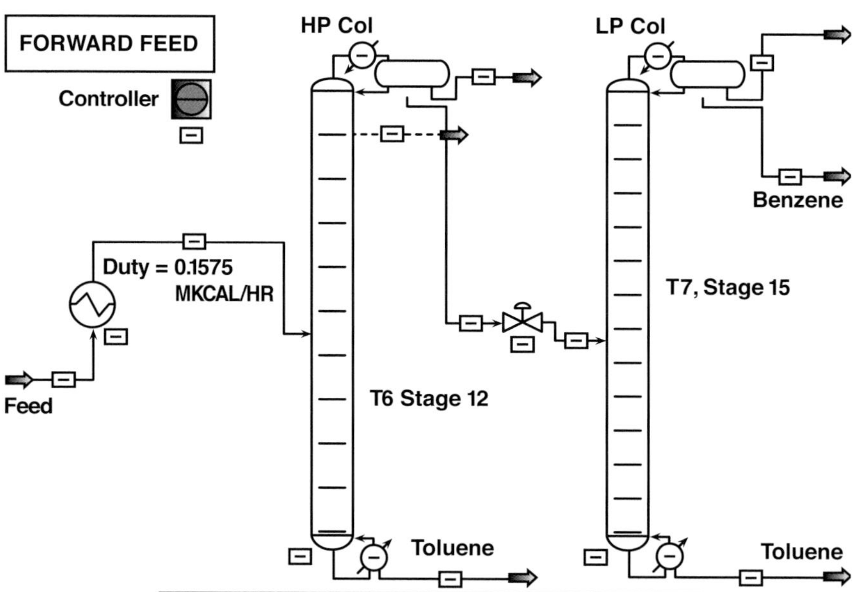

Column Name Column Description		T6	T7
Condenser Duty	M*KCAL/HR	−1.2497	−1.4705
Reboiler Duty	M*KCAL/HR	1.6513	1.2497
Column Reflux Ratio		0.3001	1.1669

Fig. 1.8.12 B–T system with a forward feed arrangement

Simulation features

A controller is used to make the HP column condenser duty equal to the LP column reboiler duty by varying the composition of the HP column overhead. Thus, the HP column reboiler duty must be compared with the reboiler duty of a conventional column.

HP column pressure is chosen to deliver the required differential temperature for the transfer of overhead heat duty from the HP column to the LP column reboiler. The feed point in the first column (HP) is preferred at upper zone and is based on the composition of the feed.

3. **Backward feed with a two-stage distillation: First column with 12 stages and second column with 15 stages**

The feed is first introduced in the LP column, and the bottom of the LP column is fed to the HP column. Part of the distillate (benzene) is withdrawn from the overhead of the LP column while the remaining distillate is withdrawn from the top of the HP column. The bottom (toluene) is withdrawn from the bottom of the HP column. A conservation of around 6% reboiler duty is seen with this arrangement. This scheme is likely to conserve even more reboiler duty when the feed is rich in a high-volatile component (more benzene).

An example of a benzene–toluene system with backward feed is given in Fig. 1.8.13.

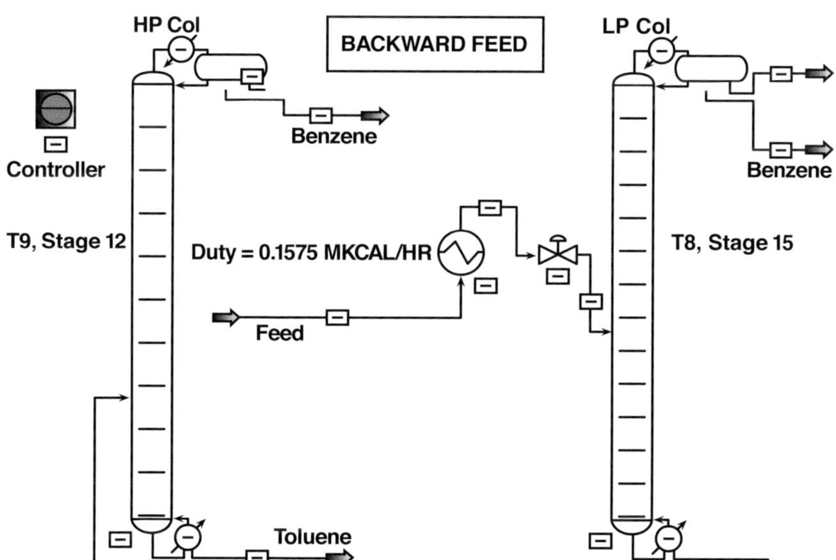

Column Name Column Description		T9	T8
Condenser Duty	M*KCAL/HR	−1.3204	−1.2623
Reboiler Duty	M*KCAL/HR	1.5750	1.3196
Column Reflux Ratio		6.8006	1.5530

Fig. 1.8.13 B–T with backward feed

Simulation features

A controller is used to make the HP column condenser duty equal to the LP column reboiler duty by varying the composition of the LP column bottom. Thus, the HP column reboiler duty must be compared with the reboiler duty of a conventional column. HP column pressure is chosen to deliver the required differential temperature for the transfer of overhead duty from the HP column to the LP column reboiler.

4. **Pre-splitter scheme where pre-splitter has 12 stages preceding splitter (15 stages)**

In this scheme (shown in Fig. 1.8.14), the feed is introduced in the HP pre-splitter column. The overhead vapor transfers its heat to the LP column reboiler, part of the overhead liquid is refluxed into the HP column, and the balance is fed to the LP column at an upper location. The bottom of the HP column is fed into the LP column at a lower location. The HP column operates at 4 kg/cm^2 to provide the differential temperature for heat transfer to the LP column reboiler. The LP column top is distillate and LP column bottom is the bottom product. A reboiler heat saving of around 20% is achieved.

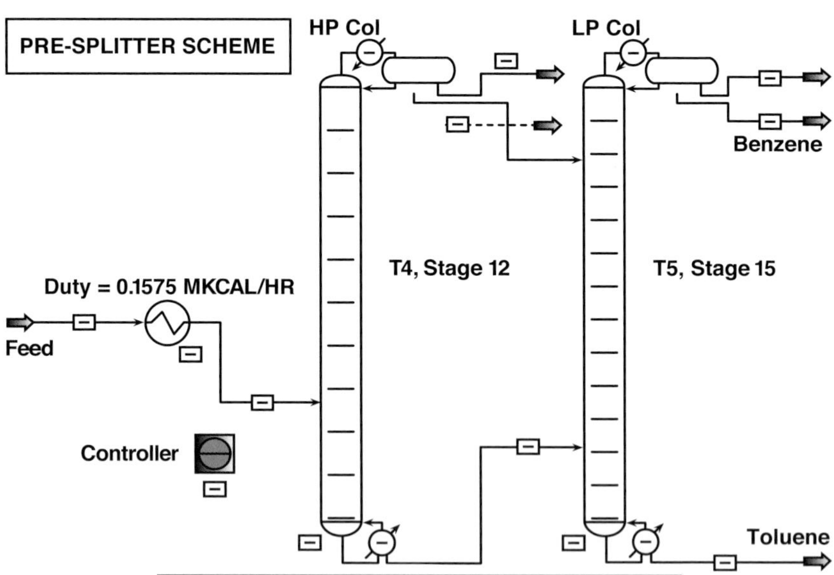

Column Name Column Description		T4	T5
Condenser Duty	M*KCAL/HR	–0.9997	–1.3483
Reboiler Duty	M*KCAL/HR	1.4108	0.9993
Column Reflux Ratio		1.2223	0.9868

Fig. 1.8.14 B-T with high-pressure pre-splitter

An example of a benzene–toluene system with a pre-splitter and heat coupling is presented in Fig. 1.8.14.

Simulation features

A controller is used to make the HP column condenser duty equal to the LP column reboiler duty by varying the composition of the HP column overhead. Thus, the HP column reboiler duty must be compared with the reboiler duty of a conventional column. HP column pressure is chosen to deliver the required differential temperature for the transfer of overhead duty from the HP column to the LP column reboiler. A partly pure (68%) toluene specified as the bottom purity of the HP column is controlling the reboiler duty of the HP column.

Essentially, a pre-splitter scheme using a reboiler and a condenser is similar to a column having an inter-reboiler and intercooler and an increased number of stages. The pre-splitter's reboiler serves as an inter-reboiler to the main column while the condenser serves as an intercooler to the main column without actual installation of an intercooler and inter-reboiler in the main column. Thus, through the installation of a pre-splitter, modification of the main column can be avoided and some heat savings can be obtained through the net increase in stages. For all practical purposes, heat savings are obtained by coupling the overhead of the pre-splitter (operated at higher pressure) with the reboiler of the main column.

This may be a very useful scheme to debottleneck existing columns for higher capacity without doing much modification to the existing column aside from adding a column in series upstream. The addition of more stages in the pre-splitter reduces the reflux ratio in the main column along with the condenser and reboiler duties. This in turn reduces the load of the main column. Substantial heat savings are also obtained through heat coupling.

A summary of the results of the different heat coupling schemes from Example Problem 1.8.1 is presented in Table 1.8.1 for the benzene–toluene separation. This compares the reboiler/condenser duties of schemes such as the split feed, forward feed, backward feed and pre-splitter feed schemes against the conventional single column scheme. The net heat duty of the condenser and reboiler is the heat duty required after coupling the overhead of the HP column with the reboiler of the LP column.

Table 1.8.1 Summary of net reboiler and condenser duties for various heat coupling schemes

Configurations	Net condenser duty, Gcal/hr	Net reboiler duty, Gcal/hr
Conventional single column	−1.6329	1.6949
Split feed in two columns	−1.1668	1.3968
Forward feed two columns	−1.4705	1.6513
Backward feed two columns	−1.2624	1.5746
Pre-splitter and splitter	−1.3484	1.4108

Example Problem 1.8.2

A refinery naphtha splitter producing ISOM feed and CCRU feed is considered as an example of how a splitter can be debottlenecked and heat can be saved using this pre-splitter scheme. The CCRU feed has a specification of C6 precursors content of around 3.5% maximum weight, and the ISOM feed has a limit of 8% weight for C7+ content. The feed preheat duty in all cases is considered the same and feed rate set at 200 MT/hr.

The conventional scheme is presented in Fig. 1.8.15 with 30 stages.

The pre-splitter scheme is shown in Fig. 1.8.16 with no overhead coupling. The reflux ratio in the splitter reduces considerably and thus provides cushion for capacity increase with the addition of the pre-splitter upstream.

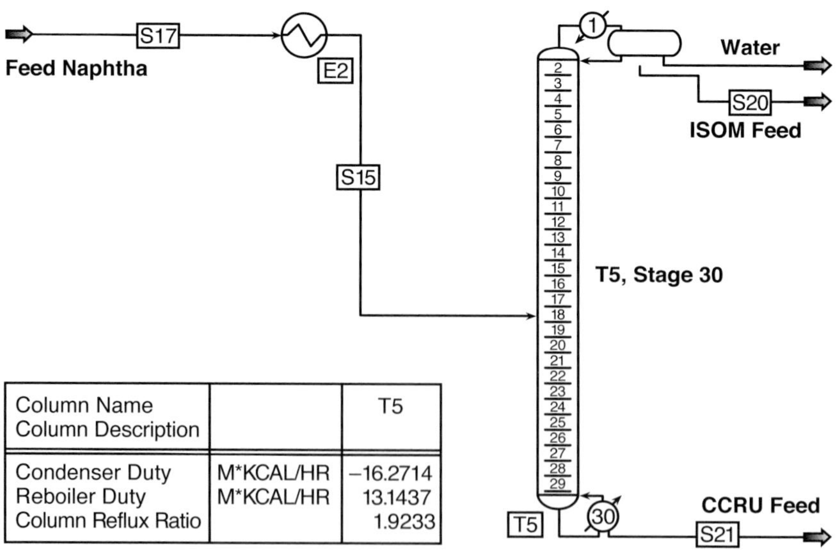

Fig. 1.8.15 ISOM/CCRU feed separation using a conventional scheme with 30 stages

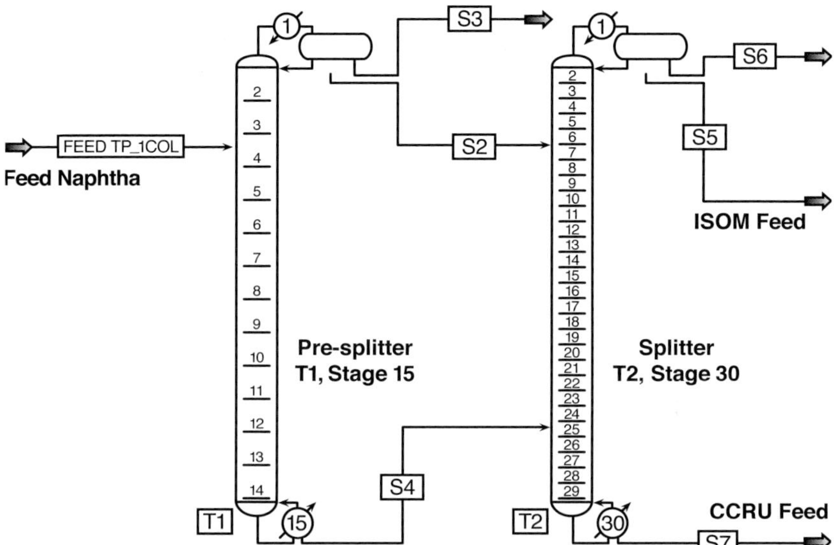

Column Name Column Description		T1	T2
Condenser Duty	M*KCAL/HR	–6.8770	–12.4723
Reboiler Duty	M*KCAL/HR	3.0000	13.2017
Column Reflux Ratio		0.3000	1.2192

Fig. 1.8.16 ISOM/CCRU feed separation with pre-splitter and no coupling

In the pre-splitter with an overhead integration scheme, the pre-splitter is operated at 8 kg/cm^2(a). The pre-splitter overhead is coupled with main splitter reboiler (Fig. 1.8.17).

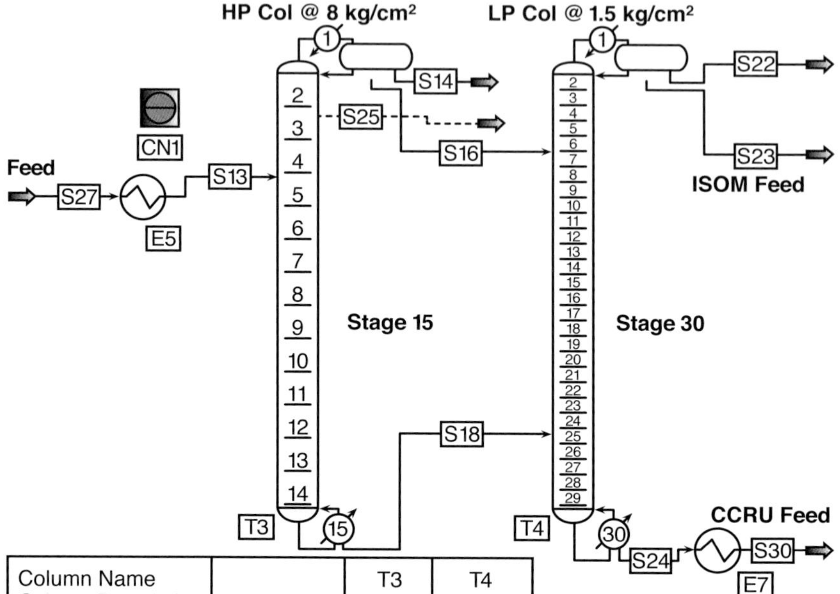

Column Name Column Description		T3	T4
Condenser Duty	M*KCAL/HR	−7.0004	−15.1963
Reboiler Duty	M*KCAL/HR	12.5000	7.0022
Column Reflux Ratio		0.5948	1.8016

Fig. 1.8.17 ISOM/CCRU feed separation with pre-splitter and overhead coupling of pre-splitter top with splitter bottom

Multi-effect scheme compared with two-enthalpy scheme

A two-enthalpy feed can also be tried in naphtha splitting for ISOM feed generation and CCRU feed generation. ISOM/CCRU feed generation using two-enthalpy feed is shown in Fig. 1.8.18.

ISOM/CCRU feed separation using a conventional scheme with 45 stages is presented in Fig. 1.8.19 for better comparison.

A summary of the condenser/reboiler duties of naphtha splitter schemes is presented in Table 1.8.2.

Net reboiler duty requirement in a pre-splitter with overhead coupling (Fig. 1.8.17) is found to be 12.50 Gcal/hr, as opposed to the 13.14 Gcal/hr of a conventional scheme with 30 stages (Fig. 1.8.15) and the 12.62 Gcal/hr of one with 45 stages (Fig. 1.8.19). The pre-splitter scheme without overhead coupling (Fig. 1.8.16) reduces the reflux ratio in the main splitter (debottlenecks the diameter limitation) and provides flexibility for capacity augmentation. It can be seen in Fig. 1.8.18 that a two-enthalpy feed using the same preheater duty can also partially reduce the reboiler duty, condenser duty, and reflux ratio of the splitter. Process engineers will need to examine the alternatives and decide which scheme is best based on individual requirements.

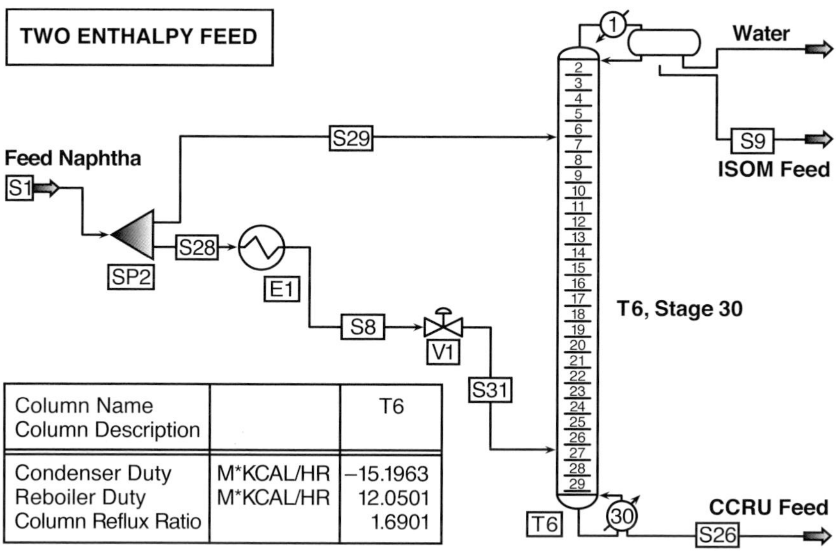

Column Name Column Description		T6
Condenser Duty	M*KCAL/HR	−15.1963
Reboiler Duty	M*KCAL/HR	12.0501
Column Reflux Ratio		1.6901

Fig. 1.8.18 ISOM/CCRU feed generation using a two-enthalpy scheme

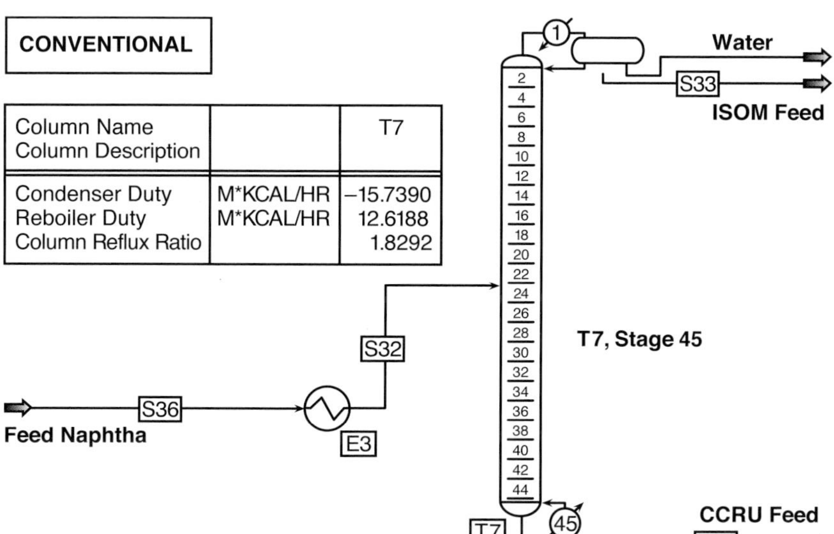

Column Name Column Description		T7
Condenser Duty	M*KCAL/HR	−15.7390
Reboiler Duty	M*KCAL/HR	12.6188
Column Reflux Ratio		1.8292

Fig. 1.8.19 ISOM/CCRU feed separation using a conventional scheme with 45 stages

Table 1.8.2 Summary of condenser/reboiler duties of naphtha splitter schemes

Type of scheme	Net condenser duty, Gcal/hr	Net reboiler duty, Gcal/hr	Splitter reflux ratio
Conventional (30 stages)	−16.2714	13.1437	1.9223
Conventional (45 stages)	−15.7390	12.6188	1.8292
Pre-splitter with no overhead coupling (Fig. 1.8.16) (45 stages)	−(6.8773 + 12.4646) or −19.3419	3.0 + 13.2 = 16.2	1.217
Pre-splitter/splitter with overhead coupling (Fig. 1.8.17) (45 stages)	−15.6435	12.5	1.8016
Two enthalpy feed, 30 stages (Fig. 1.8.18)	−15.1963	12.0501	1.6901

A process engineer may choose the scheme based on the requirements.

1.9 Thermally Coupled Distillation Schemes (TCDS) or Distributed Distillation

1.9.1 Introduction to TCDS

In order to help explain the schemes, a sharp-cut direct scheme is presented in Fig. 1.9.1.

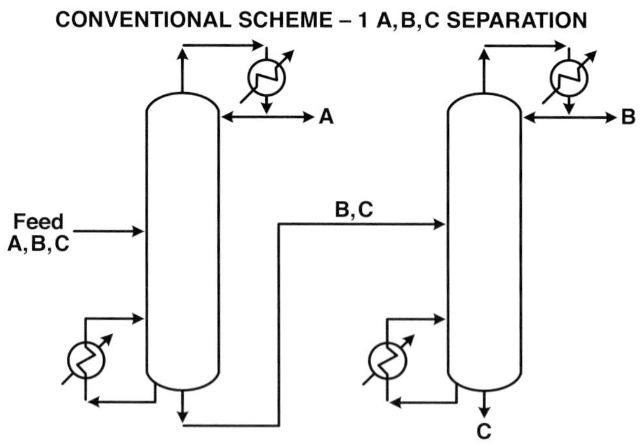

Fig. 1.9.1 Sharp-cut direct scheme

A three-component mixture (A, B, C) is separated into pure components A, B, and C by deploying two columns in a conventional sharp-cut distillation using the schemes practiced in Fig. 1.9.1 or in Fig. 1.9.4. Each column produces a pure product at one end and a mixture of products at the other end for feeding into the next column, in order to produce two pure products, one from the top and one from the bottom. Component A is recovered as top product of the first column in Fig. 1.9.1. But some repeated vaporization and condensation of one component is unavoidable in these schemes and may result in higher energy consumption. For instance, in a direct sharp-cut distillation scheme, the concentration of component B reaches its maximum below the feed stage and again decreases at the bottom of the first tower.

This phenomenon is referred by distillation experts as remixing. Remixing occurs in most conventional distillation schemes and demands higher energy. In the second tower, the components B and C are separated, with B as the top product and C as the bottom product.

A typical distribution of composition of toluene in a benzene–toluene–*para*-xylene system using a 30-stage column is shown below in Example Problem 1.9.1 This specific mixture of 40% wt (46 mole %) benzene, 20% wt (20 mole %) toluene, and 40% wt (34 mole %) *para*-xylene is fed to the first column T1.

Change in the toluene concentration in the first column (T1) can be seen in Fig. 1.9.2, which shows a peak at about the 24th stage and then a drop at the bottom. The toluene concentration reaches almost 68 mole % before dropping to almost 36 mole % at the bottom of the first column (T1). The toluene concentration of 36% is then fractionated in the second tower to 99 mole % toluene, which is the desired top product. This is a typical phenomenon in a multi-component system where the concentration of heavy key component at the bottom of the tower is reduced by heavier-than-heavy key component. The effect is more pronounced when there is a greater amount of the heavier components in the feed. The change in the toluene concentration in the first column (T1) is shown in Fig. 1.9.2.

Change in the toluene concentration in the second column (T2) is given in Fig. 1.9.3.

An indirect sharp-cut distillation scheme is presented in Fig. 1.9.4.

Similarly, in the indirect sharp-cut distillation scheme (the mixture of components A, B, and C in Fig. 1.9.4), the concentration of component B reaches its maximum between the feed tray and the top, then reduces when it reaches the top. The component C is recovered from the bottom of the first column. The mixture of components A and B from the first column top is then fed to the succeeding column for the separation of A and B. In Example Problem 1.9.1, the toluene concentration increases to around 72 mole % at the 6th tray in the first column, then at the top reduces to almost 30 mole % after being diluted with benzene. The change of the toluene mole fraction through the stages is shown in Fig. 1.9.4A.

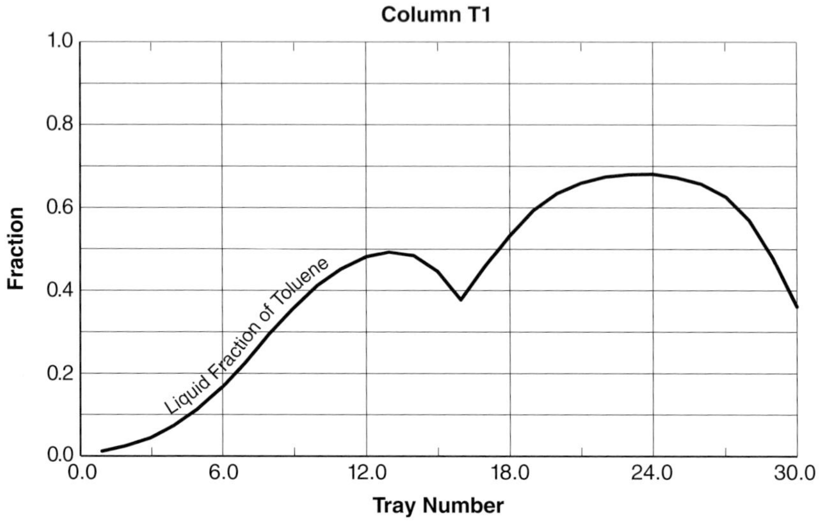

Fig. 1.9.2 Toluene liquid concentration profile (in mole fraction) in the stages of the first column in a direct scheme

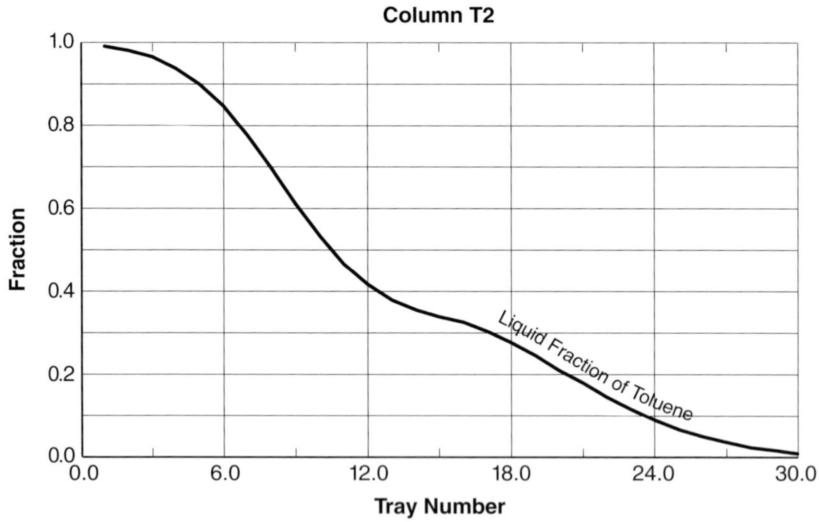

Fig. 1.9.3 Toluene liquid concentration profile in the stages of the second column

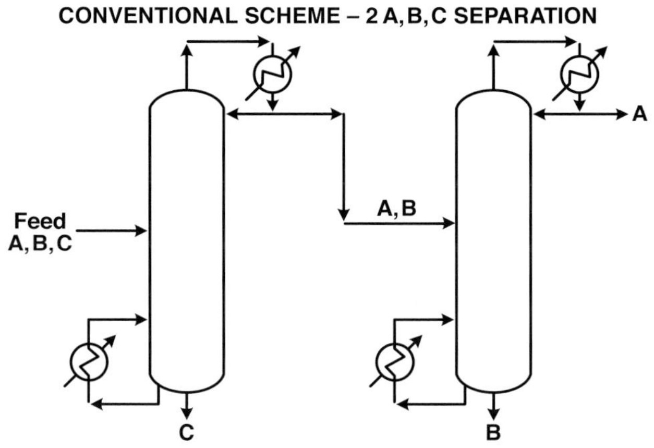

Fig. 1.9.4 Indirect sharp-cut distillation scheme

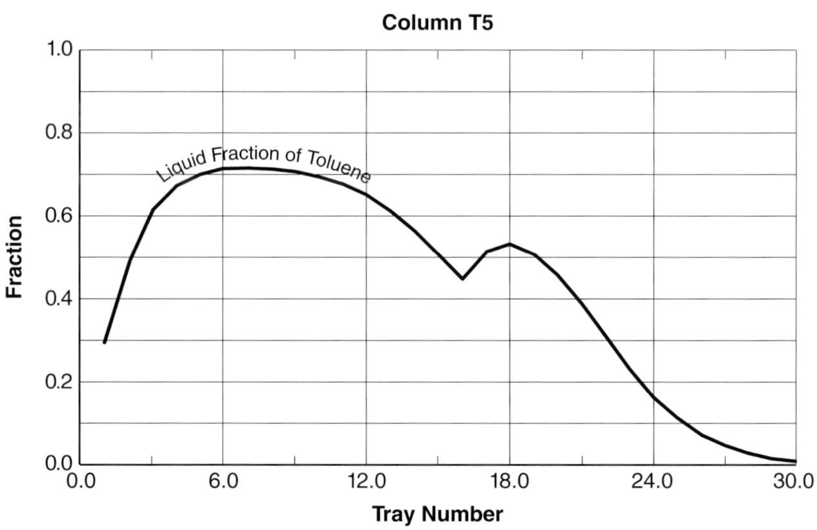

Fig. 1.9.4A Toluene liquid concentration profile (in mole fraction) through the stages in the first column in an indirect scheme

1.9.2 Distributed distillation/TCDS

a) Side rectifier/Side stripper

The concept of distributed distillation or thermally coupled distillation came about with the emphasis on energy conservation in distillation. Two TCD schemes evolved from the above observation of concentration of component B reaching maximum concentration below the feed tray in a direct scheme (Fig. 1.9.1), and above feed tray in an indirect scheme (Fig. 1.9.4).

Fig. 1.9.5 shows how the maximum concentration of component B is drawn from the main column and fed to the side stripper/side rectifier and can produce the component B of desired purity.

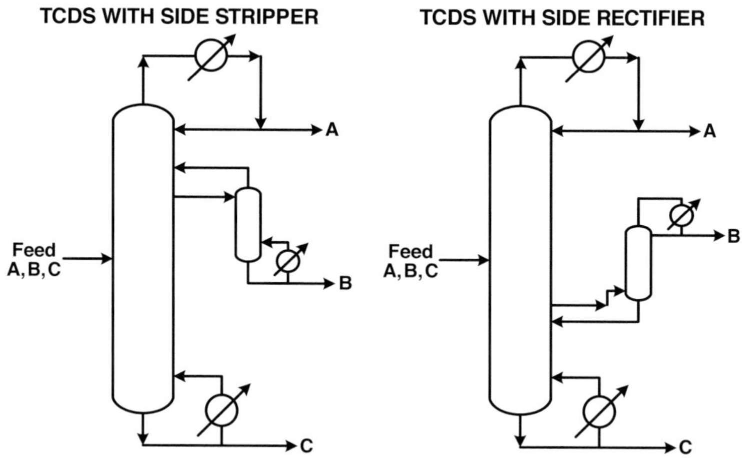

Fig. 1.9.5 TCD with side stripper/side rectifier

b) Brugma-type schemes

In the Brugma-type scheme given below, the first column produces a top comprised of components A and B, and a bottom comprised of B and C. These mixtures are fed to the second column at different locations to produce three pure components of A, B, and C. This is equivalent to adding a pre-fractionator or pre-splitter in the upstream. A Brugma scheme is shown in Fig. 1.9.6.

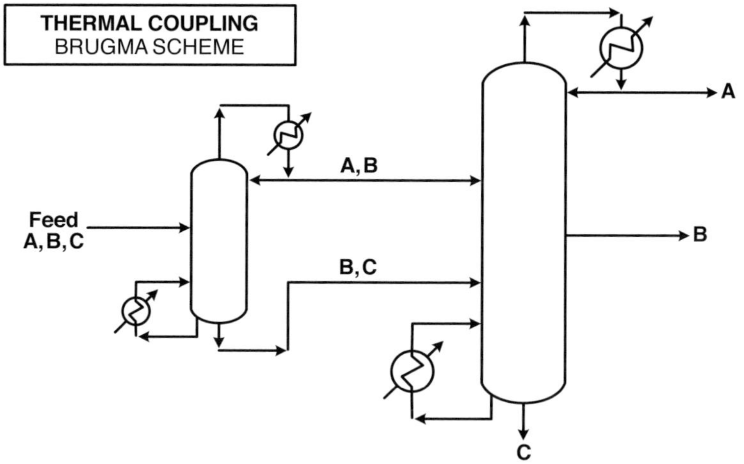

Fig. 1.9.6 Brugma scheme

c) Classic Petlyuk scheme

In the classic Petlyuk scheme, the first tower is supplied with reflux and stripping vapor from the succeeding column. The first tower produces streams rich in A, B and B, C, while the second column produces A and C as two end products and B as side draw. No reboiler or condenser is used in the first column. The classic Petlyuk scheme is shown in Fig. 1.9.7.

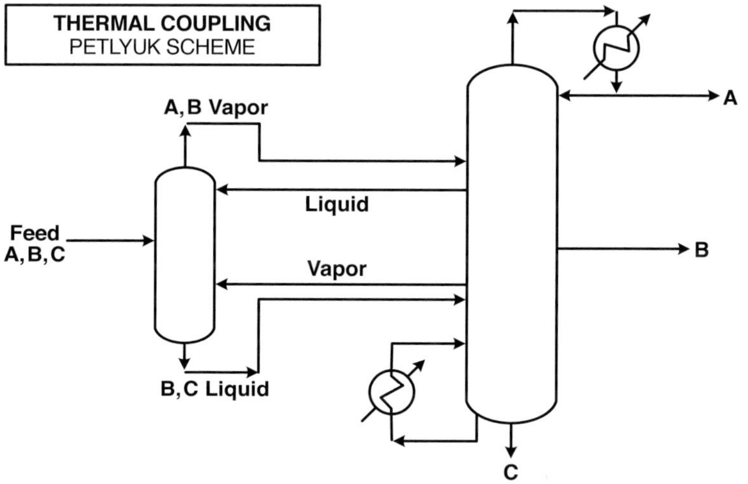

Fig. 1.9.7 Classic Petlyuk scheme

The difficulty with this scheme is to ensure the flow of vapor from one tower to the other when the both are operating at a similar pressure. Liquid can be introduced from one tower to the other by pumps, but vapor flow would need ejectors or compressors.

d) Modified Petlyuk schemes

The classic Petlyuk scheme has necessitated modifications for easier application. This particular modification uses two sequential towers. The first one can operate at higher pressure, thus ensuring the flow of vapor from first column to the second while pumps provide the flow of liquid from the second tower to the first. The reboiler is provided in the first column and the condenser in the second column. This modified Petlyuk scheme is shown in Fig. 1.9.8.

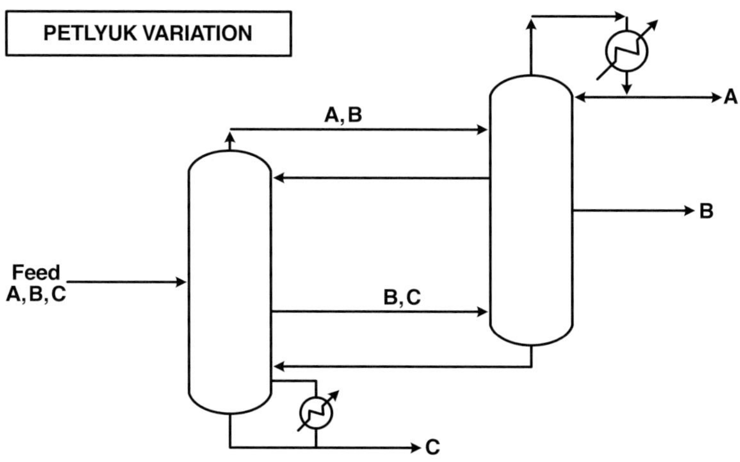

Fig. 1.9.8 Modified Petlyuk scheme

1.9.3 Benefits of using a TCDS

The TCDS is explained using the previous example of the benzene–toluene–*para*-xylene system and are compared with a sharp-cut distillation scheme.

Conventional scheme (sharp-cut)

The B–T–PX system is conventionally operated with two columns (sharp-cut). Benzene of the required purity is recovered as the top product of first column. The bottom of the first column is then fed to the second column. Toluene is recovered from the second column top with the required purity, and *para*-xylene is recovered as the bottom product of the second column.

This conventional separation consumes extra energy due to repeated vaporization. Part of the toluene is vaporized in the first column, condensed, and then revaporized again in the second to be recovered as top product.

Example Problem 1.9.1

Feed composition: 40% wt (46 mole %) benzene, 40% wt (20 mole %) toluene, and 40% wt (34 mole %)

Feed rate = 100 MT/hr

B–T–PX is separated with different schemes (with the same number of stages) for the purity of benzene, toluene, and *para*-xylene (99 mole % each). The example problem considers operation of the columns at pressure of 1.2 kg/cm^2(a). The feed preheat is considered 80°C at a pressure at 3 kg/cm^2 in all the schemes.

A conventional B–T–PX system is simulated and presented in Fig. 1.9.9.

The total reboiler duty of both the columns is 10.4902 + 7.6179 = 18.11 Mkcal/hr or Gcal/hr in the conventional configuration in Fig. 1.9.9.

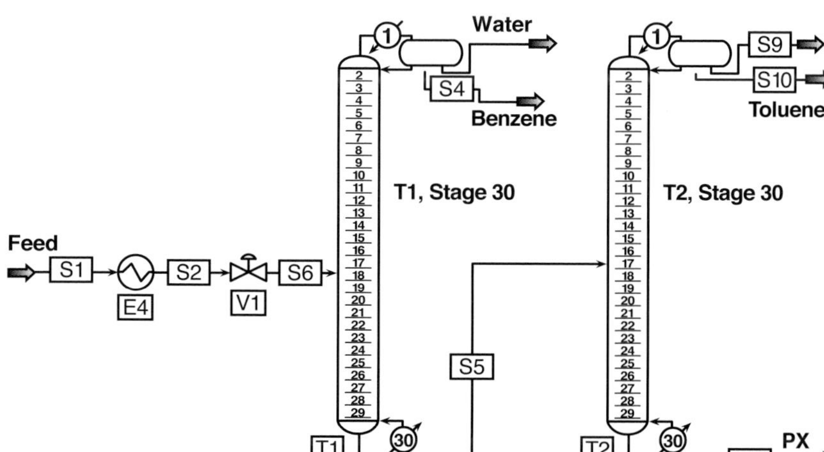

Column Name Column Description		T1	T2
Condenser Duty	M*KCAL/HR	-8.6800	-7.6012
Reboiler Duty	M*KCAL/HR	10.4902	7.6179
Column Reflux Ratio		1.3408	3.5996

Fig. 1.9.9 Conventional B-T-PX scheme simulation

TCDS for B–T–PX separation

In the TCDS with a side stripper scheme, toluene of the same purity is recovered as side product with stripper. This scheme has lower reboiler and condenser duties than the conventional scheme.

A TCDS with side stripper for a B–T–PX system with (50 + 10 stages) is given in Fig. 1.9.10.

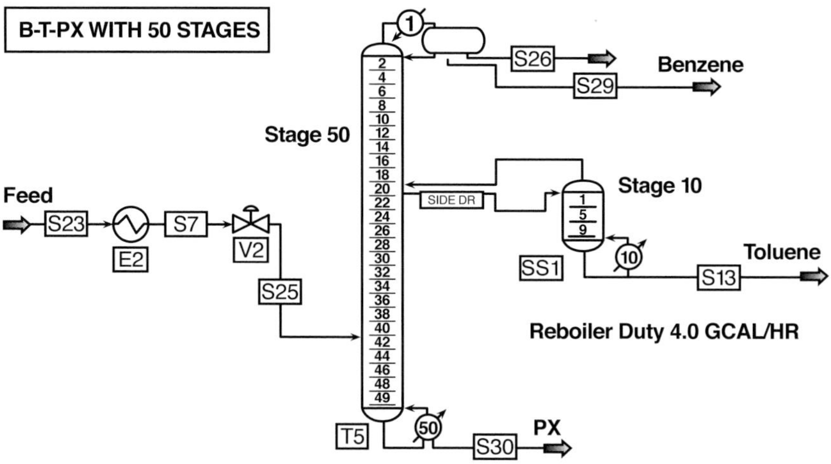

Column Name Column Description		T5
Condenser Duty	M*KCAL/HR	−15.2034
Reboiler Duty	M*KCAL/HR	13.0977
Column Reflux Ratio		1.1002

Fig. 1.9.10 B-T-PX side stripper scheme

Total reboiler duty with above configuration is 13.097 + 4.0 = 17.097 Gcal/hr as opposed to the 18.11 Gcal/hr reboiler duty of a conventional scheme (Fig. 1.9.9). Around 6% savings in reboiler duty can be achieved. In this scheme, toluene at nearly its highest concentration is fed to the side stripper. A benzene–toluene–*para*-xylene system with a side rectifier scheme simulation is shown in Fig. 1.9.11.

A TCDS with a side rectifier also consumes less energy than a conventional scheme. The reboiler duty requirement in a side rectifier scheme is 16.04 Gcal/hr, as opposed to 18.11 Gcal/hr in a conventional scheme. A savings of around 11% in reboiler duty is achieved. In this scheme, toluene at its highest concentration is fed from the main column to the side rectifier to repurify it to the desired purity. This scheme is not much used in the industry. For comparing the efficiency of TCDS schemes with multi-effect schemes, the same problem is configured in multi-effect pattern as shown below (Fig. 1.9.12).

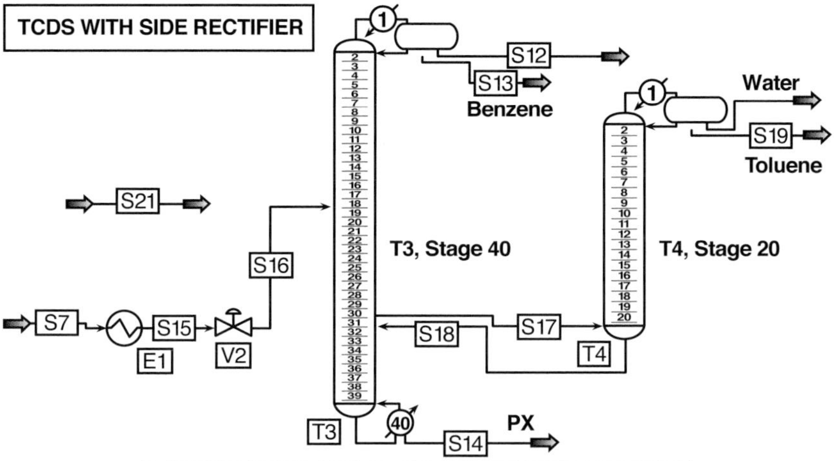

Column Name Column Description		T3	T4
Condenser Duty	M*KCAL/HR	–8.9407	–5.2905
Reboiler Duty	M*KCAL/HR	16.0435	n/a
Column Reflux Ratio		1.4124	2.1928

Fig. 1.9.11 Simulation of a B–T–PX system with a side rectifier scheme

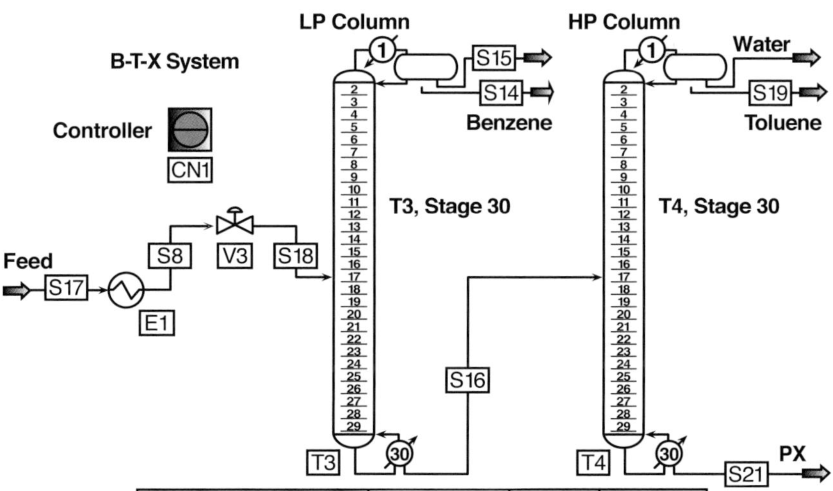

Column Name Column Description		T3	T4
Condenser Duty	M*KCAL/HR	–8.4614	–10.2717
Reboiler Duty	M*KCAL/HR	10.2717	11.5795
Column Reflux Ratio		1.2818	4.7241

Fig. 1.9.12 Simulation of a B–T–PX multi-effect scheme

Multi-effect comparison with TCDS

For comparing the efficiency of TCDS with the multi-effect scheme, the same problem (Example Problem 1.9.1) is configured in the multi-effect scheme as below. A multi-effect scheme (B–T–PX) system is shown in Fig. 1.9.12. The HP column is operated at 3 kg/cm^2(g) and the HP overhead supplying the LP column's reboiler duty (at a pressure of 1.2 kg/cm^2(a)). This shows appreciable savings in the reboiler and condenser duties. The net reboiler duty required after the HP column overhead is coupled with the LP column reboiler is 11.6 Gcal/hr, versus the 18.11 Gcal/hr reboiler duty of a conventional scheme. Substantial savings are obtained in the multi-effect (approximately 30%), particularly when compared with the TCDS in the example of this specific B–T–PX system. The savings in reboiler or condenser duty will be much greater if additional stages are considered for an additional HP column. This scheme is often found to be more efficient than TCDS schemes, even those with the same number of stages as the conventional.

Some other TCDS schemes are analyzed in the example problem below using the same B–T–PX system, but with small changes in feed composition and operating pressure.

Example Problem 1.9.2

Feed composition: 40:20:40 mole % of B:T:PX is distilled in order to recover 99 mole % of each

Feed rate: 100MT/hr

The reboiler/condenser duties shown for different schemes are compared to those of a conventional scheme with the same feed conditions and same total number of stages (60). The column is operated at 1.9 kg/cm^2 (a). A conventional scheme is simulated and presented in Fig. 1.9.13.

The total reboiler duty requirement is 11.05 + 10.43 = 21.5 Gcal/hr.

The same is analyzed below in a Brugma scheme with 15 + 45 stages (Fig. 1.9.14).

The above Brugma configuration requires a reboiler duty of around 21.7 Mkcal/hr, as compared with the requirement of 21.5 Mkcal/hr in a conventional scheme. However, the reflux ratio in the second column has reduced significantly. This may provide flexibility for capacity augmentation in the unit through installation of a pre-splitter. The reboiler condenser duties of the second column are very sensitive to the reboiler duty of the first column and the feed locations of the second column. The classic Petlyuk scheme is analyzed in Fig. 1.9.15 using the same example problem as above.

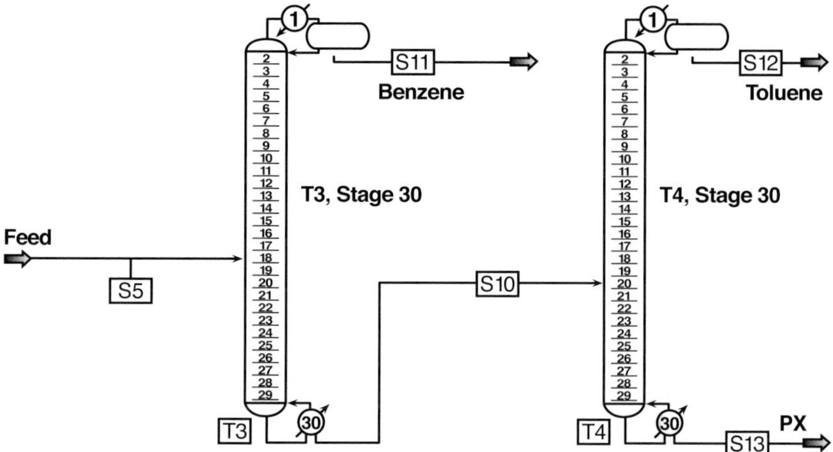

Column Name Column Description		T3	T4
Condenser Duty	M*KCAL/HR	–8.3007	–10.3606
Reboiler Duty	M*KCAL/HR	11.0519	10.4254
Column Reflux Ratio		1.6669	5.3547

Fig. 1.9.13 Simulation of a B–T–PX conventional direct sharp-cut scheme

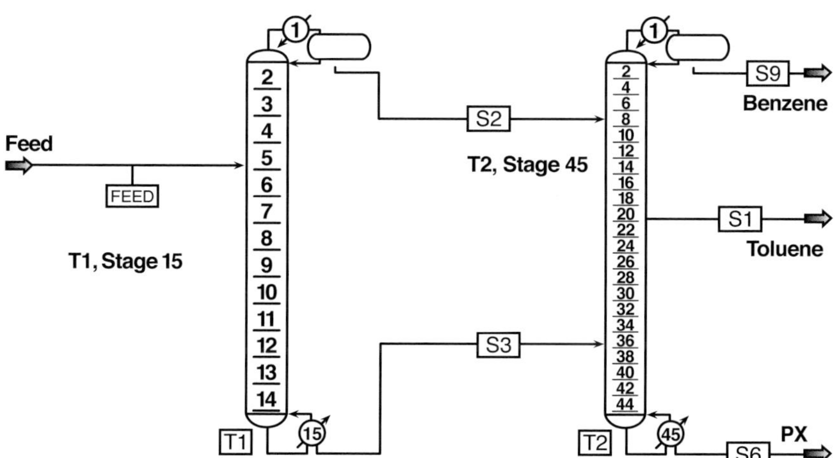

Column Name Column Description		T1	T2
Condenser Duty	M*KCAL/HR	–6.8196	–12.1172
Reboiler Duty	M*KCAL/HR	9.4032	12.3664
Column Reflux Ratio		0.9386	2.8883

Fig. 1.9.14 Simulation of a B–T–PX Brugma scheme

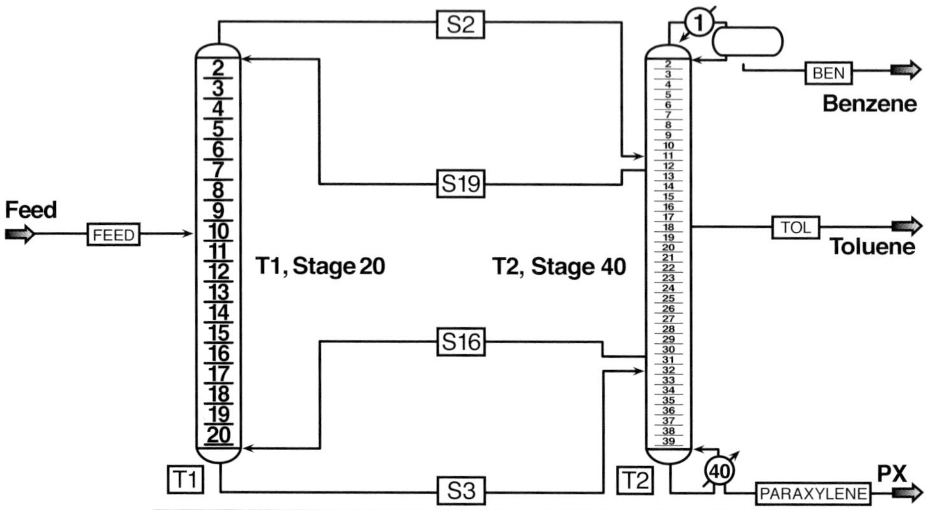

Column Name Column Description		T1	T2
Condenser Duty	M*KCAL/HR	n/a	−19.0533
Reboiler Duty	M*KCAL/HR	n/a	21.8861
Column Reflux Ratio		0.6712	5.1138

Fig. 1.9.15 Simulation of a B–T–PX classic Petlyuk scheme

The summary of observations are presented in Table 1.9.1.

Table 1.9.1 Condenser and reboiler duties of schemes from Figs. 1.9.13, 1.9.14, and 1.9.15

Scheme	Total reboiler duty
Conventional	11.05 + 10.4254 = 21.5
Brugma	9.4032 + 12.3664 = 21.77
Petlyuk	0 + 21.887 = 21.887

The Brugma and Petlyuk schemes show almost the same reboiler duty for the same feed condition, composition, and total stages as given in Example Problem 1.9.2

Effect of feed composition on reboiler duty of the TCDS schemes

The toluene proportion was only 20 mole % in Example Problem 1.9.2 where savings in reboiler/condenser duty were not observed. To address the effect of feed composition in TCDS schemes, the feed composition is altered. A feed with higher proportion of toluene (60 mole %) is considered in Example Problem 1.9.3.

Example Problem 1.9.3

The proportion of toluene which remixes in a conventional scheme is increased from 20% to 60%. The feed composition changed to 20:60:20 mole % of B:T:PX (proportion of toluene increased) and the simulation is carried out to analyze the savings in reboiler duty compared to a conventional scheme (simulation not shown). A comparison of the results of the simulations of the Brugma and Petlyuk schemes with the conventional scheme from Example Problem 1.9.3 is shown in Table 1.9.2.

Table 1.9.2 Results of Example Problem 1.9.3

Scheme	Total reboiler duties in Gcal/hr
Conventional	13.7 + 11.6 = 25.3
Brugma	23.2
Petlyuk	23.3

Table 1.9.2 shows a savings of around 8%–10% in reboiler duty with the Brugma and Petlyuk scheme from the example. Thus the savings in reboiler/condenser duty are dependent on the components, their proportions, operating conditions, etc.

This classic Petlyuk scheme is difficult to execute even if it is found to be the most efficient, as the vapor flows cannot be established from one column to the other. Methods have been developed to make two compartments in the same column work as two columns. Feed is introduced into one compartment, constituting the first column. The split of vapor and liquid flows is established internally by proprietary internals designed for splitting the vapor and liquid flows in the compartments. This design is known as a dividing wall column (DWC). This design is finding increased application in the field of distillation (see Fig. 1.9.16.).

The dividing wall columns can have arrangements such as top-dividing, bottom-dividing, and middle-dividing columns and can be simulated like the Petlyuk scheme.

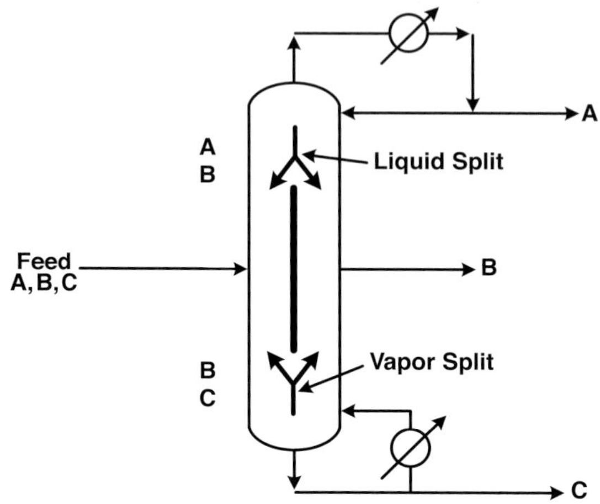

Fig. 1.9.16 Dividing wall column (middle divided)

1.9.15 Application of distributed distillation in an olefin plant

The concept of distributed distillation can be easily applied for the debottlenecking of demethanizer (DM) and deethanizer (DE) in an olefin plant. To illustrate this application, a conventional scheme is worked out and presented in Fig. 1.9.17, considering an ethylene loss of less than 5 mole % along the top of the demethanizer and methane loss of 0.05 mole % along the bottom of the DM.

The DM can be preceded by a column operating at a higher pressure (around 12 kg/cm^2), where partial separation can be achieved. The temperature level in the higher-pressure column will be higher than in the DM, the partially separated top and bottom products can be fed to the original demethanizer, and the feed can be purified to the required level with a specification for ethylene and methane loss and methane slippage from the bottom as in the conventional scheme. The reboiler and condenser duties in the scheme are as below in Fig. 1.9.18. (For simplicity, the demethanizer has been configured with only two feeds rather than three or four.)

A second scheme with a pre-splitter of 20 stages is tried in order to debottleneck the deethanizer (see Fig. 1.9.19). In this scheme, the pre-splitter bottom product, which contains very low methane, is fed to the deethanizer instead of to the demethanizer. The top vapor of the pre-splitter, which contains most of the methane, is fed to the DM.

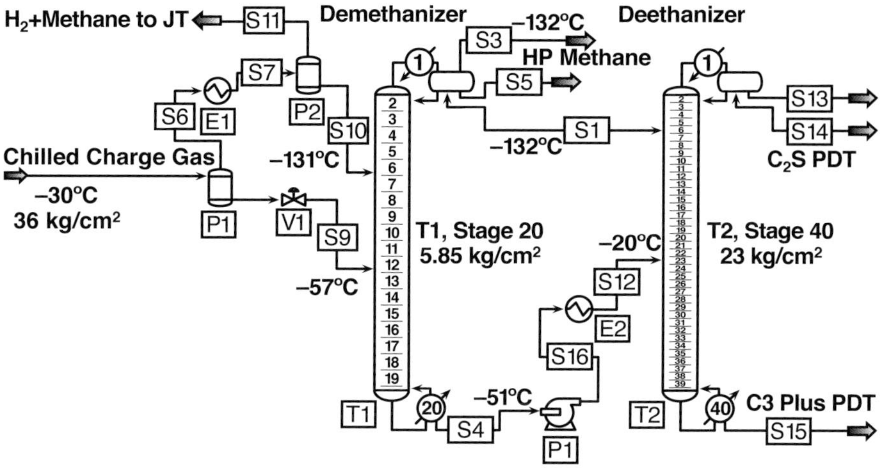

Column Name Column Description		T1	T2
Condenser Duty	M*KCAL/HR	–0.8544	–15.2988
Reboiler Duty	M*KCAL/HR	5.8553	20.0746
Column Reflux Ratio		0.1053	0.5952

Fig. 1.9.17 Simulation of conventional scheme in DM, DE

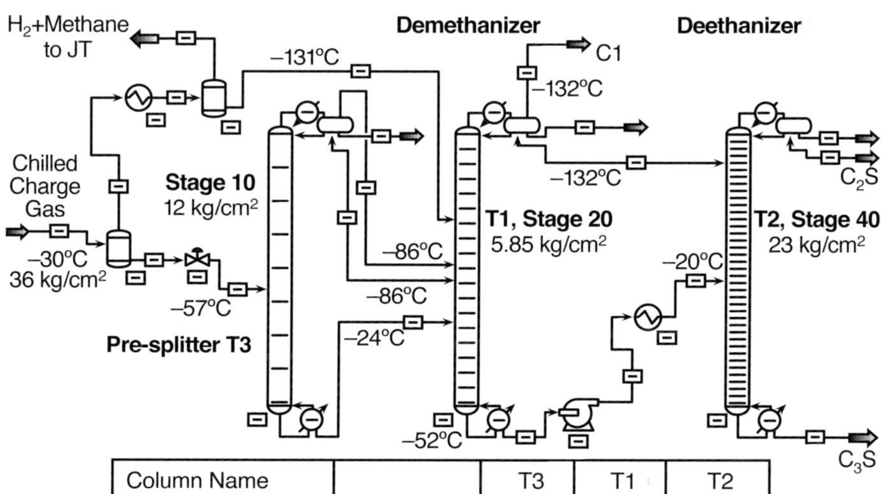

Column Name Column Description		T3	T1	T2
Condenser Duty	M*KCAL/HR	–2.2222	–0.8452	–15.4027
Reboiler Duty	M*KCAL/HR	3.0000	4.8646	20.0750
Column Reflux Ratio		1.1282	0.1051	0.5881

Fig. 1.9.18 Simulation of distillation with a pre-splitter (10 stages) ahead of DM, DE

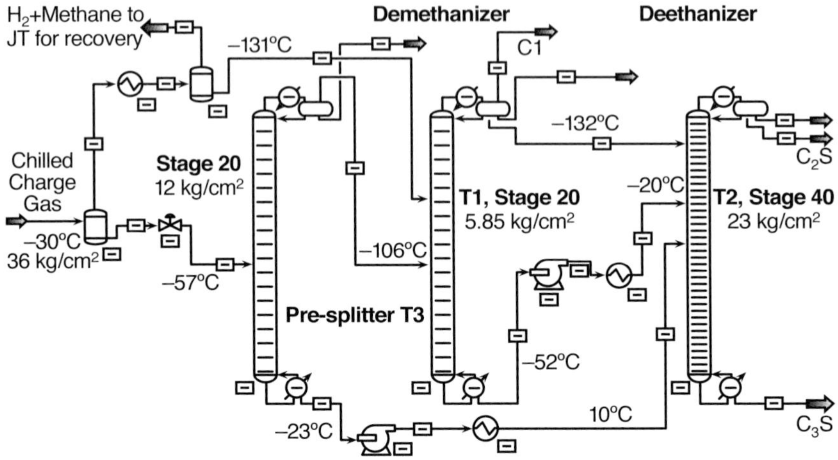

Column Name Column Description		T3	T1	T2
Condenser Duty	M*KCAL/HR	-2.8572	-0.8508	-14.4436
Reboiler Duty	M*KCAL/HR	3.4918	7.1364	15.3792
Column Reflux Ratio		1.7625	0.1049	0.5060

Fig. 1.9.19 Simulation of alternative scheme in DM, DE with a pre-splitter of 20 stages

Table 1.9.3 shows that the demethanizer is partly relieved by the installation of a pre-splitter 10 stages upstream. The pre-splitter operates at 12 kg/cm^2 pressure, an overhead temperature of −86°C, and a bottom temperature of −24°C. Thus, the demethanizer in the scheme (Fig. 1.9.18) requires a refrigerant of lower quality than in the conventional scheme. The demethanizer condenser and reboiler duties and the reflux ratio reduce marginally, though the total reboiler duty of the pre-splitter and the demethanizer increases.

Table 1.9.3 shows that the deethanizer load reduces significantly with the scheme shown in Fig. 1.9.19. In the scheme, the pre-splitter bottom product is fed to the deethanizer with the same minimum specification of methane as the demethanizer in the conventional scheme. The pre-splitter bottom is heated from −23°C to +10°C before it is fed to the deethanizer. Similarly, the demethanizer bottom is also heated from −52°C to −20°C before it is fed to the deethanizer at the upper location.

The colder feed is introduced at the upper location and warmer one to the lower location. The cold feed can be utilized for cooling any stream. In the scheme, the pre-splitter overhead drum temperature is −106°C and the pre-splitter bottom temperature is −23°C. The demethanizer load below the feed tray reduces significantly as a result of pre-splitting. Thus, the scheme can be used for capacity augmentation applications. A summary of the results of the schemes found in Figs. 1.9.17, 1.9.18, and 1.9.19 is given in Table 1.9.3.

Table 1.9.3 Summary of the results of schemes from Figs. 1.9.17, 1.9.18, and 1.9.19

Parameters	Scheme Fig. 1.9.17 without pre-splitter			Scheme Fig. 1.9.18 with pre-splitter 10 stage			Scheme Fig. 1.9.19 with pre-splitter 20 stage		
	PRSPL	DM	DE	PRSPL	DM	DE	PRSPL	DM	DE
Q_c	0	−0.854	−15.299	−2.222	−0.845	−15.403	−2.857	−0.851	−14.444
Q_R	0	5.856	20.075	3.000	4.865	20.075	3.4918	7.1365	15.379
R.R.	0	0.106	0.595	1.128	0.105	0.588	1.726	0.105	0.506

The appropriate scheme needs to be selected based on the specific requirements. A two-enthalpy feed in a deethanizer is also used industrially to reduce reflux ratio and reboiler and condenser duties, and to provide a capacity cushion.

1.10 Vapor Compression in Cryogenic Distillation

1.10.1 Description of vapor compression

Super fractionators operate with very high reflux ratios. Thus, a large amount of heat is rejected in the condensers. If the temperature difference between the column top and bottom is not large, a compressor is installed to compress the top vapor in order to raise the bubble point of the compressor discharge stream to around 10°C higher than the bottom temperature. This is done so that the heat can be transferred to the bottom, providing the reboiler duty.

Thus, in the process the overhead vapor also cools/condenses and a trim condenser is deployed in the overhead circuit to completely condense the vapor before routing it to the reflux drum. An example of the application of this process can be seen in the propylene/propane fractionators. It is also used in butene-1 production. A typical scheme for vapor compression is shown in Fig. 1.10.1.

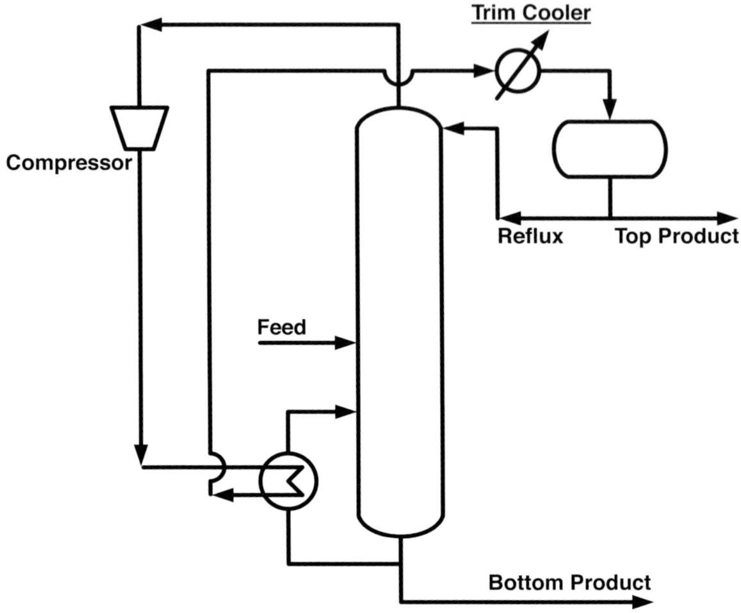

Fig. 1.10.1 Vapor compression, open-top scheme

Typical vapor compression in distillation

A butene-1 separation with two vapor compressors (C1 and C2) is shown in Fig. 1.10.2. Q1 and Q2 are the reboiler duties, and E1 and E2 are the trim cooler duties.

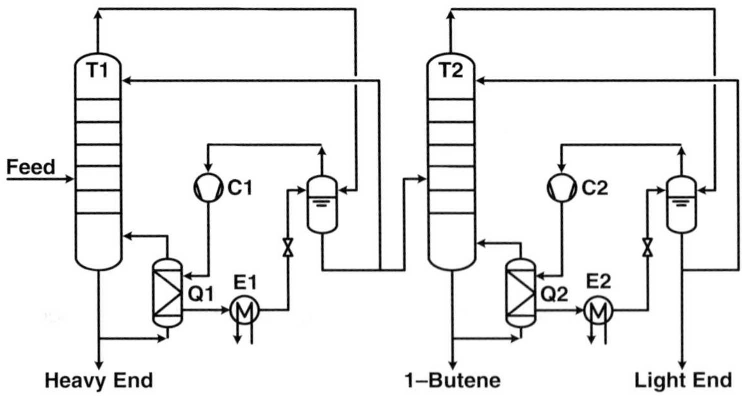

Fig. 1.10.2 Vapor compression for butene-1 separation with two compressors

Fig. 1.10.2A shows the butene-1 production scheme using a single vapor compressor (C1). Q1 and Q2 are the reboiler duties; E1 is the trim cooler, and E2 is the condenser of column T2. This scheme is often used in the industry.

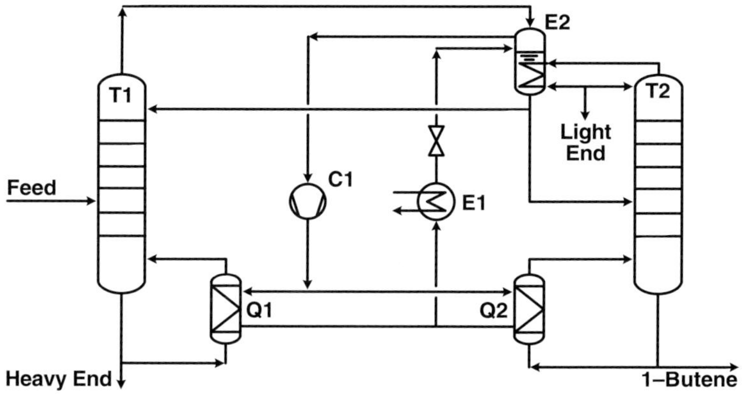

Fig. 1.10.2A Vapor compression for butene-1 with a single compressor

1.10.2 Tentative Application Guidelines

1. The temperature difference between top and bottom should be maximum 50°C for vapor compression to be economical in water-cooled columns.
2. For refrigerated columns, the literature states that for a temperature difference of up to 65°C between the top and the bottom, the application of vapor compression can be economical.
3. This is an attractive option when:
   a. the condenser and reboiler duty requirements are high;
   b. refrigeration is used for condensation.
4. This option is not attractive when cheaper fluid is available as the heat source for the reboiler of water-cooled columns. An example of such a system is a propylene/propane fractionator where tempered water or steam condensate is available for use in the reboiler at 60°C–70°C. The tempered water is generated by the cooling of several hot streams, and needs to be cooled further before it can be rerouted and used to cool hot products.
5. When installing a vapor compressor, the compressor suction line needs to be as short as possible to eliminate undue pressure drop in the suction line (caused by a higher specific volume of vapor at the suction line). A high pressure drop in the suction line may require a

higher compression ratio. Similarly, the discharge line pressure drop should also be limited.

6. Low-pressure drop column internals are desirable for reducing top and bottom differential pressure and consequently reducing the required temperature difference between compressor discharge and bottom. This will also reduce the compression ratio.
7. Often low-pressure drop exchangers are used to further decrease the compression ratio.
8. Often a high flux tube reboiler is beneficial, as it will simultaneously reduce surface area and pressure drop.

Some other vapor compression schemes (such as the open-bottom and closed heat pump schemes) are given in Fig. 1.10.3.

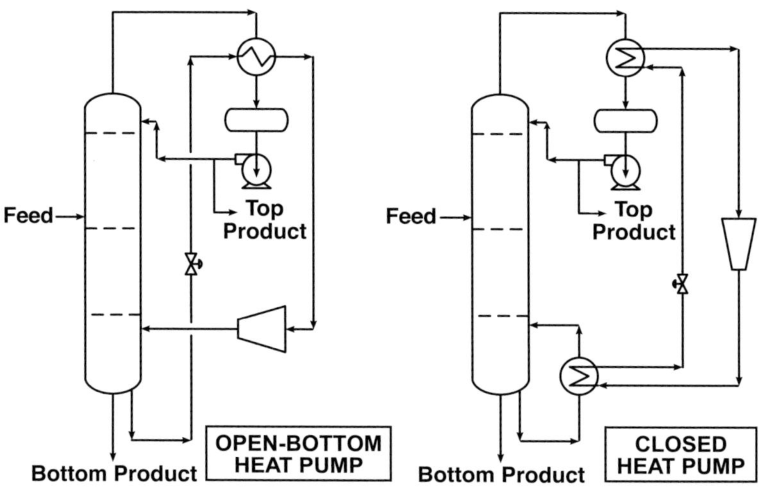

Fig. 1.10.3 Open-bottom and closed heat pump schemes

1.10.3 General guidelines for such applications

A conventional simulation of the column is to be done for achieving the required purity and yield. This will also provide the condenser and reboiler duties. Once the column converges, a pseudo top product is to be drawn in simulation to represent the vapor from the column at the column top temperature and pressure.

The pseudo-vapor will be fed to a compressor. The compressor discharge pressure is fixed so that the vapor dew point is higher by about 7°C–10°C above the dew point of the bottom product. The difference in temperature must be maintained between the dew point of the compressed overhead vapor and the dew point of the column bottom vapor and between the bubble point of the bottom liquid and the bubble point of the overhead condensed liquid. The reboiler needs to be configured as a heat exchanger performing the same duty as a column reboiler. A trim cooler needs to be placed downstream from the reboiler in order to completely condense the compressed overhead stream to saturated liquid.

Propylene/propane splitter

A propylene–propane mixture can be separated through conventional distillation using low-grade heat as a heating agent for a reboiler, and cooling water or air as a condensing agent for overhead at a higher operating pressure of around 18.5 kg/cm^2(g).

A heat pump has been proved to reduce the operating cost by lowering the operating pressure of the column, despite the input of compressor power. Refrigeration from common pool can also be used, if provided in a complex plant like a steam cracker. To illustrate vapor compression in propylene fractionators (PF), the example problem is given below.

Example Problem 1.10.1

A propylene–propane mixture from a refinery residue fluid catalytic cracking unit (RFCCU) having approximately 72 mole % of propylene and approximately 28 mole % of propane (around 41.7 MT/hr of propylene in the mixture) was processed to produce only 25 MT/hr of 99.7 mole % polymer-grade propylene as an overhead product, and the balance of propylene is not recovered due to limited demand.

This simulation can be carried out with or without vapor compression to demonstrate the differences in reboiler duty, condenser duty and reflux ratio in both cases, which are shown below. The example problem is solved using vapor compression and operating the column at 11.2 kg/cm^2 pressure (Fig. 1.10.4).

A propylene fractionator overhead system can be configured as in Fig. 1.10.4. The vapor compressor takes suction from the column overhead at a pressure of 11.2 kg/cm^2 and discharges it at a pressure of 18.5 kg/cm^2. It then delivers the heat to its own reboiler and rejects the remaining balance of the heat to a water cooler to be converted into a saturated liquid and finally sent to the reflux drum.

In the simulation, a water cooler is placed in the upstream to remove the superheat and convert it to saturated vapor before feeding it to the reboiler. The water cooler can also be placed in the downstream to the reboiler. The reflux drum operates at a compressor discharge pressure of 18.5 kg/cm^2(g). From the reflux drum, the propylene is drawn as product and part is refluxed. The drum operates at 46°C. Part of the reflux vaporizes downstream of the reflux control valve (not shown), and the reflux stream cools to less than the top temperature and in turn cools the overhead. The vapor generated due to the flashing of the reflux is released to the top dome of the column and joins the overhead vapor, which is then fed to the compressor suction.

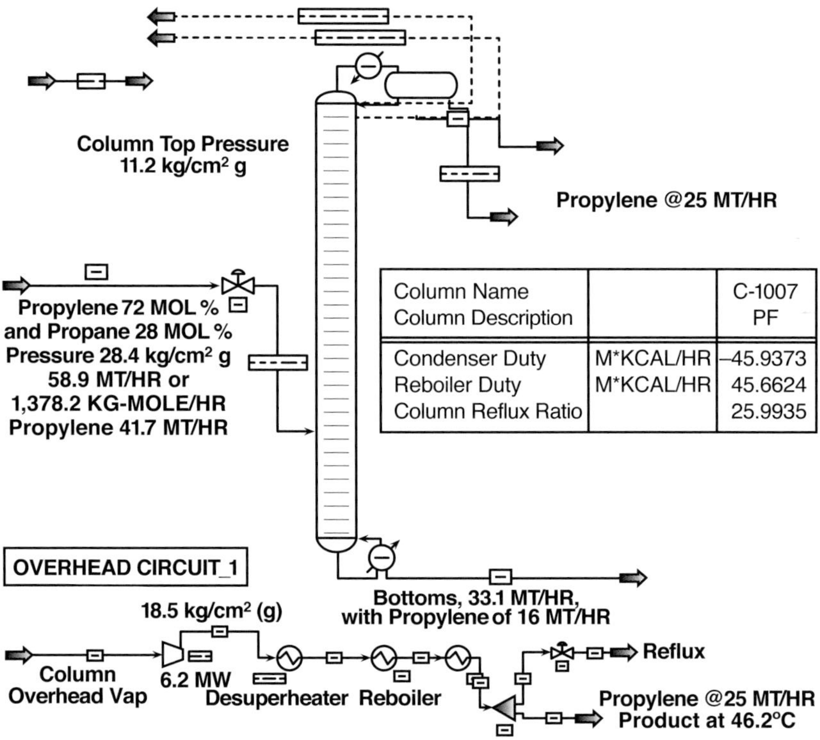

Fig. 1.10.4 Propylene fractionators (PF) at 11.2 kg/cm^2 using vapor compression

This operation may not require a pump for reflux, as the pressure of the drum may be adequate to introduce the reflux to the column. The propylene product is recovered at a normal temperature of 46°C. The power requirement of the compressor is calculated at around 6.2 MW.

Now the scheme from Example Problem 1.10.1 is worked out without a vapor compressor. This will require that the column be operated at a higher pressure in order to condense the overhead using cool water with the same purity and recovery of the top product. In the configuration in Fig. 1.10.5, the propylene column is operated at 18.5 kg/cm^2(g). The pressure is sufficient to condense the overhead vapor by circulating water in the condenser. No vapor compressor is used. The reboiler duty is supplied by the circulation of hot demineralized water

(DMW). The hot (70°C) DMW is generated by cooling the hot products from other units in multiple coolers, a process referred to as hot belting.

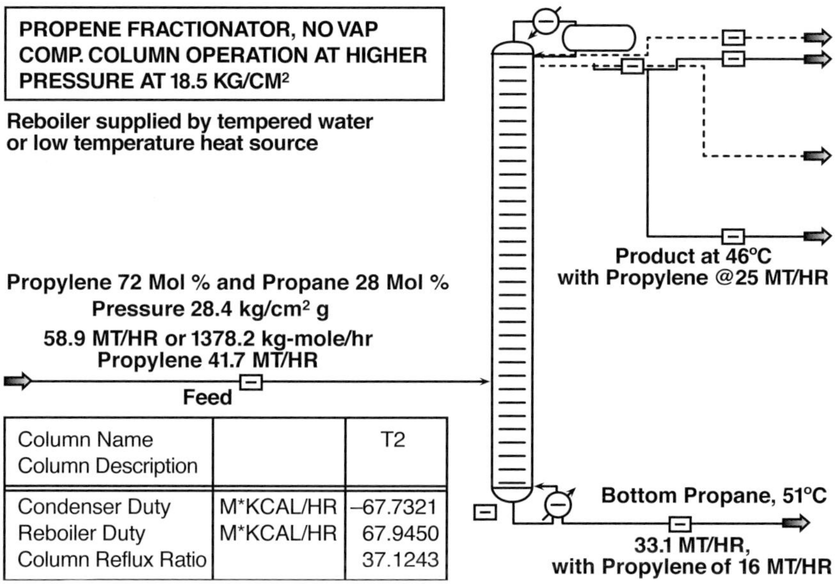

Fig. 1.10.5 Propylene splitter operating at 18.5 kg/cm^2(g) with no vapor compression

Thus, when the reboiler and condenser duties can be supplied or rejected using cheaper utilities, a vapor compressor can be avoided. The reboiler and condenser duties for the two configurations (Figs. 1.10.4 and 1.10.5) are presented below in Table 1.10.1. They are found much higher than operation at 11.2 kg/cm^2(g) with a vapor compressor.

The data in Table 1.10.1 shows that vapor compression is an efficient scheme, as both the condenser and reboiler duties were reduced significantly. However, when a low-temperature, very cheap heat source is available, vapor compression can be avoided through fractionation at a higher pressure, as seen in Fig. 1.10.5.

Table 1.10.1 Comparison of results of PF column with and without vapor compression

Parameters	Units	With vapor compression, PF at 11.2 kg/cm^2	Without vapor compression, PF at 18.5 kg/cm^2
Condenser duty	Mkcal/hr	–45.9373	–67.7421
Reboiler duty	Mkcal/hr	45.6624	67.9614
Column reflux ratio	ratio	25.9935	37.1314
Compressor power (polytropic eff. 70%)	(MW)	6.2	Nil

A butene-1 plant simulation is presented in Fig. 1.10.6 with a single vapor compressor providing reboiler heat to two columns.

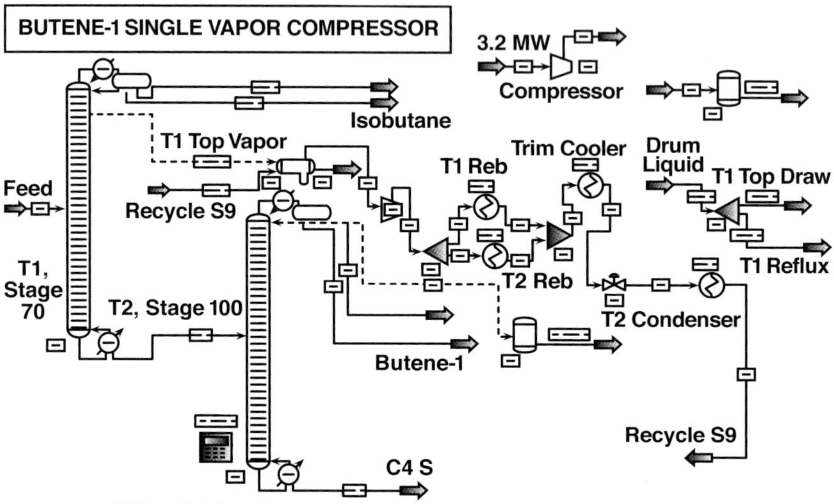

Column Name Column Description		T1	T2
Column Total Wt. Feed	KG/HR	7,500.0000	6,578.4383
Column Reflux Ratio		0.0000	13.0100
Drum Temp	°C	37.8919	41.0284
Condenser Duty	M*KCAL/HR	–9.9515	–5.0178
Reboiler Duty	M*KCAL/HR	10.0000	5.0000

Fig. 1.10.6 Simulation model of a butene-1 plant

In the simulation, the condenser duties of both the columns are taken up by the two reboilers and a common trim cooler is used to completely condense the overhead. The data on the feeds to the two columns and their different product compositions are given in Table 1.10.2.

The results of the simulation in Fig. 1.10.6 are given in Table 1.10.3.

In the vapor compression scheme in Fig. 1.10.6, a single vapor compressor is used. The compressor takes suction from the T1 (the isobutane column) overhead and the discharge is routed to the reboilers of both T1 and T2 (the butane-1 column) in parallel. The majority of the compressor discharge vapor gets condensed. A trim water cooler/condenser is used at the downstream to completely condense the isobutane vapor from the first and second columns to a saturated liquid.

Table 1.10.2 Column feeds and product compositions of a butene-1 plant

Stream name	S1	T1_TOP	T1_BOT	T1_BOT	T2_TOP	T2_BOT
Stream description		Feed to T1			Feed to T2	
Stream phase	Liquid	Liquid	Liquid	Liquid	Liquid	Liquid
Temperature, °C	35.00	37.89	47.47	47.47	41.03	48.98
Pressure, kg/cm (g)	6.00	4.00	4.01	4.01	3.50	3.50
Total mass rate, kg/hr	7,500.00	921.56	6,578.44	6,578.44	4,316.58	2,261.86
Total weight comp. rates, kg/hr						
Propane	25.00	25.00	0.00	0.00	0.00	0.00
Ibutane	643.00	642.18	0.82	0.82	0.82	0.00
Ibutene	17.00	2.91	14.09	14.09	14.09	0.00
1butene	4,180.00	251.38	3,928.62	3,928.62	3,928.09	0.53
Butane	870.00	0.05	869.95	869.95	278.29	591.67
C2butene	900.00	0.01	899.99	899.99	4.96	895.03
T2butene	865.00	0.03	864.97	864.97	90.34	774.63
H_2O	0.00	0.00	0.00	0.00	0.00	0.00

Table 1.10.3 Operating parameters of a butene-1 simulation

Parameter	T1 (isobutene column)	T2 (butene-1 column)
Colum top temperature, °C	39	41
Condenser duty, Gcal/hr	–9.95	–5.01
Reboiler duty, Gcal/hr	10.0	5.0
Reflux ratio	138.08	13.0
Reflux rate, kmoles/hr	2,229.83	998.03
Column top Pressure, kg/cm^2 (g)	4.0	3.5
Condenser temperature, °C	37.89	41.03
Compressor power, MW, 70% efficiency, polytropic		2.3
Trim cooler duty, Gcal/hr		1.6

The liquid is then throttled to partially vaporize and cool the mixture to at least 7°C cooler than the T2 overhead in order to condense it (the T2 overhead) and produce liquefied butene-1 as a product and reflux to the column. The revaporized isobutane (S9) is recycled to the compressor suction and recompressed along with fresh vapor from T1.

For simulation purposes, both columns are operated conventionally, where the first column removes isobutane (IB) with a bottom product specification of minimum isobutane. The bottom product of T1 is then fed to T2, and the recovery of butene-1 with the required purity is made possible by manipulating the reflux in T2. The butene-1 loss to the bottom is kept at minimum by regulating the reboiler duty in T2. The columns are converged conventionally, meeting the purity and recovery specifications of the products.

For the vapor compression circuit, pseudo-vapor from the T1 overhead is drawn in full and introduced to a compressor. Compressor discharge pressure is fixed at a differential temperature adequate for heat transfer to the reboilers of both column T1 and T2. The compressed isobutane is routed to two parallel heat exchangers performing the duties of reboilers in T1 and T2 in order to condense it almost fully.

The condensed isobutane is further cooled to fully condense it and throttled to a pressure low enough to generate the temperature differential required for condensation of T2 overhead vapor. It is next routed to a heat exchanger with a duty equal to that of the T2 condenser so that the T2 overhead can be condensed and the throttled T1 overhead can be further vaporized. It is then recycled to the compressor suction.

1.10.5 Cryogenic distillation

In cryogenic distillation, condenser duty is given more emphasis than reboiler duty. The reboiler duty can be supplied using cheaper utility heat, whereas condenser duty usually requires refrigeration. Examples (taken from a steam cracker for the production of ethylene and propylene) are considered below.

An olefin plant (steam cracker) uses a fair number of cryogenic distillation applications in the distillation units listed here:

1. Demethanizer (DM)
2. Deethanizer (DE)
3. Ethylene Fractionator (EF)
4. Depropanizer (DP)
5. Propylene Fractionator (PF)

The typical distillation scheme of an olefin plant is given in Fig. 1.10.7.

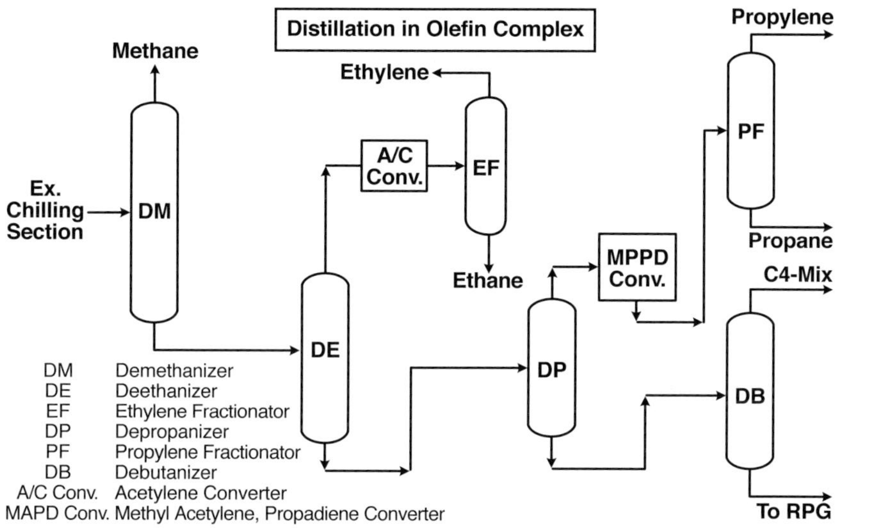

Fig. 1.10.7 Typical fractionation scheme of an olefin plant

1.10.6 Process description of an olefin plant

In an olefin plant, the cracked gas obtained from a steam cracker (which contains the desired olefins) is compressed in a charge gas compressor, treated first with amine and then with caustic. The gas is then further compressed in charge gas compressor (CGC) to around 36–40 kg/cm^2. The compressed charge gas is dried prior to chilling.

In the compressed cracked gas chilling train, all the hydrocarbon is condensed using progressive chilling, and the hydrogen is separated. The hydrogen is selectively rejected (using the Joule–Thomson cooling effect) from the charge gas after chilling it and condensing the majority of the methane and heavier hydrocarbons.

The feed to the demethanizer is charge gas compressor discharge condensate stream containing most of the lighter hydrocarbons along with the olefins produced. The condensate is fed to the demethanizer, where the top temperature requirement is around –134°C at a pressure of approximately 6 kg/cm^2—almost the dew point of pure methane (–134.5°C at 6 kg/cm^2) at that pressure.

The charge gas compressor discharge gas is chilled at different temperature levels and the condensate is fed at different locations of the column. The chilling is achieved through heat exchange with cold products, then refrigeration and self-chilling.

The feed to the demethanizer is generated at different temperatures during the process of chilling, and separators are used to collect condensates at the different temperatures. The liquid from the separators is fed back to the demethanizer through control valves that drop the downstream pressures almost to the demethanizer pressure of 6 kg/cm^2. The condensates further cool before being fed to the column through self-refrigeration. The topmost feed is fed at about −132°C and the intermediate feeds at around −102°C and −75°C. Often more feeds are used at different temperatures. This reduces the condensing load/duty.

A methane refrigeration (either open or closed type) is used to cool the overhead product of the demethanizer. The methane content of the demethanizer bottom must be low enough that the slippage of methane to the downstream deethanizer is minimal, in order to maintain the purity of ethylene from the ethylene fractionator (EF). The demethanizer application is a great example of ways in which low-temperature refrigeration duty can be conserved, both by using the suitable integration of cold streams, and by feeding at different locations in the demethanizer column (shown in Fig. 1.10.8). A simplified demethanizer/deethanizer section simulation is shown in Fig. 1.9.17 of Chapter 9, with feed from only two stages of chilling. A tentative process flow scheme of a demethanizer column is given in Fig. 1.10.8.

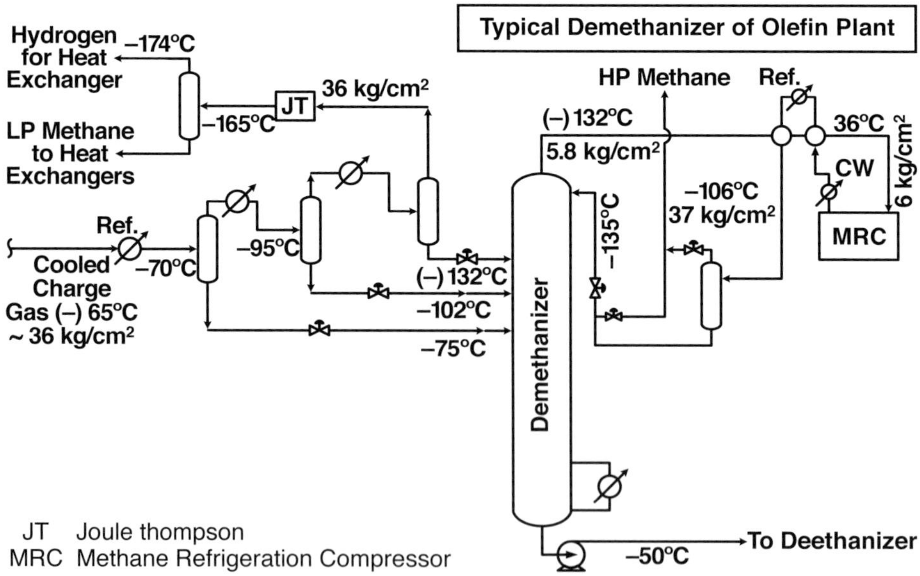

Fig. 1.10.8 Typical feed arrangement in a demethanizer

Deethanizer

Demethanizer bottom product is fed to the deethanizer after heating it and recovering the cold at about –20°C. The condenser needs refrigeration (temperature around –19°C), while the bottom needs cheap hot utility (temperature around 70°C). Top condenser duty is traditionally removed by propylene refrigeration from appropriate stage of the propylene refrigeration compressor. A typical demethanizer/deethanizer simulation is shown in Chapter 9, Fig. 1.9.17.

Ethylene fractionator

To illustrate the relative merits of the open-top, open-bottom, and closed-loop vapor compression schemes shown in Figs. 1.10.1 and 1.10.3, Example Problem 1.10.2 examines an ethylene fractionator using all the three schemes.

Ethylene fractionators typically require around 170 stages in order to separate a mixture of 81.8% wt of ethylene and 18.2% wt of ethane into ethylene with an overhead purity of 99.5% wt or higher and an ethane purity of 99.5% wt at the bottom. The feed temperature in the example is –28.6°C at a pressure of around 20 kg/cm^2(a), which is its bubble point. The simulation output along with the condenser and reboiler duties are given below for the live problem.

Example Problem 1.10.2

The example problem is worked out with different schemes. An ethylene–ethane fractionator from a steam cracker with a feed composition of ethylene: ethane of 81.8:18.2 wt/wt. The feed temperature is –28.6°C. This is tried with different schemes to obtain 99.5% wt ethylene purity. A summary of the data from Example Problem 1.10.2 is given in Table 1.10.4.

Table 1.10.4 Summary of Example Problem 1.10.2

Stream name	Feed	Overhead	Bottoms
Total mass rate in kg/hr	144,162.0	118,387.6	25,774.4
Ethylene %wt	81.8	99.5	0.5
Ethane %wt	18.2	0.5	99.5

In Fig. 1.10.10, Example Problem 1.10.2 and all of its critical parameters are shown with 170 stages and a top pressure of 16.5 kg/cm^2(a).

Conventional scheme

The conventional column operates at 16.5 kg/cm^2 pressure at the top. The condenser and reboiler heat duties are met by propylene refrigeration. The required top temperature is –36.6°C, and the bottom temperature is –14.6°C in order to obtain the purity specified by the example problem. A conventional simulation of an ethylene fractionator is presented in Fig. 1.10.9.

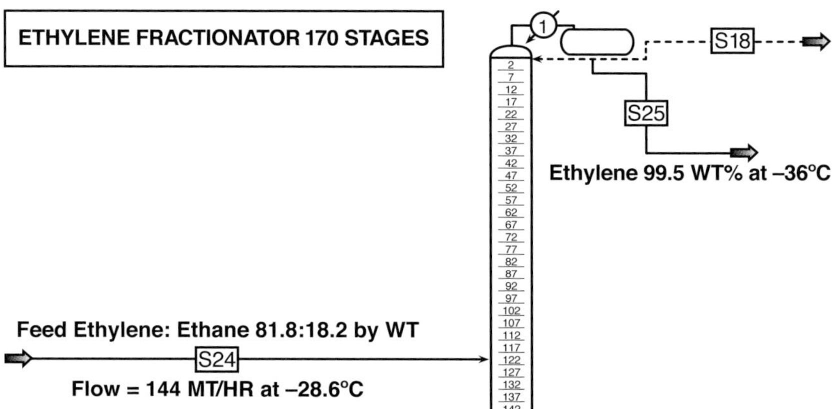

Column Name Column Description		T4
Condenser Duty	M*KCAL/HR	–33.8372
Reboiler Duty	M*KCAL/HR	33.3224
Column Reflux Ratio		2.4078

Fig. 1.10.9 Conventional simulation of an ethylene fractionator

It has been found that the reflux drum temperature for the ethylene splitter is –36.6°C and bottom temperature is –15°C. The column top pressure is 16.5 kg/cm^2(a). Thus refrigeration for overhead and reboiler is required. In an actual olefin complex, the heat duties of the EF reboiler and condenser are supplied and removed by common external propylene refrigeration system. The refrigeration system has a propylene compressor and other accessories such as a water condenser, receiver, etc., and uses propylene as refrigerant fluid. Typically, the propylene refrigeration compressor needs four stages to compress the propylene vapor from 0.33 to 18 kg/cm^2(*g*). The dew point of propylene, calculated at 0.30 kg/cm^2(*g*), is around –43°C; thus, it is possible to condense the overhead of an ethylene fractionator. Accordingly, the propylene vapor produced during condensation of the EF column overhead goes to the first-stage suction of the propylene compressor (0.33 kg/cm^2(*g*)) and the propylene vapor produced after supplying the reboiler duty goes to the second-stage suction of the propylene compressor. The stage suction pressures of the propylene compressor are around 0.33, 1.5, 4.0, 8.2 kg/cm^2(g) and may serve to meet the various requirements for chilling and refrigeration in the complex at temperature levels of –43°C, –27.8°C, –6.5°C, and 15.6°C.

It may be pertinent to mention that in a steam cracker complex, methane refrigeration condenser heat duty is removed by ethylene refrigeration, and the ethylene refrigeration condenser heat duty is rejected to the propylene refrigeration. The entire cooling load of the refrigeration of methane and ethylene is thus transferred to the propylene refrigeration, which is finally removed in cooling water by compressing propylene to 18.5 kg/cm^2 and attaining a temperature higher than the cooling water temperature (with the required temperature differential).

In propylene refrigeration, a stream can be cooled to around –36°C based on the dew point of propylene at the refrigeration compressor minimum suction pressure of 0.3 kg/cm^2(*g*) as –43°C. In ethylene refrigeration, a stream can be cooled to around –93°C based on the dew point of ethylene, which, at the refrigeration compressor minimum suction pressure of 1.3 kg/cm^2(*g*), is –99.1°C. In methane refrigeration, a stream can be cooled to around –133°C based on the dew point of methane, which is around –138°C at 6 kg/cm^2.

Application of the vapor compression scheme in a cryogenic system

The EF with vapor compression is chosen only to illustrate different schemes. An EF is used for the analysis and evaluation of features of different configurations such as open-top, open-bottom, and closed-loop vapor compression schemes.

1. Conventional vapor compression (open-top)

If a vapor compressor is used in this scheme (see Fig. 1.10.11) and develops discharge pressure of around 35 kg/cm^2 for transferring heat to the bottom reboiler (with a compression ratio of 35.0/16.5 = 2.18), the reflux drum pressure can be increased to 35 kg/cm^2.

The overhead vapor of the column will be routed to the vapor compressor, and the compressor discharge vapor will be routed to the reboiler and then returned to the reflux drum as saturated liquid after ensuring condensation through a cooler/condenser (trim cooler). The discharge pressure of the compressor is fixed to give a minimum differential temperature (7°C–10°C) across the reboiler at the column bottom.

Part of the liquid from the reflux drum will be drawn as ethylene product and part will be returned as reflux through the control valve. Part of the reflux liquid will vaporize downstream of the reflux control valve and bring the temperature of the remaining liquid lower than the top temperature of the column (–36°C).

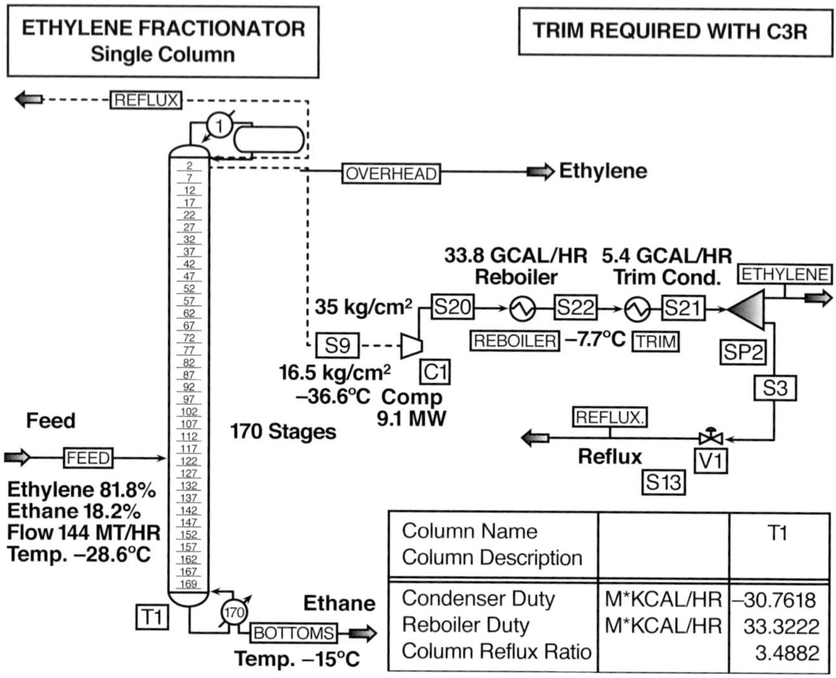

Fig. 1.10.10 Ethylene fractionator with a vapor compression open-top scheme

The temperature of the compressed vapor at the reboiler outlet is still –7.7°C and requires refrigeration to condense the stream in the trim cooler. It should be noted that the enthalpy of the overhead vapor increases upon compression with the addition of work equivalent to the heat of the compressor. At this point, imparting its heat to the reboiler by itself is not sufficient to allow condensation. In this case, some refrigeration is still required in the overhead circuit for the trim (which can be supplied by C3R). An ethylene fractionator with a vapor compression open-top scheme is shown in Fig. 1.10.10. In this scheme, there is a likelihood that some compressor lubrication oil will get into the overhead system and adversely affect the quality of ethylene produced unless proper care is taken with oil traps.

2. Open-bottom heat pump in an EF

In this scheme (shown in Fig. 1.10.11), part of the column bottom ethane is recovered as product with the desired purity of 99.5 mole % and the balance routed through the throttle valve to partly vaporize and cool it to a temperature of –43°C (lower than the bubble point of the 99.5 mole % ethylene overhead by 7°C–10°C). The pressure of the downstream throttle valve is kept at 7.1 kg/cm^2 to get the required cooling. This mixed phase is routed to the condenser and the outlet is specified as full vapor, which is then routed to the suction of the vapor compressor. The compressor boosts the pressure to 17.2 kg/cm^2 so that it can enter the column bottom with a pressure of 17.1 kg/cm^2, then a trim cooler is used to bring it to its dew point and inject it back into the column as near-vapor generated by the reboiler. The compression ratio of the compressor will be 17.2/7.1, i.e., 2.42. There is no reboiler in this scheme; the heat content of the condenser provides the reboiler duty. A controller is used to regulate the flow of bottom liquid (95 mole % ethane) through the throttle valve to remove the condenser duty of 33.8 Gcal/hr. A worry about open-bottom schemes is that there is a likelihood that the compressor lubrication liquid will get into the bottom of the system, but that is unlikely to spoil the quality of the desired top product. Open-bottom schemes also use about the same amount of power as open-top schemes.

A simulation of an EF with an open-bottom heat pump is given in Fig. 1.10.11.

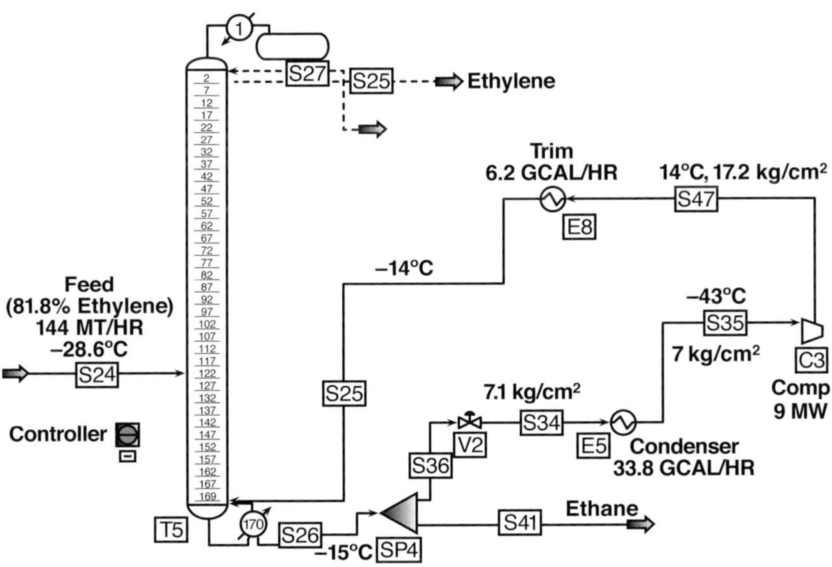

Column Name Column Description		T5
Condenser Duty	M*KCAL/HR	–33.8482
Reboiler Duty	M*KCAL/HR	0.0029
Column Reflux Ratio		2.4089

Fig. 1.10.11 Simulation of an EF with open-bottom heat pump

3. Closed-loop heat pump in an ethylene fractionator

A heat pump system can use propylene in a closed-loop in an ethylene fractionator (Fig. 1.10.12). The compressor discharge will exchange heat with the reboiler, be cooled and converted to liquid, expanded in the throttle valve, and then chilled to at least 7°C cooler than the overhead dew point of –36.6°C.

Another heat sink is required to convert the propylene completely to liquid. The throttle valve's downstream pressure is fixed at 1.3 kg/cm^2(a) in order to achieve a temperature of –43°C. The compression ratio required is 4.7/1.3, i.e., 3.6. Consequently, the power requirement may be little higher.

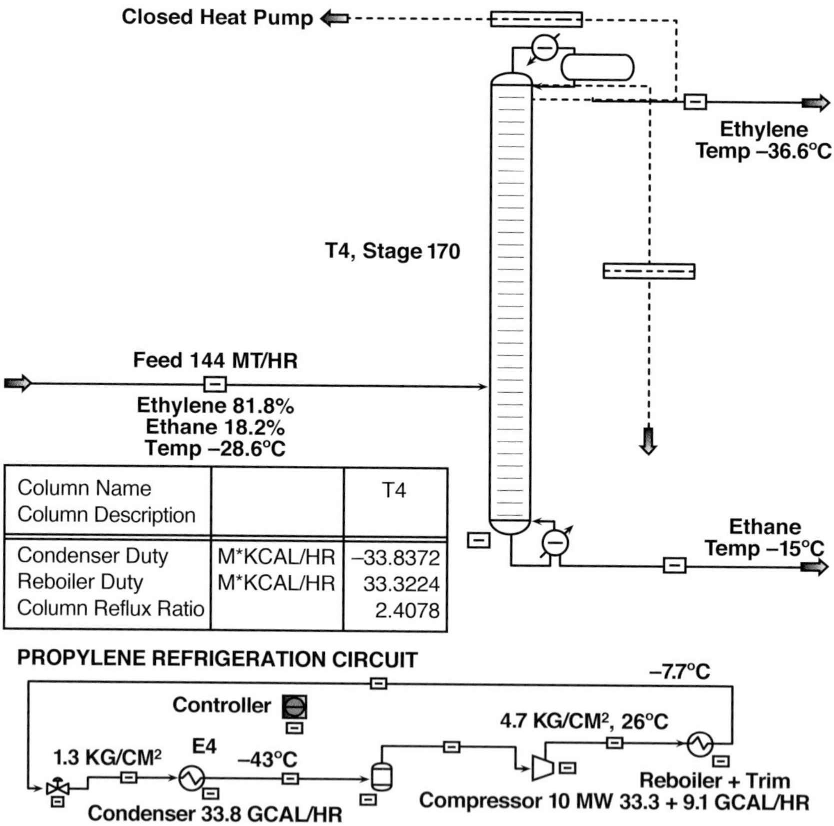

Fig. 1.10.12 Simulation of an EF with a closed-loop heat pump

In Fig. 1.10.12, a controller is used to regulate the flow of closed refrigerant in order to remove the entire condenser duty of 33.8 Gcal/hr. The refrigerant vapor is routed through a suction knockout drum to eliminate condensate, then to compressor. The compressor discharge is then routed to the reboiler. Some additional duty needs to be removed (over and above the reboiler duty of 33.3 Gcal/hr) in order to convert the compressed vapor to a saturated liquid. Heat (around 9 Gcal/hr) needs to be rejected to another refrigerant. A simulation of an EF with a closed-loop heat pump is given in Fig. 1.10.12. There is no likelihood that the compressor lubricating oil will get into the system, and it is similar to common propylene refrigeration.

Summary of the comparison of open-top, open-bottom, and closed-loop vapor compression

- The power requirement in the open-top scheme (Fig. 1.10.10) is comparable to that in the open-bottom scheme but marginally less than the closed-loop scheme.
- The power requirement in the open-bottom scheme (Fig. 1.10.11) is comparable to that in the open-top scheme but marginally less than the closed-loop scheme.
- The power requirement in a closed-loop scheme (Fig. 1.10.12) is slightly higher than the other two schemes, as it must overcome two differential temperatures for heat transfer (i.e., for the condenser and the reboiler). It also has the highest compression ratio. This scheme is preferred when compressor-oil ingress in the system is not at all desirable.

Often columns have inter-reboilers supplied by different stages of the vapor compressor in order to optimize the energy requirement.

1.11 Azeotropic Distillation

Azeotropes are primarily two different types:

1. Low-boiling azeotropes: in this type of azeotrope, the boiling point of the azeotropic mixture is lower than the boiling points of both the components.
2. High-boiling azeotropes: in this type of azeotrope, the boiling point of the azeotropic mixture is higher than the boiling points of both the components.

Within those two types, there are two subgroups: homogeneous and heterogeneous. Most of the low-boiling azeotropes are heterogeneous.

The homogeneous azeotropes are often distilled with pressure swing distillation. A scheme for the separation of a mixture of A and B is shown in Fig.

1.11.1. The first column distills the feed to its azeotropic composition through the top at that pressure and pure A-product is recovered through the bottom. The top is fed to a second column at a higher pressure, and distills a product with an azeotropic composition (for the pressure) from the top, and pure B-product is recovered from the bottom. The overhead of the second column is fed to the first.

The pressure swing distillation scheme is given in Fig. 1.11.1.

In refinery/petrochemical applications, the majority of the azeotropes are low-boiling and heterogeneous. Methyl ethyl ketone (MEK)–Water and Furfural–Water systems are examples of low-boiling heterogeneous azeotropes and are commonly encountered in the solvent dewaxing units and furfural aromatic extraction units of a lube oil production complex. Furfural and water form azeotropes with a composition of approximately 8% furfural and 92% water at atmospheric pressure.

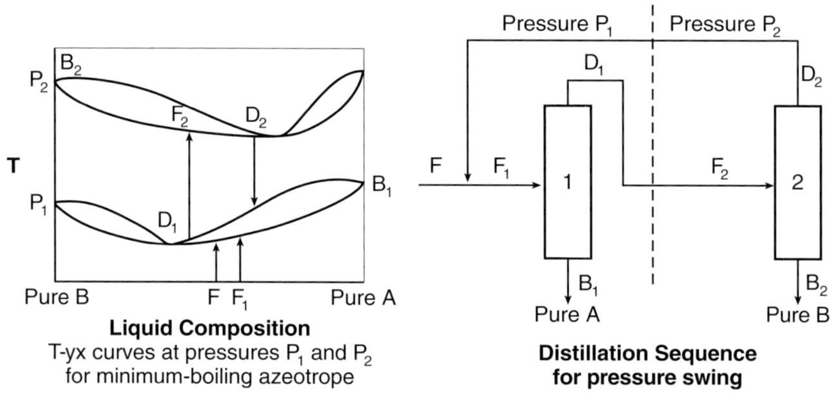

Fig. 1.11.1 Pressure swing distillation

In the furfural aromatic extraction of lube oil base stocks, the furfural from both the extract and raffinate phase is recovered through successive evaporation and stripping. The furfural vapor, along with the water, comes to the furfural (drying) column (Fig. 1.11.2).

From the furfural (drying) column bottom, pure furfural is recovered. The near-azeotropic composition is achieved in the overhead and cooled/condensed and collected in the drum. The condensate separates in two layers in the drum at 45°C:

1. the water-rich phase: (9.7% wt furfural + 90.3% wt water);
2. solvent-rich (furfural) phase: (93% wt furfural + 7% wt water).

The solvent-rich phase is fed to the top of the furfural (drying) column as the reflux and water-rich phase is fed to the water column. Pure furfural is obtained from the furfural (drying) column bottom and near-azeotropic composition from the top (76.8% wt furfural + 23.2% wt water). Pure water is recovered as the bottom product of the water column and near-azeotropic composition (13.5% wt furfural + 86.5% wt water) is distilled to the top. It then joins the overhead of the furfural (drying) column at the upstream of the cooler condenser and is routed to the drum along with the furfural (drying) column condensate. The solvent-drying section of a furfural extraction unit (FEU) for a lube complex is given in Fig. 1.11.2.

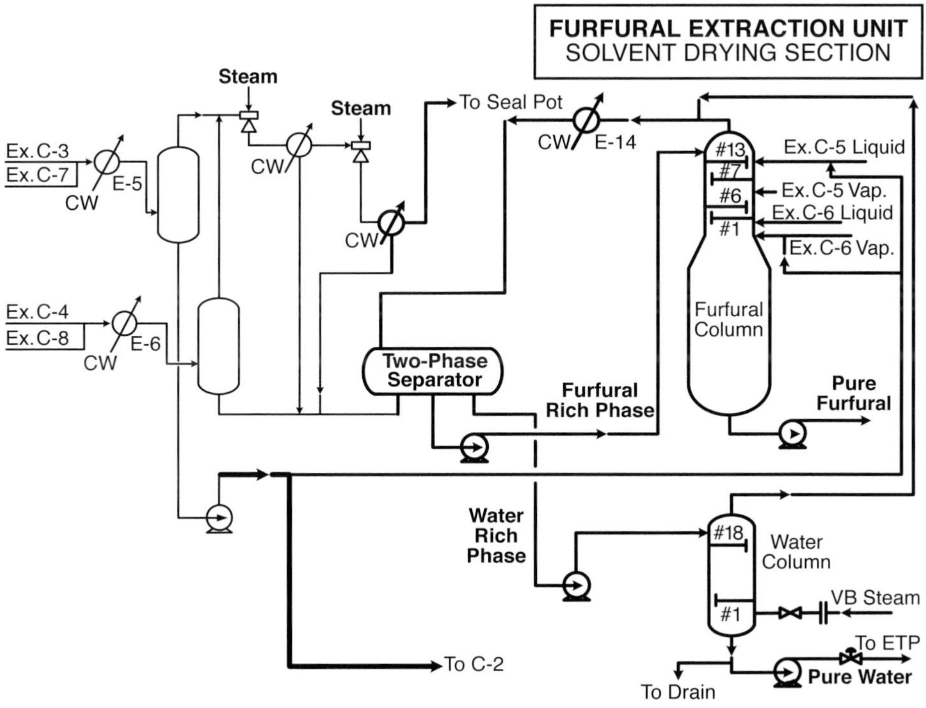

Fig. 1.11.2 Solvent-drying section of a furfural extraction unit (FEU) for a lube complex

In the industry, a mixture of acetic acid and water is distilled to get pure acetic acid and pure water using either ethyl acetate or N–Propyl–acetate as the entrainer. The feed rate in the example problem is considered as 10 MT/hr with a composition of 50:50 mole:mole of acetic acid and water. The ethyl acetate is fed as the entrainer externally at the top of the first column. The overhead vapor is cooled and separated in a drum, where the liquid phases separate in two layers. The ethyl-acetate-rich phase is recirculated in the first column while the water-rich phase (around 95 mole %) is fed to the second column to recover water as a bottom product. Stripping steam is injected at the bottom of the second column to strip ethyl acetate from water. The vapor from the second condenser is recirculated at the upstream of the cooler/condenser.

Simulation Guideline

Initially, fresh ethyl acetate is added to the top of first column as a stream to control the top temperature of the column. The bottom product is pure acetic acid. The ethyl acetate generated by separation in the drum is added to the first column as a second feed (stream name: S18). This brings down the required quantity of external fresh ethyl acetate to almost nothing. A simulation using ethyl acetate as the entrainer is shown below in Fig. 1.11.3.

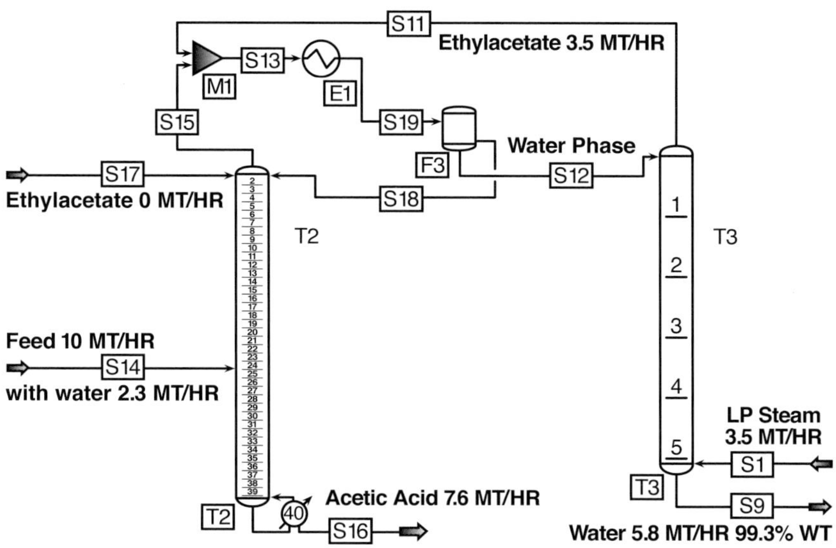

Column Name Column Description		T2	T3
Condenser Duty	M*KCAL/HR	n/a	n/a
Reboiler Duty	M*KCAL/HR	5.4575	5.0000
Column Reflux Ratio		0.8366	2.1097

Fig. 1.11.3 Simulation of acetic acid–water separation with ethyl acetate as the entrainer

Pressure swing in ethanol drying

A specimen sample case of ethyl alcohol (EtOH) drying is considered with a pressure swing.

The X-Y diagram of EtOH–water is prepared at 0.13 kg/cm^2(a) pressure. The X-Y curve shows an azeotropic composition of around 92 mole % EtOH (Fig. 1.11.4). Thus, the mixture can be separated into around 90 mole % of EtOH (near azeotrope) from the top and pure water from the bottom in the first column at 0.13 kg/cm^2(a) pressure. Subsequently, the curve is developed at 4 kg/cm^2(a). The azeotropic composition becomes around 82 mole % EtOH (Fig 1.11.5). The boiling point of the azeotropic mixture at 4 kg/cm^2(a) (116°C) is lower than the boiling point of pure EtOH at 4 kg/cm^2(a), as it is a low-boiling azeotrope. So when the top of the first column at around 90 mole % of EtOH (near azeotrope) is fed to the second column at higher pressure, the second column overhead will be around 80% mole EtOH (near azeotrope at 4 kg/cm^2(a) pressure), and the bottom will be pure EtOH. When the top of the second is again recirculated to the first column, it can be purified to 90 mole % of EtOH (near azeotrope at 0.13 kg/cm^2(a) pressure) from 80 mole % of EtOH. Thus it becomes feasible to separate the mixture into pure EtOH and pure water with this scheme. The pressure swing distillation of EtOH–water can be also understood from the T-X diagram of the two components given in Fig. 1.11.1.

The X-Y diagram at 0.13 kg/cm^2(a) is given in Fig. 1.11.4.

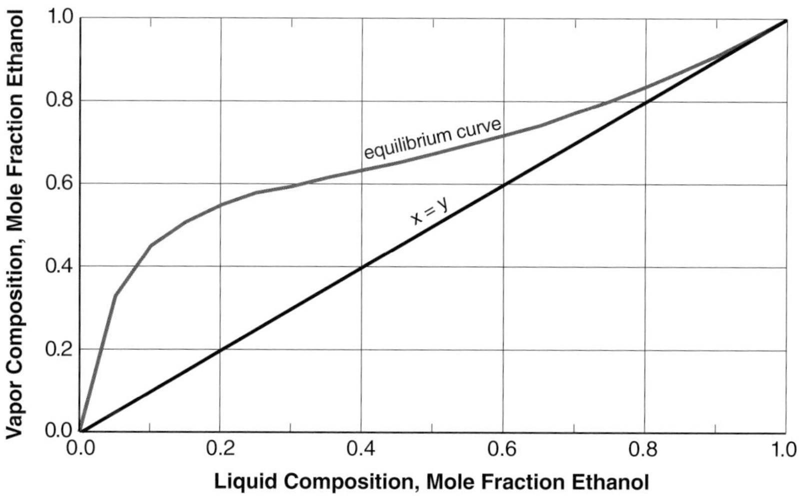

Fig. 1.11.4 X-Y plot of ethanol-water system at 0.13 kg/cm^2(a)

The X-Y diagram of an ethanol–water system at 4 kg/cm^2(a) is given in Fig. 1.11.5.

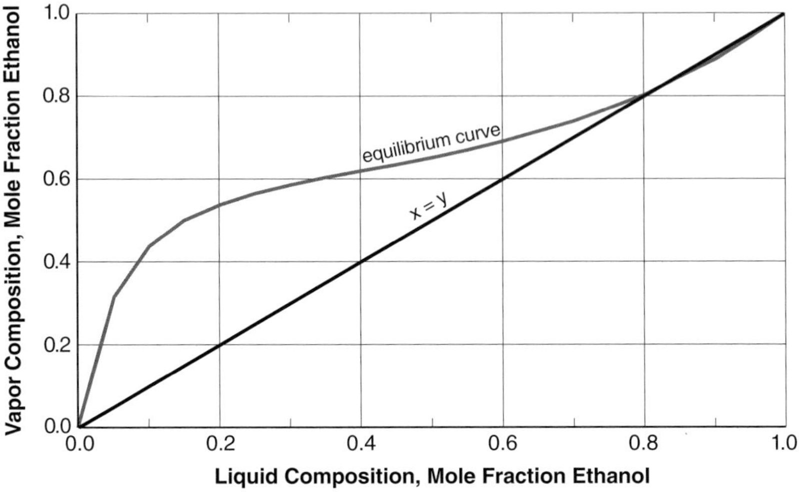

Fig. 1.11.5 X-Y plot of an ethanol–water system at 4 kg/cm^2(a)

Example

A 67:33 mole % EtOH–water mixture is first fed to the LP column (Fig 1.11.6) under a vacuum. The feed rate is 1,000 kmole/hr. Pure water separates from bottom and near-azeotropic composition is recovered from top and fed to the second column at 4 kg/cm^2 pressure.

Pure EtOH is recovered from the bottom and a mixture of EtOH and water (rich in EtOH and close to azeotropic composition) is obtained from top and recirculated to the top of the first column.

The vapor–liquid equilibrium (VLE) data generated using the parameters of the thermodynamic model are used for the simulation in both the columns in Fig. 1.11. 6.

This process consumes very high energy and is not industrially practiced. The EtOH drying is achieved using either azeotropic distillation with cyclohexane $(CH_2)_6$ as the entrainer, by extractive distillation, or through solid bed drying.

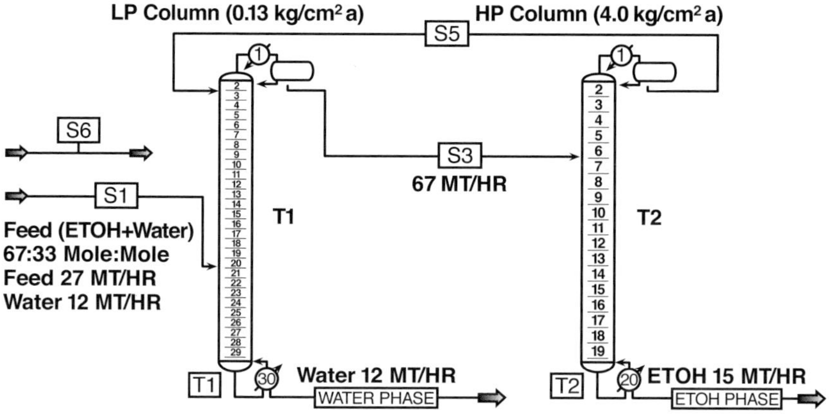

Column Name Column Description		T1	T2
Condenser Duty	M*KCAL/HR	–57.7711	–61.7657
Reboiler Duty	M*KCAL/HR	54.7364	65.8455
Column Reflux Ratio		2.7678	5.0000

Fig. 1.11.6 Simulation of the pressure-swing distillation of EtOH and water

Extractive distillation using ethylene glycol (EG) as the entrainer

The EtOH–water mixture can be separated into pure EtOH and pure water using EG as the extractive agent. Typically, EG is fed to the first column. The EG captures water and, because it has a higher boiling point than EtOH, it remains in the bottom of the first column. The reboiler heat provided in the first column preferentially vaporizes EtOH so that pure EtOH is recovered from the top of first column. The water and the EG are fed to the second column, and pure water is recovered from top and pure EG from the bottom of the second column.

Example

The example from Fig. 1.11.12 is worked out to purify 10 MT/hr of feed with 50:50 mole % of ETOH and water. Initially, a column (T1) is used to enrich the 50:50 mole % mixture toward an overhead product with an azeotropic composition of around 74 mole % EtOH. The overhead is then fed to the extractive column T2. The overhead of T2 is pure ETOH and the overhead of a third column (T3) is pure water with a bottom of pure EG. The recovered EG is circulated back into extractive column T2.

A nonrandom two-liquid (NRTL) VLE is used in T1. NRTL VLE parameters are used in T2 and T3. The reboiler duties are low in this configuration.

The feed (50:50 mixture) can be also directly fed to the extractive column T2, but this would require a higher circulation of EG and create higher reboiler duties. The simulation of EtOH–water separation using ethylene glycol is given in Fig. 1.11.7.

The application of EG in drying natural gas (NG) prior to refrigeration is very common. The EG with water is then dried in an EG dryer (glycol regenerator) and recirculated so that moisture from the NG can be reabsorbed. However, a glycol regenerator can only be used with a reboiler and condenser that have a refluxing arrangement on the wash trays due to the vast boiling point difference between EG (197.3°C) and water (100°C) at atmospheric pressure. Commercial glycol regenerators use small columns with an evaporator and condenser at the top, and a maximum of four wash trays to contain the entrainment of glycol to the overhead water stream.

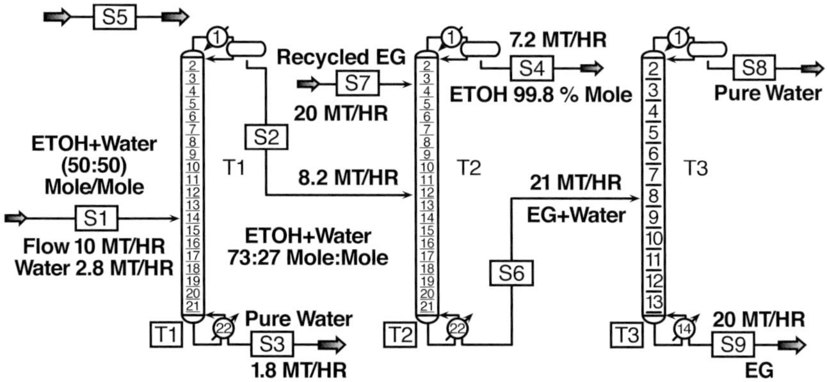

Column Name Column Description		T1	T2	T3
Condenser Duty	M*KCAL/HR	–5.8213	–4.2568	–2.7150
Reboiler Duty	M*KCAL/HR	6.0000	6.0000	3.0000
Column Reflux Ratio		2.0000	2.0000	6.0000

Fig. 1.11.7 Simulation of EtOH–water separation using EG

1.12 Extractive Distillation

1.12.1 Introduction

Extractive distillation (ED) is often performed when the relative volatility of the keys are close to unity (1.0). A third component can be added to modify the relative volatility of the light or heavy key so that the key is separated as almost pure component. The other key remains, along with the third component added. This mixture is then sent to another column that separates the key and the third component.

In the alcohol–water system above (Fig. 1.11.7), the third component used is ethylene glycol, which alters the relative volatility of EtOH (one key) and separates pure EtOH from the top and a mixture of water and EG from the bottom (see Fig. 1.11.7). The water and EG phase is fed to the next column where it is separated, with water as the top product and EG as the bottom product. In the above depiction (Fig. 1.11.7) of an EtOH–water separation, EG is used as the extractive distillation solvent. Often in extractive distillation, salts are used as the entrainer to make the separation more energy efficient; for example, potassium acetate is used as the entrainer for ethanol–water systems.

The third component actually modifies the activity coefficients of the keys to make one more volatile and the other less volatile. The third component alters the relative volatility of a two-component system when the components have different polarity. If the third component (the entrainer) is highly polar, the activity coefficient of the polar component in the mixture will be close to unity (one). The other nonpolar (less polar) component will have a higher activity coefficient. Conversely, if a nonpolar entrainer is used for the same mixture of components, it will increase the activity coefficient of the polar component.

A very high proportion of entrainer (third component) is normally used to raise the relative volatility of one key in respect to the other key. The high proportion of entrainer used in the process allows the activity coefficient of the key components to approach the activity coefficient at infinite dilution.

Correlation of activity coefficient (γ) and relative volatility (α)

Partial pressure: $(p_i) = \gamma_i \times P_i X_i = Y_i P_t$ or $Y_i P_t = \gamma_i P_i X_i$

or $Y_i = \gamma_i (P_i / P_t) \times X_i = \gamma_i K_i X_i$

Relative volatility: $\alpha_{ij} = (Y_i/X_i)/(Y_j/X_j) = (\gamma_i/\gamma_j)(P_i/P_j)$

P_i, P_j = vapor pressures of ith and jth components

P_t = total pressure

Even if the ratio of vapor pressure (P_i/P_j) is high, the ratio of activity coefficients (γ_i/γ_j) can still be low, which may bring relative volatility α close to unity and make the separation impossible.

Extractive distillation is based on introducing an entrainer that modifies the ratio of activity coefficients to increase the value of α and make the separation easy. In other words, if γ_{1E} is much higher than γ_{2E}, the ratio is likely to become high unless it is reduced by vapor pressure ratios (P_i/P_j) or (P_1/P_2).

The equation for an ideal mixture at low pressure and relative volatility is $\alpha^* = P_1/P_2$, where P_1 and P_2 are the vapor pressures for components 1 and 2 respectively. This can be rewritten as $\alpha = (P_1/P_2)(\gamma_1/\gamma_2)$ at infinite dilution. The ratio (γ_1/γ_2) at infinite dilution is often called the selectivity of the entrainer. So, if the ratio of the activity coefficients at infinite dilution is high, then α will be high. When a large amount of entrainer or solvent is used, γ_1 and γ_2 reduce to activity coefficients at infinite dilution γ_1^∞ and γ_2^∞. This can be calculated for the given system from correlations given in the standard literature.[1]

Alternatively, γ_1^∞ and γ_2^∞ are determined experimentally, using ebulliometry and chromatography. The entrainers commonly used in industrial practice can be seen in Table 1.12.1 in descending order of polarity.

Table 1.12.1 Polarity of industrial solvents

Solvent	Polarity
Water	1.0
Ethylene glycol	0.79
NMP	0.36
Sulpholane	0.41
DEG	0.72

Water can be a very good entrainer due to its high polarity; however, it forms azeotropes with many compounds, which makes the separation of water from one component difficult, and restricts its use as an effective entrainer. The effects of water as an entrainer are given below in Table 1.12.2.

Table 1.12.2 Effects of water as an entrainer

Component desired	Component rejected	Normal relative volatility	ED relative volatility
DIPE	Ethanol	1.04	5.7
MEK	Ethanol	1.05	3.0

The activity model equation parameters need to be provided to the designer from experimental data for the systems that are not available in the simulator database. Experimental activity coefficients at infinite dilution can be fed into a simulator to generate VLE data for the application in ED. The UNIFAC thermodynamic model can be chosen to determine the behavior of the components in presence of ED solvent for initial estimation. Any activity coefficient correlation (Wilson equation, NRTL equation, etc.) can be chosen for simulation. For common systems, data is available in some simulators or VLE data collections by DECHEMA and others. For uncommon systems, data of the activity coefficient at infinite dilution needs to be generated experimentally for feeding into the simulator to work out applications with ED. At this level of high entrainer concentration, the interaction of a ternary system can be ignored. It may actually be considerably more accurate to consider two binaries (component 1 and the entrainer, and component 2 and the entrainer) to determine the values of the activity coefficients. Additionally, in extractive distillation, the entrainer is introduced in large amounts in order to alter the relative volatility of the components; thus, the liquid load increases in the section below the solvent feed in the column, and the downcomer area needs to be large enough to handle the huge liquid load. In practical ED column design, there should be a few trays above the entrainer entry in the upper section to act as wash stages and reduce entrainer loss along with top product.

The use of ED is becoming increasingly popular. One of its applications is in the recovery of benzene from a mixture of paraffins and cyclo-paraffin from pyrolysis gasoline in ethylene units (gas/naphtha cracker). The solvent used is sulpholane or NMP (*N*-Methyl-2-pyrrolidone). Extractions of 1,3-butadiene are carried out from the C4 mix streams of an ethylene plant with acetonitrile (ACN), NMP, or another proprietary solvent.

A simulation of a benzene extractive distillation using NMP is presented in Fig. 1.12.1.

Typically, in an olefin plant, pyrolysis gasoline is first treated for removal of diolefins, and subsequently deheptanised in a deheptaniser column. Then the top products of the column—benzene, cyclohexane (CH), hexanes, 2,4-dimethylpentane (2,4-DMP), 2,3-dimethylpentane (2,3-DMP), and 2,2-dimethylpentane (2,2-DMP)—are fed to the extractive distillation column (EDC) (Fig. 1.12.1). NMP (or often sulfolane) may be used as a solvent in quantities of almost three to four times the benzene in the feed in order to extract the entire quantity of the benzene along with the NMP.

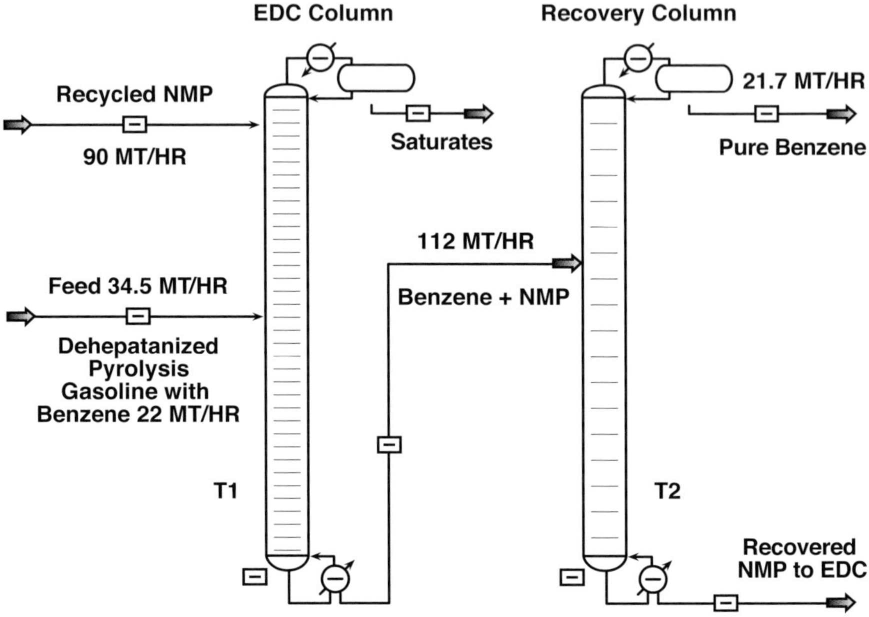

Column Name Column Description		T1	T2
Condenser Duty	M*KCAL/HR	–4.0331	–37.3569
Reboiler Duty	M*KCAL/HR	5.4000	40.0000
Column Reflux Ratio		3.0000	18.0000

Fig 1.12.1 Simulation of a benzene extractive distillation using NMP

The top is a mixture of saturates that are rejected as raffinate. The bottom (NMP and benzene) is fed to the NMP recovery column to recover NMP as bottom product and pure benzene as top product. A high purity of benzene is obtained. The NMP recovered is then recycled to the EDC column three to four trays below the top. The ED solvent takes an appreciable amount of reboiler duty in the recovery column. A better ED solvent can reduce the quantity of solvent required and can reduce the reboiler duty.

The NMP recovery column operates at 0.3 kg/cm^2(g) top pressure and the bottom temperature becomes 145°C to 147°C. Thus, medium-pressure steam can be used as a reboiler heating agent instead of operating in vacuum (with sulfolane as solvent).

Desired properties of ED solvents

1. Chemical stability at operating temperatures.
2. Miscibility with components 1 and 2 in all proportions.
3. A large difference in boiling point with components 1 and 2.
4. Low molar volume so that the downcomers are not overloaded by the large volume of entrainer used in the ED column.
5. Nontoxic.
6. Noncorrosive.
7. A boiling point that allows the stripping/recovery of the ED entrainer in the recovery column without the application of a vacuum is preferred.
8. The ED entrainer may not form any azeotrope with any components.

1.13 Case Studies

Case Study 1

1.13.1 Paraxylene feed production from a C8 precursor rich, straight-run naphtha

A unit producing *para*-xylene feed had an overhead coupling arrangement. The first column (LP column) was operated at 1.8 kg/cm^2(g) and the second column (HP column) was operated at 4.5 kg/cm^2(g). The overhead of the second column was supplying part of the reboiler duty (5.0 Gcal/hr) for the LP column. (See Fig. 1.13.1).

The LP column had an auxiliary reboiler supplied by high-pressure steam. The heat duty requirement of the first column with this scheme was 10.5 Gcal/hr, comprised of 5.0 Gcal/hr supplied by the overhead of the high-pressure column and 5.5 Gcal/hr from an external source. The heat duty requirement of the second column with this scheme was 12 Gcal/hr. Total reboiler duty requirement from external source was 5.5 + 12 = 17.5 Gcal/hr.

The recovery of C8 precursors were 85% by weight and their purity in a PX feed required was a minimum of 60% by weight. The original scheme for PX feed preparation is presented in Fig. 1.13.1.

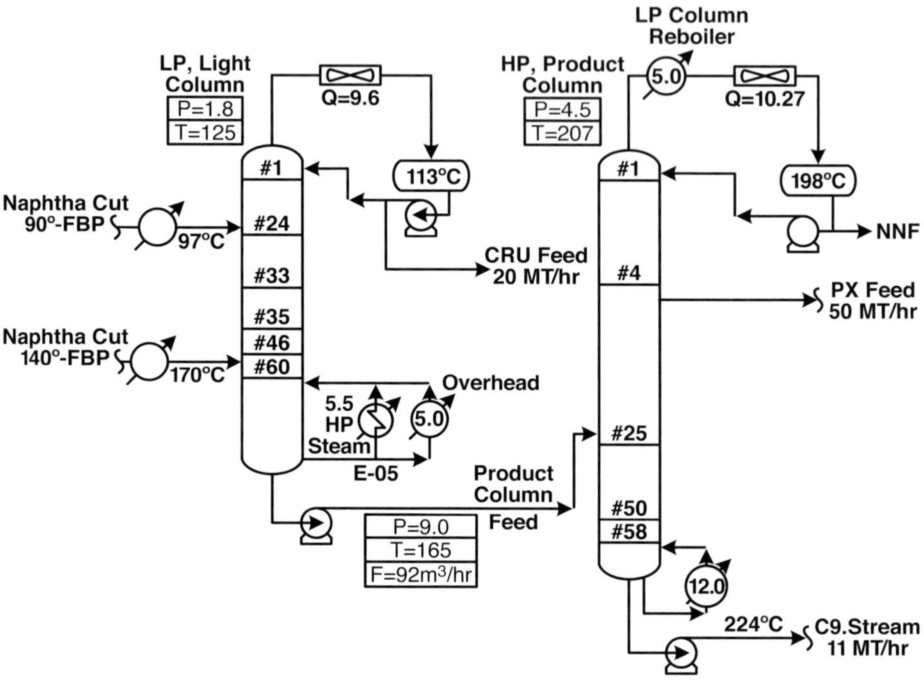

Fig. 1.13.1 Original scheme for PX feed preparation

The unit was modified to optimize further energy using TCDS/distributed distillation principles. Essentially, an additional side cut was drawn from the LP column to capture almost all the C8 precursors, and fed to the HP column. The bottom of the LP column was also fed to the HP column bottom to recover any slippage of C8 precursors from the LP column's bottom stream (Fig. 1.13.2). The total external duty requirement was reduced to 3.3 + 8.5 = 11.8 Gcal/hr. A conservation of around 6 Gcal/hr was obtained, keeping the same recovery and purity percentages for the C8 precursors as in original scheme (Fig. 1.13.1).

The scheme in Fig. 1.13.2 is a combination of a multi-effect distillation scheme with the features of a distributed distillation scheme. The bottom product of the LP column could have been drawn from the LP column itself instead of feeding to the HP column. Feeding the LP column bottom stream to the HP column was introduced to increase flexibility and ease of execution.

The modified scheme for PX feed preparation is presented in Fig. 1.13.2.

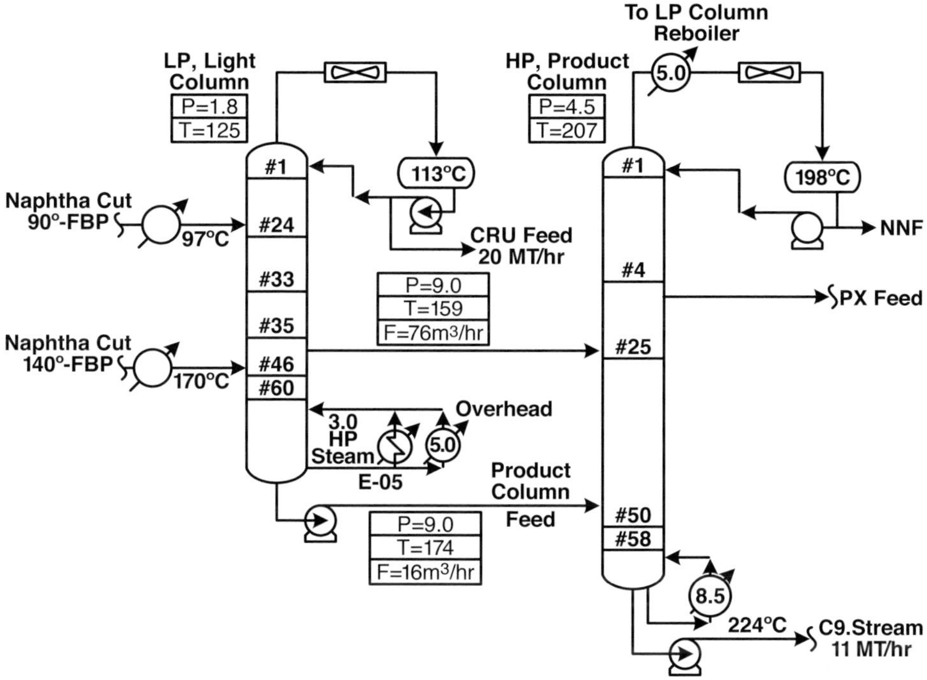

Fig. 1.13.2 Modified scheme for PX feed preparation

Case Study 2

One case study of energy minimization/marginal-capacity increase with the features of a two-enthalpy feed is presented in Section 1.6 for a debutanizer (Figs. 1.6.1 and 1.6.2). The capacity of the debutanizer was increased by around 15% using the method of introducing cold and hot feed at two different locations in the column.

Case Study 3

A case study of energy minimization in the separation of isomerization feed and catalytic reformer feed is presented in Section 1.6 for a naphtha splitter (Figs. 1.6.3 and 1.6.4) using the principle of a two-enthalpy feed. The purity of ISOM feed in respect to C7 content improved.

Case Study 4

A propylene fractionator unit was originally designed to produce 93 mole % propylene with a higher feed rate and overhead propylene product. There were 130 trays provided in the column, with feed point originally envisaged at the 53rd tray from the top. But there was no demand for 93 mole % purity propylene. It was converted to 99.7 mole % purity simply by lowering the feed location from the 53rd to the 95th tray and recovering only the required quantity (25 MT/hr) of polymer-grade propylene, leaving the rest of the propylene at the bottom with propane to be sold as LPG.

In the plant, no cheap heat source was available, so the original design considered the use of vapor compression. The reboiler duty and condenser duty were reduced, thus the overhead vapor compressor flow and, consequently, the power requirement, were reduced.

Abbreviations

ACN: acetonitrile

ASTM D-86: standard method of distillation specified by American Standard Test Method

C3R: propylene refrigeration

CCRU: continuous catalytic reformer unit

DMW: demineralized water

ED: extractive distillation

EDC: extractive distillation column

EtOH: ethyl alcohol

FUEM: Fenske–Underwood–Erbar-Maddox correlation

FUG: Fenske–Underwood–Gilliland correlation

ISOM: isomerization unit

LPG: liquefied petroleum gas

NMP: n-methylpyrolidone (a solvent)

NRTL: nonrandom two-liquid

UNIFAC: liquid activity model

VLE: vapor–liquid equilibrium

Note

1. Bruce E. Poling, John M. Prausnitz, and John P. O'Connell, *The Properties of Gases and Liquids*, 5th ed. (New York: McGraw-Hill Professional, 2000).

Distillation Tower Auxiliaries 2

2.1 Estimation of Efficiency

2.1.1 Murphree tray efficiency

$E_{mv} = (y_n - y_{n-1})/(y_n^* - y_{n-1})$ for the tray, where y_{n-1} is the vapor phase mole fraction of the more volatile component reaching the nth tray, y_n is the vapor phase mole fraction of the more volatile component leaving the nth tray, and y_n^* is the vapor phase mole fraction of the more volatile component in equilibrium with the liquid composition of the nth tray.

Overall column efficiency (E_{oc}) can be calculated from the Murphree tray efficiency equation (E_{mv}) using this relationship, developed by Lewis.

Overall efficiency

$$E_{oc} = \ln(1 + E_{\text{mv}} (\lambda - 1))/\ln\lambda$$

For $\lambda = m \times Gm/Lm$, m is equal to the slope of the equilibrium line, and Gm and Lm are molal vapor and liquid flow rates respectively. It is very difficult to correctly determine the slope of equilibrium line. Normally, one average slope of the equilibrium line from above the feed tray and one average slope from below the feed tray are considered for an estimation of overall efficiency.

One ratio of the Gm/Lm tray is considered in the trays above feed tray. Similarly, one ratio of Gm/Lm tray is considered in the trays below feed tray. Thus, two different efficiencies are normally calculated for the rectification section and stripping section. E_{mv} is normally determined experimentally.

2.1.2 Tray efficiency (empirical prediction methodology)

Overall efficiency can be calculated through the following steps using either the O'Connell Correlation of Efficiency ($E_{oc} = 0.492 \times (M_u \times \alpha)^{-0.245}$) or the O'Connell Correlation in chart form.

1. The average temperature is to be calculated as the arithmetic mean of the top and bottom temperatures of the column.
2. The tray, with a temperature close to the average temperature calculated is to be chosen for the estimation of relative volatility and viscosity.
3. The viscosity of the liquid in the tray is to be evaluated in centipoise (cP) based on the composition and temperature of the tray and is chosen as M_u.
4. Relative volatility between the key components to be evaluated at the temperature of the chosen tray as α.
5. Calculate ($M_u \times \alpha$)
6. Calculate $E_{oc} = 0.492 \times (M_u \times \alpha)^{-0.245}$

The overall tray efficiency is closely linked with the liquid distribution on the trays.

Calculations of efficiency using the O'Connell Correlation for the benzene–toluene system from Example Problem 1.1.1 in Chapter 1.1, and the benzene–toluene–*para*-xylene system from Example Problem 1.2.1 from Chapter 1.2 are given in Table 2.1.1.

Table 2.1.1 Tray efficiency calculation for B–T and B–T–PX systems

	B–T System	B–T-PX System
Column top temperature	85.04°C	85.1°C
Column bottom temperature	125.2°C	126.8°C
Average temperature	105.1°C	105.9°C
The average temperature is found very close to tray-22 temperature.		
Liquid viscosity (μL) at tray 22 in centipoise, CP	0.248	0.25
Equilibrium constant (K_1) for benzene in tray 22	1.4811	1.452
Equilibrium constant (K_2) for toluene in tray 22	0.655	0.641
Relative volatility – α (K_1/K_2)	2.261	2.265
Product of liquid viscosity and relative volatility ($\mu L \times \alpha$)	0.561	0.566
Tray efficiency from O'Connell correlation, $E_{oc} = 0.492\ (\mu L \times \alpha)^{-0.245} \times 100$	56.7%	56.6%

Often this calculation is performed in two sections. Separate efficiencies are calculated for the section above the feed zone and the section below the feed zone. Experts observed that the efficiency calculated using this method is conservative (less than actually experienced).[1]

2.1.3 Influence of other factors on tray efficiency

1. **Weir height** increases efficiency. This is particularly true of H_2S, CO_2 absorption with amines, caustic scrubbers, etc.
2. **Length of flow path** increases efficiency until it causes channeling.
3. Reduction of the **fractional area of holes** increases efficiency.
4. **Vapor/Liquid loads** have almost no effect on efficiency.
5. Efficiency increases as the **liquid viscosity** is lowered.
6. **Lower relative volatility** increases efficiency.
7. At higher **pressure**, tray temperature (bubble temperature) increases, which lowers the viscosities of tray liquids and increases efficiency. This also reduces relative volatilities, which in turn contributes to higher efficiency.

2.1.4 Overall efficiencies for conservative design

Table 2.1.2 lists overall efficiencies of some systems commonly encountered. The tray efficiencies given are conservative and can be used safely for simulation and rating. In the ethylene fractionator and the propylene fractionator, there are often higher efficiencies observed than those mentioned in the table.

Table 2.1.2 Conservative efficiencies of industrial columns

Process	Service	Overall Efficiency	
		Above feed	Below feed
Catalytic reformers	Hydrotreater stripper	85	75
	Stabilizer	95	80
	Reformate splitter	80	70
	Debutanizer	90	80
Hydrotreater	Kerosene stripper	75	35
	Gas oil stripper	95	20
	Hydrocracker fractionator	70	65
	Hydrocracker stripper	70	65

Process	Service	Overall Efficiency	
		Above feed	Below feed
Cat cracker	Main fractionator	70	–
	Absorber-deethanizer	60	60
	Stabilizer	80	70
	Sponge oil absorber	–	60
	Cat naphtha splitter	80	70
Aromatic complex	Reformate splitter	70	65
	Benzene column	70	70
	Toluene column	65	60
	Xylene column	80	80
	C8–C9 splitter	80	70
	Deisohexanizer	80	70
Crude distillation	Crude column	66	30 wash section 30-50
	Vacuum column	66	30 wash section 30-50
	Stabilizer	80	70
	Naphtha splitter	80	70
	Depropanizer	80	70
Steam cracker	Demethanizer	100	100
	Deethaniser	100	100
	Ethylene fractionator	95	95
	Depronanizer	90	75
	Propylene-propane fractionator	95	90
	Gasoline fractionator	70	65
	Gasoline stripper	70	65
	Depropanizer	90	75
	Debutanizer	90	75
Amine absorbers	Gas/DEA	–	10
	Gas/MEA	–	15
	Gas/MDEA		15
Amine regenerators	MDEA	–	50
	DEA	–	50
Sour water strippers		–	50 (25%–35% for grass root designs)
Solvents	Hexane column	75	80
	Heptanes column	75	80
Other	Ethanol/water	60	60
	Iso-propanol/water	70	70

2.1.5 Effect of percent flood on tray efficiency

Flooding reduces the efficiency as entrainment increases. The flow parameter F_{LV} is defined as $F_{LV} = (L_W/V_W)(D_v/D_L)^{1/2}$ where L_W = liquid mass flow rate in kg/s, V_W = vapor mass flow rate in kg/s, and D_v and D_L = vapor and liquid densities in kg/m^3, respectively.

2.1.6 Effect of entrainment on plate efficiency

Dry-tray efficiency can be corrected for the effect of entrainment using the Boles equation: $E_a = E_{mv}/(1 + E_{mv}\ [\psi/(1-\psi)])$, where E_a is the actual plate efficiency, and ψ is the fractional entrainment, which is equivalent to the entrained liquid/gross liquid flow. Fractional entrainment of up to 10% is allowed, as this entrainment is not likely to affect the tray efficiency much. Fractional entrainment variation with a flow parameter is given in Fig. 2.1.1.

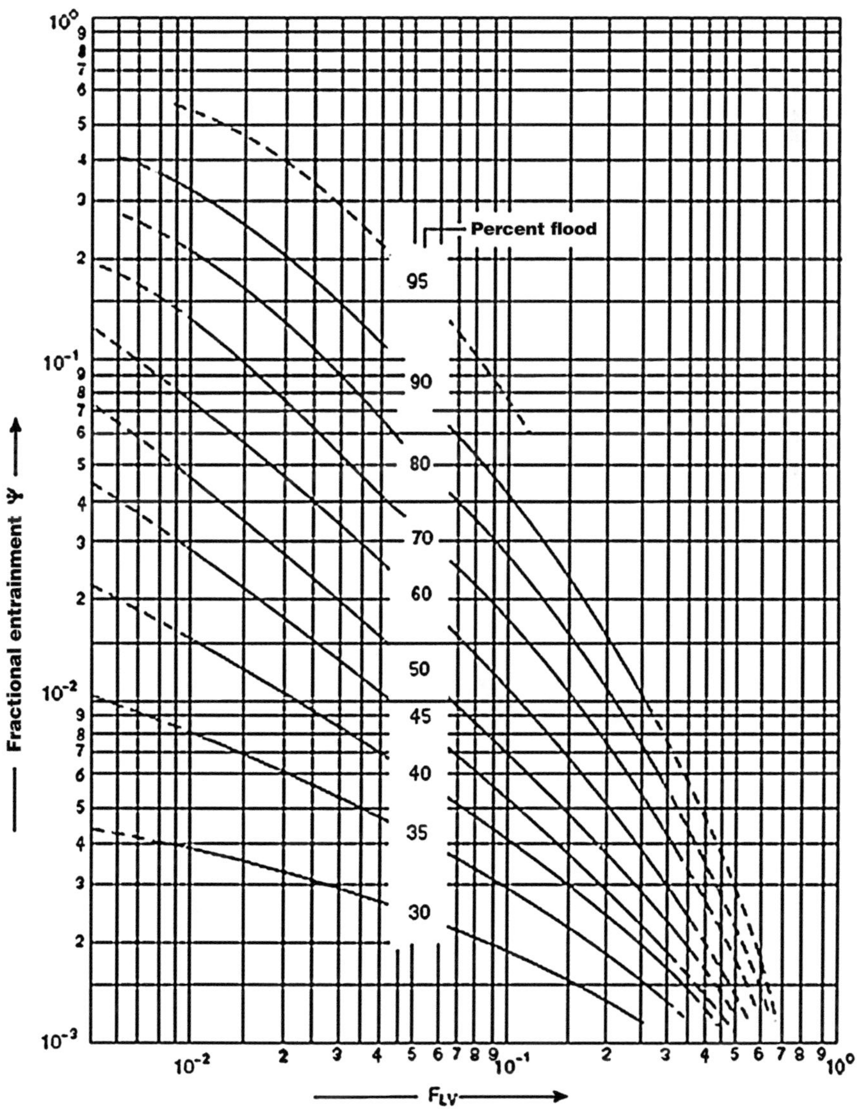

Fig. 2.1.1 Fractional entrainment variation with a flow parameter ($F_{LV}{}^{1}$)

2.2 Reboilers in Distillation Columns

There are two types of reboilers primarily used in distillation columns:

1. Once-through
2. Recirculation type

Each may have two subtypes of configurations:

1. Thermosiphon
2. Forced circulation

2.2.1 Once-through reboiler

In a distillation column with a once-through reboiler, the internal reflux is collected in a trough from the tray and fed to the reboiler. The reboiler outlet vapor and liquid are delivered to the column bottom. The reflux liquid is not directly routed to the column bottom without reboil. In the once-through arrangement, the reboiler can be considered as one theoretical stage. These reboilers are deployed only for low boil-up systems such as in strippers. Once-through reboiler arrangements are given in Fig. 2.2.1.

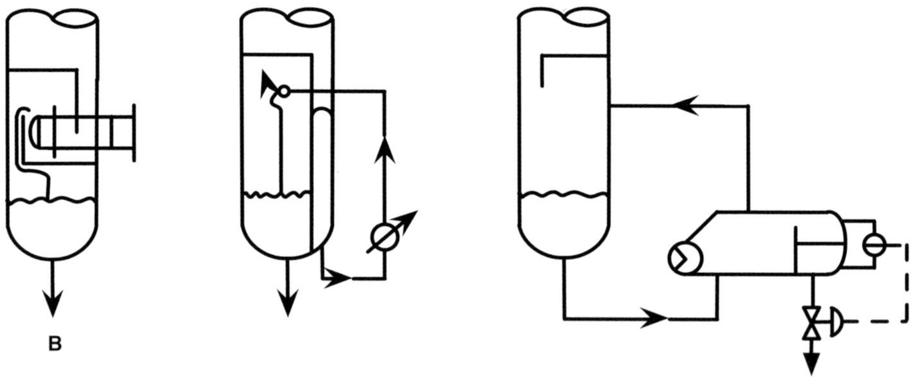

Fig. 2.2.1 Once-through reboiler arrangements

2.2.2 Recirculation-type reboilers

In a recirculation-type reboiler, liquid from the bottom spout is drawn and passed through the reboiler. The liquid from the bottom to the reboiler is constituted by the internal reflux from the tray above and part of the liquid from the reboiler outlet. In this type of reboiler, the liquid recirculates through the reboiler. The recirculation-type reboilers are used in columns operating with high reflux ratios, as in splitters. In this type of reboiler there can be arrangements with baffles and without baffles. The reboilers with baffles can deliver a fraction of a stage, and the ones without baffles cannot.

Often, a column bottom reboiler with a baffle is used. The partition baffle separates the bottom spout into two compartments. The internal reflux is routed into one chamber and the liquid is then sent to the reboiler either by using a thermosiphon, or forcing circulation to the top of the same compartment, where the liquid will overflow to the next compartment before being sent out. This ensures a constant head of liquid at the inlet of the reboiler, which is necessary for consistent good performance and is referred to as *baffles with overflow*. The other arrangement has a reboiler outlet returning to the product side compartment with a large hole in the baffle for recirculation of part of the reboiler outlet liquid to the reboiler inlet, and the arrangement is referred to as *baffles with underflow*.

Vaporization in reboiler

Vaporization in a thermosiphon reboiler is kept at around 15%–30%, and the critical heat flux is never crossed in order to avoid film boiling in the reboiler. A rule of thumb to prevent this film is to restrict the differential temperature between hot and cold fluids to 30°C–40°C. However, critical heat flux needs to be calculated for the rigorous design of a thermosiphon reboiler.

A forced circulation reboiler is often designed as a normal heat exchanger without phase changes. This is achieved by keeping the control valve at the outlet of the reboiler and elevating the pressure to arrest vaporization at the upstream of the control valve. Where a furnace is used as the reboiler, a vaporization percentage of greater than 50 should not be considered at the furnace outlet to avoid high tube metal temperature and coking in furnace tubes. The fractional vaporization may be controlled by increasing the liquid circulation through the pump.

2.2.3 Kettle reboiler

A kettle-type reboiler is shown in Fig. 2.2.2. With this type of reboiler, the column liquid is always kept in the shell side. The reboiler shown in Fig. 2.2.2 is a once-through type, as the tray liquid is directly fed to the column bottom and then is routed through the reboiler. Only the vapor from the reboiler is returned to the column, and liquid from the kettle is drawn as product.

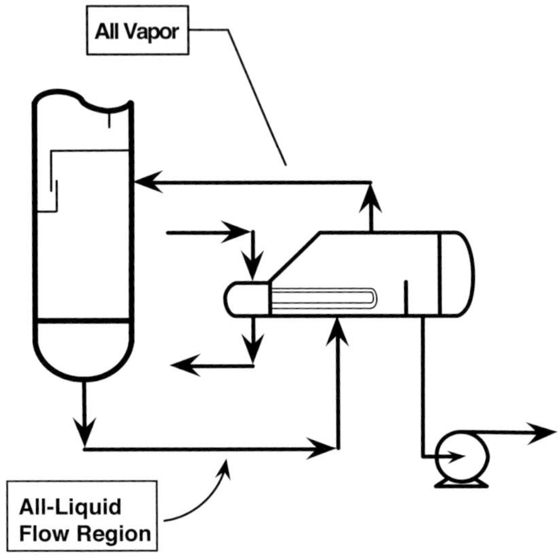

Fig. 2.2.2 Kettle reboiler

Features

1. Vapor–liquid disengagement occurs in the vapor-disengaging space of the reboiler; thus, return piping is sized for total vapor flow.
2. Proper elevation is critical in a kettle reboiler operation.
3. Kettles need a large plot space, but require less elevation than a thermosiphon.
4. Circulation works through differences in gravity.
5. The boiling mechanism is primarily pool boiling.
6. Prone to fouling due to low velocities, high fractional vaporization, and high retention time.
7. Needs a blow down facility to purge out degradation products and avoid residue accumulation.

2.2.4 Forced circulation reboiler

A forced circulation reboiler scheme is shown in Fig. 2.2.3.

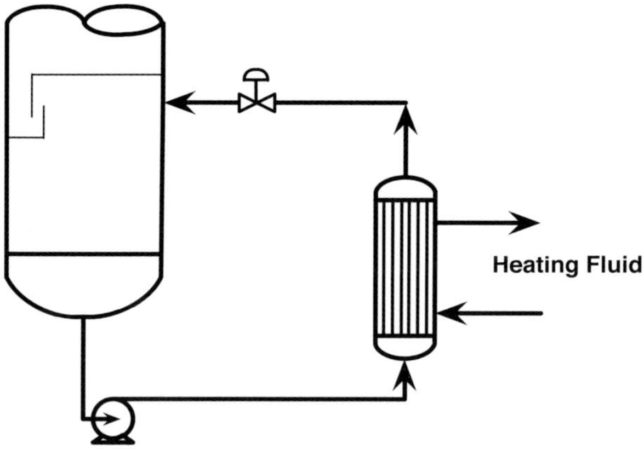

Fig. 2.2.3 Forced circulation reboiler

Features

1. Some forced circulation reboilers are designed to operate full of liquid (see Fig. 2.2.3).
2. Flashing occurs at the downstream of the control valve.
3. Valve should be located close to the column shell at the outlet of the reboiler.
4. Forced circulation reboilers offer greater operational flexibility than thermosiphons.
5. Needs a smaller plot space than kettle reboilers.
6. Elevation is less critical due to the availability of a pump.
7. Commonly operated with high circulation rates and no vaporization before the control valves.
8. Can be horizontal and vertical. Boiling fluid is kept in the tube side for a vertical type and in the shell side for a horizontal type.

However, when a fired furnace (see Fig. 2.2.4) is used as a reboiler, the control valves are located at the inlet and vaporization takes place in the reboiler. The flow of column bottom liquid through the reboiler is controlled by the inlet flow control valves on the flow passes. The maximum vaporization to be considered in a furnace reboiler is 50%. The arrangement in Fig. 2.2.4 is a recirculation type.

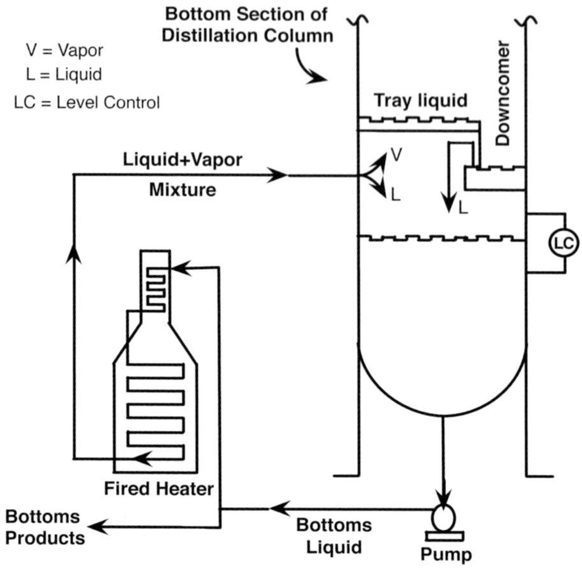

Fig. 2.2.4 Forced circulation reboiler using a furnace

Furnace reboilers are most commonly used where column bottom temperature is high and can be used with or without a partition plate at the column bottom (recirculation type).

A thermosiphon reboiler is shown in Fig. 2.2.5.

In Fig. 2.2.5, the reboiler is a vertical thermosiphon (recirculation type), so the column bottom liquid is in the tube side.

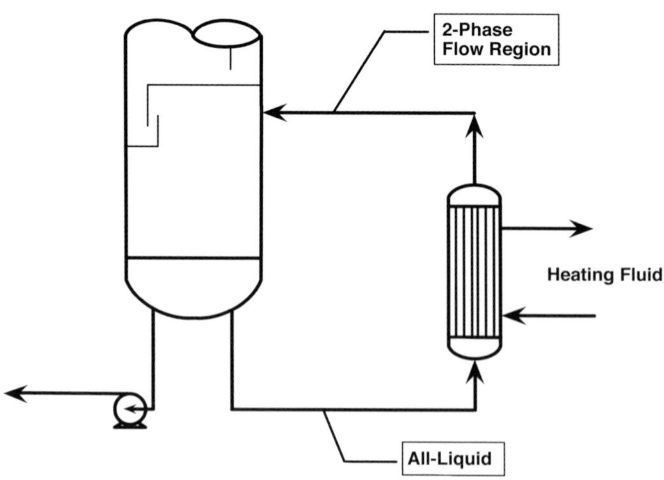

Fig. 2.2.5 Vertical thermosiphon reboiler

Features

1. Return piping to the column must be sized for two-phase flow.
2. Thermosiphons operate based on thermodynamic principles, rather than gravity; therefore, design detail is important.
3. The design of thermosiphon reboilers is often difficult for a wide operating range.
4. Proper elevation is critical.
5. Film boiling should be avoided in thermosiphon reboilers.
6. The differential temperature across the wall should not exceed 35°C in order to avoid film boiling and avoid exceeding the critical heat flux (a function of the critical pressure of the boiling fluid).
7. These reboilers can be horizontal and vertical. The boiling fluid is kept in the tube side for a vertical type and in the shell side for a horizontal type.
8. For horizontal thermosiphon type reboilers, liquid must be uniformly distributed in the shell, particularly in the case of multi-component mixtures. To ensure good liquid distribution, two liquid inlets and two vapor outlets are often used. Horizontal longitudinal baffles are used in TEMA H-type shell arrangements.
9. Thermosiphon reboilers need additional column skirt height to provide the static head required for circulation. Vertical thermosiphons require a greater column skirt height than horizontal thermosiphons.

Experts describe three zones in a reboiler. When column liquid first enters the reboiler it is subcooled because of the liquid head and heat loss from lines. It is first heated to boiling temperature in the reboiler (preheat zone). Heat transfer rate is low in this zone. After boiling point is reached, vaporization starts, a two-flow regime is established, and the zone has a high heat transfer rate. The zone above nucleate boiling is often referred to as the mist zone and has a low heat transfer rate. So, at least 80% to 85% of the reboiler tube bundle needs to be submerged in liquid or else there may be severe loss of heat transfer. Thus, a vertical thermosiphon reboiler often needs a partition baffle inside the column to ensure submergence of the tubes and constant head to the reboiler. When a partition baffle is used, it may be advisable to provide a 1.5- to 2-inch interconnecting line with valves across the lowest points of the two chambers. This helps drain water and heavies from the reboiler side compartment to the outlet and ensures smooth startup and operation. A temperature pinch can also occur in a reboiler when the liquid inside the column is a mixture of wide boiling components and is more likely above the nucleate boiling zone.

A horizontal thermosiphon reboiler with the column bottom liquid in the shell side is shown in Fig. 2.2.6.

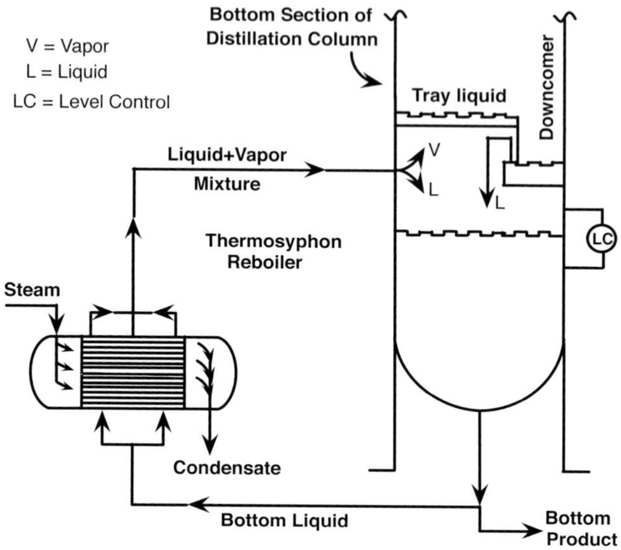

Fig. 2.2.6 Horizontal thermosiphon reboiler

Each thermosiphon reboiler can have two configurations:

1. Once-through
2. Recirculation type

These are shown in Fig. 2.2.7. In a once-through type, the reboiler liquid inlet comes from a sump below the downcomer of the tray above.

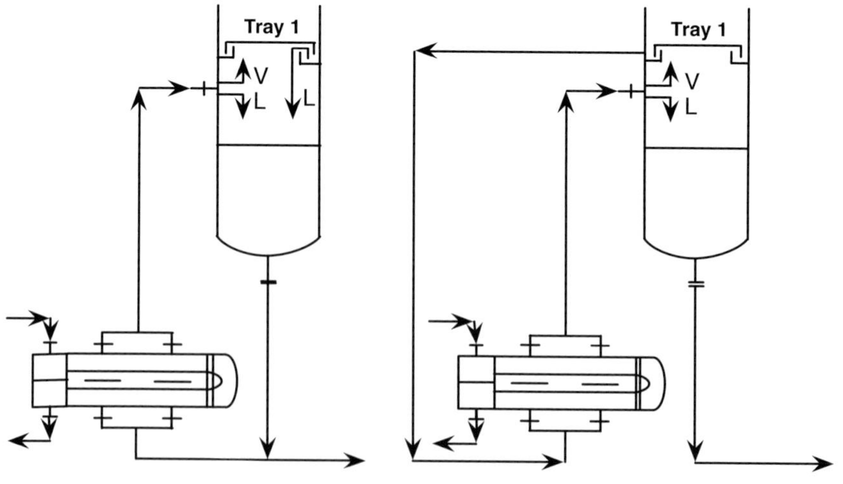

Fig. 2.2.7 Horizontal thermosiphon recirculating and once-through types

Fig. 2.2.7 shows the schematics for horizontal thermosiphon reboilers, both recirculating and once-through types. A scheme of stab-in and forced-type reboilers is provided in Fig. 2.2.8A. Vertical once-through and recirculating types are shown in Fig. 2.2.8B.

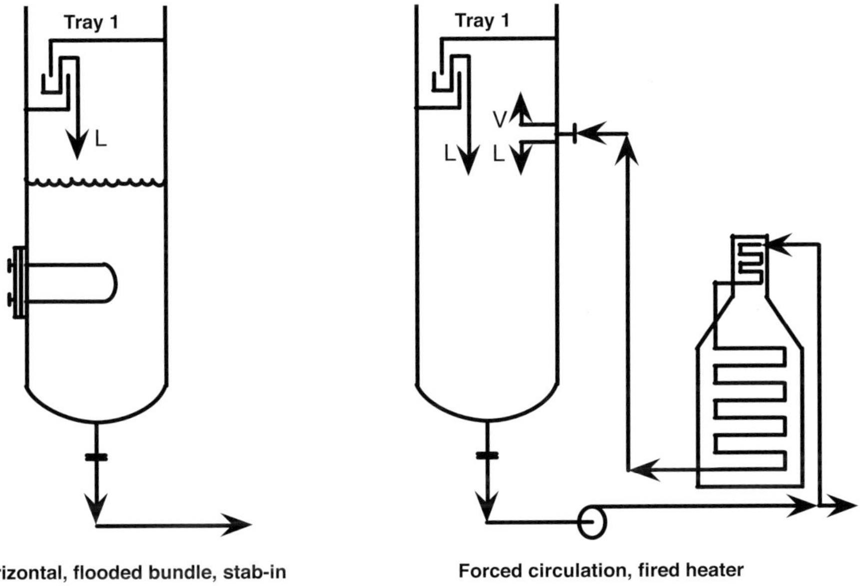

Fig. 2.2.8A Stab-in and forced-type reboilers

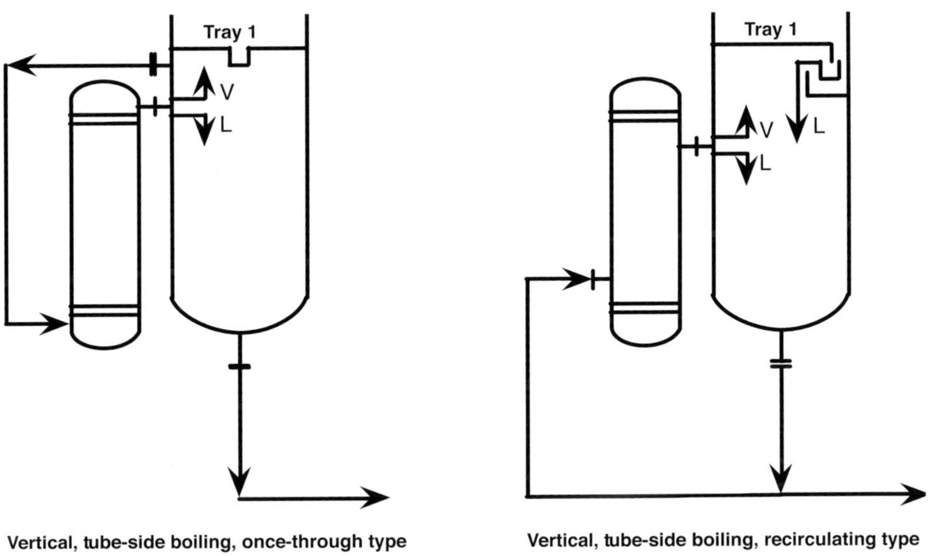

Fig. 2.2.8B Vertical once-through and recirculating types

The tentative criteria used to choose the reboiler type are described in Table 2.2.1.

Table 2.2.1 Tentative criteria for choosing reboiler types

Attributes	Preferred type of reboiler
Low relative volatility systems	Recirculating
High relative volatility systems	Once-through
Large exchanger size or high duty requirements	Horizontal
Small exchangers	Vertical or stab-in
Hazardous liquid	Stab-in
Solids present	Kettle
Fouling service	Forced circulation
Low plot area available	Vertical
Requirement of high temperatures	Fired heaters
Reboilers with high heat fluxes	Forced circulation or flooded bundles
For high viscosity liquids (>25 cp)	Forced circulation
Availability of low ΔT	Kettle type
Requirement of low capital cost	Vertical thermosiphon
Reliability	Kettle/forced circulation
Boiling fluid in tube	Vertical thermosiphon/forced circulation
Boiling fluid in shell	Horizontal thermosiphon/kettle/internal (stab-in)

2.2.6 Common reboiler problems

In a thermosiphon type, the circulation rate of the bottom liquid through the reboiler may be inadequate due to incorrect estimation of the size of the inlet/outlet lines and/or the elevation difference between the column bottom and the reboiler. In some services, decomposition products accumulate in the legs and reduce the circulation rate. These types of problems are quite common.

Case studies of common reboiler problems

1. A hydrocracker stripper preceding the main fractionator had a furnace reboiler duty of around 16 Gcal/hr. The reboiler was possibly designed as a recirculation type with a partition baffle and pump, like the forced circulation recirculation type reboiler in Fig. 2.2.8. Accordingly, the column bottom had a baffle to create two chambers. The internal reflux from the tray above was entering through the right-side chamber and the liquid was pumped through the reboiler furnace.

There was another pump in the bottom of the left-side chamber to pump the liquid out to the downstream equipment (the charge furnace and fractionator) under the level control of the left chamber (product side). However, due to possible incorrect installation, engineering, and/or execution, the reboiler return stream was routed to the left-side chamber, rather than returning to the right-side chamber top and overflowing to the left-side chamber only after maintaining constant levels up to the baffle top in the right chamber (reboiler side). However, theoretically, the once-through arrangement provided may be correct where the boil-up is low, as in this case of a stripper (possibly with a thermosiphon reboiler), but not with a pump feeding the reboiler that needs minimum suction head. Thus, there was frequent level loss in the right chamber, which gave rise to suction loss in the furnace recirculation pump. This endangered the reboiler furnace due to a loss of flow in the tubes, and could result in overheating and coking of the furnace tubes.

Practically, the right chamber (reboiler side) should be always full (constant level) to provide constant suction head to the recirculation pump, and liquid should overflow to the left chamber to be pumped to a downstream section under level control. A modification was executed to interconnect the suction lines of the furnace recirculation pump and the bottom pump-out pump as shown in Fig. 2.2.9. Thus, the original plan was converted almost to a full recirculation type and the unit started operating without further problems. The correct configuration (recirculation type with baffle) is shown in Fig. 2.2.8.

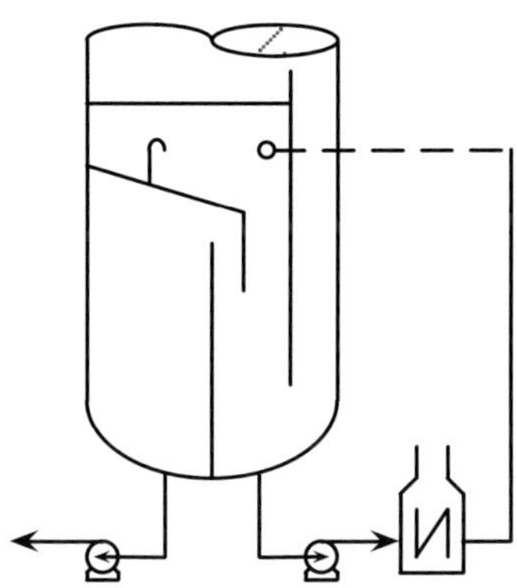

Fig. 2.2.8 Correct configuration (recirculation with baffle)

Actual configuration showing modification is shown in Fig. 2.2.9.

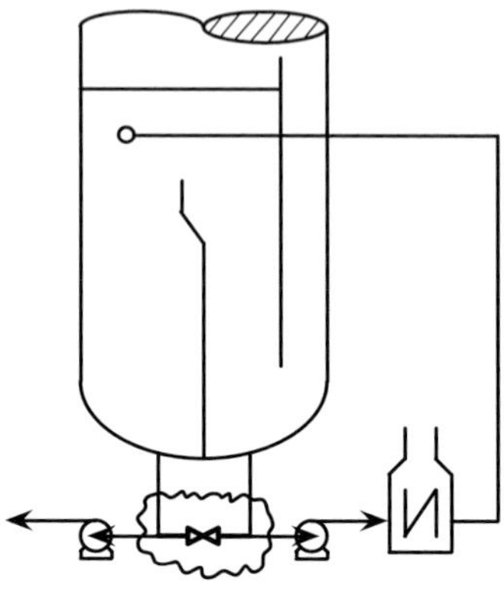

Fig. 2.2.9 Actual configuration showing modification

2. The author has experienced plenty of similar problems in thermosiphon recirculation-type reboilers with a baffle. In some cases, the manway in the partition plate had gaskets (manway covers should always be seal welded) and developed gasket leaks, which resulted in a migration of liquid from the right to the left chamber (product side) and consequently, a level loss in the right chamber (reboiler side). This reduced the circulation through the thermosiphon reboiler and coked up the column side of the reboiler, affecting the performance of the column.

   Often in columns with two thermosiphon reboilers heated by the same hot fluid or different hot fluids, one reboiler gets higher circulation than the other due to errors that can occur during the installation of the internals of the column bottom spout.

3. Another case deals with a column with two reboilers: one vertical thermosiphon and the other forced circulation, both with a baffle.

   The forced circulation reboiler (with a pump) was added in revamp. The circulation through the vertical thermosiphon reboiler was greatly reduced and the reboiler tube side became completely coked up.

   This resulted in only one reboiler being actually available. The column (a refinery debutanizer) performance dropped due to the reduced reboiler heat input.

4. A refinery sour water stripper was not performing satisfactorily to bring down the H_2S level in the stripped water. It was found that the bottom temperature of the stripper was low and the top pump-around reflux flow that maintained the column top temperature was also less than desirable. The stripper had a kettle-type reboiler where the liquid

to the reboiler was routed from the chimney tray located above and the reboiler outlet vapor returned below the chimney tray. The liquid from the kettle reboiler was routed to the column bottom spout. It was suspected that the chimney tray pan was dislodged and some of the liquid from the chimney tray was dropping directly into the spout. Thus the heat input to the column had reduced. This was reflected by the low bottom temperature and low pump-around flow. The column was opened and it was found that the chimney tray pan was partly dislodged. The pan was rectified and unit began functioning properly.

2.3 Condensers and Column Pressure Control

Essentially, there are three types of condensers:

1. Partial condensers (with noncondensable bleed)
2. Bubble temperature
3. Sub-cooled

Common column pressure control schemes are described below.

In the scheme (Fig. 2.3.1), the pressure of the column is controlled by manipulating the noncondensable bleed. These are usually deployed in partial condensers. The scheme is applicable when the net vapor rate is always positive.

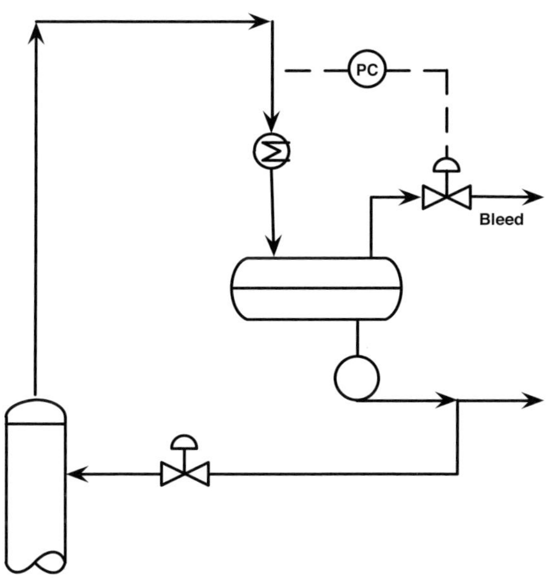

Fig. 2.3.1 Noncondensable bleed control

A scheme with a compressor with spillback control is shown in Fig. 2.3.2.

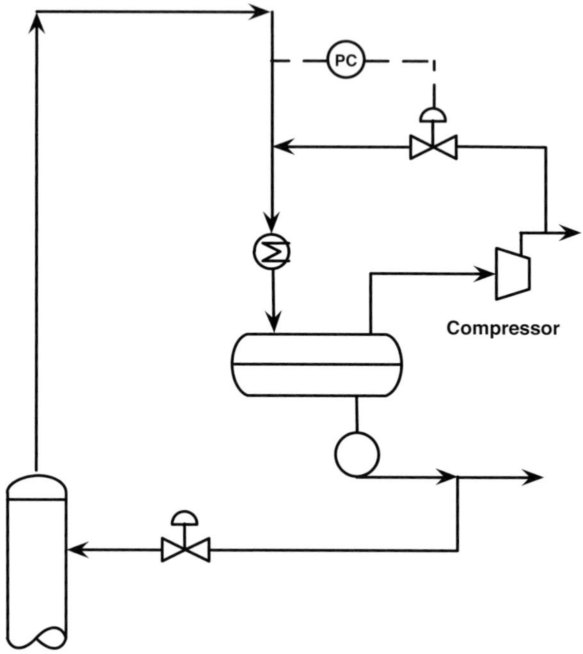

Fig. 2.3.2 Compressor with spillback control

The scheme shown in Fig. 2.3.2 is applicable when the net vapor rate is always positive and vapor is moving to higher-pressure equipment or systems. In a refinery FCCU or coker unit fractionator, overhead gas is picked up by a compressor to be fed to the gas concentration unit at a higher pressure under the pressure control of the fractionator. The pressure controller output is on the compressor spillback valve and often on the speed control device of the compressor drive. If the drive is a steam turbine, the output manipulates the speed by varying the opening of a steam governor valve. This method usually works in a split-range mode manipulating the spillback valve and turbine speed. If the drive is a motor, pressure control is achieved by manipulating the control valve at the compressor suction and spillback.

A scheme with condenser inlet line restriction is shown in Fig. 2.3.3.

In the scheme in Fig. 2.3.3, the condenser is located above the reflux drum and condensates enter from the bottom. This scheme is often used in naphtha splitters when the vapor line size is not very big.

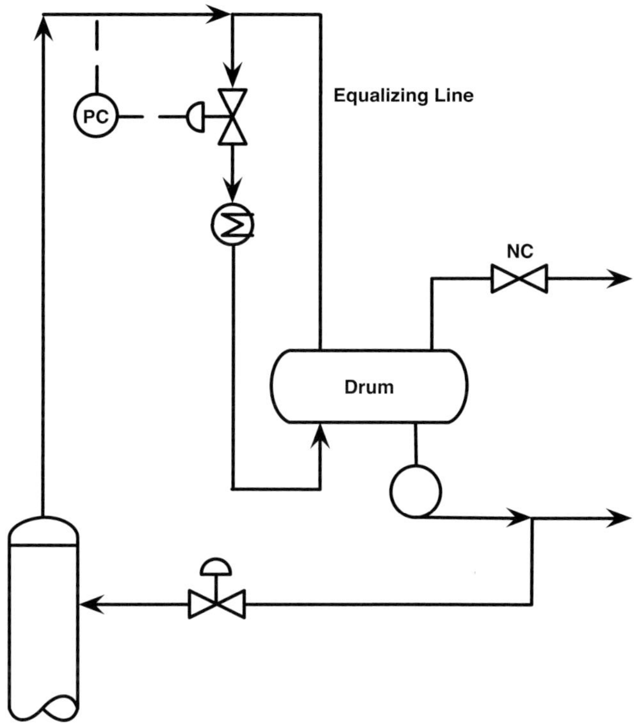

Fig. 2.3.3 Condenser inlet line restriction

In the scheme in Fig. 2.3.4, the condenser is also located above the reflux drum and condensates enter from bottom. However, the condenser duty is adjusted by changing the submergence, and column pressure is controlled through the condensate temperature. An equalizing line is essential in this scheme.

A scheme with condenser outlet line restriction is shown in Fig. 2.3.4.

In the scheme in Fig. 2.3.5, a condenser is at the same level as the drum and condensates enter from bottom. The column pressure is controlled using a hot vapor bypass line to the drum.

In the scheme in Fig. 2.3.6, submergence is altered through the liquid outlet flow rate from the reflux drum. The scheme does not find much industrial application.

In the scheme in Fig. 2.3.7, pressure control is achieved by varying submergence with condensate entry from the top. Drum pressure is controlled by the valve on the hot vapor bypass line.

A pressure control scheme with an air-cooler blade-angle variation is shown in Fig. 2.3.8.

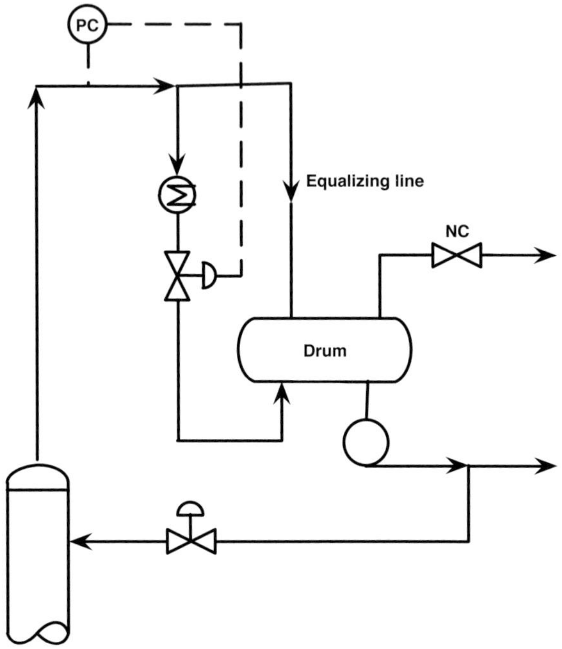

Fig. 2.3.4 Pressure control with submergence variation

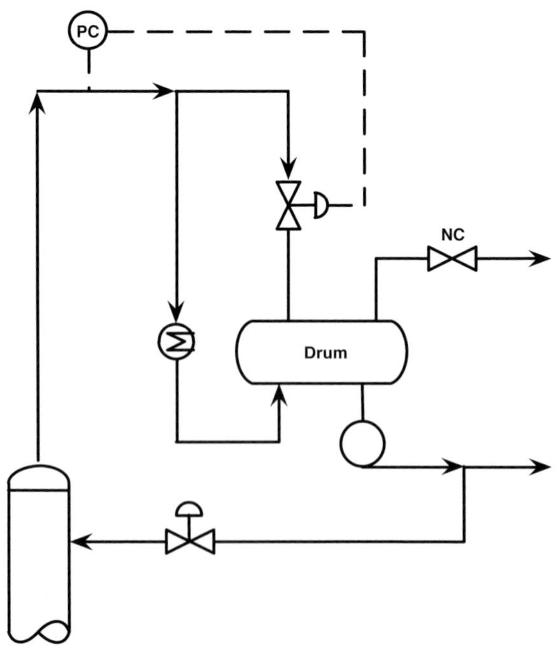

Fig. 2.3.5 Pressure control using a hot vapor bypass

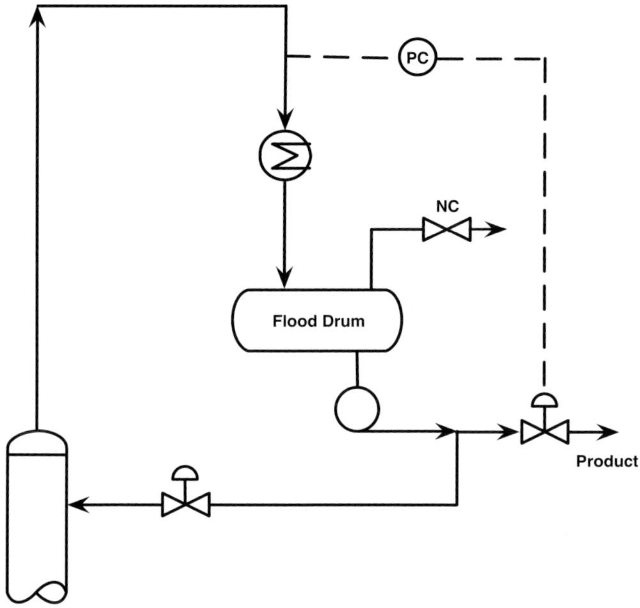

Fig. 2.3.6 Pressure control with liquid withdrawal variation

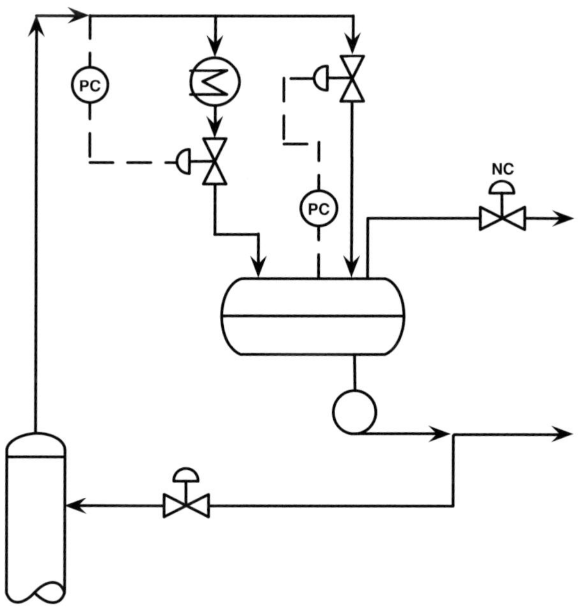

Fig. 2.3.7 Pressure control with submergence variation

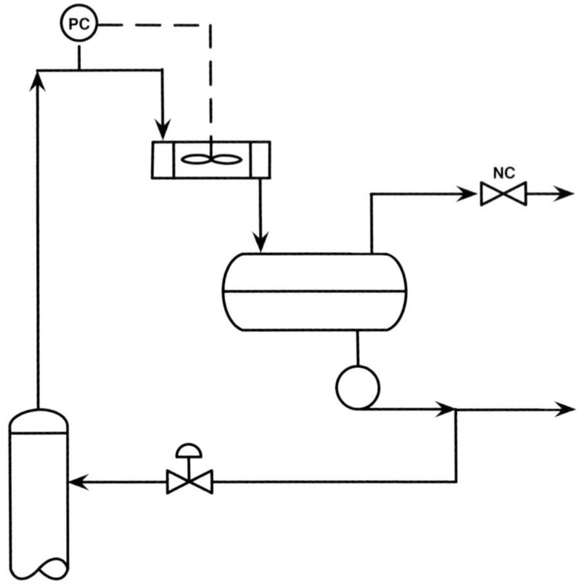

Fig. 2.3.8 Pressure control with air-cooler blade-angle variation

It is common practice not to have a valve in the condensate outlet in case the overhead condenser is an air cooler. If a valve at the condensate outlet is required, the air cooler needs to be installed at a downward incline of 15°.

A pressure control scheme with a vapor line restriction is shown in Fig. 2.3.9.

The scheme in Fig. 2.3.9 is usually applied in high-pressure columns where the vapor line is not so big that it limits the size of the vapor line control valve. This is often used with a hot vapor bypass line and control valve to control the drum pressure (Fig. 2.3.10).

The scheme in Fig. 2.3.10 is used when the vapor line is relatively small, in order to keep the control valve from becoming big in size. Normally, this is also applicable in high-pressure columns where there is no noncondensable. A scheme with makeup and bleed split-range control is presented in Fig. 2.3.11.

In Fig. 2.3.11, the pressure of the reflux drum is controlled by makeup gas flow to the drum and a bleed of gas from the drum. The control valves typically operate in a split range.

The disadvantage of the system is that a certain amount of makeup gas is dissolved in the drum liquid, so when the overhead is processed in another distillation column, the dissolved gas finds an entry to the next column and adds to the load of that column.

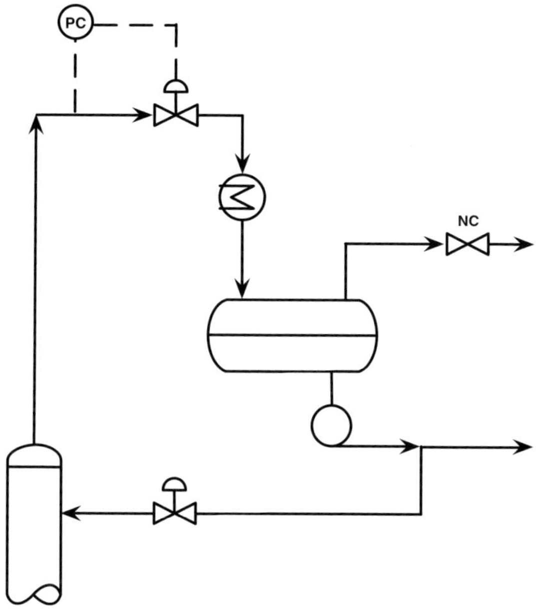

Fig. 2.3.9 Overhead throttle pressure control

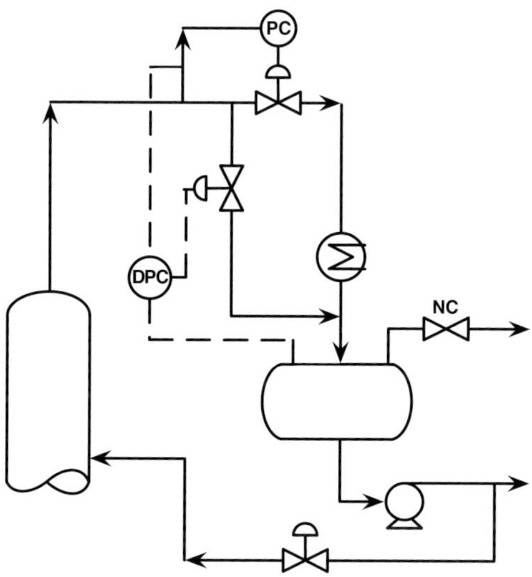

Fig. 2.3.10 Overhead throttle pressure control with differential control for drum

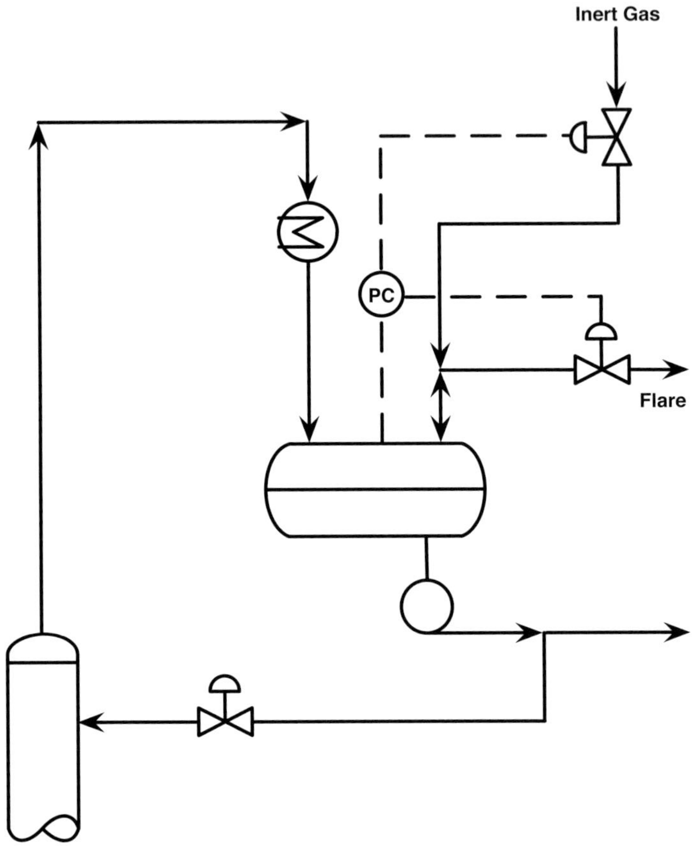

Fig. 2.3.11 Makeup and bleed split-range control

If the makeup gas is fuel gas from refinery fuel gas header, the dissolution of the gas may adversely affect the quality of the overhead liquid. In one refinery, the makeup gas was from refinery fuel gas header. The overhead naphtha often failed copper corrosion tests due to the dissolution of hydrogen sulfide in naphtha from fuel gas used as makeup.

In one refinery, the split range control was used to control the pressure of a crude distillation column preceded by pre-topper, as in Fig. 2.3.11. The makeup gas was from refinery fuel gas header. The fuel gas used as makeup in the crude column overhead can dissolve at low reflux drum temperatures and get into the downstream stabilizer column and partly in the LPG stream. This LPG stream, when fed to the depropanizer/deethanizer, used to overload the columns with the dissolved fuel gas components.

2.4 Vapor–Liquid Contacting Devices

Generally two types of contacting devices are used:

1. Cross-flow
2. Counterflow

2.4.1 Cross-flow

Cross-flow occurs when the vapor flow is perpendicular to the liquid flow path. Sieve trays, valve trays, and bubble cap are examples of cross-flow devices and the arrangement is shown in Fig. 2.4.1.

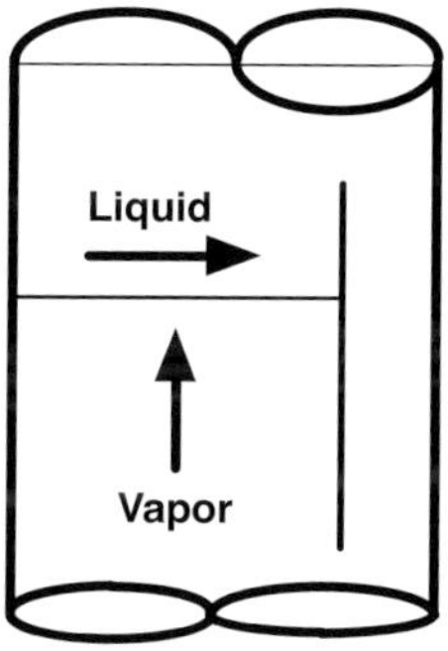

Fig. 2.4.1 Cross-flow device arrangement

2.4.2 Counterflow

Counterflow occurs when the vapor and liquid flows in a counter-current fashion. Packing and heat transfer trays are the examples of counterflow devices presented in Fig. 2.4.2.

These types of trays are normally used as heat transfer trays in the FCCU and coker desuperheating sections. These trays are categorized as counterflow devices.

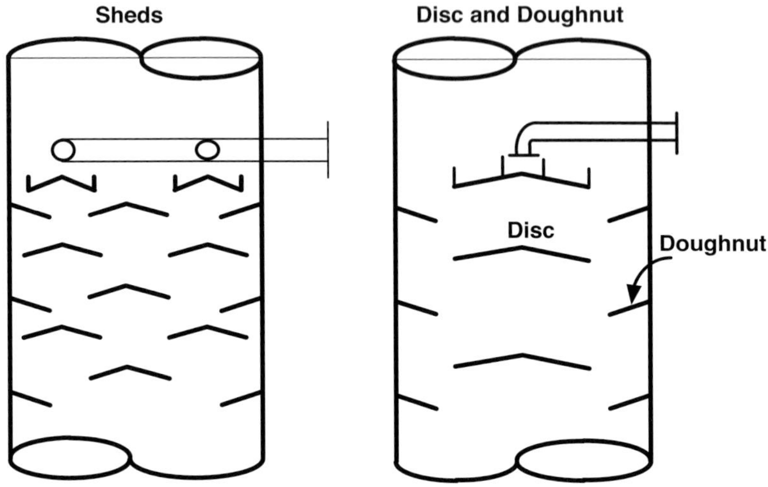

Fig. 2.4.2 Counterflow devices

Tray-fitted columns

The main requirement of a tray is that it provides intimate mixing between liquids and vapors. Common types of cross-flow trays are the bubble cap, valve, or ballast trays.

Two layouts are generally adopted for the liquid flow.

1. Cross-flow with a good length of single-pass liquid path.
2. A multi-pass arrangement, which is suitable for handling high liquid rates.

The liquid flow arrangements in columns with trays are shown in Fig. 2.4.3.

Packed columns

Generally two types of packing are used:

1. Random packing
2. Structured packing

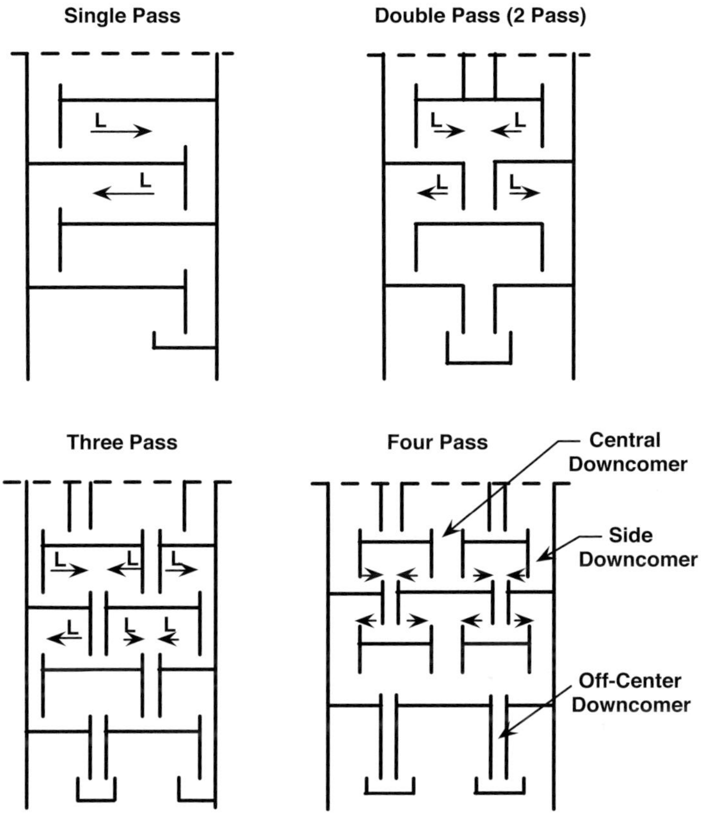

Fig. 2.4.3 Liquid flow arrangements in columns with trays

2.4.3 Selection of internals—plate or packed

Features of plate and packed columns

1. Columns with trays (plate columns) can be designed to handle a wider range of liquid and gas flow rates than packed columns.
2. Packed columns are not suitable for very low liquid rates because minimum wetting liquid rates are required and misdistributions of liquid can adversely affect performance.
3. Columns with trays can be designed more reliably than packed columns. There is always some likelihood of improper liquid distribution throughout a packed column under all operating conditions, particularly in large columns.
4. It is easier to make provisions for inter-reboilers and intercoolers in columns with trays.

5. It is easier to make provisions for the withdrawal of side-streams from plate columns. In packed columns, chimney trays are required for side withdrawal.
6. If the liquid causes fouling or contains solids, it is easier to make provisions for cleaning in a plate column.
7. Manways can be installed on the trays.
8. Packing should always be considered for small diameter columns (columns less than 0.6 m) where the installation of plates would be difficult. It may also be cheaper to use packing in such columns, replacing it when it becomes fouled.
9. For corrosive liquids, a packed column will usually be cheaper than a plate column, as the latter will need expensive metals to arrest corrosion for plates. Packing can be made of cheaper material like plastic or ceramic to avoid corrosion.
10. The liquid holdup is appreciably lower in a packed column than a plate column. This may be important when the inventory of toxic or flammable liquids needs to be kept as small as possible for safety reasons.
11. Packed columns are more suitable for handling foaming systems.
12. The pressure drop per equilibrium stage (HETP) is usually lower for packed columns than plated ones.
13. Typically, packing is not used in high-pressure columns where the density difference between vapor and liquid is lower. This creates problems in the vapor–liquid separation in the net area of the column. Furthermore, packing is not preferred where liquid surface tension is high (lighter hydrocarbons), as the wetting of the packing may be poor and may result in poor contact between vapor and liquid.

2.4.4 Flow regime in trays

Distillation experts label flow regimes in trays based on their vapor/liquid rates. The most commonly encountered are

1. Froth regime: normal liquid rate
2. Spray regime: low liquid rate
3. Emulsion regime: encountered in high-pressure systems with relatively high liquid loads. Most of the gas is emulsified within the liquid, and the mixture behaves as a uniform two-phase fluid.

2.5 Tray Flooding Mechanisms

Experts describe flooding mechanisms as below.[2]

2.5.1 Mechanisms

Spray entrainment flooding

At low liquid flow rates, trays operate in a spray regime. As vapor velocity is increased, a point is reached where a considerable amount of the liquid droplets are carried (entrained) to the tray above along with the vapor.

Froth entrainment flooding

At higher liquid loads, when vapor velocity is raised, froth height increases and approaches the tray above, leading to froth entrainment flooding.

Downcomer backup flooding

Liquid is backed up in the downcomer due to pressure drop in the tray, liquid height on the tray, and frictional losses in the downcomer. When the backup in the downcomer exceeds the tray spacing, the downcomer liquid finds a way to the tray above, which causes downcomer backup flooding.

Downcomer choke flooding

This is caused if the frictional losses in the downcomer, particularly at the downcomer entrance, are high due to high liquid flow rates at the inlet. The liquid accumulates in the tray instead of flowing down, which results in downcomer choke flooding. At the inlet of the downcomer, the volumetric flow of liquid is always higher than at the outlet, as the liquid is more aerated and has a lower density than the liquid at the outlet and consequently requires a larger flow area.

2.5.2 Correlations for flooding

The flooding correlations commonly used are:

1. Kister and Haas correlation
2. Fair's correlation
3. Koch correlation of downcomer choke
4. Smith correlation

The Kister and Haas correlation is known to have been derived from a large database. This uses physical properties, operating variables, and tray geometry and is applicable to a sieve and valve tray. Fair's correlation has been standard for the industry but may tend to be conservative. The Koch correlation finds wide application for predicting downcomer backup.

The basic equation used for predicting entrainment due to spray is the Souder–Brown equation.

$$U_{S,\,flood} = C_{SB}\sqrt{\frac{\rho_L - \rho_V}{\rho_V}}$$

$$C_{SB} = U_{S,\,flood}\sqrt{\frac{\rho_V}{\rho_L - \rho_V}}$$

As an improvement over this equation, the entrainment flooding correlation expresses C_{SB} as a function of other variables, such as tray spacing, vapor/liquid load, fractional hole area, etc.

A process engineer generates the vapor/liquid flow rates through the column stages and the relevant properties like vapor/liquid densities, viscosities, surface tension, etc., which are then supplied to the column internal expert for designing the particular internals that will deliver the number of stages and meet the pressure drop specified.

For determining the dimensions of the column, the process designer needs to carry out a tentative estimation of column diameter and height. Usually the determination of the conservative diameter can be done by a steady state simulator with the ability to calculate diameter using a calculation algorithm based on tray or packing. For valve trays, the Glitch 4900 is most commonly used, while the Smith method is used for bubble cap and sieve trays.

Column diameter is estimated based on the type of internals, the system factor, the tray spacing (normally 600 mm), and the percentage of flooding specified.

1. 70% for columns with a diameter of up to 2 meters
2. 80% for columns with a diameter greater than 2 meters

The estimation of column diameter needs to be conservative. The column internal expert will match the pressure drop and the stages specified during detail design of the internals.

Rating of columns

A process engineer is required to work out a conservative rating of an existing column. This is typically done by providing the actual diameter, spacing, and expected system factor, then calculating the flood percentage. Commercial simulators often have the capacity to detail-rate the trays by inputting the diameter of the column, the hole diameter, the tray spacing, the number of liquid passes, etc. The percent flood, pressure drop, and downcomer backup are also calculated.

If the percent flooding is less than or equal to 70% for a column with a diameter of less than 2 meters, then the column diameter is considered satisfactory. The downcomer backup is also checked, and must be at maximum 40%–50%. For columns with larger diameters, a flood percentage of up to 80% is considered acceptable for the preliminary rating of a column. The downcomer backup is also checked and must be at maximum 40%–50%. The final rating of the internals is preferably done by internal experts. At times, it is necessary that the process designer or engineer perform this rating, for the purposes of troubleshooting and preliminary adequacy checks. Therefore, it is necessary for process designers and engineers to have some understanding of the fundamental principles of column diameter estimation.

2.5.3 Estimation of tower diameter

The flooding condition fixes the upper limit of vapor velocity. The flooding velocity is estimated using Fair's correlation:

$$U_f = K_1((D_L - D_v)/D_v)^{½}(\sigma/20)^{0.2}$$

where σ is surface tension in dynes/cm. K_1 (often called C_{SB}) is the constant obtained from the correlation between K_1 and the liquid–vapor flow factor (flow parameter) at flooding velocity. $(\sigma/20)^{0.2}$ is the correction factor for any liquid surface tension other than 20 dynes/cm. $F_{LV} = (L_W/V_W)(D_v/D_L)^{1/2}$ where L_W is the liquid mass flow rate in kg/sec, V_W is the vapor mass flow rate in kg/sec, and D_v, D_L are vapor and liquid densities in kg/m^3, respectively. There are other correlations that can be used, like the Smith–Dresser–Ohlswager correlation, or the Kister and Haas correlation.

The liquid–vapor flow factor (K_1) can be obtained from Fig. 2.5.1.

Percentage flooding = u_a/u_f, where u_a is actual velocity based on net area and u_f is flooding velocity on net area.

Based on this vapor velocity and the vapor flow rate obtained from the column vapor–liquid profile, approximate tower diameter is calculated based on the vapor–liquid separation. Net flow area $(A_n) = (VG/(u_a \times 3{,}600))$ m^2, where VG is vapor flow rate in m^3/hr at actual pressure and temperature (ACMH), and u_a is vapor velocity m/s.

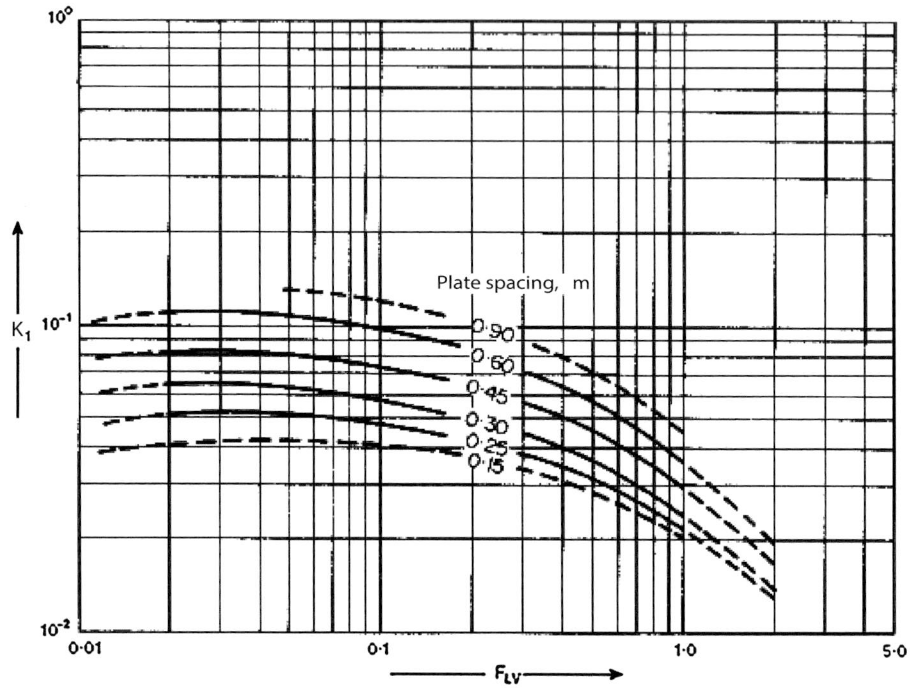

Fig. 2.5.1 Flow parameter with coefficient for flooding[3]

Estimation of a column cross-sectional area

A typical downcomer area is around 10%–12% of the column cross-sectional area. Thus, the column cross-sectional area (*CA*) is defined as *CA* = ((net area)/(1 – 0.10)) m^2 or ((net area)/0.9) m^2.

Often net area and total cross-sectional area are estimated separately for the section above the feed and the section below the feed based on their individual vapor/liquid rates.

There are several alternative methods available based on certain graphs and standard literature. A rule of thumb for the quick calculation of estimated column diameter is given in Table 2.5.1.

Table 2.5.1 Constant (*C*) for the Souder-Brown equation with different tray spacing

C in ft/sec	Tray spacing, inches
0.122	18
0.150	21
0.167	24

The Souder–Brown equation, used for finding the maximum allowable vapor velocity in vapor–liquid separation vessels, is as follows:

$$V_{max} = C((D_L - D_v)/D_v)^{0.5}$$

V_{max} = the velocity through the net area in ft/sec. The column diameter calculated is based on the vapor velocity, not the liquid load, and is conservative by at least 20%. If the liquid load is very high, the column diameter can be governed by the liquid load.

Another quick calculation for estimating the column diameter is to assume that the actual velocity through the net area is 0.6–0.8 m/sec. The vapor flow in actual cubic meters per hour (ACMH) can be divided by 0.6 or 0.8, which will give the net area. The net area for vapor/liquid disengagement will be equal to (flow) m^3/sec. divided by 0.6 or 0.8. The total area (including downcomer) = the net area/(1 – 0.10).

Components of tray fitting

1. Tray support ring (TSR). This circular ring is welded with the column shell.
2. Bolting bars. These are welded with the column shell.
3. Downcomer apron. These are fitted on bolting bars.
4. Tray decks. These are fitted on TSR and support beams.
5. Tray support trusses/ beams. These support tray segments.

Types of downcomers

Downcomers can be of different types:

1. Segmental downcomers
2. Stepped apron downcomers
3. Pipe downcomers
4. Envelope downcomers
5. Recessed inlet downcomers

2.5.4 Downcomer backup

The backup of liquid (the liquid height in the downcomer) is caused by the pressure drop over the plate and the resistance due to liquid flow in the downcomer itself. The height of liquid in the downcomer balances the differential pressure in two successive trays and the loss of pressure due to the flow of liquid in the downcomer.

Fig. 2.5.2 shows a downcomer backup as the height of the liquid (expressed as h_{dc}), where h_w is the weir height, h_{ow} is the crest of liquid above weir, I_t is tray spacing, and h_{fc} is the clearance beneath the downcomer. The downcomer backup is shown schematically in Fig. 2.5.2.

The downcomer area and plate spacing must be such that the level of the liquid and froth in the downcomer is well below the plate above.

If the level rises in the downcomer, the column will flood due to downcomer backup.

A higher pressure drop in the trays will also manifest in higher downcomer backup and eventually lead to downcomer backup flooding. Similarly, higher head loss in the flow of liquid in the downcomer will also result in higher downcomer backup.

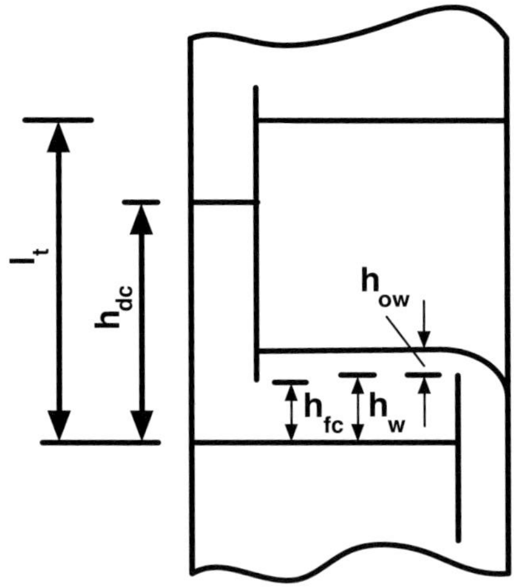

Fig. 2.5.2 Downcomer backup estimation

In the event of inadequate downcomer area availability at the inlet, premature flooding can take place. This is termed as downcomer choke. If the downcomer outlet area is restricted, this will manifest in higher downcomer backup and may result in downcomer backup flooding. Some fundamentals may be valuable and worth remembering for troubleshooting purposes. As a rule, liquid flow of 10–12 gallons per minute (gpm) per linear inch of weir length is seldom exceeded; when liquid flow crosses this value, a multi-pass tray is normally selected.

Design velocity in downcomers

The maximum velocity of clear liquid in the downcomer is limited by choking in the downcomer and the disengagement of vapor bubbles from the liquid.

Typically, two approaches for sizing downcomers are used by experts:

1. Residence time approach
2. Maximum allowable velocity approach

Experts recommend a maximum downcomer velocity in the range of 0.2–0.5 ft/sec and a residence time from 3–7 sec, depending on the foaming behavior of the liquid.

Downcomer pressure drop

There is a head loss in the downcomer caused by the liquid flow. Experts recommend the following correlation to estimate the head loss:

$H_{ud} = 0.65\ (V_{ud})^2$ where H_{ud} is head loss under the downcomer in inches of liquid, and V_{ud} is velocity under the downcomer.

Typical clearance under the downcomer is 40 mm (with a 50 mm weir height). Weir height is to be 5–10 mm above the height of the bottom edge of the downcomer from the deck in order to maintain downcomer liquid seal.

Estimation of dry tray pressure drop as head in inches

H_{dry} → dry tray pressure drop in inches where $H_{\text{dry}} = C_1 + C_2\ V^2/2g_C$.

C_1 and C_2 are proprietary constants.

Alternatively, $DP_{\text{dry}} = K(V_H)^2 \times (D_V/D_L)$ can be used for valve trays.

DP_{dry} is expressed in inches of liquid. V_H = hole velocity in ft/sec. D_v and D_L are vapor and liquid densities in lb/ft^3.

$K = 1.05$ for a 2 mm-thick tray deck, 0.82 for a 3.25 mm-thick tray deck, or 0.58 for a 0.58 mm-thick tray deck.

Estimation of total tray pressure drop

$DP_T = DP_{\text{dry}} + 0.4(\text{gpm}/L_w)^{2/3} + 0.4H_w$

H_w = weir height in inches

Some designers also add a residual head term to take care of potential errors for surface tension, froth density, and froth height.

Estimation of downcomer backup

$Hdc = H_{\text{dry}} + H_w + H_{ow} + H_{ud}$

Estimation of H_{ow} (crest above the weir)

$H_{ow} = 0.4(\text{gpm}/L_w)^{2/3}$ is the Francis formula for weirs, where L_w = weir length in inches, H_{ow} = weir crest in inches of liquid, *gpm* = liquid flow rate in gallons per minute. To ensure an even flow of liquid along the weir length, experts recommend that the crest be at least 10 mm at the lowest liquid rate.

Typical dimensions of internals

1. Typical weir height: 50 mm (For vacuum applications, a lower weir height of 6–12 mm is practiced to reduce pressure drop).
2. Hole diameter: 5 mm.
3. Plate thickness: 3 mm (For SS–410 S tray deck metallurgy).
4. Height under downcomer: 40 mm.
5. Downcomer liquid seal: 8–10 mm. The downcomer should always have a liquid seal to avoid vapor short circuiting through downcomer. Due to installation errors, the downcomer seal is often found absent. This is a serious error in fitting.
6. Weir length: normally around 60%–85% of the column diameter. A good initial value recommended by experts is 76%, and the equivalent downcomer area is 10%–12% of the column cross-sectional area. For double pass trays, the typical width of the central downcomer is 200–250 mm.
7. Perforated area: the area covering the holes or dispersers. Calming zones are unperforated strips of plate at the inlet and outlet. Distillation experts recommend that the width be 75 mm for any diameter less than 1.5m and 100 mm for any higher diameter.
8. Hole area and hole size: area is approximately 10%–14% of active area; size of holes varies from 2.5–12 mm. The hole is either drilled or punched depending on the plate thickness. Typical plate thickness is 5mm for carbon steel decks and 3mm for stainless steel decks. Typical metallurgy used for a tray deck is SS–410S.
9. Hole pitch: typically twice the hole diameter; normal range is 2.5–4 times the diameter. Square and equilateral triangular patterns are preferred by internal designers.
10. Number of valves: maximum is approximately 120 per square meter of active area.

Check of entrainment

After the preliminary design of the column diameter and design of the tray are finished, a check for fractional entrainment is typically done. However, if the flooding percentage is low, this is not required.

As a rough guide, the upper limit of ψ can be understood to be 0.1. Fig. 2.1.1 shows the effect of flow parameter on efficiency.

Dumping

As the vapor rate is lowered below the weep point, the fraction of liquid dropping below increases and reaches a point where all the liquid on the tray drops below and no liquid reaches the downcomer. This condition is referred as dump point.

Turndown

This is the ratio between the minimum allowable vapor flow rate and the normal operating (intended) vapor throughput. The turndown can be increased by the following:

1. Partially blanking tray holes. Weeping at lower vapor rates can be reduced by partially blanking off the dispersers, but in such cases the maximum capacity of the column is reduced.
2. Using tray decks with a low fractional hole area at the expense of lowering maximum capacity.

Typical sieve tray turndown is low (about 1:2), whereas a valve tray turndown is typically higher (1:4–1:5). Valve trays offer very low turndown. In valve trays, the valve starts to lift as the vapor rate increases or the pressure below the plate increases and becomes equivalent to the valve weight plus the liquid head. This opens the orifice and gas starts to flow through. Lower flow closes the orifice. Typically two types of valves with different weights are used. The heavier ones close at a lower vapor flow and lighter ones keep the gas going through the orifices.

System factors

Traditional flooding equations are corrected with a derating factor popularly known as the system factor (SF). The actual backup (with foam) is calculated by dividing the calculated backup by the system factor. Some designers also discount flooding velocity with the system factor (calculated using Souder–Brown types of equations). System factors of some common services are listed in Table 2.5.2.

Table 2.5.2 System foaming factors for common services

System	Foaming factor	Remarks
Crude/vacuum	0.85/0.9	For grassroots design/revamp
Pre-topper in CDU	0.8–0.85	Preferred. Some of the crudes may result in considerable foaming in the service.
Gas amine absorber	0.65–0.7	
Glycol absorber	0.65–0.7	
Amine regenerator	0.85	
Glycol regenerators	0.85	
Sour water strippers	0.7	0.65 for grassroots application
Caustic regenerators	0.3–0.6	
Oil absorbers	0.85	Preferred 0.8

2.5.5 Estimation of the height of a column with trays

A process simulator can calculate the diameter of a column but height of the column needs to be calculated by a process engineer, who prepares the data sheet for the column.

The following can be considered when estimating the overall height for columns with trays.

1. Number of actual trays based on stages in simulation.
2. Tray spacing (based on the spacing considered in the simulation for diameter estimate; typically 600 mm).
3. Provide at minimum twice the regular tray spacing height (minimum 1,200 mm) anywhere a change in the column diameter is expected.
4. Provide a minimum of 1,500 mm anywhere changes in the number of liquid passes are planned.
5. Provide a maximum space of 1.5 times regular tray spacing (or 750 mm) where draw-off pans are provided and/or where intermediate feeds are introduced.
6. Provide a height of at least 1,500 mm between the chimney tray and the tray above. Chimney height is based on residence time in the chimney tray and controllability of the chimney tray level (usually 6 inches from HHL to LLL). Clearance above the chimney hat to the upper tray deck or bottom of the packed bed is to be maintained at a minimum of 12 inches.
7. Provide a minimum space of tray spacing plus 150 mm where the downcomer is provided with a separate seal pan.

8. The space required above a feed entry with a disengaging device is minimum 16 inches from the upper edge of the disengaging device to the pan or chimney tray above. (Normally, internal pipes are provided in the crude or vacuum column for the distribution of the overflash liquid to the stripping section tray incomers. If not, external pipes with seal and metering devices are used.)
9. The top dome (TL) height may be maintained at twice the spacing of the tray or 1,200 mm (the greater of the two numbers is normally selected) from the topmost tray deck.
10. One manhole is to be provided after 8 to 10 trays. The tray spacing is to be at least 1,200 mm where manholes are provided.

Number of liquid passes

1. The number of ballast units that can be placed on a tray decreases as the number of passes increases. This may also cause an increase in pressure drop and downcomer backup. Tray efficiency will decrease as the number of passes increases due to a smaller flow path length (FPL). Experts say that flow path lengths smaller than 18–22 inches may severely lower tray efficiency.[4]
2. Typically, changes in the number of liquid passes are often required
   a. when a circulating reflux stream is introduced in petroleum distillation, or
   b. below the feed tray due to higher liquid loads.

   Experts recommend a design involving a 90° downcomer rotation.[5] Experts also recommend increasing the tray spacing by 1.5–2 ft in these locations.[6] Some designers prefer to avoid an odd number of passes.

Special features/fittings

A splash baffle or vapor hood is often used for the spray regime design when the liquid flow rate is very low. This is both to avoid the liquid droplets being flung directly to the downcomer and to collect the droplets in order to induce the froth regime. This should be used only with low liquid flow rates and never in high liquid flow rates as splash baffles/vapor hoods choke the liquid flow to the downcomer and may lead to premature flooding. High-capacity trays are used for capacity revamps. The trays normally convert the area under the downcomer to an active area. The proprietary multiple downcomer tray (MD tray) is a type of high-capacity tray widely used in industrial applications like propylene–propane fractionators (PFs). These trays are suitable for high liquid flow services and have small downcomers spread over the tray area and normally deploy lower tray spacing (450 mm) than is conventional (600 mm).

2.6 Common Problems Encountered in Distillation Columns

Problems commonly encountered in distillation are not usually caused by the actual theories of distillation adopted in design, but rather by difficulties or errors with the internals. The following are some of the most common complications:

1. Incorrect fitting of the tray decks with tray support rings (TSRs) and support beams, which results in heavy weeping.
2. Liquid overloading on trays (in gpm/inch of weir), which results in downcomer choke flooding.
3. Inadequate downcomer bottom clearance, which leads to high backup and flooding.
4. Sticking or clogging of tray dispersers (like valves, etc.), which causes a high pressure drop in the column or section of the column.
5. Heavy foaming, which results in poor separation.
6. Often short downcomer residence times result in the recirculation of vapor (unreleased) from the tray above to the tray below along with the downcomer liquid. This results in severe loss of efficiency due to back mixing.
7. Water in reflux leads to pressure fluctuations in a column and results in poor separation. The most common cause of water ingress in a distillation column is improper separation of water in the reflux drum. Sometimes a small leak in the water condenser can result in more water entering the reflux drum (for low-pressure columns) and can get into the column along with the reflux. An incorrect measurement of the water–hydrocarbon interface level can also lead to water in the reflux.
   Steps to avoid water in the reflux:
   a. Reflux draw pipe should always be extended 6 inches inside the reflux drum
   b. Adequately size the reflux drum for water–oil separation
   c. Adequately size the water boot

8. Dislodgement of a tray deck or sections can be caused by the following:
   a. Condensate in stripping steam: Tray decks/segments often get dislodged due to an improper original fitting, water shots, or pressure surges during operation and startups. The author has personally experienced a chronic problem with the dislodgement of the atmospheric and vacuum columns' stripping section trays. The stripping agent is superheated steam. Typically, 4 to 9 trays are provided in the bottom stripping section. The trays can become dislodged during start up when the stripping agent is introduced in the column, as the stripping steam may contain condensate due to a stagnant condition of the lines during shutdown. If this is the case, the stripping agent will put water in the hot column, inducing a surge and causing consequent tray damage.
   b. High liquid level in column bottom: If the liquid level becomes high in the bottom of the column and goes above the reboiler vapor return or stripping steam inlet nozzle, the vapor may travel through the liquid pool in the form of slugs. This may damage the bottom trays and tray supports, and can result in tray dislodgement.

      This can be resolved by stopping the reboiler return vapor/stripping steam flow in case of high levels in the bottom sump by way of terminating reboiler hot fluid or shutting off the stripping steam valve. Another reason for dislodgement of column bottom trays is quick pump out of the built-up level, a phenomenon often referred to by experts as a vapor gap. The liquid from upper trays cannot drain out as fast as it is pumped out from the sump. A vapor gap is generated below the bottom tray, and when the load of accumulated liquid comes on the tray, it fails to support and dislodges.[7] Cross-flow trays are susceptible to this kind of failure where downcomer clearance and downcomer flow area are small.

      Often bottom trays (particularly those in crude and vacuum columns) are changed to counterflow trays without downcomers to alleviate the problem. It may be better to use counterflow trays (baffle trays, disc and doughnut trays), as there is a bigger flow area for the accumulated liquid to flow down and reduce the likelihood of vapor gap and tray dislodgement. However, repeated failures of counterflow trays in the column bottom are still observed and also experienced by the author. It is likely that failure can be avoided by strengthening the trays and lowering the rate of pump-out.

9. Absence of self-venting nozzles and lines in product and pump-around draw-offs: Draw-off nozzles and lines must meet self-venting criteria. The liquid drawn from the column trays still has some amount of gas (aerated liquid) dissolved in it. If the same liquid is fed to pumps, they are likely to cavitate with the release of gas from the liquid at pump suction. The nozzle and connecting line size must be adequate to allow the release of the gas back into the column. After around 10 ft of vertical travel, the withdrawal line can be swaged in order to reduce the diameter. This will help release vapor back into the column.
10. Incorrect construction of draw-off channels/draw-off sumps: This is a very common problem encountered in distillation columns in refineries. The author has observed quite a few instances of the problem in crude and vacuum distillation units where channels and draw-off sumps were too small to handle the liquid load. This resulted in the unavailability of pump-around reflux flows. The problems were subsequently corrected as per the guidelines below.
    Guidelines for designing a draw-off box:
    a. The depth of the draw-off box should be 150 mm or 1.5 times the diameter of the withdrawal nozzle, whichever is greater.
    b. The draw-off box area should allow a maximum downward velocity of 1 ft/sec.
    c. The velocity in the channels should be at maximum 1 ft/sec. The author has observed several cases where noncompliance with the above criterion led to unavailability of flow for pump-around refluxes in refinery columns.
    d. It is also very important that the longer sides of the chimney be parallel to the liquid flow direction toward the withdrawal box or collector channel (which eventually lead to the draw-off box); otherwise the flow of liquid may be restricted.
11. Inadequate residence time at column bottom/draw-offs: Residence time at column bottom sumps and draw-off trays is a very important parameter, especially when the bottom liquid is fed to a furnace. Loss of suction of the pump due to unreleased gas may coke up the furnace. If residence time in the spout is low, this problem can be resolved by external vessel breathing with the column vapor space above the liquid.

    Another option is to provide a liquid gallery so that liquid from the bottommost tray can flow through it (while releasing vapors) and reach the bottom. People often use a self-venting nozzle and line for the bottom products going to pump in the absence of an adequate bottom residence time. However, this may not be a fool-proof arrangement.

Of course, when sufficient residence time is possible, the nozzle size can be smaller than the self-venting size. In revamps, such nozzles are commonly encountered. In one naphtha splitter producing ISOM feed and CCRU feed with a furnace reboiler, the nozzle size was allowed to remain the same after it was established that the residence time at the bottom was adequate.

12. Hydrates in trays: Hydrates are likely to form in natural gas processing plants in the distillation section. The hydrates can choke the tray dispersers. This increases the pressure drop and leads to column flooding due to the partial plugging of the dispersers.
13. Low clearance between tray deck and downcomer/downcomer edge with inlet weir during fitting: Often tray inlet weirs are used by internal experts to ensure a better distribution of liquids, particularly where liquid flow rates are low. The horizontal clearance between the downcomer and the inlet weir needs to be maintained such that the downcomer backup is within allowable limits of 40%–50% of tray spacing.
14. Coking in trays and choking due to the deposition of salts: Coking can occur in trays when hydrocarbons degrade, polymerize, and form deposits. Often components present can combine to produce salts and deposit on trays. Sometimes particulates carried from upstream deposit on tray dispersers, reducing the flow area for the vapor. The liquid channels may also become blocked due to the deposits.

2.7 Case Studies

Case Study 1

The author and team designed two amine regenerator columns that experienced a lot of amine carryover to the reflux drums. However, the lean amine H_2S loading was found to be in order, so it was difficult to find the root cause of the problem. It was observed that the liquid flow rate at the wash section was around 0.8 gpm per linear inch at design capacity and the internal designer has used a splash baffle, presumably for turndown consideration.

A problem was suspected in the wash section of the columns, as the section helps to reduce carryover to overhead. It was felt that in the wash section there might be a great deal of weeping from the tray, and as a result tray liquid was not reaching the outlet downcomer, which was adversely affecting vapor–liquid contact and defeating the purpose of washing. The regenerator columns were opened. It was found that segments of the tray deck had been fitted improperly. The fittings were rectified. Further, the trays were made leakproof using gaskets and seal welding to eliminate liquid weeping from the trays. The problem was permanently resolved.

For bigger amine regenerator columns, the wash section diameter is often reduced to resolve the problem of low liquid loading. The liquid load in the wash zone of an amine regenerator is normally low, thus the diameter is often reduced in some designs. This case study is also mentioned in Chapter 6 (on amine regenerators). The wash section of the amine regenerator is shown in Fig. 2.7.1.

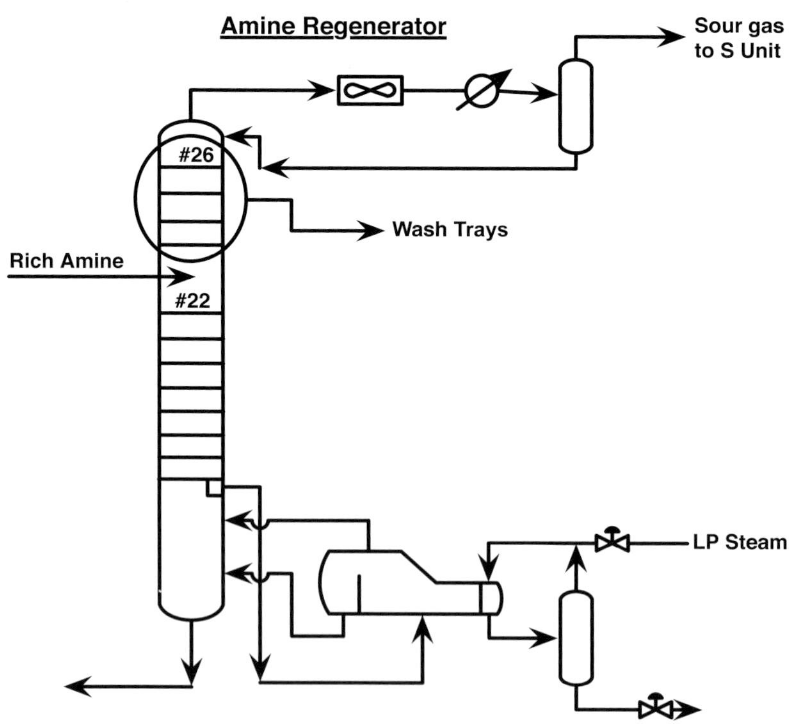

Fig. 2.7.1 Amine regenerator column

Case Study 2

This problem was experienced in a refinery crude unit after capacity augmentation. The flash point of heavy naphtha side cut (110°C–140°C) was not coming up to the required levels despite adequate stripping in side stripper of a crude distillation unit.

This is likely to occur in revamp applications where the downcomer area has not been increased and does not meet the norm for minimum residence time in the downcomer. It is always advisable to ensure minimum residence time even if the velocity in the downcomer and downcomer backup is adequate. In shutdown of the unit, the modification was carried out to increase downcomer area, and the problem was resolved.

Case Study 3

Often proprietary Z bars are used to increase the downcomer area without cutting and removing the existing bolting bars and rewelding the new bolting bars in revamp applications. This type of fitting reduces the work done inside the column to increase the downcomer area. A Z bar is fitted (without welding) on the existing bolting bars and a downcomer apron is fitted on it, increasing the downcomer area and weir length. The reduction in bubble area on the tray deck is compensated by increasing the number of dispersers on the active area. The capacity of a large naphtha splitter in a refinery was augmented by around 40% by fitting Z bars to increase downcomer area without cutting the existing bolting bars. The splitter was deployed for production of ISOM feed and CCRU feed. A schematic diagram of a Z bar is shown in Fig. 2.7.2.

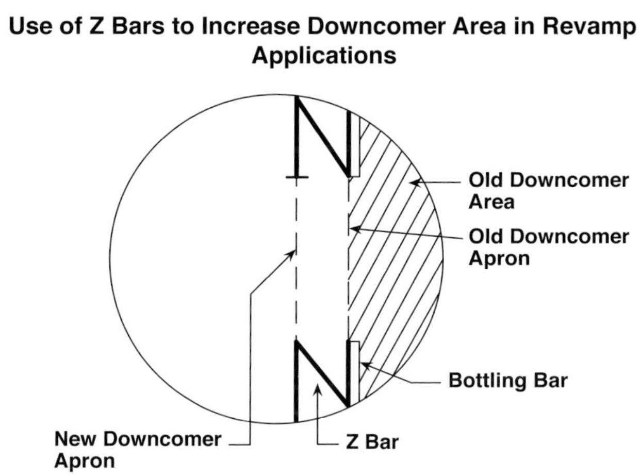

Fig. 2.7.2 Tentative schematic drawing of Z bars

Case Study 4

The required flow of HGO pump-around (around 250 MT/hr) was not achieved in a crude unit during startup after revamp, though the trays were designed and supplied by a reputed vendor. The column did not have a chimney tray to draw HGO pump-around and draw-off to the side stripper; instead, the draw was provided from the withdrawal pan located at the bottom of the central downcomer of the valve tray (4-pass). In the pump-around section, there were four liquid passes, and in the sections above and below, there were two tray passes. The trays were rotated by 90° in the pump-around zone. The pump-around was returned to three trays above (four-pass). The drawings were checked and inadequate channel width and depth were found. The same was corrected in the shutdown. The arrangement for the 90° rotation of the downcomer was created in line with the guidelines in *Chemical Engineering* by Henry Z. Kister.[7] The problem was resolved by modifying in line with the referred recommendation, and the desired flow of pump-around was achieved.

Case Study 5

The crude distillation column of an atmospheric vacuum distillation unit was facing a problem since the capacity revamp of the unit: the required ATF pump-around flow was not achieved, and any attempt to increase the flow was resulting in loss of pump suction.

There was a packing bed for the pump-around, and after cooling, the pump-around liquid was returned at the top of the bed with a distributor. Due to the low ATF pump-around, the cooling load was shifting to the top condenser (air cooler). Thus, the vapor velocity through the overhead air cooler was becoming alarmingly high and resulted in the thinning of the air cooler tubes. Furthermore, due to low pump-around flow and duty, less heat was transferred to the crude feed, resulting in low feed preheat. ATF yield was also low. Tray configuration was an overflow and partial draw-off type (see Fig. 2.7.3).

The top view of the existing chimney orientation is shown in Fig. 2.7.4.

The cause of the low ATF flow problem was narrowed down to faulty design in the ATF chimney tray, which was critically examined. There was inadequate cross-sectional area in the central collector channel, directing the liquids to draw-off sumps. There was also less area in the draw-off sump than necessary.

Withdrawal nozzle sizes were also found to be inadequate (not consistent with the self-venting size requirement discussed later in this chapter). The segmental weir below the packing did not have a hat. However, such a downcomer arrangement is theoretically correct if the liquid falling through the open segmental downcomers of the chimney tray is less than the desired internal reflux (IR).

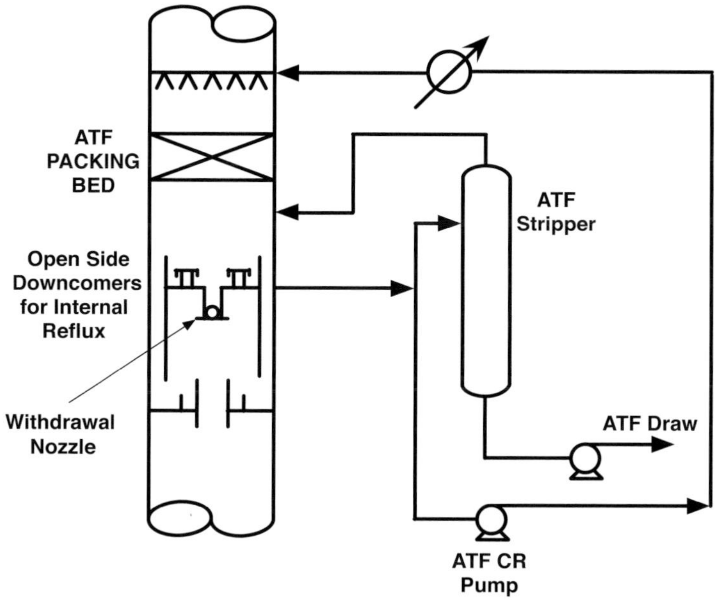

Fig. 2.7.3 Existing withdrawal arrangement for ATF circulating reflux (CR)

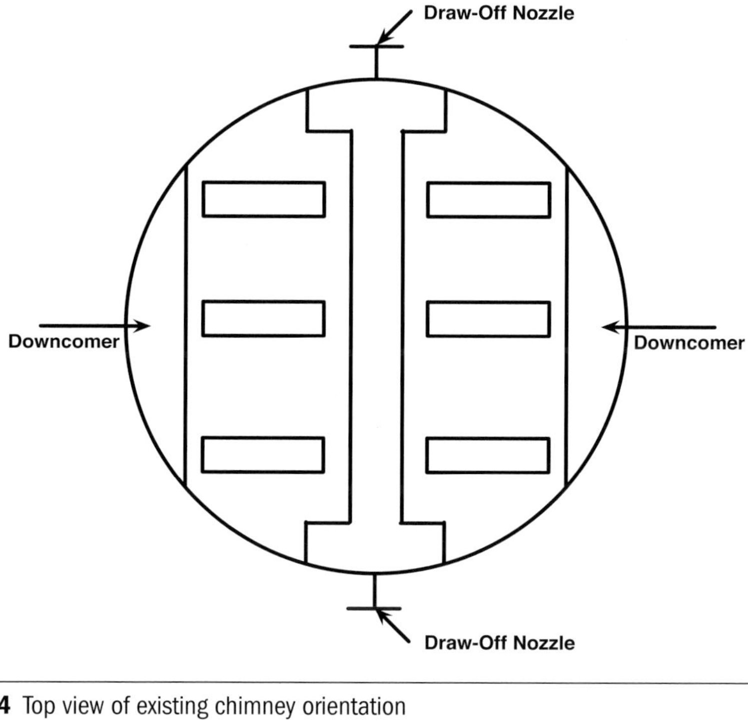

Fig. 2.7.4 Top view of existing chimney orientation

It was checked and found that 10% of the liquid falling on the chimney tray from the packing above (which can drop through the open area of the downcomers) was less than the IR required for the tray below. Thus, theoretically, it should not need hats over the downcomers; however, the likelihood of liquid channeling in the packing above cannot be eliminated. Due to channeling in the packing above, more than 10% of liquid can drop through the downcomers of the chimney tray. This could result in a higher-than-desirable IR through the packing to the downcomers of the chimney tray, and may result in additional reflux below the chimney tray. The additional IR in turn may result in more vapor condensation than desired in the tray below and lead to lesser availability of ATF.

The chimney tray in Fig. 2.7.5 is modified in a shutdown based on these findings. The chimneys are oriented transversely and the central collector is eliminated, routing the liquids directly to the draw-off boxes. Pipe downcomers have been introduced in place of segmental downcomers. Hats are still provided on the top of downcomers. The size of the withdrawal nozzles was increased to 18 inches to ensure self-venting flows, as shown in Fig. 2.7.5.

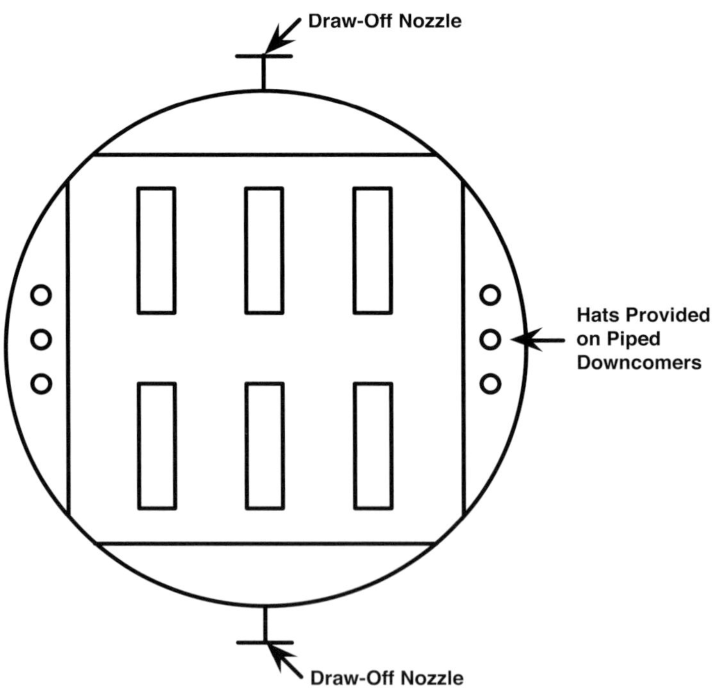

Fig. 2.7.5 Modified chimney tray, draw-off box, and piped downcomers

All the problems for this case were resolved. Pump-around flow increased to around 350 m^3/hr, top reflux was reduced, preheat increased by 15°C, and the ATF yield increased by around 4% wt. The unit has been operating for almost three years.

Case Study 7

This issue was observed in one crude unit's wash zone (area below the gas oil tray). The gas oil draw would often become dark-colored once the unit was restarted after a revamp shutdown. The column was opened during shutdown and the inspection carried out starting from the wash zone, as this section contributes to the gas oil product color.

An inlet weir (provided where liquid load is low and vapor load is high for better liquid distribution, as in the case of wash zone) was in the tray. In the first tray above the flash zone, it was found that the horizontal clearance between the downcomer and the inlet weir was not uniform. At one side, the downcomer was almost touching the inlet weir, leaving no gap for the downcomer liquid to flow and resulting in the buildup of liquid in the downcomer. This was causing the coloration of the gas oil. This was corrected by setting a uniform gap between the downcomer and inlet weir as specified by proper alignment of downcomer and inlet weir. The gas oil color improved.

Case Study 8

A capacity revamp was done on a crude unit, which had a low pre-topper bottom residence time, potentially due to incorrect design. This was causing fluctuations in pre-topped crude flow through the downstream section during operation at higher capacity, which was in turn causing occasional low feed flow trips in the furnace. (Low feed flow trips protect furnace tubes from overheating and coking.) An auxiliary vessel was used to improve the residence time, which resolved the problem. A schematic diagram is shown in Fig. 2.7.6.

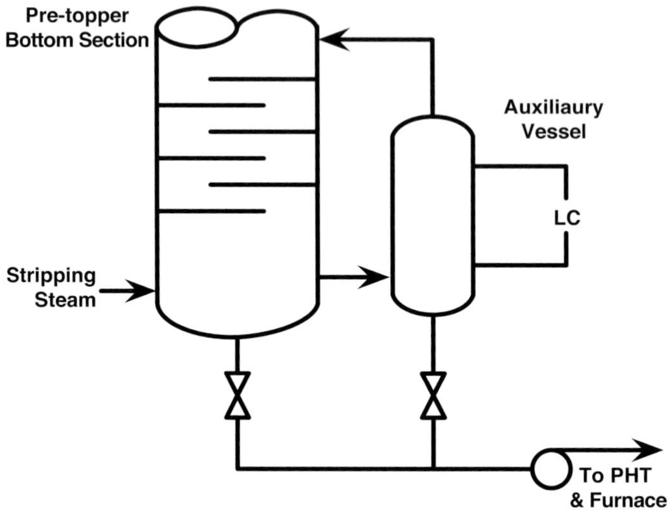

Fig. 2.7.6 Use of auxiliary vessel to improve bottom residence time

Case Study 9

An FCCU main fractionator (which uses RCO as feedstock) in a refinery was frequently experiencing high pressure drop across the column during operation. The pressure drop would increase quite quickly after start up of the unit.

It was reported that after the column was water-washed during shutdown, the pressure drop would reduce after the startup and increase after a short period of operation. It was also found that the performance of the desalter was not satisfactory. Thus, chlorides in the feed can combine with ammonia and eventually deposit in the colder parts (top trays) of the column as NH_4Cl. To resolve the problem, an increase in column top temperature was first tried, but the refinery was unable to increase the fractionator top temperature due to the limitation of the endpoint of the FCCU gasoline. Thus the solution could not be continued. The refinery was advised to lower the pump-around duties to increase the top reflux ratio of the fractionator limited by the reflux drum temperature. This was suggested in order to increase the hydrocarbon partial pressure in the upper section of the column and thereby reduce the partial pressures of NH_3 and HCl, which would lower the deposition temperature of NH_4Cl. The refinery was further advised to improve desalter operation to arrest chloride slippage, the condition improved gradually, and the problem was resolved. (This is also mentioned in Chapter 15.)

Case Study 10

In a VGO hydrotreater, the stripper column preceding the fractionator was getting repeatedly choked. It was discovered that the company had started routing visbroken naphtha along with the VGO for the purpose of its disposal. It was suspected that the deposition of primarily NH_4CL (possibly part NH_4HS) on the upper (colder) part of the stripper was causing the choking of the dispersers of the fractionator. This was corrected by a modification in the upstream, namely adding an additional water injection at the inlet of the air cooler (located at the overhead of the hot low-pressure separator (HLPS)) for washing of the NH_4Cl of the effluent stream. The unit already had a water-wash upstream of the cold high-pressure separator (CHPS). The problem was resolved by implementation of the above modification. (This is also mentioned in Chapter 15 and Chapter 4.)

Case Study 11

In one FCCU main fractionator, frequent loss of bottom level was observed, with an increase in reactor temperature (increasing conversion). The bottom level was brought back to normal by decreasing the reactor temperature.

It was suspected that the main column bottom (MCB) pump-around return liquid (used to desuperheat the superheated feed vapor) might not be able to flow down through the openings of disc and doughnut trays, and was thus building up and not reaching the bottom, resulting in a lowering of the bottom level. The wash liquid below the HCO tray (wash zone) could also have been building up.

The vapor–liquid profile for the entire column was developed at highest conversion. This was given to the column internal designer, particularly so that they could check the adequacy of disc and doughnut trays used at the desuperheating section. The designer ascertained that disc and doughnut trays are not adequate to allow the flow of liquid and suggested changing them to grids and installing coarse packing in the HCO wash zone and sacrificing one wash-zone stage to allow the free flow of liquid down. This was implemented and the problem completely disappeared. The capacity of the unit has even improved beyond the original design capacity with the use of available cushion in the other sections of the unit (this is also mentioned in Chapter 15).

The schematic drawing of desuperheating section is shown in Fig. 2.7.7.

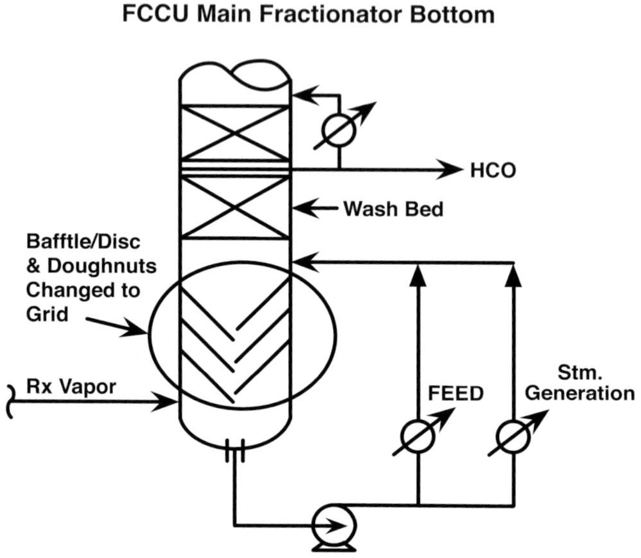

Fig. 2.7.7 FCCU column desuperheating section

2.8 Packed Columns

Packings are counter-flow devices.

Advantages of packing

1. Typically, the open area for vapor flow in the tray is around 10%–14% of the column cross section. This results in high velocity and pressure drop during vapor flow through tray dispersers. Open area in a packed column is approximately 50% or more of the column cross section. This provides a much lower pressure drop compared to columns with trays.
2. Liquid height in each tray is around 2 inches, which creates additional pressure drop for columns with trays. This additional pressure drop due to tray levels is not relevant for packing.
3. The low liquid holdup of packing trays is often considered an advantage, as it reduces polymerization/degradation and is safer.
4. Packing has greater applicability for use with corrosive products.
5. Packing can be installed in small diameter columns where the installation of trays is difficult.

The following are two broad classes of packing:

1. Structured packing: stacked rings, grids, and proprietary structured packing with a regular geometry.
2. Random packing: rings, saddles, and proprietary shapes, which are dumped in the column and take up a random arrangement.

Structured packing

One advantage of structured packing is that it has a lower height equivalent to theoretical plate (HETP), typically less than 0.5 m, and a lower pressure drop than random packing.

Application of structure packing

1. High vacuum distillation
2. For column revamps to increase capacity

Packing in packed columns is not designed by the process designers engaged in overall unit design. It may be designed by packing experts based on the inputs of vapor–liquid flows and properties like vapor–liquid densities, viscosities, etc. However, the process engineer is required to calculate the diameter and height of the column, keeping in mind that it must be adequate for column internal suppliers to install the internals without any problem at a later date. Both the height and diameter of the column need to be estimated conservatively by a process designer before the column internals are designed by column internal experts. Diameter can be estimated by a steady state simulator that takes trays or packing into account, or more conservatively, diameter can be calculated only taking trays into account. The height estimation requires a calculation of height equivalent to a theoretical plate (HETP).

2.8.1 HETP estimation

HETP is required by the process designer for calculating the height of the columns. A conservative approach is used so that the internal expert can provide the required number of stages during internal design.

HETP can be estimated by this rule of thumb (the Frank–Ludwig–Vital rule): HETP in ft = 1.5 *dp*; *dp* = packing diameter in inches for pall rings, with a minimum of 2 ft for columns with a diameter less than 2 ft. The HETP is estimated more often using this rule than it is predicted using models.

Another rule is HETP = $(1200/a_p) + 4$, where a_p = packing surface area per unit volume in ft^2/ft^3. There are other rules, but the above two equations are conservative enough to estimate HETP and calculate bed height to get the estimated overall column height. The purpose is to determine the column height in order to accommodate packing and other internals that will provide the desired stages considered in design. Often HETP estimation is made more conservative by introducing the effect of viscosity. The HETP increases with an increase in liquid viscosity.

Once the HETP for the pall ring is determined, it is multiplied with a prorating factor (*fp*) for the selected packing. Commercial charts and tables are available that list the *fp* for different commercial packings. The *fp* for a 2-inch pall ring from a particular supplier is 1.0, and that of the 2-inch Flexipac from Koch-Glitsch is 0.71.

Example:

HETP in ft = 1.5 × *dp* = 1.5 × 2 = 3.0 ft = 36 inches for a 2-inch pall ring. The packing selected is the Flexipac with an *fp* of 0.71.

Thus, the HETP of the Flexipac is 0.71 × 36 inches = 25.6 inches (approximately).

The height of the Flexipac bed will be equal to (number of stages considered in the bed) × (HETP).

Estimation of overall column height for packed column

Once the bed height is estimated, the heights required for the distributors, bed supports, etc., are estimated conservatively and total height of the column is then estimated.

1. For each distributor, around 24–36 inches are considered (as an estimate) for determination of column height.
2. A maximum packing height of around 8 m is deployed in a single bed. If additional height is required to deliver the specified number of stages, another packed bed is added with a redistributor between the beds.
3. The redistributor can be spray type or gravity type. The gravity type redistributor is essentially a chimney tray with a liquid distribution facility (i.e., with sieves or drip tubes). The height of the redistributor is used to estimate for overall column height determination. A minimum height of 1,500 mm may be considered for installation of the redistributor.
4. For liquid collection from beds and vapor distribution, chimney trays are used in packed columns. A height of around 1.5 meters is

considered for each chimney. Minimum chimney height considered is 9 inches. Chimney height is based on residence time in the chimney tray and controllability of chimney tray level (HLL to LLL is normally 6 inches). Clearance from the chimney hat to the upper tray deck or the bottom of the packed bed is to be maintained at a minimum of 12 inches.)

5. A manhole is to be provided after each packing bed. The clearance preferred is 1,200 mm for a 24-inch manhole.

The estimate should be conservative so that column internal vendor does not encounter difficulties in accommodating the stages specified.

Column diameter of a packed column

Column diameter with packing can be estimated using a simulator with specified packing or even trays assumed in order to get a conservative diameter. For purpose of preliminary rating, a simulator can be used, but it needs to be confirmed by internal experts.

Some special features and devices:

1. Mounted condensers are sometimes used in column overheads when the overhead product is vapor.
2. Water trap-out trays are used to collect water from trays and drain.
3. Total draw-off trays in this context are the same as chimney trays.

   A chimney tray with side downcomer is used as a partial draw-off tray, mostly below a packed bed. A chimney tray arrangement with side downcomer is shown in Fig. 2.8.1.

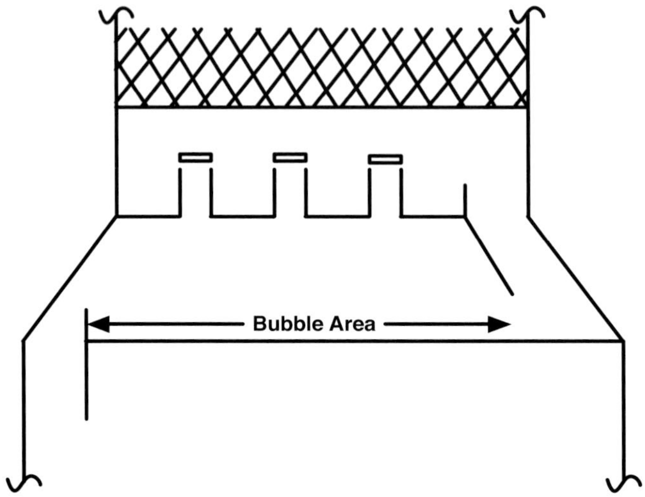

Fig. 2.8.1 Chimney tray with downcomer below packing

Liquid distribution on a packed bed

Liquid is introduced to the bed through liquid distributors. Two types of distributors are commonly used.

1. Spray distributors
2. Gravity distributors

2.8.2 Common problems experienced in packed columns

1. Packing blowing up: Packing is often installed with hold-down grids at the top to avoid lift or blow-up. But with a pressure surge in columns, the packing can blow up in full or part.
2. Breakage of packing: Packing may break after prolonged use. This is more pronounced with structured packing, and is often caused by corrosion, as structured packing is made with thin metal sheets.
3. Coking of packing: Coking is common problem in packing in high-temperature services aggravated by low liquid flow rates.
4. Minimum wetting: A minimum liquid rate is always required to keep the packing surface covered with film of liquid that gives it the desired efficiency.

5. Channeling of vapor and liquid: Vapor and liquid channeling is the most common problem in a packed column, and affects the separation between adjacent cuts. The likelihood of channeling increases with the height of the packing. Often packed beds are fragmented and redistributors are used to avoid channeling.
6. Malfunctioning of liquid distributors: Spray distributors deploy nozzles with small holes in the spray header. Often, the holes will get plugged, or the nozzles dislodged, which affects the performance of liquid distribution. To avoid particulates, duplex filters are used at the upstream of the distributors. The filtering element needs to be cleaned and checked frequently as suggested by pressure drop across the filters. To assess the health of the distributors, it is always advisable to provide pressure gauges at the upstream of distributors and downstream of the duplex filters.

   It is also very important to take readings of the back pressure in spray distributors corresponding to the flow. An increase in the back pressure (normalized for the flow) indicates distributor choking. A sudden decrease in back pressure indicates the dislodgement of nozzles in the distributors or the rupture or leak of gaskets in the distributor header or laterals. After fitting the nozzles in the distributor header or laterals, an internal vendor and their client needs to check nozzle performance for gasket leaks and irregular spray patterns by circulating demineralized water (DM water).

   In a gravity-type distributor, the liquid is either pumped or introduced to the troughs using elevation differences. From there, it is distributed in the packed bed through a sieve or drip tubes by gravity. When pumps are used to feed the liquid, it should have a duplex filter to arrest the particulates before the liquid is introduced to the troughs. The most common problems in gravity distributors are holes in the tubes getting choked by corrosion products or external particulates, and the failure of the tubes.

2.9 Common Installation Errors

In many cases, the problem is caused by errors in installation. They are described in this section along with the correct arrangement.

Often, strip-out from the side stripper is introduced within the pump-around trays (typically three trays) (Fig. 2.9.1). This arrangement has been seen in many columns and rectified by the author during revamp by rerouting the strip-out above the PA return. If the strip-out is incorrectly introduced, the strip-out vapor will condense and recirculate. The draw-off nozzle for pump-around trays should be flushed with the bottom of the pan; otherwise, a weep hole is required on the pan. The desired elevation of stripper and seal in a two-pass tray column is presented in Fig. 2.9.2.

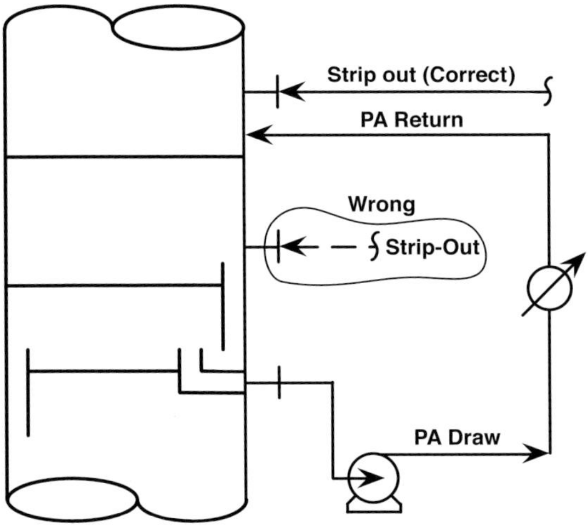

Fig. 2.9.1 Strip-out wrongly entering between PA sections

Similar arrangements have been seen in other columns, where the strip-out enters the tray above the two-pass tray. It can be seen that the liquid drawn from the tray is actually the liquid from the tray above. If the strip-out is returned to the location shown, this would be introducing it below the tray and would recirculate the strip-out.

Furthermore, for two-pass trays, the strip-out will flow through half of the tray's active area as the central downcomer will not allow the vapor to bubble through the other half of the tray.

It should be noted that the central and off-center downcomers are always provided with vapor-equalizing tunnels to equalize the pressure in each section constituted by the downcomers.

For side-stripper installation, the difference in elevation between the column draw tray and stripper top should be a minimum of 4 meters (Fig. 2.4.2). This elevation difference provides the necessary head to overcome the pressure loss in the product draw line and draw-off control valve. A liquid seal of 1,500 mm is typically provided so that the strip-out vapor does not blow through the liquid draw line due to pressure drop in the stripper overhead line joining the column. Normally, the control valve loop provides the seal required (Fig. 2.9.2).

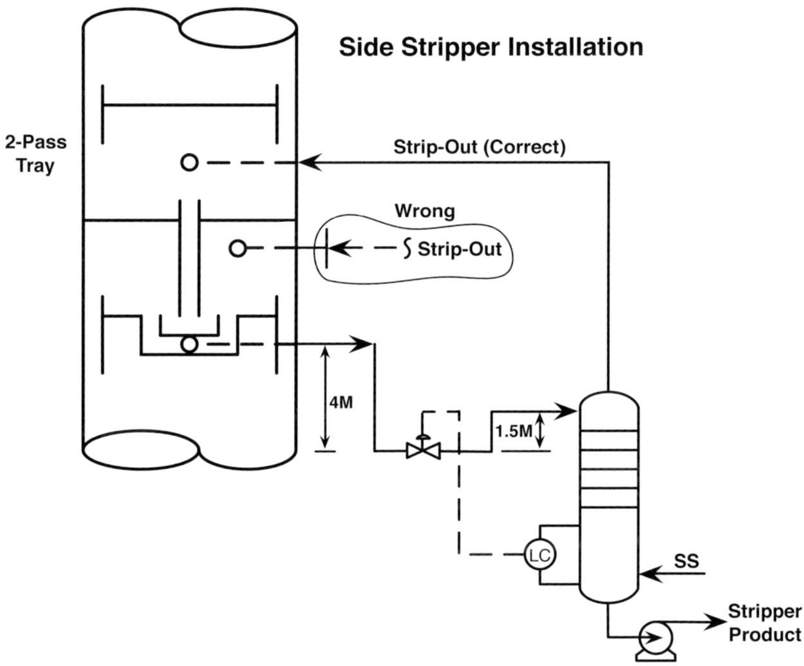

Fig. 2.9.2 Desired elevation of stripper and seal in a two-pass tray column

2.10 Calculation of Nozzle Size

Self-venting withdrawal nozzles and lines

Liquid outlet line from the column should meet the self-venting criteria. The liquid draw from the distillation column is partly aerated. The size of the self-venting line should provide enough free area for the released vapor to return to the column; otherwise, it is likely to go to the pump and cavitate it.

Appropriate self-venting nozzle size is estimated based on the velocity in the nozzle, which typically ranges from 0.6–1.2 m/sec for side draw-off. For a flow below 300 m^3/hr, a nozzle velocity of 0.6–0.7 m/sec can be considered. For flow between 300–500 m^3/hr, a velocity of 0.9 m/sec can be considered. For a flow between 500–1,000 m^3/hr, a velocity of 0.9–1.0 meter/sec can be considered. For a flow of 1,000 m^3/hr and higher, a velocity of 1.0–1.2 m/sec can be considered. These may be little conservative. Some designers use a velocity of 0.9 m/sec at all ranges of flow, but a conservative sizing may better. Some generalized correlations are suggested by experts and practitioners.

Normally, size reduction (swage down) is not allowed before a 3-meter vertical drop of the withdrawal line from the draw-off nozzle. The draw-off line may have bends/elbows after a minimum vertical descent of 3 meters. Bends may have a long radius elbow.

Column inlet nozzle

The diameter of the column inlet nozzle (the flash zone of the vacuum distillation column in particular) is calculated with a maximum gas velocity of 0.6–0.7 times the velocity of sound in the media, or as a maximum of 350 ft/sec. The velocity of sound is calculated as $31.6((KP)/\rho)^{0.5}$ m/sec, where P is the absolute pressure in KPa and ρ is the density of the gas in kg/m^3. K is Cp/Cv (specific heat ratio) of the vapor. Alternatively, it can be calculated by the expression $Vs = 6{,}160(K/1.1)^{0.5} \times (T/M)^{0.5}$, where Vs is velocity of the sound in meter/min, T is temperature in Kelvins, and M is the molecular weight. Often for vacuum transfer lines or nozzles the diameter thus calculated is divided by 0.98 to account for the flow of liquid.

2.11 Typical Residence Time in Vessels, Tower Bottom

1. The preferred time for column bottom sump being pumped to furnace is 10 minutes. This amount can be less if the nozzle is self-venting. The author has used 3–5 minutes in several applications without issue.
2. Column bottom pumped to storage is 2 minutes.
3. Reflux to column is 5 minutes.
4. Column bottom pumped to reboiler furnace is 5 minutes.
5. Residence time for withdrawal from chimney trays is normally less than 1 minute in order to contain the height of chimney. External vessels can be provided if higher time is desired for the pump.

Residence time is calculated based on the volume of the vessel/tower between the low-level alarm point and a high-level alarm point. The low-level alarm point is normally set at 200–300 mm above the bottom tangent line of the towers. If residence time is known, the high-level alarm point can be estimated using the residence time.

Abbreviations

gpm: gallons per minute

Notes

1. Reproduced from J.R. Fair, "How to Predict Sieve Tray Entrainment and Flooding," *Petro/Chem. Eng.* 33, no. 10 (1961): 45.
2. Henry Z. Kister, *Distillation Operation* (New York: McGraw-Hill, 1990).
3. Fair, "How to Predict," 45.
4. Kister, *Distillation Operation.*
5. Ibid.
6. Ibid.
7. Ibid.

Crude, Vacuum, and Condensate Distillation 3

3.1 Crude Distillation Unit

3.1.1 Introduction

Crude oil distillation is an open-art technology. The crude oil is distilled at almost atmospheric pressure and separated into various desired fractions in the crude oil distillation unit. At the bottom of the atmospheric distillation column, reduced crude oil (RCO) is produced and is further fractionated under vacuum to produce vacuum gas oil (VGO) as a feed to the hydrocracker or fluid catalytic cracking unit (FCCU) in a fuel-type vacuum unit, or distilled to produce lube oil feedstocks in a lube-type vacuum unit.

Fractions produced from the atmospheric distillation column are given below.

Typical products of crude oil distillation with true boiling point (TBP) cut ranges:

1. Unstabilized naphtha: IBP–120°C
2. Heavy naphtha: 120°C–140°C
3. Kerosene: 140°C–270°C
4. Light gas oil: 270°C–320°C
5. Heavy gas oil: 320°C–370°C
6. Reduced crude oil: 370°C+

The unstabilized naphtha is stabilized in a stabilizer and produces LPG and gas as overhead. A naphtha splitter is often used to split unstabilized naphtha into light naphtha/ISOM feed and catalytic reformer unit (CRU) feedstock.

3.1.2 Configurations

There are several configurations commonly deployed in crude distillation units:

1. Without pre-topper
2. With pre-topper
3. With pre-flash drum

A crude unit without a pre-topper or pre-flash drum is shown in Fig. 3.1.1.

A crude unit with a pre-flash drum is shown in Fig. 3.1.2.

A crude unit with pre-flash tower (pre-topper) is shown in Fig. 3.1.3.

The unit without a pre-topper and pre-flash drum (shown in Fig. 3.1.1) is a simpler configuration. It typically has two preheat trains: one up to the desalter, the other from the desalter downstream to the furnace inlet. Sufficient pressure is maintained at the upstream furnace inlet control valves so that no vaporization takes place at the upstream of the valves.

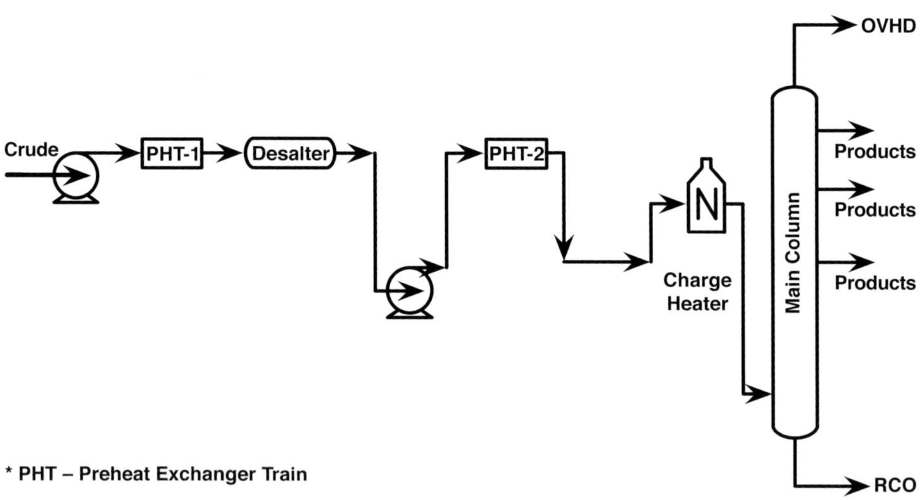

Fig. 3.1.1 Crude unit without a flash drum or pre-topper

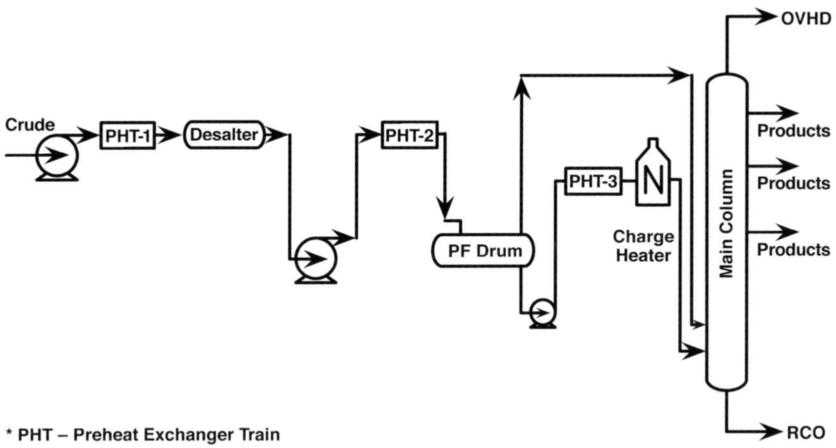

Fig. 3.1.2 Crude unit with a pre-flash drum

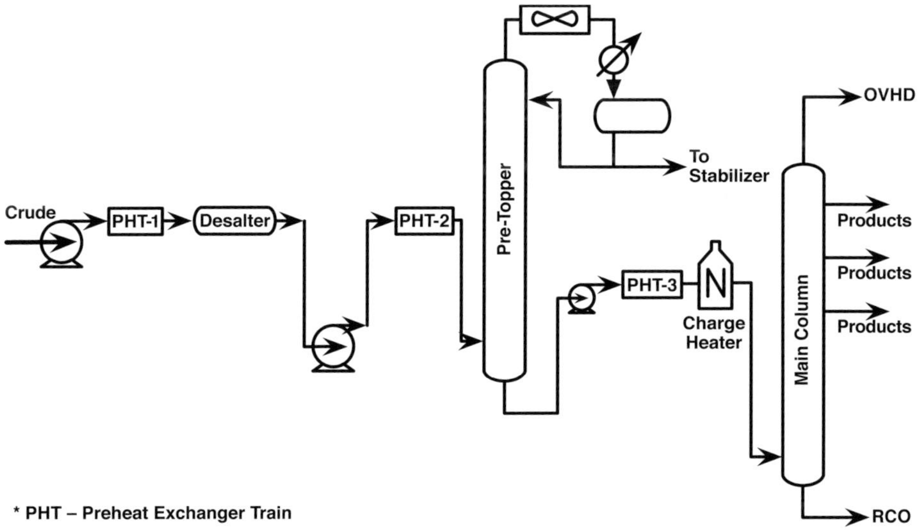

Fig. 3.1.3 Crude unit with pre-topper drum

There is always the likelihood that the crude will vaporize at the inlet of the furnace feed control valves if a very high preheat temperature is targeted. A substantially high pressure would be required in the last heat-exchanger train in order to arrest vaporization. This would increase the pump pressure, require a higher design pressure for the heat exchangers, and consequently require a higher cost for the last heat-exchanger train. For lighter crudes with a larger amount of naphtha fraction, a configuration with a pre-topper or pre-flash drum is more relevant. This eliminates the likelihood of vapor formation at the inlet of the furnace at a relatively lower pressure and the highest achievable preheat, as some of the naphtha fraction is removed through the pre-flash drum or pre-topper before being fed to the charge heater (furnace).

Use of pre-flash drum

A pre-flash drum is often necessary for units with higher capacity where a bigger heat-exchanger train is deployed to provide more surface area for crude preheating. In this case, the pressure drop is likely to increase in the heat-exchanger trains, and residual pressure at the furnace inlet control valve will reduce.

The pre-flash drum is provided as a break drum, and pre-flashed crude from the drum is fed by a pump to the heat-exchanger train, charge heater (furnace), and main fractionator. This reduces the furnace mass velocity and pressure drop, as approximately 3%–10% (depending on the temperature of the inlet to the pre-flash drum and lighter fraction present in the feed) of the crude vaporizes in the pre-flash drum and finds its way to the main fractionator by passing the downstream heat-exchanger train and charge heater (furnace). This reduces the likelihood of crude vaporization at the furnace inlet. Further, the pre-flash drum bottom pump provides the pressure to arrest vaporization at the furnace feed control valve upstream at the highest achievable crude preheat temperature. Using a pre-flash drum is a much cheaper option than installing a pre-topper; however, it does not debottleneck the stabilizer and main fractionator.

Both pre-toppers and pre-flash drums help keep the unit running smoothly, even when some water ingresses along with the crude. Part of the water vaporized from the crude gets removed through the overhead, which reduces the water content of the pre-flashed or pre-topped crude going to the third heat-exchanger train (operating at higher temperature) and the furnace, and decreases the emergencies caused by gasket failures, leaks through the flanges, etc. For crude units designed for high preheat, a pre-flash drum is a minimum necessity.

The pre-flash drums can experience severe foaming and can lead to carryover of liquid to the column flash zone and lowering of flash zone temperature. It can also pass vapors to the bottom pump suction and can lead to cavitation of the pump. Horizontal pre-flashed drums are preferred with two inlet nozzles compared to vertical drums. If a vertical flash drum is used, a tangential feed entry is necessary and four anti-swirl (vortex) baffles are essential. The baffles need to extend almost up to bottom TL. The cross-sectional area should be designed for 40% of critical velocity (V_c), where $V_c = 0.048((DL - DV)/DV)^{0.5}$ in meters per second, and DL and DV are liquid and vapor densities respectively in kg/m^3. The vertical flash may be designed very critically to avoid the likelihood of liquid carryover and vapor gassing. The author has experienced incidents of severe lowering of flash zone temperature where a vertical pre-flash drum was used. Installation of anti-vortex baffles reduced carryover and temperature drop to some extent.

Use of pre-toppers

Units with pre-toppers are somewhat more energy efficient than conventional units (those with only a topper and a pre-flash drum), as the pre-topper operates at a lower reflux ratio and removes part of the naphtha as pre-topped naphtha so it does not have to be heated to the furnace coil outlet temperature. Only the pre-topped naphtha is fed to the stabilizer, which results in a lower reboiler duty requirement in the stabilizer; thus, this configuration calls for marginally lower heat requirement.

Benefits of using pre-toppers

Furnace heat duty and pressure drop reduces due to the reduction of pre-topped crude flow by around 8%–12% through the furnace. Furnace pressure drop also reduces due to the reduction of lighter stock (naphtha) in the furnace feed. Only pre-topped naphtha is fed to the stabilizer instead of total naphtha; thus, the reboiler heat requirement in the stabilizer reduces. All the LPG components are captured in pre-topped naphtha.

Further, if pre-topping is performed with stripping steam, the pre-topped crude temperature drops by 10°C–15°C as compared to the inlet. This allows the absorption of low-temperature heat at the downstream of the pre-topper. A pre-topper is actually very useful for revamp cases where the furnace passes mass velocity or heat duty limits and the main column or stabilizer diameter limits operation at a higher capacity.

Pre-toppers typically have two configurations:

1. Reboiler at the bottom
2. No reboiler and stripping steam used at the bottom

When a reboiler furnace is used at the bottom, the crude is fed in around the middle and the column typically has 30–35 trays, 12–15 of which make up the stripping section, which is important for ensuring no slippage of LPG components along with the pre-topped crude.

When steam is used in the bottom of the pre-topper, the crude feed is introduced 6–9 trays above the bottom. The column typically has 24–27 trays, 6–9 of which constitute the stripping section. A preheat train is provided downstream of the pre-topper to heat the pre-topped crude to the furnace inlet temperature. The pre-topper pressure is fixed at a specific value which ensures that no LPG components are bled from the reflux drum at 40°C–45°C and the reflux drum pressure. In simulation, the pre-topper condenser needs to be configured as a subcooled condenser, whereas the topper condenser should be a bubble condenser when preceded by a pre-topper. The temperature of the topper reflux drum should set the pressure of the reflux drum. It is also important to note that the overhead heat of the pre-topper can be effectively transferred to the crude, as the top temperature of the pre-topper will be higher than the topper (due to high-pressure operation) and topper pressure can be kept lower.

Typically, crudes with higher naphtha content need a pre-topper to limit the topper and stabilizer diameter and also to save energy. An additional preheat train is provided at the downstream of a pre-topper. Flow through this train also reduces with the reduction of lighter stock in the charge.

A pre-topper or pre-flash drum is thus essential for the design of gas condensate distillation units. Adding a pre-topper is the best configuration for debottlenecking or revamping a crude distillation unit's capacity, as it reduces the diameter requirement of the topper and stabilizer. Furthermore, introducing a pre-topper provides flexibility for integrating some low-temperature heat with pre-topper bottom as the bottom cools by around 10°C–15°C if stripping steam is used in the bottom. Additionally, if furnace heat duty and mass velocity are limiting, installing a pre-topper or pre-flash drum helps in debottlenecking. Installing a pre-topper for revamp also reduces the condenser duty requirement of the topper.

Pre-topper inlet temperature

Pre-topper feed needs to be heated to a certain temperature to recover part of the naphtha as pre-topped naphtha from the pre-topper overhead. For a fixed cut point and flow rate of pre-topped naphtha, if the feed temperature is increased, the reflux ratio will increase and consequently more heat will be rejected into the overhead condenser of the pre-topper. This will lead to heat loss from the system, so the pre-topper inlet temperature needs to be fixed based on wise judgment. It is covered in earlier chapters that the preheat efficiency of a distillation column drops with an increase of feed preheat or feed q. A similar case has been seen in a crude distillation unit. In one unit, the preheat to the pre-topper was around 260°C and the pre-topper had a furnace deployed as a reboiler. There was no heat-exchanger train to further heat the pre-topped crude before routing it to the charge heater. It was found that a feed preheat of around 225°C was adequate for recovering the pre-topped naphtha of the required quantity and cut point. The condenser duties in both the above preheats of 225°C and 260°C were compared. The condenser duty in the higher preheat case was found to be higher by around 3 Gcal/hr, making it a loss.

In the revamp of the unit, the feed temperature was reduced, stripping steam was introduced at the bottom, and a third heat-exchanger train was added at the downstream of the pre-topper. The pre-topped crude preheat increased by the equivalent of nearly 3 Gcal/hr.

Pre-topper inlet temperature can be increased and a larger quantity of pre-topped naphtha can be drawn to reduce the load on the topper if it is limiting. This will lead to increased pre-topped naphtha production and consequently lower topped naphtha production. If the pre-topper inlet temperature is increased too much, the following problems may occur and will need to be addressed.

1. Vaporization of the crude at the pre-topper inlet control valve upstream.
2. Higher reflux ratio in pre-topper and consequent loss of heat to the pre-topper condenser.
3. Partial pressure of hydrocarbon vapors in the overhead of the topper will try to reduce, which can increase the dew point of water and condense water at the top section of the topper.
4. Partial pressure of H_2S in the overhead of the topper is likely to increase, and this may increase H_2S-accelerated acid corrosion in the topper overhead.

Design approaches to arrest vaporization at pre-topper inlet control valve upstream

The problem of vaporization at the pre-topper inlet control valve in a grassroots unit can be resolved by increasing the discharge pressure of the desalter outlet pump and maintaining a higher pressure at the valve inlet.

In case of revamps, the crude preheat train at the upstream of the pre-topper can be configured with parallel trains so that pressure drop in the crude side can be reduced to make more pressure available at the pre-topper inlet and suppress vaporization.

Alternatively, the inlet control valve can be installed at the inlet of the last one or two heat exchangers in the train, since the temperature is lower at that location and the crude is likely to be in liquid state, corresponding to the pressure and temperature at the location.

The second option may be applicable for revamp cases where pump pressure cannot be increased, as it is limited by the design pressure of the heat exchangers in the train. The last exchangers where crude is likely to vaporize should have crude in the shell side with vertical-cut baffles to allow two-phase flow of crude through the heat exchangers.

In this option, the vaporizing crude will provide a higher differential temperature in the last heat exchangers between the hot and cold fluid. This higher differential temperature is often considered an advantage if the design of the two-phase exchanger is not a problem. Similarly, when a pre-flash drum is used, a higher inlet temperature is also likely to result in vaporization at the pre-flash drum inlet control valve upstream, and the problem of vaporization at the pre-flash drum inlet can be resolved in the same way as in case of pre-topper.

The likelihood of a higher reflux ratio in the pre-topper can be avoided by withdrawing more pre-topped naphtha as an overhead product. It is always advisable to install a balancing heat exchanger at the upstream and the downstream of a pre-topper heated by the same hot fluid.

In case the pre-topper inlet temperature increases beyond the requirement, the extra heat can be passed to the downstream instead of causing a higher reflux ratio in the pre-topper and the rejection of extra heat to the condenser.

An actual heat exchanger synthesis scheme showing a buffer exchanger at the upstream and downstream of a pre-topper to maintain pre-topper inlet temperature and reflux ratio, and reduce heat loss to the pre-topper condenser is given in Fig. 3.1.4.

The light gas oil pump-around (LGO PA) is used as a heat source at the upstream and downstream. In case the pre-topper inlet temperature needs to be lowered, the bypass on the LGO PA exchanger downstream will close, the pre-topper inlet temperature will reduce, and the heat will be transferred to the pre-topped crude instead of being rejected to the pre-topper condenser.

The installation of a pre-topper can lead to reductions in the partial pressure of hydrocarbons, increases in water partial pressure, and increases in the partial pressure of H_2S in the topper overhead. The water partial pressure can be lowered by decreasing the stripping steam to side strippers, lowering pump-around duties marginally to increase reflux ratio, or by reducing the top pressure of the topper. Similarly, H_2S partial pressure can be reduced by increasing the reflux ratio and lowering topper pressure.

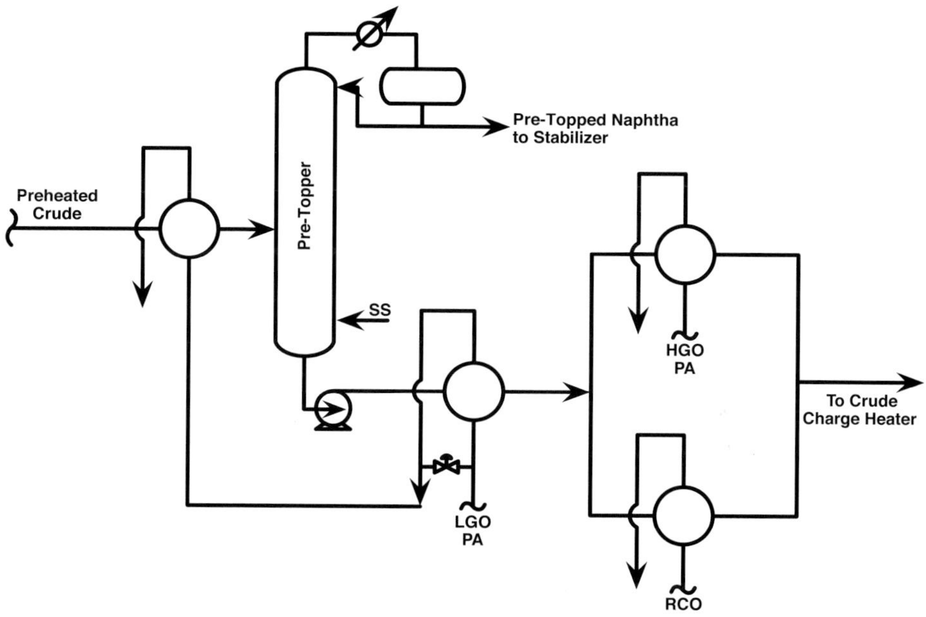

Fig. 3.1.4 Crude unit with balancing heat exchanger before and after pre-topper

3.1.3 Maximizing efficiency of crude units

Crude and vacuum units in a refinery are energy-intensive units and a major part of that energy is spent in crude vacuum units. A crude unit can be made energy efficient by the following means.

1. Maximizing crude preheat/increasing steam generation from hot streams
2. Operating at lowest furnace outlet temperature
3. Utilizing low-temperature heat sources for reboil in side strippers instead of stripping steam
4. Operating at lowest pressure
5. Maximizing pump-around duties
6. Use of pre-toppers with low reflux ratios
7. Low-temperature heat recovery schemes
8. Direct coupling of the crude distillation section with the vacuum distillation unit by feeding hot RCO directly to the vacuum distillation unit from the crude column bottom instead of cooling and sending it to storage and refeeding to the vacuum unit from storage

9. Maximizing the number of side draws and pump-around refluxes and using them to preheat crude or generate steam
10. Using pre-topper or topper overhead for crude preheating. This may increase crude preheat to some extent, but provides opportunity for steam generation from other hot streams.
11. Using a top pump-around to heat crude and lower condenser duty.

Heat integration in a crude distillation unit

The art of designing a crude distillation unit lies in designing a unit that will not only produce the desired products with the desired separations, but also one that will have efficient heat integration for achieving maximum heat recovery from hot products and result in high preheat and steam generation. Crude units with vacuum-unit integration are heat-surplus environments mainly because the heat supplied by the two furnaces is available to the crude.

Typically, the maximum preheat attainable with this configuration on gulf crude is around 295°C–300°C. The remainder of the heat from hot products is typically used for utility heating or steam generation. It is felt by many that crude with a low RCO yield would result in a much lower preheat, but the recoverable heat from pump-around refluxes is often high, which provides an opportunity for the designer to configure the heat-exchanger train to recover the full potential of pump-around heat and attain high preheat. With overhead heat integration, it may be difficult to increase the preheat temperature by a substantial amount if the existing heat-exchanger train synthesis is very efficient. However, steam generation is possible using the hot streams which would have otherwise gone to heat crude in the pre-desalter heat-exchanger train in absence of overhead heat integration.

Column overhead heat integration

Overhead integration may or may not increase the crude preheat. It depends on the heat recovery pinch temperature. If the pinch temperature is higher, then no preheat gain will be achieved by integrating overhead heat, but the cooling duty for hot products will increase. However, additional steam generation may be possible with overhead integration. So this needs to be introduced only after a heat recovery pinch study and critical examination of the requirements. Overhead heat integration can be used with the following options.

1. Introduce a top pump-around in the topper
2. Integrate pre-topper overhead heat with the crude if the pre-topper exists in the configuration

3. Integrate main topper heat along with crude

It is worth mentioning that introducing a top pump-around is the easiest and cheapest way to integrate part of the overhead heat with the crude. This is achieved by reducing the reflux ratio, which lowers heat rejection to the overhead condensers or coolers.

Some designs maintain the top temperature of the main fractionator solely by a top pump-around and only have a provision for reflux as and when required. This scheme is shown in Fig. 3.1.5.

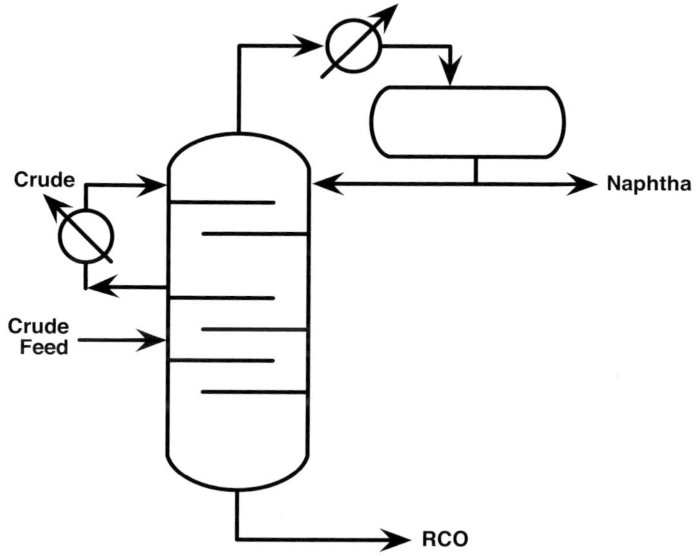

Fig. 3.1.5 Basic scheme of overhead integration

Pre-topper overhead integration is a good option and safe from the point of view of corrosion in crude and overhead heat exchangers.

An actual pre-desalter heat-exchanger train is shown where the pre-topper overhead is exchanging heat with the crude at a lower temperature (shown in Fig. 3.1.6). The first exchangers are pre-topper overhead and crude. The pre-topper overhead is split and the crude is heated to 67°C from a temperature of 30°C, with a heat transfer of around 14 Gcal/hr of overhead heat to the crude. This also reduces the overhead trim condenser duty.

If the topper overhead is integrated, it is likely to be coupled with the crude at almost the same location as the pre-topper overhead. If the overheads of both the topper and pre-topper are integrated with the crude, there may be an excess of low-temperature heat.

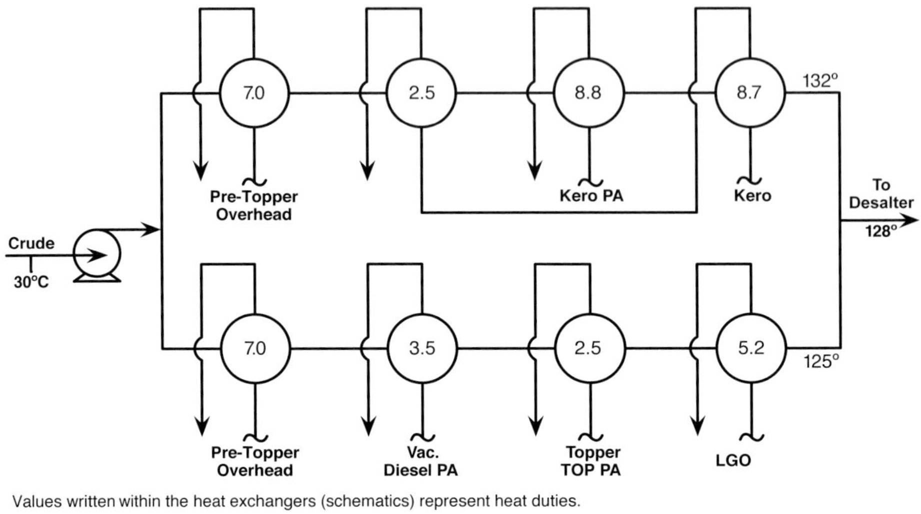

Fig. 3.1.6 Pre-topper overhead coupled with crude for preheat

If this is done, some hot streams should be deployed to generate steam and then be routed to preheat train 1, so that the full benefit of overhead integration is harnessed. Some steam generation using the LGO stream (Fig. 3.1.6) can be done prior to routing this stream to the preheat train before the desalter. Other hot streams such as vacuum residue and heavy vacuum gas oil (HVGO), which were likely to deliver heat in this train to maintain the desalter temperature, can be used for the generation of steam and subsequently can be routed to the preheat train at lower temperatures. Thus, column overhead heat integration increases the efficiency of a unit through increased preheat, additional steam generation, or by preheating other cold streams. At locations where cooling water availability is a problem, lighter products that need cooling to 40°C–45°C are finally cooled with the cold crude or condensate (feed to the unit). The hot products first exchange heat with crude then are cooled with air coolers to around 65°C–700°C, followed by cold crude versus product exchangers in place of trim coolers (water coolers).

Integrated atmospheric and vacuum unit

The other method for heat integration is the use of an atmospheric and vacuum unit (AVU), where the RCO is fed directly to the vacuum furnace before being fed to the vacuum column. This is considered an energy efficient configuration, as no RCO heat is delivered to the air-cooler or the tempered water coolers and lost. An integrated crude vacuum unit scheme is shown in Fig. 3.1.7.

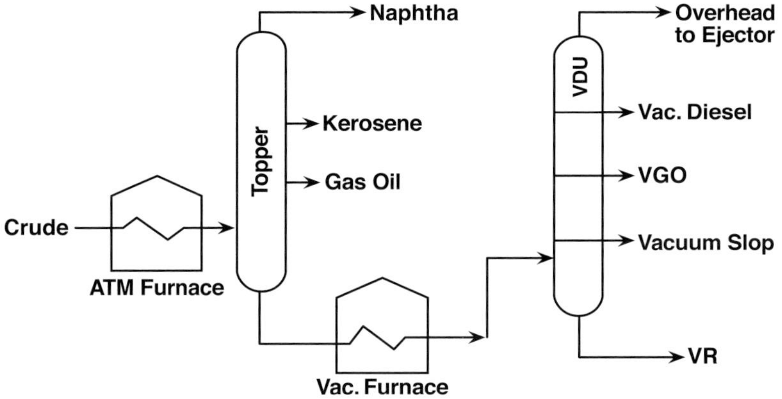

Fig. 3.1.7 Integrated crude vacuum unit scheme

Optimum furnace coil outlet temperature

Coil outlet temperature (COT) must be adequate to vaporize the distillates plus around 2%–3% of the crude, which provides wash liquid below the gas oil draw to maintain the quality and color of gas oil.

If a higher COT is maintained, pump-around refluxes and top reflux are likely to increase. Additional pump-around heat can be recovered in the preheat train through heat exchange of crude with pump-around refluxes, but some can get lost to condensers through the increased reflux; thus, a higher COT may result in a loss of heat.

There should be at least two stages below the gas oil tray. The coil outlet should be sufficient to vaporize enough crude to ensure that there is an ample amount of liquid below the bottom of the lowest wash tray or at the bottom of wash section packing bed. The internals of the wash section should be adequately irrigated to maintain the quality of gas oil.

If there are more than two or three stages in the wash, a larger amount of wash liquid is required to keep the bottommost stage irrigated. This would consequently require a higher furnace coil outlet temperature. The flow rate of wash liquid from the bottommost tray or packing of the wash section will be shown by the external overflash measurement, if available.

The coil outlet should typically provide 3%–5% of the overflash. This ensures that all the trays in the wash zone are adequately irrigated. In case of an internal overflash (no measurement), the flow of overflash can be determined using simulation of the column. Often a furnace recycle is provided, where a part of the overflash liquid is withdrawn and mixed with the furnace feed to eventually act as wash liquid.

However, for a grassroots design, a higher furnace coil outlet temperature is preferred. This necessitates provisions for higher heat removal using a heat-exchanger train and requires a higher furnace duty, which provides cushion in the surface area of the heat exchangers and furnace duty, and depicts a higher preheat in design.

Operation of crude towers at low pressure

Operating crude towers at low pressure provides flexibility for maintaining the lower coil outlet temperature and conservation of heat. In addition, due to the improvement of relative volatility at lower pressures, the tower can operate with a lower reflux. However, when the column is close to flooding, lowering the pressure may induce flooding due to the higher vapor volumetric flow.

Designers often prefer to operate crude towers at a lower pressure and deploy compressors to increase the pressure of the gas, condense or absorb the gases in the naphtha, and feed the liquid to a stabilizer in order to recover LPG and stabilize the naphtha.

The scheme below (Fig. 3.1.8) shows the operation of a crude unit at low pressure. A gas compressor is used to compress the gases from the reflux drum, which is now operating at a low pressure of around 0.5 kg/cm^2(g). The compressor compresses the gas to around 6.0 kg/cm^2(g). The compressed gas is then mixed (recontacted) with naphtha from the reflux drum and cooled through an air or water cooler to around 45°C and collected in the drum. The gas from the drum, if there is any, is routed to the fuel gas system and the liquid is fed to a stabilizer after heating. The unit may have a top pump-around to contain the top temperature and reduce condenser duty.

It may be noted that when the design is required for a unit to process a wide variety of crudes, then this may be a preferable option.

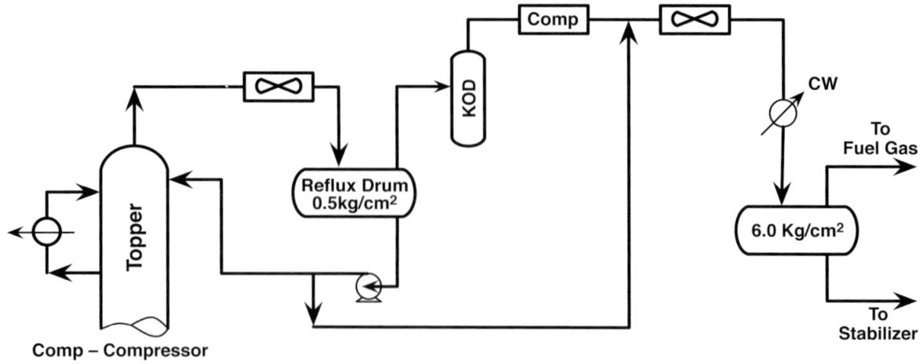

Fig. 3.1.8 Low-pressure operation using compressor

Side stripper heat optimization

Use of low-temperature heat sources for reboiler heat is shown below in Fig. 3.1.9.

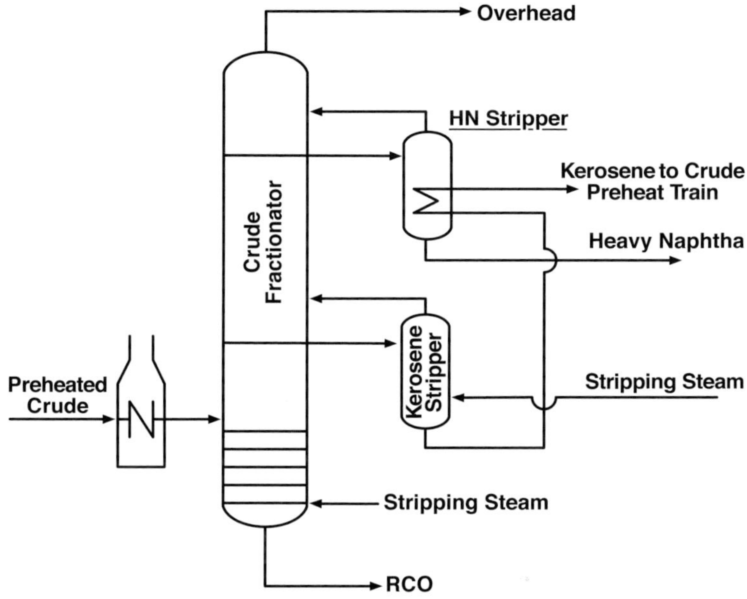

Fig. 3.1.9 Kerosene providing heat to heavy naphtha stripper reboiler

The low-temperature heat sources are normally used as reboiler fluid in heavy naphtha and kerosene strippers, particularly when the low-temperature heat requirement is not there to heat the crude up to the desalter temperature when the overhead heat is coupled with crude.

The stripping operation of light gas oil and heavy gas oil side strippers continues using stripping steam, as these components call for high-temperature heat sources as reboiler fluid.

LGO PA can be used as a reboiler heat source in a kerosene stripper, as it provides a sufficient temperature difference between the stripper liquid and the hot stream if the heat of the LGO PA is not necessary for increasing the preheat of the crude. This may be applicable in an atmospheric and vacuum integrated unit where part of the LGO PA heat can be given to the kerosene stripper reboiler without sacrificing crude preheat.

Stripping steam may be minimized in LGO and HGO side strippers. It may be noted that even if the stripping steam isn't introduced in the stripper, the stripper can still improve the flash point. A long stripper helps in attaining flash point and is preferred, especially for kerosene strippers or aircraft turbine fuel (ATF) strippers, as it requires less stripping steam or a lower reboiler heat. Thus, a long stripper is always preferred for obtaining flash point and achieving energy conservation.

Different column overhead schemes

A typical overhead scheme of a crude tower is shown in Fig. 3.1.10.

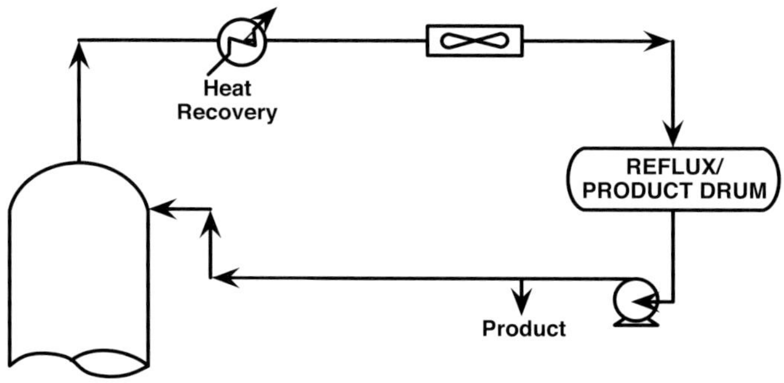

Fig. 3.1.10 Conventional column overhead system scheme

The most common overhead circuit with crude heat integration followed by air cooling is shown in Fig. 3.1.10 with a single overhead drum. The overhead of the column may or may not have integration with the crude. Often two drums are used in the overhead of a topper where overhead heat is integrated with crude. A two-drum overhead scheme is shown below in Fig. 3.1.11.

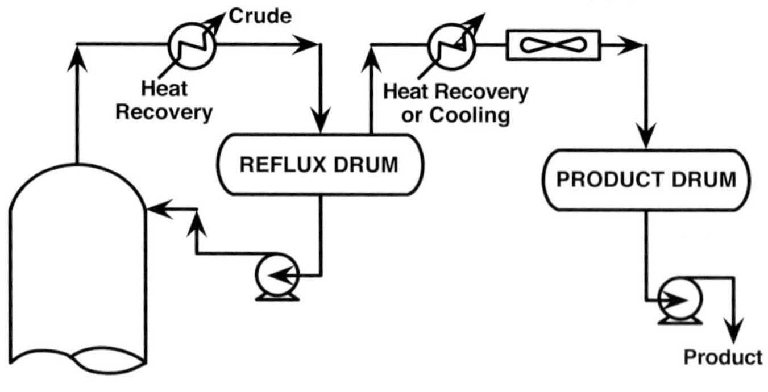

Fig. 3.1.11 Two-drum overhead system

The hot drum is typically maintained at 70°C–85°C so that the majority of the heat is recovered in overhead condensation at a higher temperature, and can be delivered to the crude. In this configuration, the reflux liquid is returned to the column from the hot drum and does not contain the lighter fraction, as it escapes the hot reflux drum and is condensed in the water cooler before being sent to the cold or product drum.

Thus, the reflux liquid containing lower proportions of lighter components revaporize and join the overhead naphtha vapor. When the lighter components in overhead vapor have been reduced, its dew point and bubble point increases. This provides a higher temperature differential between the condensing vapor and the cold crude and decreases the surface area required for heat transfer. Two drums can be also used for decreasing the required surface area for the air coolers placed at the upstream of the hot drum.

3.1.4 Steps involved in the design of a distillation column

The design of a distillation column typically involves the following steps.

1. Specify the degree of separation or set the product specifications/purities.
2. Select the operating conditions (pressure of the column, the reflux drum, etc.).
3. Determine the stages and reflux requirements.
4. Select the type of contacting device (plates or packing).
5. Size the column based on diameter, number of actual trays, etc.
6. Design the column internals (plates, distributors, packing supports, etc.).

Column top pressure

The column reflux drum pressure is determined to start with. The reflux drum pressure must be such that the bubble point of the overhead (for the composition of top product) is greater than the temperature of the cooling water by around 15°C. For example, for cooling water with an inlet temperature of 33°C, a condensate temperature of 40°C–45°C can be used, with consideration of a 10°C–15°C differential over the cooling water inlet temperature.

The reflux drum pressure needs to be such that the bubble point of the top product is 40°C–45°C at reflux drum pressure. Once the reflux drum pressure is fixed, the column top pressure is calculated by adding typically 0.5 to 0.8 kg/cm^2 to the drum pressure to account for the pressure drop in the overhead condenser and overhead line.

Column top temperature

Column top temperature is the dew point of the overhead vapor at the partial pressure of the hydrocarbons at the top. The hydrocarbon partial pressure in turn is estimated from the hydrocarbon vapor phase mole fraction at the column top multiplied by the total pressure at column top.

The manual methods are no longer used for design or rating because the heat and material balances performed to derive the vapor–liquid loading in each section are very tedious and not practical for design. Instead, steady state simulators are most often used.

3.1.5 Crude column simulation and design

Inputs required

1. Crude true boiling point (TBP) data (essential)
2. Crude density/American Petroleum Institute (API) gravity, ((141.5 – (131.5/sp. gr.)) (essential)
3. Crude molecular weight (optional)
4. Viscosity (optional)

Specifications required

1. Column pressure/reflux drum pressure
2. Fixed product draw rates or distillation points
3. Pump-around duties and return temperature or temperature differentials
4. Column top temperature or overhead product rate or distillation endpoint of overhead product

Typical number of stages in the sections of a crude unit

1. Between top naphtha and superior kerosene (SK): 5–8 stages. (It is a function of naphtha endpoint and SK flash.)
2. Between SK and diesel: 4 to 5 stages. (It is a function of the specified gap or overlap between SK and diesel.)
3. Wash section (above flash zone and lowermost draw): 2–3 stages.
4. Bottom stripping section: 2–3 stages.

Tray efficiencies and system factors to be considered for design

Pre-topper

Efficiency: 65%–70%

System factor: 0.8–0.85

Topper

Efficiency above wash zone: 60%–65%

Efficiency of wash zone: 30%–40%

Efficiency of stripping zone: 30%

System factor: 0.85–0.9

Typical simulation inputs

Thermodynamics (K-value): GS/SRK

Transport property: PETRO/SIMCI heavy oil correlation. The viscosity of crude and heavy fractions may be estimated from a mid-percent viscosity curve given along with crude TBP and can be entered in the simulation.

Density: API

Kerosene PA differential temperature (DT): 55°C–60°C

LGO PA (DT): 55°C–60°C

HGO PA (DT): 55°C–60°C

Typical reflux ratio: 1.5–1.8

DT is the differential temperature between pump-around supply and return.

Pump-around reflux heat duties and differential temperatures

Pump-around duties need to be maximized for energy efficiency. This is limited by gap or overlap specifications between adjacent products, and a minimum internal reflux between trays. The internal reflux in a section controls the separation between the cuts with a fixed number of trays in the section.

A differential temperature between pump-around supply and return temperature is typically specified as around 55°C–60°C. A larger temperature difference would lower the flow of pump-around refluxes but also reduce the differential temperatures across crude preheat exchangers, which can render the heat unavailable for the crude at higher temperature ranges.

Sometimes when the pump-around section of column is overloaded due to high flow in the section, the DT of the pump-around is increased to reduce pump-around flow, and part of the pump-around heat is passed to the lower-temperature section of the crude preheat train and may result in an inefficient design.

Location of pump-around refluxes

Pump-around refluxes can cool the vapors and condense them to be drawn as products from below the pump-around return. A pump-around intended to condense a product can be drawn from the same tray (Type B in Fig. 3.1.12), then cooled and typically returned three trays above. Some amount of internal reflux typically joins the pump-around return from the top and condenses the cuts below.

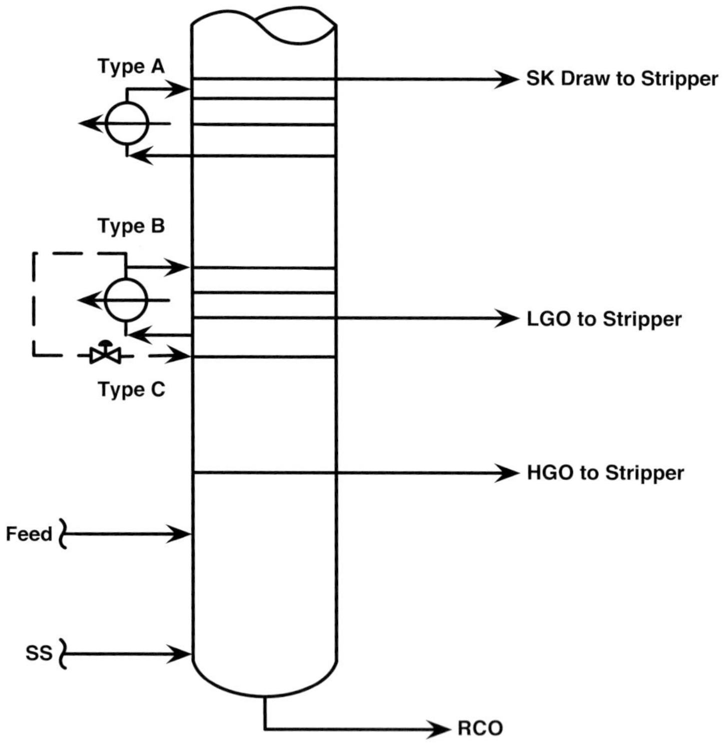

Fig. 3.1.12 Different arrangements of pump-around refluxes

A pump-around can also be located above the draw tray below the upper side cut. In this case, the pump-around will generate higher internal reflux to flow down the stages to the draw tray (Type A in Fig. 3.1.12). This arrangement helps create a better separation of the adjacent cuts. Separation is a function of the internal reflux multiplied by the stages in the section; thus, separation between the cut and the cut above improves in this arrangement. However, if the pump-around is placed at an upper location, the pump-around draw temperature reduces and the quality of pump-around heat becomes inferior. This may affect the heat recovery and thermal efficiency. A combination of both types can also be practiced, if sufficient stages are available in the section.

A pump-around can also be configured as a conservative intercooler (Type C in Fig. 3.1.12). In this situation, a part of the pump-around can be cooled and then returned below the draw tray to cool and condense the products located below. This has been practiced by the author in one design when the pump-around section was loaded heavily and there was no scope to increase it further.

Typical stabilizer design parameters

Typical number of actual trays in the stabilizer: 44–50

Reflux ratio: 2–2.5

Efficiency of trays: 65%–70%

System factor: 0.9

Thermodynamics, VLE (K-value): PR, SRK, GS

Transport property: PETRO

Density: API

Typical naphtha splitter design parameters

Typical number of actual trays in splitter: 40–60

Reflux ratio: 1.5–3.5

Efficiency of trays: 65%–70%.

Thermodynamics (K-value): GS

Transport property: PETRO

Density: API

It is better not to use black oil as a heat source for the stabilizer or the reboiler of naphtha splitters, as leakage in the reboilers can contaminate the naphtha and CRU/ISOM feed deploying the noble metal catalyst. The catalyst can be poisoned if leakage does occur.

Water dew point calculation of a crude column

Overhead product is D kg/hr

Reflux rate is R kg/hr

Total overhead vapor $V = (D + R)$ kg/hr

Kg moles of HC vapor at the top $(M_{HC}) = (D + R)/\text{molecular wt}$

Flow of stripping steam = $S1$ kg/hr (including side strippers)

Molecular weight of top naphtha typically ranges from 90–100

Water along with crude = $S2$ kg/hr (from desalter)

1% wt of water along with reflux = $(1 \times R)/100 = S3$ kg/hr

Total water at the top of column (S kg/hr) = $S1 + S2 + S3$

Moles of water at column top (M_{H_2O}) = $S/18$ kmoles/hr

Mole fraction of water = $X_{H_2O} = M_{H_2O}/(M_{HC} + M_{H_2O})$

Partial pressure of water at top (p_{H_2O}) = $X_{H_2O} \times P_{top}$

P_{top} = top pressure in kg/cm^2 abs

At the dew point, the partial pressure of water at the top is equal to the vapor pressure or saturation pressure of water at that temperature. The steam table can be referred to and the water saturation temperature corresponding to p_{H_2O} can be found. The saturation temperature is the water dew point at the column top. The top temperature should be kept above the water dew point by around 6°C–10°C to avoid water condensation and corrosion at the upper section of the column.

If there is a top pump-around in the column, it should return to the column top with a coalescer for water removal. Otherwise, heavy corrosion may result at the uppermost part of the column. Usually the top section of the column is Monel-lined to avoid corrosion.

3.1.6 Chloride corrosion of overhead systems

Crudes contain salt typically 10–20 part per thousand barrels (ptb). So the crude is desalted in desalters to reduce the salt content to 1–3 ptb through a single stage desalter. However, it can be reduced further to less than 1 ptb by two-stage desalting.

The salts hydrolyze to form hydrogen chloride using the following mechanisms while heated.

$MgCl_2 + 2H_2O \rightarrow Mg(OH)_2 + 2HCl$ (highest)

$CaCl_2 + 2H_2O \rightarrow Ca(OH)_2 + 2HCl$ (moderate)

$NaCl + H_2O \rightarrow NaOH + HCl$ (minimum)

The hydrolysis of NaCl is the least, $MgCl_2$ is the highest, and $CaCl_2$ is moderate.

H_2S is an almost insoluble gas at the overhead dew point temperatures, but it comes in contact with condensed water that has iron ions. Those iron ions in contact with H_2S in the gas phase will form iron sulfide (FeS) that will precipitate and release additional acid, increasing the occurrence of corrosion by the following reactions:

$2HCl + Fe + 2H_2O \rightleftharpoons FeCl_2 + 2H_2O + H_2$

$FeCl_2$ (aqueous) + H_2S (gas) $\rightleftharpoons$ FeS + 2HCl

FeS becomes a major component of the deposit of corrosion product.

H_2S in a crude unit is produced in a small amount due to inadvertent cracking in the furnace. Some crudes contain appreciable amounts of H_2S. H_2S is often introduced into the crude unit when slop (containing H_2S) is processed, which accelerates hydrochloric acid corrosion at the overhead.

Neutralizing amines or ammonia (along with filming amines) are commonly used in the overhead circuit of the column to arrest the corrosion of the overhead. However, ammonia may not be considered a good neutralizing agent as it cannot neutralize HCl in an aqueous phase due to its poor solubility in water at the overhead dew point temperature. Traditionally, NaOH was injected in the downstream of the desalter in a crude unit to convert $MgCl_2$ and $CaCl_2$ to the more stable (less hydrolysable) NaCl and to reduce the liberation of HCl in order to avoid corrosion in the overhead. In more recent days, however, this practice is being discontinued, as it has been found that the NaOH injected and NaCl produced may have a detrimental effect on the secondary units like the coker, visbreaker, residue hydrotreater, etc.

Perils of high top PA duty and very low reflux ratio in crude units

Top PA is often introduced in crude units to efficiently capture part of the overhead heat and deliver it to the crude. With the introduction of top PA, the reflux ratio of the column is reduced substantially.

The following example was observed in a crude unit operating at a reflux ratio of 0.3–0.5 with a top PA compensating the top reflux. Severe corrosion was noticed in the overhead air cooler with a deposit of iron sulfide in the air-cooler pass partition plates and corrosion products blocking the air cooler tubes. This resulted in no flow through some of the banks of the air cooler and an increased velocity through the tubes that did permit flow. This in turn dislodged the protective FeS film in the air cooler, exposing the metal to a fresh HCl attack.

Some amount of H_2S is always generated from sour crudes during the process of heating in the furnace. If the reflux ratio or reflux flow is less then H_2S partial pressure will increase in overhead vapor as a result of lower hydrocarbon partial pressure caused by the lower reflux ratio.

A high top PA duty may result in the following effects.

1. Partial pressure of water increases at the top
2. Partial pressure of H_2S increases at the top

An increase of water dew point at the top increases the likelihood of water condensation and water dew point corrosion at the top. This can be mitigated by increasing the top temperature or better, by decreasing the stripping steam using reboilers or compensating with a higher COT.

Increase in the partial pressure of H_2S may be dangerous, as H_2S is a known chloride corrosion accelerator. H_2S in its vapor phase reacts with the primary corrosion product of HCl (i.e., $FeCl_2$), produces FeS, and regenerates HCl, enabling the acid to attack the fresh iron piping and overhead equipments.

This problem was resolved by the following actions:

1. Lowering the top PA to increase the reflux ratio to 1.0–1.2.
2. Lowering the column pressure by bleeding some gas from the drum to the FCCU gas compressor suction. The bleeding of the gas will result in H_2S bleeding from the system, which will make the overhead liquid less rich in H_2S. Reduction of total pressure will also reduce the partial pressure of H_2S.

The problem was greatly reduced and all of the old or corroded parts of the air cooler banks were replaced.

So, a high top PA transfers low-temperature heat to the crude and reduces top reflux significantly, but can also cause too great a drop in the reflux ratio, which can be detrimental. However, a high top PA can still be used if the column is operated at a lower pressure, and a compressor is used to suck up the gas and route it for recontacting with naphtha at a higher pressure or to the LPG recovery section. At a low column pressure, the partial pressure of H_2S goes down and the H_2S-accelerated HCl corrosion reduces.

However, providing a top pump-around is the cheapest and least cumbersome means of recovering part of the overhead heat. It is always preferred over introducing an overhead vapor and crude exchanger.

Problems with overhead vapor integration in crude unit

Typically, for delivering overhead heat to the crude, a partial or full crude flow is taken to the condenser platform to exchange heat with the overhead vapor. The crude is heated to around 50°C–60°C. The overhead vapor is further cooled and condensed in the air cooler or water cooler, then routed to the reflux drum. The crude side pressure is normally very high (around 35– 40 kg/cm^2), while the overhead side pressure remains low (0.6–1.7 kg/cm^2) depending on the column operating pressure. Thus, the differential pressure is very high and the environment is corrosive, making the likelihood of a leak in these exchangers very high. In case of a tube or floating head gasket leak, the crude will contaminate the naphtha, which may in turn contaminate the CRU/CCRU feed.

To reduce the likelihood of leakage, the overhead and crude heat exchanger can be installed at the outlet of the offsite booster pump in the unit (which has a low crude side-pressure of 6–10 kg/cm^2) and therefore leads to less differential pressure betweenthe crude and overhead vapor side).

The crude/overhead heat exchanger can be located upstream of the unit crude booster pump to reduce the likelihood of crude leaking into the naphtha. However, at this location, temperature at the suction of the unit crude booster pump will be higher, but the pressure available at suction will keep the crude in full liquid state and avoid cavitations of the unit crude booster pump.

3.1.7 Maximization of crude preheat

Already, many studies, papers, and reviews have been published on this topic. This section will bring out some specific aspects of what can occur during the design, revamp, and operation of crude units.

It is often felt that a good heat exchanger synthesis through a pinch only increases preheat. The first thing required when carrying out a network synthesis or a pinch study is the heat source table. The heat source table is developed from the results of simulation. The heat source table of an actual atmospheric vacuum-type unit with a pre-flash drum is shown in Table 3.1.1. A hot composite, cold composite and grand composite curve may be the next step after a simulation is run and a heat source table is constructed from the results of the simulation. The hot and cold composites and grand composite can be developed including the topper overhead heat as available heat.

In a crude unit, cold streams are typically crude, demineralized water (DMW) for desalting, stabilizer feed, stabilizer reboiler heat, stripper reboiler heat, etc. The cold stream flows along with their heat capacities are used for construction of the cold composite curve.

The true composite curve (TCC), grand composite curve (GCC) and shifted composite curve (SCC) are presented in Figs. 3.1.13, 3.1.14, and 3.1.15 respectively. The GCC shown is not extended beyond the achievable preheat (upper open jaw not shown in GCC). The achievable preheat is read from TCC as 305°C, and the pinch temperature can be read as 285°C from SCC or GCC. The open lower jaw of GCC shows the possibility of steam generation equivalent to around 20 MW at 200°C.

The simulation plays the biggest role in finalizing the heat source table—the hot streams, their terminal temperatures and their (mass flow rate) × (heat capacity) values.

During simulation, one can increase the pump-around duties limited by the internal reflux in the zone above the pump-around return. Preference is given for an increase of heat duty in the high-temperature pump-around. The internal reflux (IR) must remain above the minimum to maintain the efficiency of the trays between the topmost side cut (kerosene or heavy naphtha) with top naphtha.

Table 3.1.1 Heat source table for an atmospheric–vacuum unit (AVU)

Heat Source Table				
Stream name	Flow (MT/hr)	T_{in} (°C)	T_{out} (°C)	Q (MMkcal/hr)
Crude	937.50	30.00		
Pre-flash Drum				
Feed crude	937.50	154.00		
Pre-flash vapor	13.60	154.00		
Bottom product	923.90	154.00		
Crude Column				
Feed (PF bottom)	923.90	375.00		
Feed (PF vapor)	13.60	154.00		
Topper naphtha	106.90	56.00		
Heavy naphtha	20.85	179.00		
ATF/kero	145.20	226.00		
LGO	114.20	264.00		
HGO	49.30	323.00		
RCO	501.50	362.40		
SK CR	654.70	201.10	135.10	25.90
LGO CR	495.70	282.80	216.80	21.30
HGO CR	144.20	337.30	271.30	6.50
Q_c/(reflux ratio)				32.3/(1.03)
Stabilizer Column				
Stabilizer feed	106.90	120.00		
LPG	12.10	45.00		
Stabilized naphtha	94.80	153.00		
Q_c/(reflux ratio)				5.25 (4.1)
Q_R				6.60
Vacuum Column				
Feed (RCO)	501.50	410.00		
Vac. diesel	270.70	111.50		
LVGO	100.00	248.00		
HVGO	101.70	309.50		
Slop	20.00	366.50		
VR	250.70	374.00		
Vac. diesel CR	403.50	111.50	53.50	11.64
LVGO PA	114.00	248.20	182.20	4.60
HVGO PA	450.00	309.50	243.50	19.40
Atm. column furnace duty (absorbed) (with 295°C preheat and 375°C (COT))				66.30

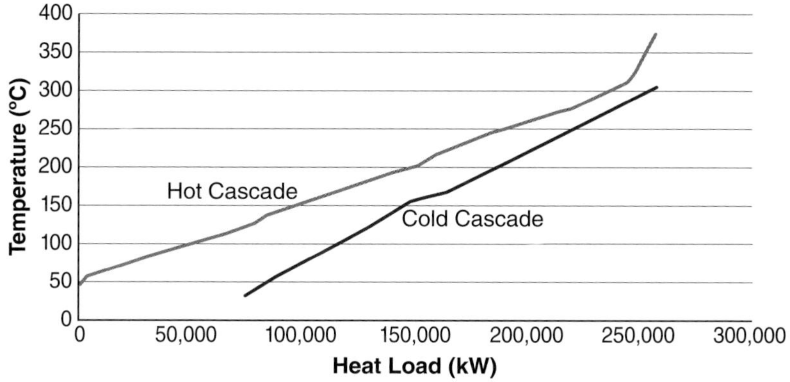

Fig. 3.1.13 True composite curves

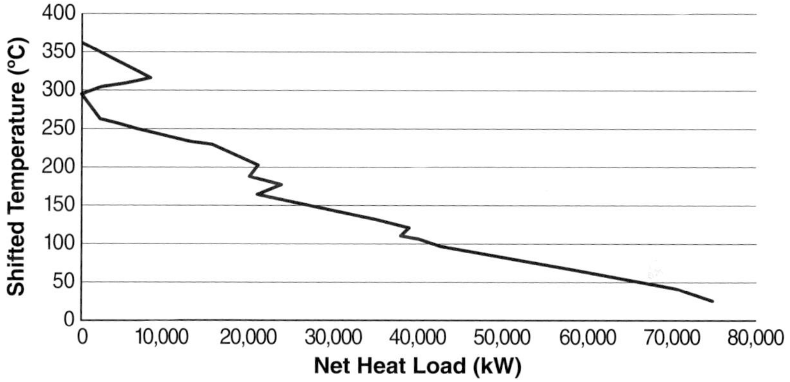

Fig. 3.1.14 Grand composite curve

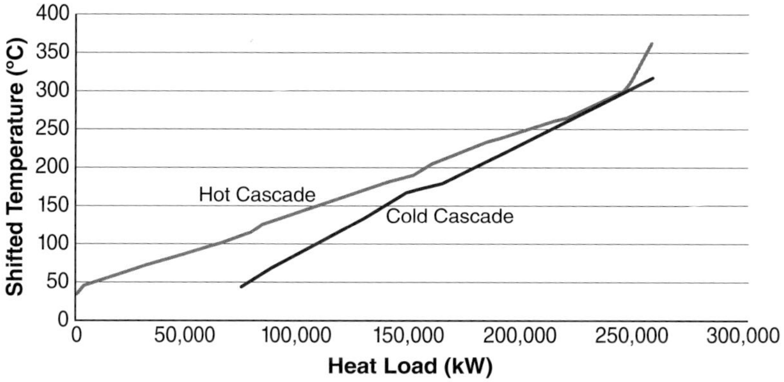

Fig. 3.1.15 Shifted composite curve

Further, a balance must be reached between the stages separating adjacent cuts and the internal reflux in the section in order to achieve optimum design. Drawing of products/pump-around refluxes at the highest temperatures is always preferred.

In a typical crude column, there is one overhead product and two to three side cuts with strippers. Two possible side cuts are kerosene/ATF (140°C–250°C/270°C), and gas oil (270°C–370°C). There can be two pump-around refluxes, usually kerosene PA and gas oil PA. The gas oil can be split as two separate cuts (LGO and HGO) from two locations with an individual pump-around. The typical draw temperature of the LGO is approximately 295°C and that of the HGO draw tray is around 330°C, as opposed to the approximately 303°C of common gas oil draws. The heat of the HGO PA and HGO draws are at higher temperatures than the LGO PA and LGO draws.

At this point, it would be prudent to maximize the HGO PA and HGO draw. The last two exchangers can be RCO and HGO PA in a parallel arrangement in a crude distillation unit to deliver maximum heat recovery with minimum surface area. A parallel arrangement scheme of HGO PA and RCO in a heat-exchanger train is shown in Fig. 3.1.16.

An HGO PA heat exchanger placed parallel to an RCO heat exchanger will have a high differential temperature across the crude vs. RCO heat exchanger.

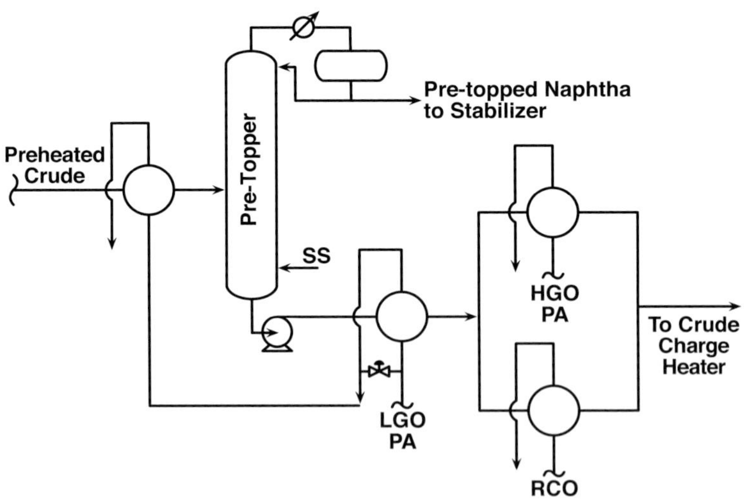

Fig. 3.1.16 Parallel arrangement of HGO PA and RCO in an HE train

3.1.8 Typical heater pressure drop in crude vacuum furnaces

1. Crude heater: 8–12 kg/cm^2
2. Vacuum heater: 5–7 kg/cm^2. Vacuum heater transfer line to column is usually designed for a pressure drop of around 150 mm Hg to lower the liquid film temperature at the furnace outlet with a maximum vapor velocity of 0.6–0.7 times the sonic velocity.

The pressure drop in a furnace can be estimated by tailored commercial software or by a program based on a steady state simulator. Pipe sections with a diameter equal to that of heater tubes of equivalent length must be considered, along with inlet and outlet temperatures, as in the main furnace simulation. This is discussed in detail in Section 13.4 of Chapter 13.

Side stripper installation practices

The liquid products drawn from a column must enter the stripper with a liquid seal (around 1,200–1,500 mm) to avoid the backup of stripper vapors in the main column through the liquid draw line. Typically, the liquid seal is automatically incorporated with installation of stripper level control valve piping (forming a U loop) in the product draws from the column to the side stripper. Typically, a side stripper is located at least 4 meters below the draw of the main column tray sump in order to provide the necessary elevation difference to compensate for the pressure loss due to flow through the draw-off line and control valve. A practical guideline for the installation of stripper is shown in Fig. 3.1.17.

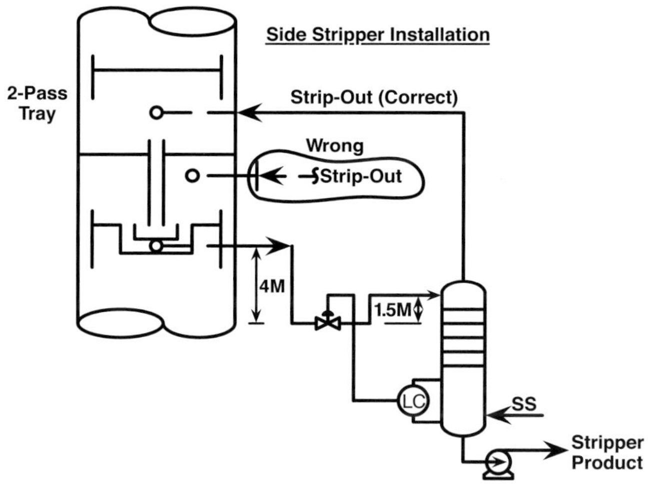

Fig. 3.1.17 Side stripper installation showing strip-out connection

3.1.9 Capacity revamp of crude units (methodologies followed)

A pre-flash drum can be added to decrease furnace mass velocity and pressure drop. The pre-flash drum inlet temperature can be increased to provide more relief.

Pre-topper can be added to decrease the mass velocity and pressure drop through the tubes of the crude charge furnace, resolve the limitation of the topper and stabilizer diameters, and resolve the limitation of the topper overhead condenser duty. Pre-topper inlet temperature can be increased to draw more distillate from the pre-topper and extend additional relief to the topper. If required, a side naphtha cut can be drawn with the side stripper to provide even higher relief.

In one revamp, part of the kerosene cut was also drawn from the pre-topper with a side stripper by increasing the pre-topper inlet temperature. The pre-topper kerosene and topper kerosene blended to constitute the total kerosene pool. However, these maneuvers can be attempted only for revamp applications.

If the furnace heat duty is limiting while augmenting crude capacity, then a parallel furnace can be added for additional flow and heat duty. A similar configuration has been adopted in two refinery units. Product splitting and pump-around splitting can be done if the diameter of the column in the section is limiting the flow required for delivering the duty.

A single gas oil draw with a single pump-around can be split into two gas oil draws (LGO and HGO) with their own pump-around refluxes and side strippers.

A single kerosene draw can be converted and split into a lighter kerosene draw and a heavy kerosene draw with individual pump-around refluxes and side strippers. However, these solutions can only be practiced if the column has an adequate number of trays and tolerates the degree of separation between two adjacent cuts. Packing or high-capacity trays can be used in toppers to resolve diameter limitations.

Stabilizer diameter limitation can be circumvented by installing a pre-topper and using only pre-topped naphtha as stabilizer feed. If the top section of the stabilizer is loaded, the feed temperature can be decreased and compensated for by a higher reboiler duty. If the bottom section is loaded, the feed temperature can be increased and reboiler duty can be reduced to give respite in bottom section loading without affecting LPG recovery. High capacity trays can be used for resolving diameter limitation to some extent. Staged reboilers can also be used. Parallel reboilers in the stabilizer can be used to augment reboiler duty.

In a naphtha splitter, stabilized pre-topped naphtha and topper naphtha need to be fed in at different locations in the naphtha splitter. Stabilized pre-topped naphtha can be fed in at an upper location and preheated topper naphtha at a comparatively lower location. This reduces reboiler duty, condenser duty, and reflux ratio, and relieves the column. A naphtha splitter reboiler duty limitation can be resolved by increasing the feed temperature. However, this will also increase the condenser duty. The capacity of a naphtha splitter (used to prepare ISOM feed and CCRU feed) with a furnace for a reboiler was augmented by 40% when the feed temperature was increased from 90°C to 125°C by installing a feed preheat heat exchanger that used low-pressure steam. An additional condenser was also installed.

The installation of a feed preheater resolved the limitation of the reboiler furnace at a higher capacity. The concept of a two-enthalpy feed can also be used in a stabilizer or naphtha splitter to reduce the reboiler and condenser duties and the reflux ratio (example given in Example Problem 1.6.1 in section 1.6 of Chapter 1).

3.1.10 Capacity revamp case studies

Case Study 1

The unit had a pre-topper with a fired reboiler. The bottom of the pre-topper was pumped through a charge heater to the topper. It was found that the reflux ratio in the pre-topper was higher than required. The configuration of the unit was changed with the following objectives.

1. Increase the capacity of the unit to 3 million metric tonnes per annum (MMTPA)
2. Introduce flexibility to process 50:50 heavy Indian origin NG crude: Bonny light or equivalent.

In the revamped configuration, the reboiler was removed from the pre-topper. The flash zone to the pre-topper was moved down, leaving 8 trays at the bottom, and the stripping steam was introduced in the pre-topper. A third preheat train was introduced to heat the pre-topped crude. The original pre-topper reboiler furnace (single-pass with a duty of 8 Gcal/hr) was used in conjunction with the existing charge heater (16 Gcal/hr), which had two passes, and a third new furnace (with an absorbed duty of 12 Gcal/hr) was added in parallel, and the combination used as a charge heater.

The new unit with the NG crude provided a preheat of 290°C as compared to the 245°C maximum achieved with the NG crude before revamp. Capacity reached 3 MMTPA and the unit was operating satisfactorily. The pre-topper inlet temperature was increased to 235°C–240°C to increase the vaporization in the pre-topper. Temperature control was provided to maintain pre-topper inlet temperature and to control reflux ratio.

The original configuration is shown in Fig. 3.1.18 and the revamped configuration is shown in Figs. 3.1.19 and 3.1.20.

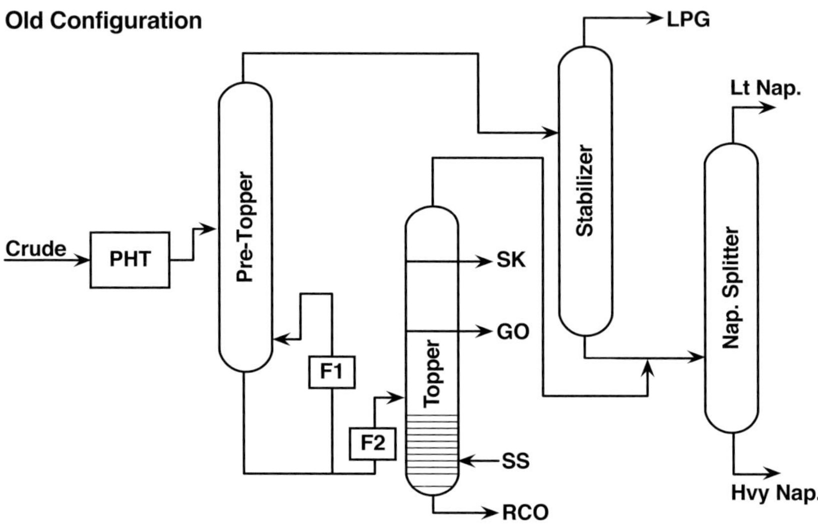

Fig. 3.1.18 Original configuration

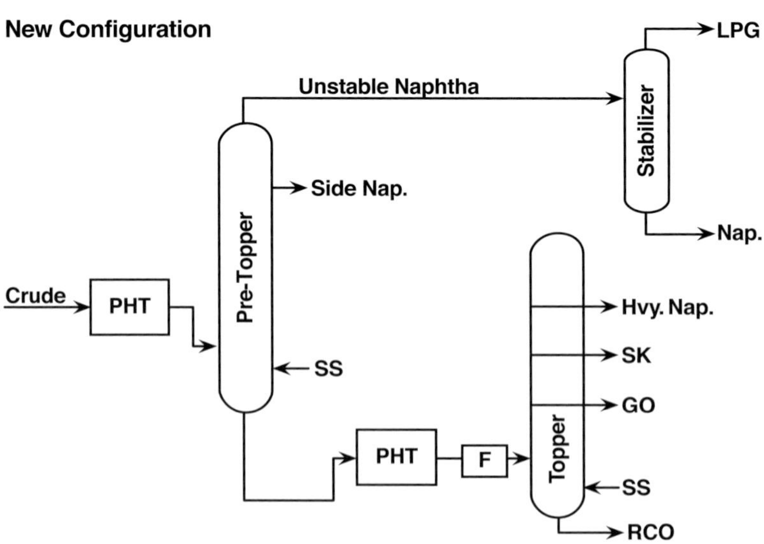

Fig. 3.1.19 Revamped configuration

A new charge furnace was constituted with the original charge furnace (F2), original reboiler furnace (F1), and new furnace (F3) all in parallel. All three furnaces had different duties and a different number of flow passes (see Fig. 3.1.20). The flows through each furnace were adjusted based on the duty available. The mass velocity and pressure drop were found to be in order for the flow and tube pass available in each furnace.

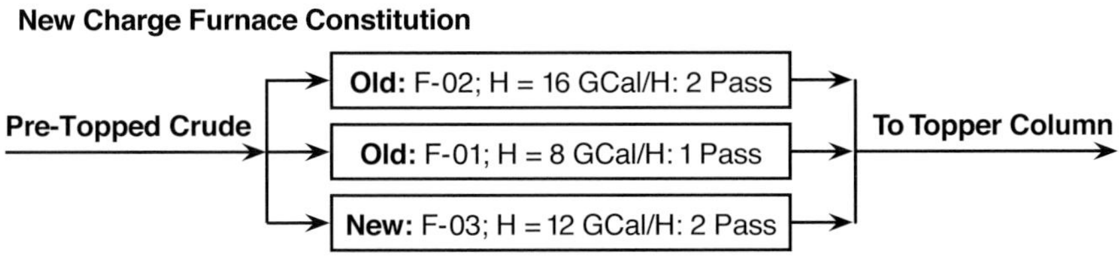

Fig. 3.1.20 Constitution of crude charge furnace using three unequal duty furnaces

It was observed that the stabilizer bottom section was flooding with the required reboiler duty. To resolve this problem, the stabilizer feed inlet temperature was increased from 90 to 115°C in the revamp design, and the reboiler duty was reduced, which resolved the problem.

Case Study 2

A similar approach was adopted when revamping another crude unit. The furnace heat duty was limited to increase the capacity by around 60%. One balancing furnace was added in parallel to the existing furnace (see Fig. 3.1.21). An old pre-topper was also added to debottleneck the fractionator and stabilizer.

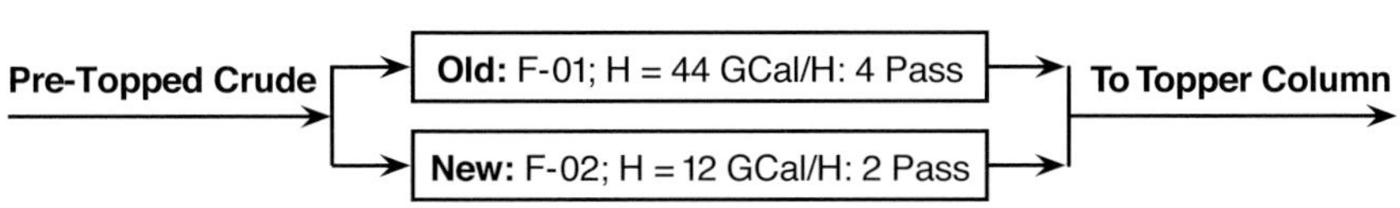

Fig. 3.1.21 Constitution of a crude charge furnace using two unequal duty furnaces

The old pre-topper bottom residence time was found to be less than required for the desired capacity. An extra vessel was added downstream to the pre-topper to increase the pre-topped crude residence time before feeding it into the pump, the downstream preheat train, and the charge furnace. Residence time at pre-topper bottom sump is very critical, as the product is fed to the pump, which then feeds it to the furnace. Any problem in the pump due to cavitation may lower the flow through the furnace, which greatly increases the likelihood of coking in the furnace.

In the unit, the stabilizer feed inlet temperature was reduced from 110°C to 90°C in an attempt to resolve the problem of flooding in the section above the feed tray, as the stripping section had some cushion. A parallel reboiler using medium pressure steam was also added to the stabilizer to compensate and supplement the reboiler heat requirement.

Case Study 3

A revamp design was carried out to increase the processing capacity of a crude unit from 3 MMTPA to 7 MMTPA on Kuwait crude. In the original configuration, there was a single topper column and a stabilizer. The unit did not have a pre-flash drum.

The topper had the following draws and pump-around refluxes:

1. Heavy naphtha cut
2. Kerosene/ATF cut
3. Gas oil cut
4. Top PA, kerosene PA, and gas oil PA

The column had 55 trays, and the widest section had a diameter of 6 meters (more than required for a 3 MMTPA unit). The stabilizer also had some cushion in terms of its diameter. The existing configuration of the crude column is given in Fig. 3.1.22.

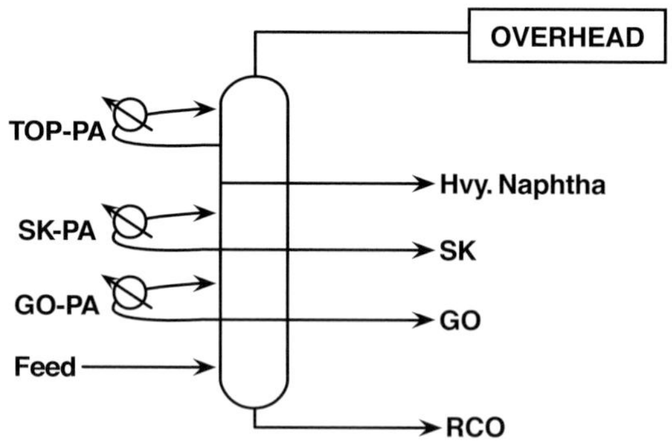

Fig. 3.1.22 Original configuration with 3 pump-around refluxes

List of modifications introduced

1. Additional kerosene cut withdrawal was introduced without a side stripper below the existing kerosene draw.
2. Additional kerosene pump-around was introduced.
3. Additional HGO draw was introduced with an additional side stripper.
4. Additional HGO pump-around was introduced.
5. Top pump-around was removed.
6. Overhead condensers were augmented.

The same column thus could be used, splitting the pump-around refluxes and product draws as per the above arrangement. The old heat-exchanger train of 3.0 MMTPA has been retained by allocating the pump-around refluxes and products of similar flows in the train.

A new parallel train was added to take the heat duties of the additional pump-around refluxes and products. The RCO stream was split into two streams. One stream was routed to the old train so that no hydraulic or surface area limitation would occur in the old heat exchangers, and the balance of the RCO was routed to the new heat-exchanger train.

Often, when pump-around duties need to be increased with an increase in throughput, but are limited by the diameter of the column, an additional pump-around can be configured instead of just one. That is the methodology adopted in this particular revamp. There was a requirement for the production of linear alkylbenzene feedstock (LABFS) from the unit. That requirement was fulfilled with the additional SK2 draw (without stripper), which was necessary to restrict the diameter of the column. The revised configuration is shown in Fig. 3.1.23.

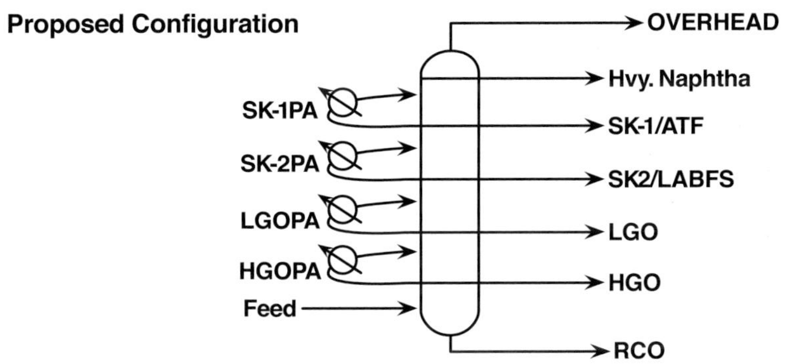

Fig. 3.1.23 Modified configuration with four pump-around refluxes

Case Study 4

One very old crude unit was in operation overseas. It had a pre-flash drum and a unique configuration, with the naphtha splitter located upstream of the stabilizer.

The off-gas from the reflux drum of the crude column was being compressed and pumped to the fuel gas system. The refinery was finding it difficult to maintain the compressor, as spare parts were not available. In addition, it was not initially understood what the compressor was pumping.

Later, it was understood that gas might be coming from the naphtha splitter overhead reflux drum as the naphtha was fractionated into the C5–90°C cut as the top product and into the 90°C–FBP cut as the bottom product going to the CRU (before stabilization). The naphtha splitter pressure was only 0.8–0.9 kg/cm^2(g), so it was inferred that due to there being less naphtha at the top of the naphtha splitter, the LPG components were being released from the reflux drum of the naphtha splitter.

Upon running a simulation, it was found that no gas was likely to bleed at the operating pressure of the crude column with a reflux drum temperature of 45°C. Later, the unit was simulated again, and it was found that the main column receiver operating at 1.5 kg/cm^2(g) might bleed at times depending upon the reflux drum temperature.

With the original configuration (Fig. 3.1.24), the gas and LPG components were released mostly from the naphtha splitter top, compressed, then sent to the fuel gas system. The pre-flash drum was operating at a low temperature of around 90°C due to low heat pickup at the upstream, which was leading to low vaporization of hydrocarbons. Thus, mass velocity through the furnace was at its limit. Consequently, the pressure drop through furnace was high and further increase in capacity was not possible. It became difficult to convince the refinery that the stabilizer should precede the naphtha splitter, and that the current location of the stabilizer was to blame for the loss of the valuable LPG components from crude to the fuel gas. Further, it was not possible to increase the operating pressure of the main column, which was further limited by the design pressure of the column and the safety valve setting.

The refinery also wanted to improve the kerosene flash point, raising it to around 50°C.

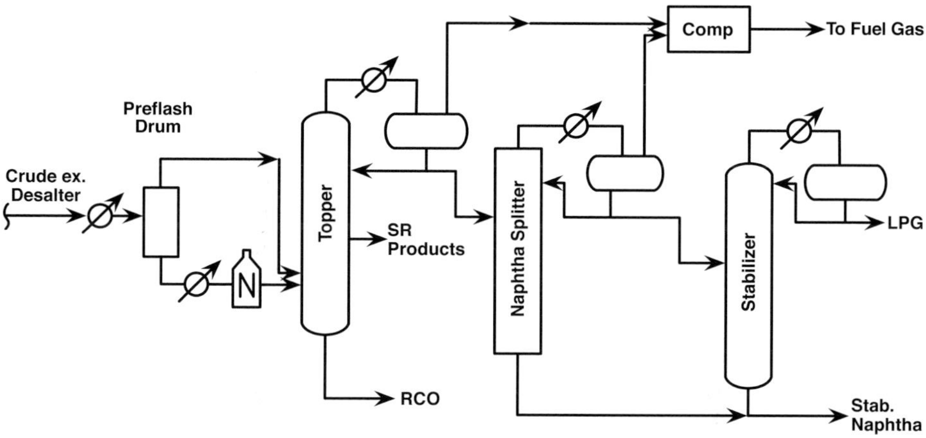

Fig. 3.1.24 Original configuration with naphtha splitter ahead of stabilizer

The solutions determined and offered are given below:

- The kerosene draw tray was relocated to the LGO draw location. Thus, a higher number of trays (15) became available between the top and kerosene withdrawal trays.
- The LGO PA was converted to kerosene PA, and the HGO draw and HGO PA were converted to combined gas oil draw and gas oil pump-around. Pump-around duties increased.
- The heat-exchanger train was reoriented to obtain a pre-flash drum temperature of around 210°C (primarily from low-temperature heat sources).
- The pre-flash drum temperature was increased to cause approximately 8%–10% vaporization in the pre-flash drum, which in turn reduced the quantity of pre-flashed crude. The lower quantity of pre-flashed crude flowed through the heat-exchanger train to the furnace, causing a reduction in mass velocity and pressure drop in the furnace, which allowed capacity augmentation.
- The stabilizer was placed upstream of the naphtha splitter and the compressor was eliminated. The feed preheat was increased and, with the available heat duty, capacity increased by around 30%.

A redundant section with a naphtha stabilizer and splitter was refurbished to stabilize the naphtha and recover around 90% of the LPG potential. The majority of the IBP–140°C cut was routed to a redundant stabilizer followed by naphtha splitter unit to generate CRU feed. The remainder of the IBP–140°C cut was routed to the existing small stabilizer and around 10% of the stabilized naphtha cut was recirculated to the upstream of the crude column's overhead air cooler, then back to the reflux drum to reduce gas bleeding from the crude column. The stabilized naphtha recirculated to the main column overhead receiver dissolved the small amount of gas that would have otherwise bled occasionally.

Case Study 5

A crude unit was originally designed without a pre-topper and had a stabilizer to produce LPG. The LPG was further fractionated into C3 and C4 to produce the propane required for its propane de-asphalting unit. The propane produced was then deethanized using a deethanizer. The unit had a submergence loop in the topper condenser outlet line to subcool the overhead and minimize LPG loss (see Fig. 3.1.25).

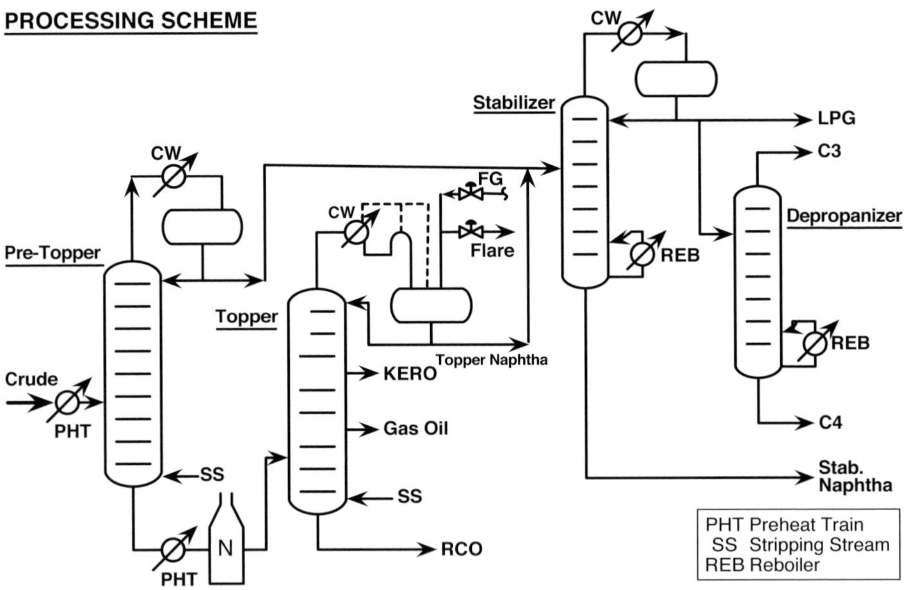

Fig. 3.1.25 Original configuration with condenser subcooling loop

A pre-topper was later added ahead of the topper for capacity augmentation, but the submergence in the topper overhead condenser was not removed. The topper's pressure continued to be controlled using fuel gas makeup and bleed to flare (split range control). The unit was studied for further capacity revamp. It was found that the stabilizer, depropanizer, and deethanizer were approaching their flooding limits in the old unit, thus no further capacity increase was possible. It was also found that due to the subcooling in the topper condenser, an appreciable amount of fuel gas was being dissolved in the topper condensate and when fed to the stabilizer to stabilize the naphtha, it was partly getting into the LPG and, in turn, loading the downstream depropanizer and deethanizer.

To solve the problem, the topper condenser submergence loop was removed and converted to a bubble condenser (with or without a submergence control valve) on the condensate line while the reflux drum liquid temperature was increased to maintain pressure. The fuel gas dissolving in topper naphtha was reduced. Routing of the topper naphtha to the stabilizer stopped, and only pre-topper naphtha continued as a feed to the stabilizer. This reduced the load in the stabilizer. The fuel gas components in LPG were reduced and this relieved the depropanizer and deethanizer. Thus, the stabilizer, depropanizer, and deethanizer were relieved, and a capacity increase was made possible.

Case Study 6

A unique overhead integration is presented in Fig. 3.1.26. Typically crude and overhead heat exchangers are provided at the upstream of the air cooler to provide the advantage of higher differential temperatures in the crude and overhead heat exchanger.

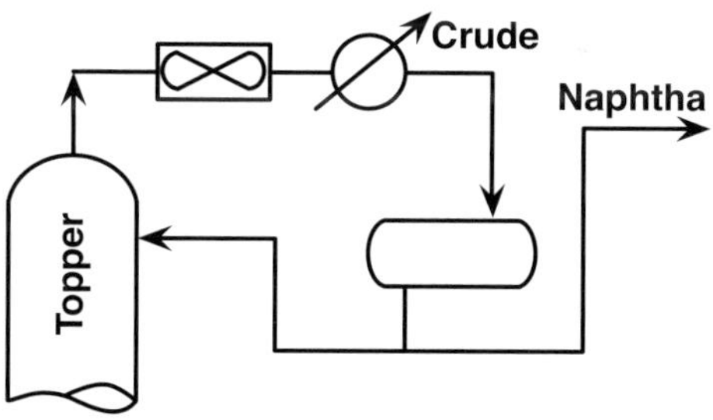

Fig. 3.1.26 Crude preheat exchanger at downstream to air cooler

In one crude vacuum distillation unit, the overhead air cooler heat duty of the topper was limiting, and the air cooler outlet temperature was around 95°C. The unit already had a pre-topper. The high reflux drum temperature was causing a large quantity of naphtha vapors to bleed to the flare.

To solve the problem, a part of the cold crude was used to cool the overhead vapor at the downstream of the topper overhead air cooler (Fig. 3.1.26). This lowered the reflux drum temperature and also increased the preheat of the crude. The crude preheat heat exchanger was not installed upstream of the air cooler, as this would lower the differential temperature across the air cooler and might further derate the air cooler.

Case Study 7

There is always a drive to reduce stripping steam to conserve energy in the distillation units. Often people reduce stripping steam in the main column or strippers and heat input to reboilers in the side strippers. This should be practiced with care and caution, as it can lead to a loss of LPG components along with lighter side cuts (products).

The author has experienced this problem in a crude unit with a capacity of 6 MMTPA with a top pump-around and a heavy naphtha draw through the side stripper.

The LPG yield was lower as compared to other units processing the same crude. It was found that the main column stripping steam was low on Kuwait crude (around 4–5 MT/hr). The conditions were simulated and it was found that the LPG components were slipping with the heavy naphtha (120°C–140°C) draw and kerosene product. The IBP of the kerosene product drawn from the stripper was low, and it was fed to the mineral turpentine oil (MTO) splitter column. The finished kerosene product flash point was improving in the MTO column located downstream for the production of MTO as a top product. The bottom product of the MTO column was finished kerosene product. The main column bottom stripping steam was increased to 8 MT/hr and the heavy naphtha stripper reboiler heat input was also increased. Heavy naphtha flash improved considerably to 12°C–13°C and the LPG yield went up by at least 0.3% on the crude.

The loss of LPG was also experienced in two other units with the reduction of stripping steam to conserve heat.

3.2 Vacuum Distillation Unit

3.2.1 Configurations and types of vacuum distillation units

These are the two primary types:

1. Fuel type
2. Lube type

Fuel type

In this type of unit, RCO is further fractionated into the following cuts.

1. Gas oil (up to 370°C)
2. VGO (370°C–565°C)
3. Vacuum residue (565°C+)

Lube type

1. Gas oil (GO)
2. Spindle oil (SO)
3. Light oil (LO)
4. Inter oil (IO)
5. Heavy oil (HO)
6. Vacuum residue (VR)

Vacuum column (fuel type) simulation

(DT) Heavy Diesel PA: 70°C–90°C or return temp 55°C– 60°C

(DT) light vacuum gas oil (LVGO) PA: 55°C

(DT) HVGO PA: 55°C

DT is the differential temperature between pump-around supply and return.

Stages in vacuum column (fuel type)

Top PA = 1–2 stages

Between heavy diesel and LVGO = 2–5 stages

LVGO PA= 1–2 stages

LVGO–HVGO = 1–2 stages

HVGO PA = 1 stage

Wash section = 2–3 stages

Stripping section = 3 stages

Efficiencies to be considered for design

Above wash zone: 60%–65%

Wash zone: 30%–40%

Stripping zone: 30%

Wash liquid at the bottom of wash bed ~ 1 M3/hr/M2 Column C.S. area

1. Thermodynamics (K-value): BK10/GS
2. Transport property option: PETRO (Method)/SIMCI heavy oil correlation. The viscosity of RCO and heavy fractions may be estimated from a mid-percent viscosity curve given along with crude TBP.
3. Density option: API (Method)

Fuel-type vacuum columns are designed for a far lower number of stages (typically 9–10 stages above flash zone). Typically 3 stages are provided in the stripping zone. Fewer stripping stages may reduce VGO production unless more stripping steam or a higher vacuum is deployed. Fuel type units are primarily used to produce vacuum gas oil as feed for fluid catalytic cracking units (FCCUs) or hydrocrackers/VGO hydrotreaters, depending on the configuration of the refinery. The VGO in both cases has Ni + V metal and a Conradson carbon residue (CCR) limit. Residue FCCUs do not have the stringent limit. Hydrocrackers have stringent feed metal content, CCR, N_2, and asphaltene specifications: anything beyond the design value of contaminants in the feed may deactivate the catalyst.

Lube type vacuum columns

1. Lube type units have a higher number of stages between the adjacent draws. They also have side strippers. They are typically designed for an absolute pressure of 65 mm Hg (a) or with 25 mm Hg (a) at top.
2. Higher numbers of stages are required for achieving higher separation between two adjacent cuts and obtaining the required viscosities of the cuts and their yields.
3. Typically operate with higher internal refluxes. Typically 3–4 stages are provided between two adjacent draws of lube oil base stocks with side strippers. At least three stages in the side strippers are preferred.

Main concerns when designing vacuum units

The economics of operation depend mainly on attaining a low flash zone pressure in the column so that the required vaporization is achieved at lower temperatures and also by using the optimal reflux liquid required for the wash bed/trays (bed/trays located below the HVGO draw tray). Typically, two stages are provided below the HVGO tray to attain the desired quality of VGO as a hydrocracker feed or FCCU feedstock. This wash bed operation and design is very critical, as a higher wash liquid rate will require higher vaporization, which will require a higher heat input through furnace, higher stripping steam, and a higher vacuum. A lower reflux may reduce the efficiency of the wash section. More serious is that this may result in the drying of liquid at the bottom section of the wash bed, resulting in heavy coking on the wash bed/trays. The coking will result in higher pressure drop across the bed and result in higher overall DP across the column top and flash zone. A minimum wash liquid is required at the bottom of the wash bed to avoid coking. Continuous developments and upgrades are coming for the wash bed to enable operation with a minimum liquid rate at the bottom of the bed. This liquid collects on slop withdraw chimney trays and the amount of vacuum slop pulled is a measure of the liquid rate at the bottom of wash bed. An improved version combines grid and structured packing in the wash section. The structured packing is placed over the grid and fixed with holding bars.

The distributor plays a major role in the performance of the wash bed, as it is responsible for uniformly spraying the HVGO internal reflux (wash liquid) on the wash bed.

Two types of distributors are normally used:

1. Pressure/spray distributors
2. Gravity distributors

Gravity distributors are in general more efficient as the irrigation of packing can be more evenly accomplished. Gravity distributors were not used in the wash zone until the 1980s, as its drip holes can eventually become coked. There is less likelihood of coking in a spray distributor, as the liquid under pressure pushes through the spray holes. The spray holes may have a 90° cut opening for covering all the areas of the packing or a 120° cut opening as decided by the packing and spray distributor suppliers. The better the distribution, the better the efficiency and the smaller the chances of coking. Normally, vacuum columns are to be provided with pressure tapping points across all the packing beds to measure pressure drop. A higher pressure drop may be due to overloading, flooding, or coking of the beds. Similarly, trayed columns should also have pressure tapping points for measuring the differential pressure (DP) across important zones (e.g., the wash zone), as the overall DP measurement is used to estimate coking or flooding.

Disengaging device

The disengaging device fitted in the flash zone helps to reduce VR entrainment in the vacuum slop cut. VR entrainment in the vacuum slop cut is typically calculated by the sodium (Na) content balance (mostly contributed by the VR entrainment) of vacuum slop, which provides a measure of entrainment at the flash zone.

In earlier days, circumferential 270° disengaging devices were used where liquid–vapor separation was achieved through angular motion with tangential inlet nozzles.

Radial-type disengagers are improved versions of conventional circumferential disengaging devices. In this type, the furnace outlet vapor and liquid is introduced in a diametrical channel covered at the top. It has radial arms with side vanes to release and separate vapor and liquid.

Ejector and overhead system

If the column is a wet type (having a pre-condenser), the overhead vapors are routed to one or more parallel pre-condensers. Condensed steam and hydrocarbons are routed through the barometric legs to a hot well and the noncondensables are routed to first-stage ejectors followed by a condenser, then to a second ejector followed by another condenser. The net off-gases are routed to a hot well, vented through the hot well water seal to the atmosphere at a safe location, and often sent to low-pressure burners with flame arrestors and supplementary fuel gas firing.

The maximum vacuum achievable in a wet type depends upon the actual vapor pressure of the condensate at the temperature of the condensate water. If the pre-condenser condensate temperature is 40°C, the maximum achievable vacuum is 55–60 mm Hg (a).

If it is a damp type, the overhead vapors go to the booster ejector then to the condenser. The booster ejector should build up enough pressure that the saturation temperature of water at the outlet pressure is at least 10°C–15°C higher than the cooling water temperature (approximately 33°C–40°C), so that steam can condense. In a damp type column vacuum, the maximum vacuum achievable is around 20–25 mm Hg (a).

Hot well

The hot well definitely plays an important role in the operation and the safe use of the vacuum column. The barometric legs are dipped in the pool of steam condensate or water in the hot well. Each barometric leg has to be separately dipped and cannot be combined. If they are combined, the off-gas from the second-stage barometric leg can get into the first-stage barometric leg (at lower pressure) and lead to the recirculation of off-gases, which will lead to the reduction of the effective capacity of the ejector system. If there are many barometric legs from pre-condensers as a result of having parallel pre-condensers, they can be combined in a vessel with a liquid seal. The vessel should be located 10 meters above the hot well. The pre-condenser barometric legs are bigger in diameter than the barometric legs of the condensers after the ejectors, as they handle a higher load of condensate as a result of the condensation of stripping steam and coil steam. The height of the seal in the hot well is typically 200–250 mm. The off-gas should have a lower seal height than the barometric legs and should be spaced at least five times its diameter from the nearest barometric leg.

There are several types of hot wells with barometric legs entering almost at the middle, with oil drawn out from one end after the overflow baffle and condensate drawn from other end. This type may result in turbulence in either the oil overflow side or the water overflow side and can result in carryover of water in oil or oil in water. It is best to put all barometric legs on one side of the hot well and have both the water and the oil come from the other side through overflow baffles constituting a water chamber and an oil chamber, as shown in Fig. 3.2.1. In the proposed hot well, there should not be any level tap or utility connection from the seal trough; the chamber should have a drain with a blind. This is to ensure that the legs are always dipped in the water and air cannot break the seal and find its way to the vacuum column to break the vacuum. Water connection must always be provided in a hot well to seal all the legs during startup. Water must also be overflowing before the ejectors start to ensure that barometric legs are always sealed when the vacuum is pulled through ejectors.

Barometric legs of the same stage (same pressure) can be combined and introduced in the hot well from the bottom with a deep U loop. This reduces the entry nozzles in the hot well and automatically incorporates the desired water seal. Care is to be taken to ensure that legs of different stages (different pressures) never combine.

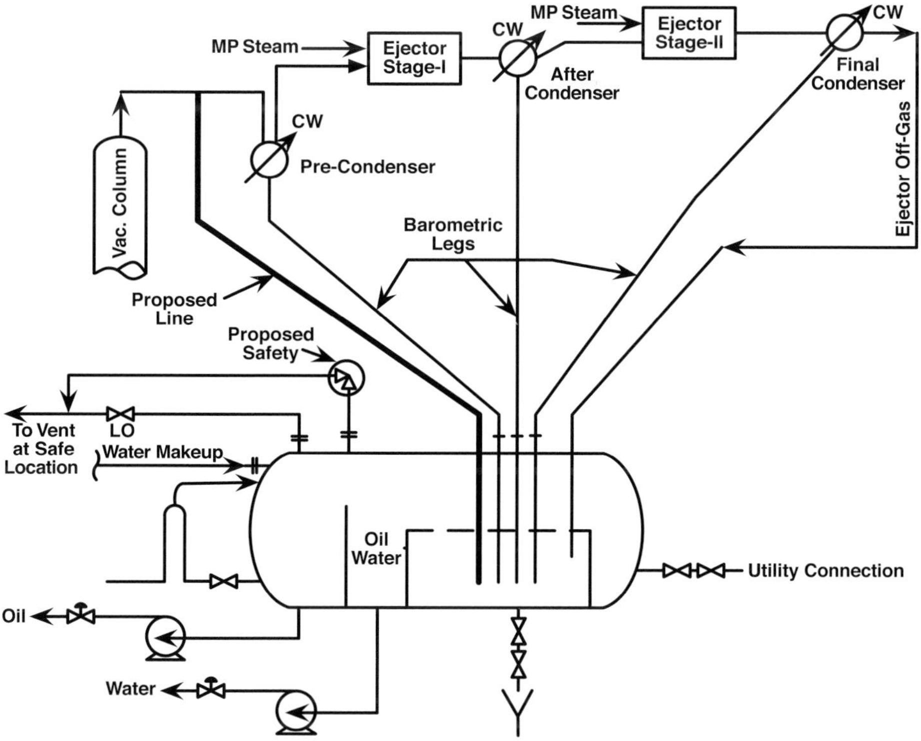

Fig. 3.2.1 Hot well with a line from overhead dipped in liquid

Elevation of the condensers

Vacuum column condensers operate in vacuum. Water drains from the condenser through barometric legs to the hot well. So the minimum theoretical height requirement is 10 meters. To compensate for DP in the barometric legs caused by flow and slightly higher-than-atmospheric pressure in the hot well, a higher elevation is provided (usually around 20 meters is preferred).

The problem of air ingress

Air ingress is likely to occur in a vacuum column through flanges, valve glands, etc. Typically, in a vacuum column, a pressure test followed by a vacuum test is performed before startup to ensure tightness of the column flanges and valve glands. Ejector off-gas is comprised of air leak into the column, gases produced due to cracking in the furnace, dissolved gases in the feed, and water vapor. Data may be generated from the startup to determine the typical composition of the gas at normal coil outlet temperature. An increase in CO_2 is a concern, as it suggests that air leak from the hot part of the column has increased (the flash zone, HVGO, LVGO zones, etc.). An increase in O_2 indicates increased air leak from the colder section of the column (column top, pre-condenser ejector flanges, etc.). It may be advisable to use a pitot tube for measuring the flow of the off-gas with almost no pressure drop and consequently no loss of vacuum. The increase in off-gas flow indicates a leak or increased production of decomposition gas and requires attention. An increase of light hydrocarbons in the ejector off-gas indicates cracking and consequent production of a higher amount of decomposition gas.

Overpressure prevention in a vacuum column

A vacuum column is likely to be pressurized in the event of ejector steam failure, pre-condenser/condenser water failure, reflux failure, blocked flow, etc. In a vacuum column, safety valves are provided in accordance with the contingency specified by API 521/API RP 520. The proper location of the safety valve discharge is a debated issue. If it is open to the atmosphere and it pops, it may discharge hot hydrocarbons into the atmosphere. If it passes, it may suck in air and load the ejectors, and the vacuum may drop. If the leak is high, the ingress of O_2 can generate sparks due to the presence of pyrophoric iron inside the overhead circuit. If it is connected to the flare and passes, it might pull the vacuum into the flare header. (There is a refinery where this did occur; the flare seal pot water rose in the flare header and the flare header supports collapsed.)

Many new ideas have been created for safety valves, such as adding a pressure switch at the downstream of the safety valve to indicate whether it is passing or not. A fool-proof scheme has been developed where the safety valve can be installed in the hot well instead of in the vacuum column. It would release the same quantity and provide for the same contingency as one in the vacuum column. The scheme mainly envisages one line of requisite diameter from the overhead vapor line directly routed to the hot well (without a valve) and dipped into the hot well liquid. The scheme is shown in Fig. 3.2.1.

During operation, almost 10 meters of water will rise in the line and, in the case of pressurization, the gas will escape through the line, blow through the seal, and go into the off-gas vent line. To make a contingency plan for blocked flow, a safety valve needs to be provided on the hot well. This was suggested to a refinery for its vacuum column for implementation, as they did not have a safety valve in the vacuum column in the design.

Vacuum column top pressure

The minimum pressure achievable is the vapor pressure of water at pre-condenser condensate temperature. If the condensate is at 40°C, then the vapor pressure of water at 40°C is 0.075 kg/cm^2(a) or 55 mm of Hg (a). So the minimum pre-condenser pressure achievable is 55 mm Hg. The pre-condenser inlet pressure will be 55mm Hg + 5–10 mm Hg (pressure drop in pre-condenser). The column overhead will be 55 + 5 + 10 (pressure drop in overhead line assumed as 10 mm Hg), which equals 70 mm Hg. Thus the minimum top pressure achievable in a vacuum column with a pre-condenser preceding the ejectors will be 70 mm Hg (a). If a top pressure much lower than 60 mm Hg (a) is desired, a booster ejector has to precede the pre-condenser. This requires substantial steam input to the booster ejector so that the noncondensables are compressed to a pressure of 65 to 70 mm Hg and the steam can be condensed by cooling water in the subsequent condenser. With a booster ejector, a column top pressure as low as 19 mm Hg (a) can be achieved. A more common top pressure encountered with a booster ejector is 20 to 25 mm Hg (a).

Normally four types of designs that have a great influence on the VGO quality and yield are used.

1. Dry (no stripping section, no coil steam)
2. Wet (pre-condensers limit column overhead pressure by cooling water temperature)
3. Damp column without stripping steam (no pre-condenser, coil steam used to adjust flash zone oil partial pressure)

4. Damp column with stripping steam (no pre-condenser, coil or stripping steam used).

Different configurations of a vacuum column

1. The first one operates typically at 8–12 mm Hg (a) at the top. The vapor goes directly to the ejectors.
2. The second type operates at 60–70 mm Hg (a) at the top and has a pre-condenser. The noncondensables are pulled from the pre-condenser by an ejector.
3. The third type of operation is done at 18–25 mm Hg (a) at top. Without stripping steam, the column is provided with a booster ejector followed by a condenser.
4. The fourth type again operates with a top pressure of 18–25 mm Hg (a) and uses both stripping steam and coil steam. This type is often considered best to increase the cut point of VGO by limiting the contaminants (vanadium, nickel, etc.) in the VGO with the same number of stages in the wash section.

Selection between four modes of operation

Out of the four modes, the quality of the HVGO product is better for a given cut point in the damp mode with stripping steam. The difference in quality between the wet and the damp with stripping is due to the lower oil partial pressure in the damp case.

Total VGO cut point is a measure of VGO yield and indicates efficiency of operation with respect to the up-gradation of the front end of the VR to VGO. The efficiency of the fuel type vacuum unit is primarily related to the VGO cut point achieved for the particular crude.

Ejector load determination

The off-gas generated in the vacuum column consists of the following:

1. Air leaking through the flanges, instrument tappings, and/or valve glands
2. Decomposition gas and/or cracked gas produced in the furnace
3. Dissolved inerts in the feed, if fed from tankage

The air leak is estimated by the absolute pressure in the column and the volume of the column as given in standard charts, often from the feed rate. Typically, 200% leak is assumed over that calculated by standard charts.

The formation of decomposition gas is estimated as a function of the furnace coil outlet temperature. In a wet type column (with a pre-condenser), the load to the first stage ejector is air leak + decomposition gases + saturated water vapor at pre-condenser condensate temperature. Similarly, the second-stage ejector load is air leak + decomposition gases + saturated water vapor at the first-stage ejector condenser condensate temperature. In a damp-type column, the first ejector (booster ejector) load is estimated as the air leak + decomposition gases + stripping steam + coil steam + hydrocarbon vapors coming out at the column temperature and pressure (available from the simulation result). It is advisable to add some cushion to the hydrocarbon vapor quantity from the simulation result.

In a dry type column (no pre-condenser, no stripping steam, no coil steam), it is only the off-gases, the air leak + decomposition gases + hydrocarbon vapors.

The final ejector outlet pressure is the specified maximum hot well pressure plus the height of the water seal used in the hot well for the off-gas. The ejector's off-gas analysis typically indicates the presence of N_2, CO_2, H_2S, O_2, C_1, and C_2, etc. The presence of these, along with the presence of CO_2, suggests that the air from the leak reacts to a certain extent with the hot hydrocarbon inside the column.

1. Air leak from chart (from commercial chart) = $X1$ kg/hr = $X1/29$ = $M1$ kmoles/hr
2. Decomposition gas (from standard chart)= $X2$ kg/hr = $X2/40$ = $M2$ kmoles/hr
3. Total noncondensables from charts = $M1 + M2$ kmoles/hr
4. Assumed noncondensables (200%) = $2(M1 + M2)$ kmoles/hr (However, some designers assume that the quantity of noncondensables is 150% of the noncondensable quantity calculated from charts for specifying ejector load.)
5. Assumed pre-condenser temperature = 40°C
6. Vapor pressure of water (steam table) at 40°C = 0.075 kg/cm^2(a) = 55 mm Hg (a)

As the vapor is saturated with water, the partial pressure of water in the vapor becomes equal to the vapor pressure of water at 40°C.

Let MH_2O = moles of water vapor at pre-condenser outlet, and then calculate MH_2O from the following equation:

$55 = ((MH_2O)/(MH_2O + 2(M1 + M2))) \times$ (Absolute pressure of pre-condenser in mm Hg)

Load to the ejector downstream of the pre-condenser is $2(M1 + M2) + MH_2O$ kmoles/hr.

Load to the ejector downstream of the pre-condenser expressed in mass units = $(2(X1 + 2X2) + (MH_2O \times 18))$ kg/hr.

Calculation with booster ejector ahead

If a booster ejector is used, the load to the booster ejector (upstream of the pre-condenser) = $(2(X1 + X2) + (MH_2O \times 18)$ + (stripping steam) + (coil steam) in kg/hr) + HC vapors. The subsequent ejector loads are calculated as previously described.

Estimation of wash stages in a vacuum column from operating data

The health and efficacy of the wash stages in a vacuum column have always been in question, particularly in view of the quality of VGO as a hydrocracker or FCCU feed.

Typically 2–3 stages are provided in the wash bed. Wash liquid VGO is supplied to the top of the bed, and liquid from the bottom of the bed is collected as vacuum slop and drawn as product; often part of the slop drawn is recycled through the furnace along with the feed. The quantity of liquid recovered from the slop tray is a measure of the quantity of liquid at the bottom of the wash bed, which typically maintains a rate of 0.5–1 m^3/hr per m^2 of column cross section. The stages available in the wash section can be roughly estimated by the ratio of slop drawn from the slop tray to the HVGO internal reflux introduced at a fixed flash zone temperature. This assumes no leak or overflow from the slop tray. A vacuum slop product to the HVGO internal reflux with a flow ratio ranging from 0.5–0.3 indicates the availability of approximately 2 stages in the wash zone for typical furnace coil outlet temperatures and vacuum maintained. A higher ratio (where the reduction of internal reflux is less) indicates the availability of fewer stages in the wash zone due to coking or channeling. However, this can be best estimated using a total simulation of the column that takes into consideration the actual HVGO IR introduced and vacuum slop produced. It is necessary to have correct flow quantities of HVGO IR and vacuum slop to estimate the actual stages.

VGO cut point determination

The VGO cut point is conveniently estimated by adding the VR and the vacuum slop flow and converting the sum to a percentage on the crude. Then subtract the percentage from 100 to obtain the percentage of crude boiling up to VGO. The same is marked on the abscissa on the crude true boiling point (TBP) plot, and the temperature is read from the ordinate as the cut point of VGO. If VR + vacuum slop yield is 25% by volume, then the yield (volume) of the lighter products up to VGO is (100 – 25) = 75%.

Refer to Fig. 3.2.2 to find the cut point of VGO, marked 75% vol. on the abscissa, and read the temperature in the ordinate. The temperature noted in the ordinate is the cut point of the VGO.

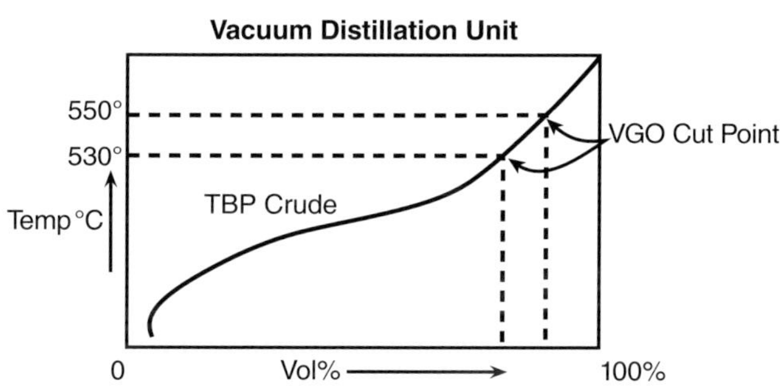

Fig. 3.2.2 VGO cut point estimation

Problems in vacuum units with different Euro IV specifications for diesel

The 95% temperature point in accordance with the American Society for Testing and Materials (ASTM) distillation temperature (T95%) for diesel has been lowered in a new diesel specification conforming to Euro III and IV, which changes the T95% distillation temperature from the previous 370°C to 360°C. This has reduced the gas oil draw from the atmospheric column and caused a small amount of gas oil to be inadvertently dropped to the RCO and fed to the vacuum unit. This has resulted in the following:

1. Increase in the RCO produced and increased gas oil content in RCO
2. Increase in the vacuum heater duty requirement
3. Vacuum column diameter being pushed to the limits
4. Vacuum column wash zone velocity reaching its limits and nearing a point where it will affect the VGO quality
5. More gas oil getting into the hydrocracker or FCCU feed as the vacuum top-heavy diesel draw is reduced

Refiners often drop gas oil components to contain T95% of vacuum diesel into the VGO to maintain the T95% of the total diesel pool. This increases the content of gas oil in VGO and results in an extra load on the hydrocracker/hydrotreater or FCCU. Alternatively, refiners often reduce the draw of atmospheric gas oils to maintain the T95% of the total diesel pool, reduce vacuum furnace outlet temperature to maintain the vacuum furnace duty and wash zone velocity, and end up with a reduction in the cut point of the VGO. This does help in resolving vacuum column diameter limitation to some extent, but lowering the VGO cut point adversely affects the economics of operation and increases the VR production. Some refiners continue to draw straight-run diesel components as heavy as earlier and routes to the DHDT and incorporating a T95% improvement catalyst in the diesel hydrodesulfurization/hydrotreating units (DHDS/DHDT) to improve the T95% distillation point of the diesel product. Typically, some liquid yield loss is associated with T95% improvement in a DHDS/DHDT unit, which means it may not be the most economic alternative.

Alternatives that reduce the vacuum furnace and column loads as well as the gas oil content in the diesel are shown below:

1. To reduce the vacuum furnace and column loads, the ability to draw an additional gas oil cut—heavy atmospheric gas oil (HAGO) or extra heavy gas oil (EHGO)—can be provided below the HGO draw with or without a side stripper. The gas oil dropped can be pulled out so that it does not go to the vacuum furnace or column. It can be blended with the FCCU feedstock, as the heavier diesel fraction is likely to crack in the FCCU. However, this may increase the sulfur content of the FCCU feed and consequently increase the gasoline sulfur if the FCCU feed has low sulfur or uses an unconverted oil from the hydrotreater or hydrocracker as feedstock. This in turn is likely to reduce the RON (Research Octane Number) of the treated gasoline in the FCCU gasoline treatment unit.

2. To reduce gas oil components in the VGO, the mass transfer stages between the vacuum diesel and the LVGO can be increased. Four to five stages will be required to significantly reduce gas oil in the VGO. Typically, vacuum columns are provided with 2–3 stages between the vacuum diesel and the LVGO; however, there may not be space available to accommodate additional stages as required.

3. The wash zone stages in an atmospheric column can be also increased to improve the T95% of straight-run gas oil (SRGO).

4. Alternatively, to reduce gas oil components in the VGO, install an LVGO side stripper with at least three stages and use stripping steam as the stripping agent. This increases the vacuum diesel yield considerably without adversely affecting the T95% of the vacuum diesel. This scheme is advised by the author and is under implementation in many refineries.
5. To reduce the vacuum furnace load and reduce wash zone velocity, a scheme has been developed to draw additional HGO (containing part of the front end of the VGO) with a side stripper from the atmospheric column at a location below the HGO tray, then cool and route it to vacuum at the LVGO/HVGO pump-around return. Using this method, the vacuum furnace and wash zone will be bypassed by the EHGO cut.

Thus, the vacuum column wash zone velocity (which in this case is a concern for maintaining the quality of the VGO) is reduced. This method has also been advised and implemented in an atmospheric-vacuum unit (AVU). The atmospheric furnace duty increased to provide a higher atmospheric furnace outlet temperature and enable EHGO draw. The unit had a provision to draw EHGO and heat exchangers to cool the EHGO with crude. The same facility commissioned EHGO cooled to 110°C through heat exchange with the crude and introduced with the HVGO pump-around return at the upstream of the filter.[1]

The unit had a problem with the VGO quality when the unit was operating at a higher vacuum capacity. This problem was aggravated by a change in the diesel specifications. The unit had one HVGO cut with an HVGO pump-around and a light diesel oil (LDO) trim cut between the vacuum diesel and the HVGO (Fig. 3.2.3). The HVGO PA section was also heavily loaded. The unit had almost five mass transfer stages between the vacuum diesel and the HVGO.

The VGO color improved with the EHGO routing to the HVGO PA return due to the decrease in the wash zone velocity. The cut point of HVGO increased and the gas oil content of HVGO came to minimum. Furthermore, the EHGO worked as a reflux for the HVGO PA zone and reduced the pump-around duty requirement and the load of HVGO PA section. With the lowering of the HVGO PA duty, a slight drop in feed preheat (around 2°C) was observed. The modified scheme is shown in Fig. 3.2.4.

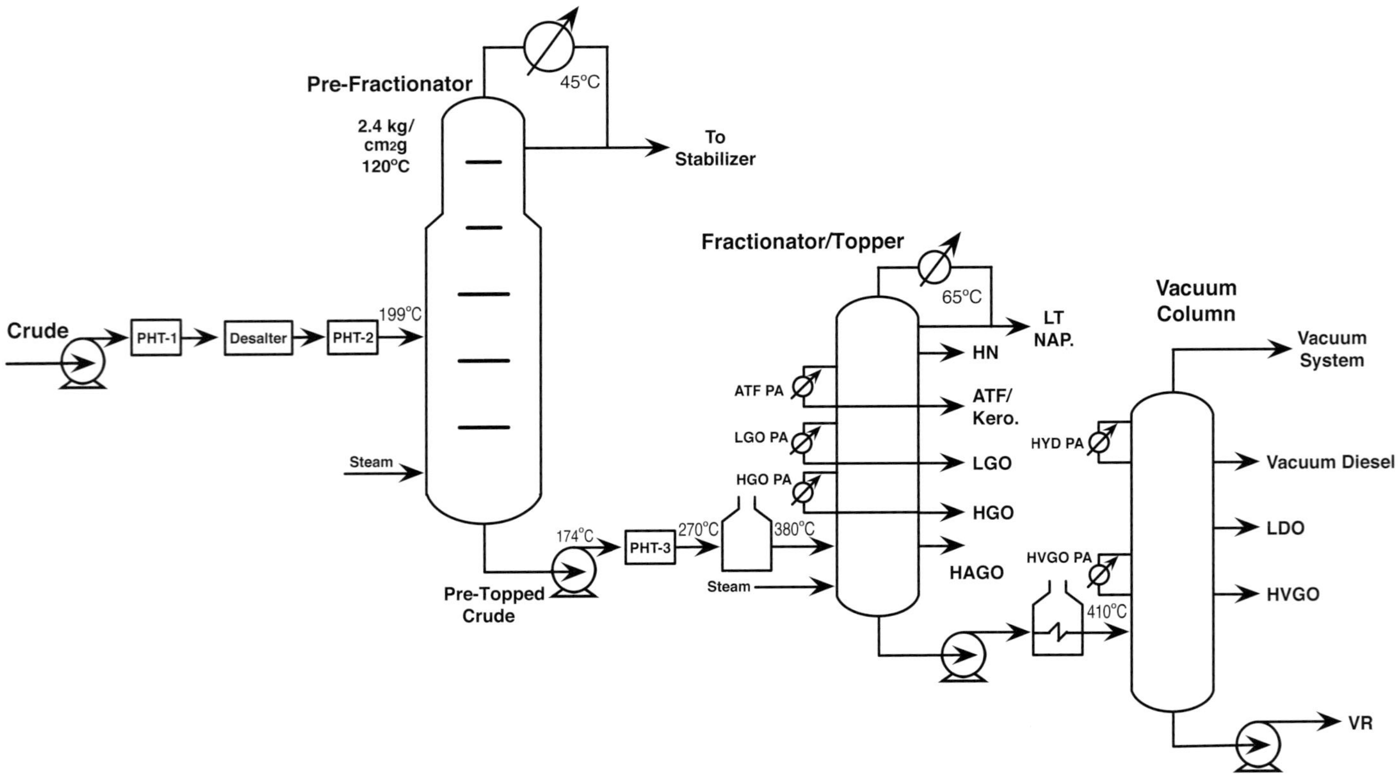

Fig. 3.2.3 Original configuration

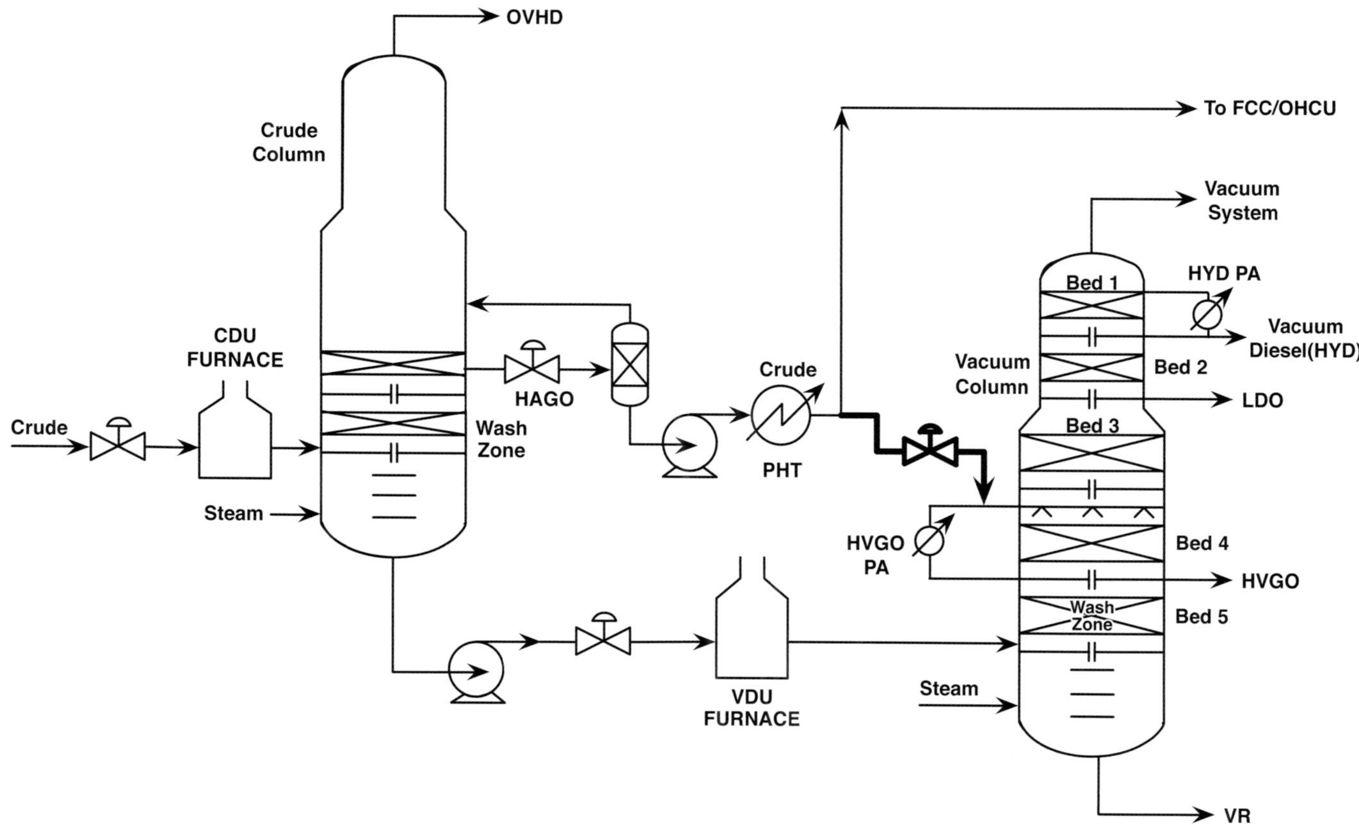

Fig. 3.2.4 Modified configuration with HAGO draw introduced in VGO PA return

3.2.2 Case studies

Case Study 1 (Increase in cut point of VGO)

It was observed that one unit was designed to produce VGO with an ASTM cut point of 565°C. However, the actual cut point was calculated to be only 525°C (see Fig. 3.2.2). This cut point was too low, causing very inefficient operation.

One of the problems was that the refinery did not reflux the minimum wash quantity, so when it was first tried, there was no flow. It was concluded that this was probably due to the spray distributor nozzles being coked up. It was decided to shut down the unit to change the wash bed.

A significant degree of coking was found in the wash bed, so the wash bed was changed and after the unit was restarted, 90 m^3/hr of HVGO IR was introduced. The wash oil flow with the new wash bed was strictly monitored, as was the vacuum slop flow, based on 0.5–1 m^3/hr per m^2 of column cross-sectional area. The diameter of the column in the wash zone was 8.9 meters. The minimum liquid at the bottom of the wash bed was calculated as $(0.5 \times \pi/4 \times (8.9)^2) \sim 20$ m^3/hr. This was maintained at a rate of 20–25 m^3/hr.

The TBP cut point increased from 525°C to 565°C. Vanadium and nickel content were reduced, asphaltene content decreased, and column DP decreased. The increase in VGO yield was around 3% of the crude, and the VR viscosity went up as a result of higher VGO yield. This just goes to show how important the wash oil flow and vacuum slop flow are for indicating the health and efficiency of a vacuum tower.

Case Study 2 (Revamp of a hydrocracker feed preparation unit)

Production of extra VGO (XVGO)

An extra VGO cut (XVGO) can be provided below the HVGO cut, where the wash zone is located below the XVGO with 2–3 wash stages and XVGO is returned to wash zone as wash liquid.

The LVGO + HVGO can be routed to the hydrocracker, and XVGO can be routed to the FCCU. In FCCU units, some quantities of catalyst are always rejected to maintain the activity and the contaminant level; thus the buildup of impurities in the FCCU feed (such as Ni, V, N_2) is reduced, allowing the XVGO to be fed to the FCCU. This is a feature of deep-cut vacuum operations where VGO cut points beyond 565°C–570°C are targeted with very high wash rates. This feature can also be implemented when contaminant levels increase in hydrocracker feed and adversely affect the VGO recovery.

When the metals (Ni + V) and asphaltenes increase in a VGO with a normal cut point, it renders the VGO unsuitable as hydrocracker feed. The only option available is to drop the back end of the VGO to the vacuum residue and increase vacuum residue production. This adversely affects profitability.

It may be recommended to provide an additional XVGO or heavy heavy VGO (HHVGO) draw at the bottom of the HVGO if the unit is likely to process a wide range of crudes in a fuel-type vacuum distillation unit.

This is particularly applicable in the following cases:

1. Processing of heavy crudes
2. Processing of crudes with high levels of nitrogen, metals, and asphaltenes

The wash zone needs to be placed below the XVGO or HHVGO draw. The HVGO draw needs to be adjusted to control the hydrocracker feed quality, and the XVGO or HHVGO draw must be manipulated to maintain the cut point or the FCCU feed contaminant level in the feed blend.

In a revamp scenario, it is difficult to provide the additional draw due to lack of a space in the column. However, some rearrangements may be possible. In one of the hydrocracker feed preparation unit revamps, there was a slop pump-around below the HVGO wash zone in the original design. It was dropped in revamp, and an extra VGO cut was incorporated.

The unit was originally designed to produce feedstock for a hydrocracker unit (HCU) and had a capacity of 2.5 MMTPA. The design preheat was 298°C. It was observed that the asphaltenes and N_2 content of the VGO would often increase with a change in the feedstock. The hydrocracker was originally designed to process a VGO feedstock with a maximum asphaltene content of 100 ppm (n-heptanes insoluble) and a maximum N_2 content of 800 ppm wt.

The refinery desired to augment the unit capacity beyond 2.5 MMTPA to the maximum level feasible without any change in the existing charge heater or the main column.

The study done showed that with an increase in preheat, more feed could be processed within the designed furnace duty. Thus, emphasis was placed on enhancing the preheat and increasing capacity while maintaining the asphaltene and N_2 contents at the stipulated levels.

The original configuration with two VGO draws (LVGO and HVGO) is shown in Fig. 3.2.5.

The new configuration is shown in Fig. 3.2.6 with the additional HHVGO draw.

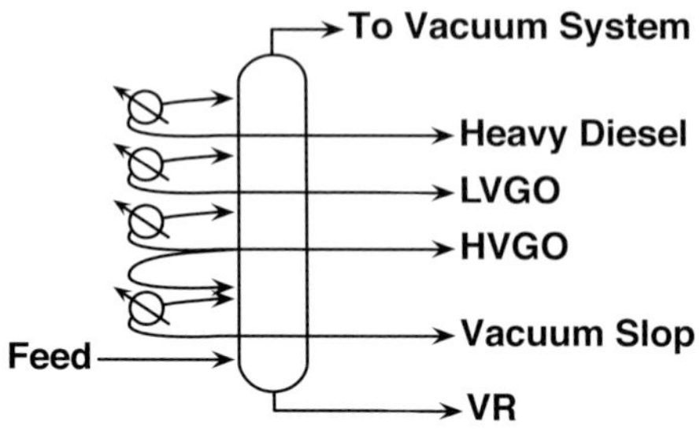

Fig. 3.2.5 Original configuration with two VGO (LVGO and HVGO)

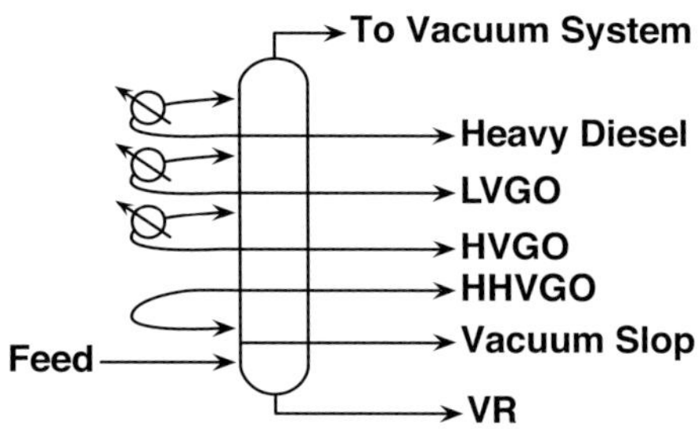

Fig. 3.2.6 Revised configuration with additional HHVGO draw

A third VGO withdrawal as HHVGO below the HVGO draw off zone was introduced in the revamp (see Fig. 3.2.6). The envisaged routing of VGO components is shown below:

VGO (LVGO + HVGO) → HCU feed

HHVGO → FCCU Feed

The wash section was relocated below HHVGO tray. Thus, it was possible to keep the VGO, asphaltene, and N_2 within the stipulated limits and reduce the production of VR by pulling the HHVGO. The high N_2 content of the HHVGO was diluted by the VGO obtained as FCCU feed from the high sulfur (HS) RCO (which is low in N_2) processed in another vacuum unit. Thus, the distillate was increased by increasing the (LVGO + HVGO + HHVGO) cut point and reducing the VR from the vacuum column.

The preheat could be increased to around 320°C by splitting the feed RCO train and recovering the HVGO rundown product heat in the feed which, in earlier configurations, was used for low-pressure steam generation. The unit now operates at 3.3 MMTPA, delivering the revamp objectives.

Case Study 3 (Capacity and preheat improvement of a fuel-type VDU)

One very old fuel-type vacuum unit with a capacity of 0.8 MMTPA was revamped to 1.2/1.5 MMTPA. The unit was operating with three side cuts: heavy diesel, VGO, and vacuum slop. It also had two pump-arounds: heavy vacuum diesel PA and VGO PA.

In the revamp, a new HVGO draw was introduced along with the pump-around, and a good amount of the heavy diesel PA heat that was previously rejected to the water cooler was shifted to the LVGO/HVGO pump-around, making it available to preheat the feed after a suitable alteration of the heat-exchanger train.

In the wash section, the packing equivalent to two stages was installed. As a result of this modification, the preheat temperature increased from 250°C to 315°C and VGO cut point increased considerably. In addition to the improvement in the quality and yield of the VGO, there was a reduction in the ΔP of the column to a level of 8–10 mm Hg. The VR quality also improved for converting the same to bitumen.

The old VDU configuration with only two pump-around refluxes and one VGO is shown in Fig. 3.2.7.

A third pump-around and an additional VGO draw were introduced. The new VDU configuration is shown in Fig. 3.2.8. The LVGO and HVGO constituted the total VGO pool.

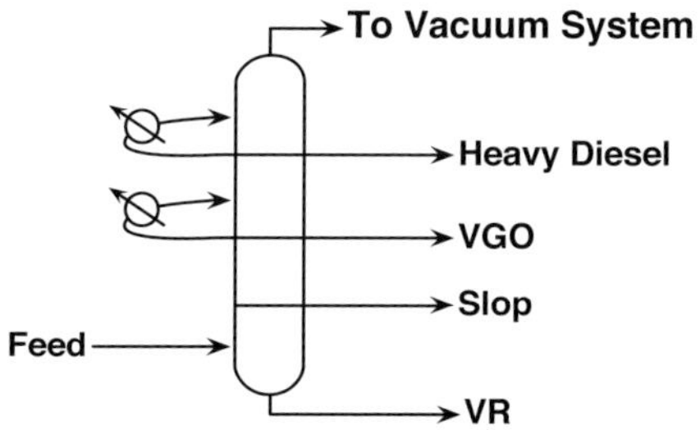

Fig. 3.2.7 Original configuration with single VGO draw

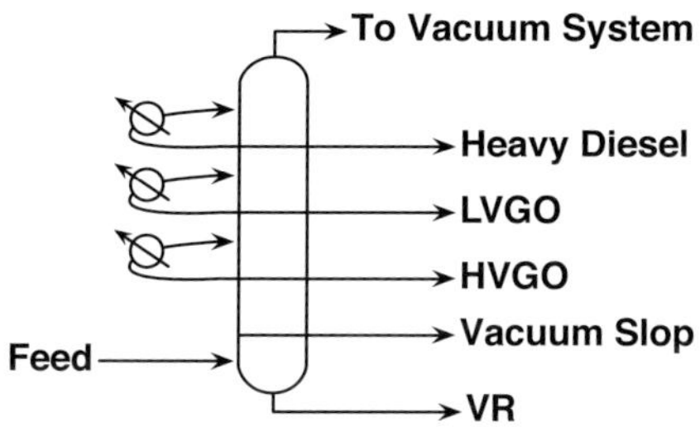

Fig. 3.2.8 Revised configuration with LVGO and HVGO draw

Case Study 4 (A lube-type vacuum unit capacity revamp)

One very old lube-oil-type vacuum unit was originally installed to produce lube-based stocks like spindle oil (SO), light oil (LO), inter oil (IO), heavy oil (HO), and vacuum residue (VR) (see Fig. 3.2.9). The feed preheat of the unit was low, around 240°C–250°C (original preheat train shown in Fig. 3.2.10). The refinery wanted to enhance the capacity by 20% and increase the preheat to the maximum possible extent.

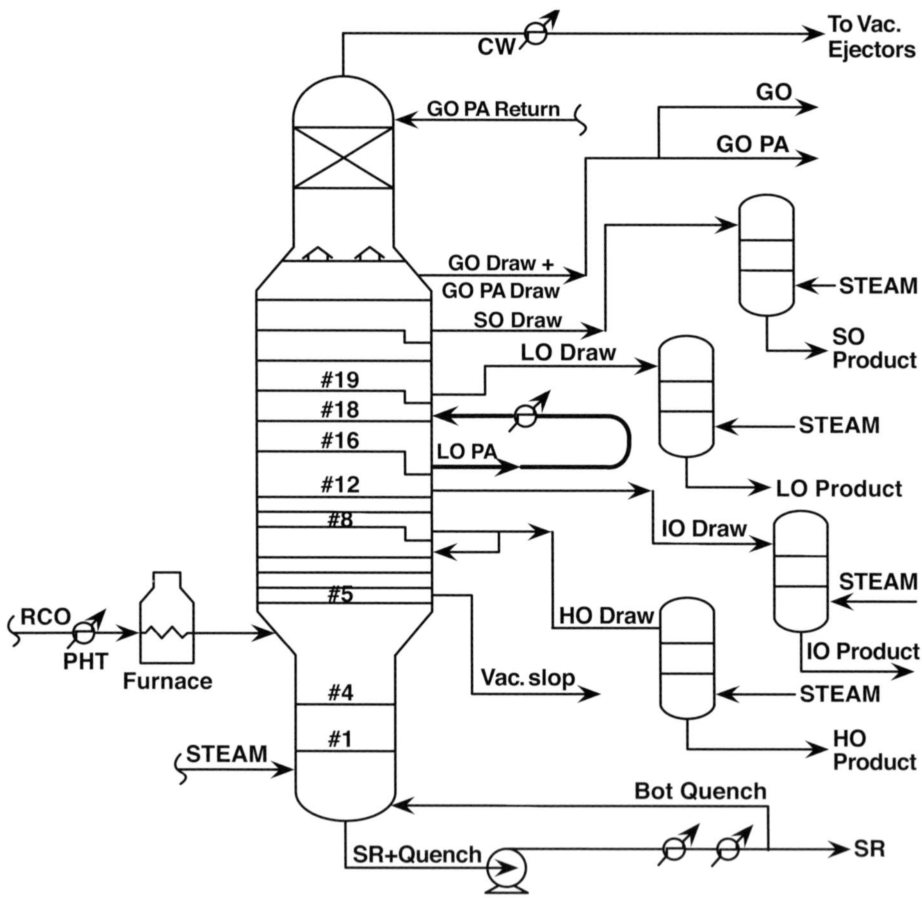

Fig. 3.2.9 Original scheme of LO pump-around below LO draw

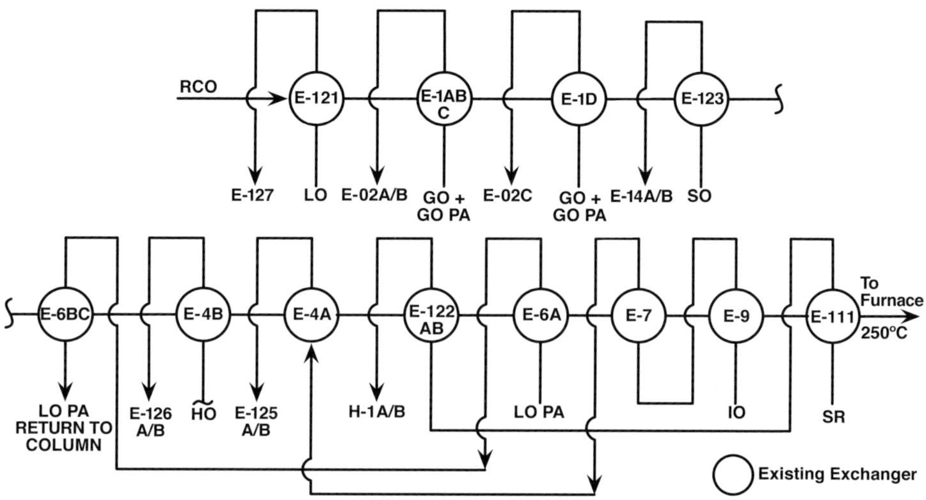

Fig. 3.2.10 Original heat-exchanger train configuration

The column had two pump-around refluxes: a top gas oil pump-around and a second pump-around between the LO and IO draws, where it was only in a position to condense IO and HO cuts. Thus, it was not possible to increase the pump-around duty to integrate it with the feed and enhance the feed preheat. Any increase in the pump-around duty at that location would have condensed part of LO in the PA section, which would have then traveled to the IO cut, affecting the viscosity of the IO. As a solution, the second pump-around was relocated to the LO tray, drawing the PA from the LO tray and returning it to a tray three above after the requisite cooling (Fig. 3.2.11).

At this new location, the pump-around was able to cool and condense the LO, IO, and HO cuts. Thus, the pump-around duty was increased without affecting product quality. Consequently, it would be possible to increase pump-around duty and transfer the heat to the feed in order to increase the feed preheat.

The draw temperature of the pump-around at the new location (upper) was lower by around 20°C, but the heat duty increased considerably. Additionally, the enhanced feed rate was also causing extra pressure drop in the feed circuit, so a parallel train was laid to resolve the problem, keeping the original train by and large intact (Fig. 3.2.12). A preheat temperature of around 290°C was achieved. Some steam generators were added to improve the general energy efficiency of the unit.

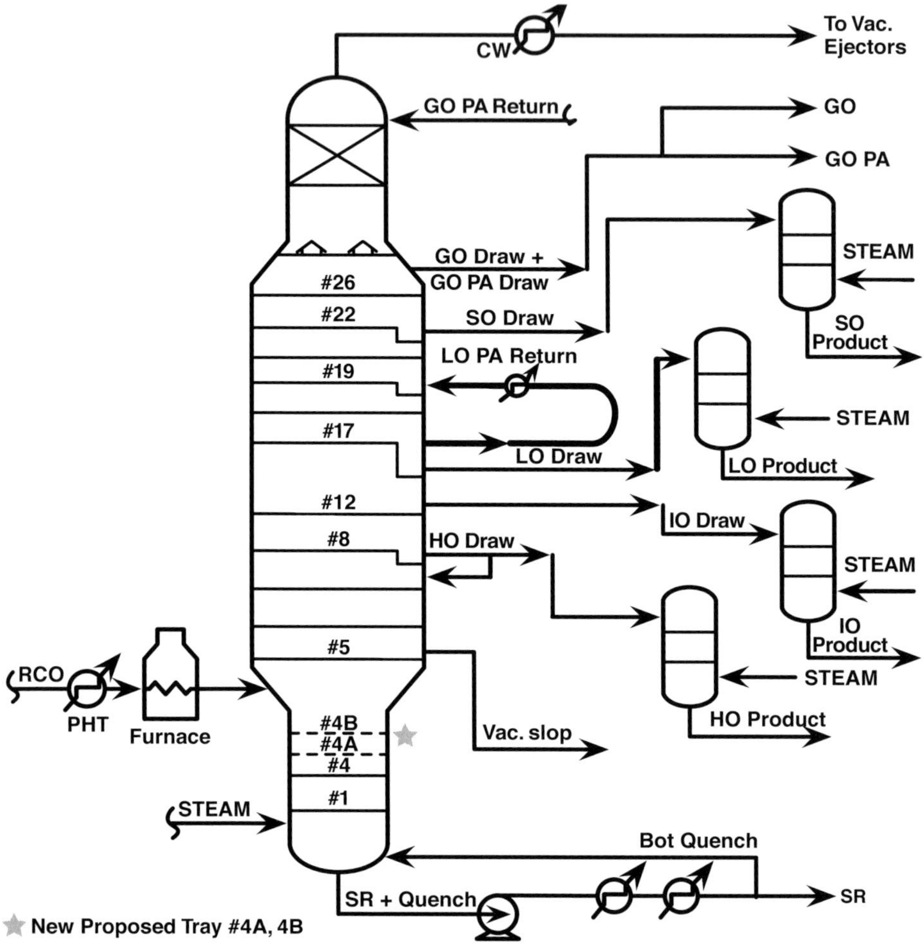

Fig. 3.2.11 Revised configuration with LO PA from LO draw tray

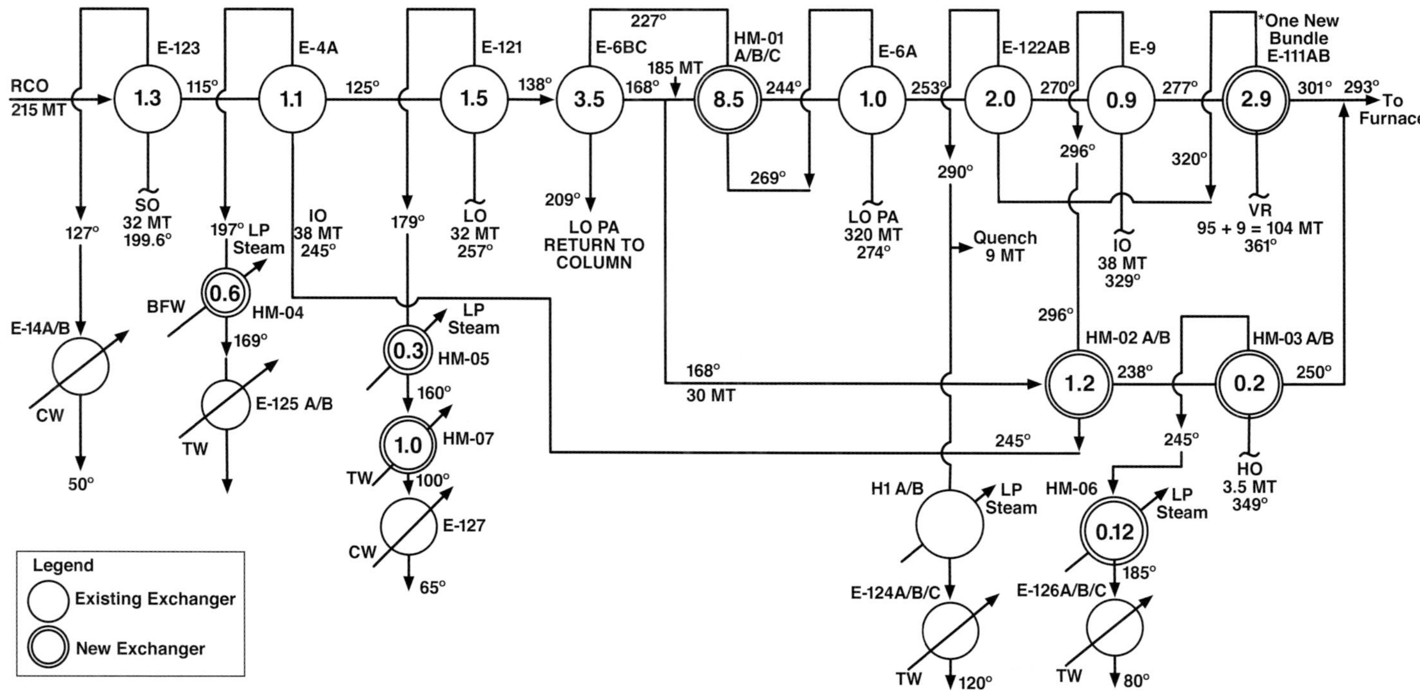

Fig. 3.2.12 Modified heat-exchanger train for preheat improvement

Case Study 5 (Production of lube oil cuts from the unconverted oil from hydrocracker grassroots unit)

In a conventional process, lube oil cuts are produced from reduced crude oil (RCO) after distillation in a vacuum distillation unit, but lube oil cuts possessing the desired viscosity can only be produced from certain crude oil stocks. Of late, lube oil cuts are also produced from the unconverted oil from hydrocrackers.

In this process, the unconverted oil is fractionated in a vacuum column, normally producing two to three side cuts along with a bottom cut. Typically, there are 4–5 stages provided between the adjacent cuts. The column top pressure generally preferred is 25 mm Hg (a). The configuration of one such unit is given below in Fig. 3.2.13.

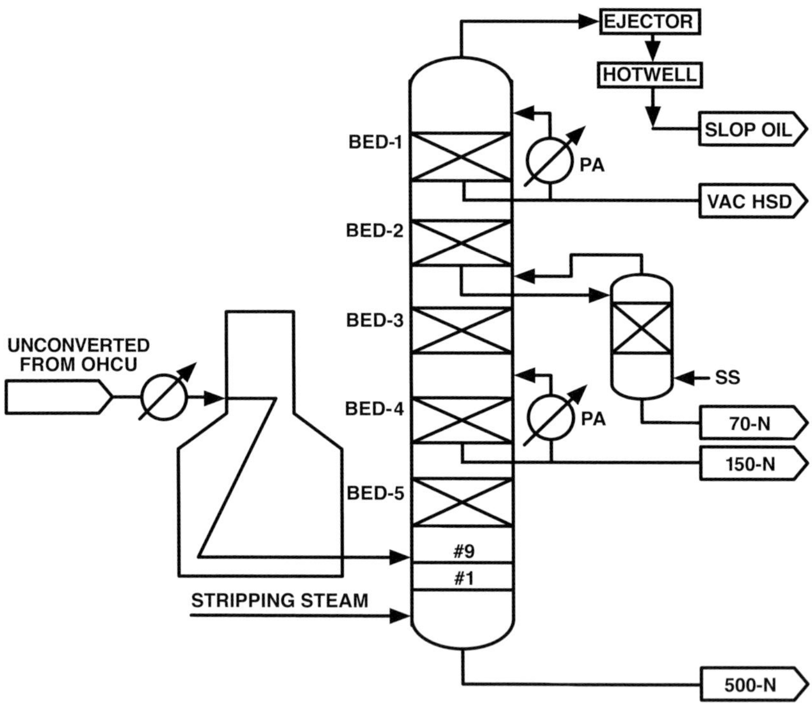

Fig. 3.2.13 Scheme of a vacuum column for a lube oil cut from the unconverted product of a once-through hydrocracker unit (OHCU)

The column does not need the side strippers required in a conventional lube-oil-cut-producing vacuum column, as the unit operates with a low furnace coil outlet temperature of around 340°C and would produce less cracked products. Five stages are already provided between the two adjacent cuts.

However, a side stripper is still provided in the 70N draw to minimize the gas oil components in the 70N draw to maintain its viscosity, as well as maintain the T95% point of the gas oil. The side stripper is helpful even with the five mass transfer stages provided between the gas oil draw and the 70N draw.

3.2.3 Case Studies of preheat improvement in crude vacuum units with the least modification

Case Study 1 (Preheat improvement in an AVU)

So many old crude vacuum units are operating with low preheat, and therefore, low efficiency. With the recent increase in energy costs, these unit designs are revisited. A big revamp to increase heat efficiency is normally not economically viable (unless capacity expansion is associated). Therefore, refiners want to improve the efficiency of the units with small modifications and with less unit outage required to implement the modification.

The unit shown below is an atmospheric-vacuum unit producing lube-oil based stocks (Fig. 3.2.14). The unit had a crude preheat of 240°C, and the refinery wanted to examine whether it was possible to increase the preheat.

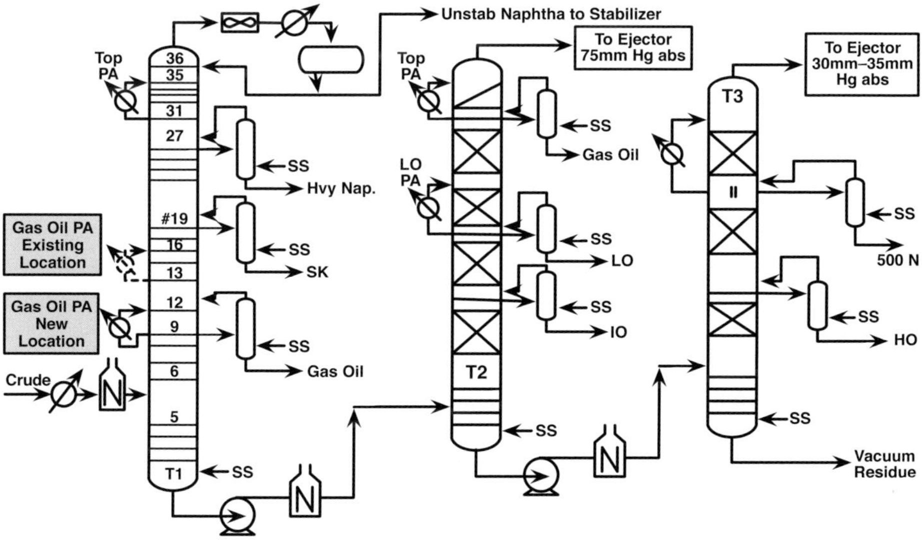

Fig. 3.2.14 Original gas oil PA location lowered

A heat source table was constructed through simulation of the existing configuration and using the true composite curve (TCC), it was found that the preheat could be increased to a maximum of around 250°C, and that the high-temperature range heat was not sufficient to increase the crude preheat beyond that temperature.

In the original configuration, the gas oil PA was below the SK draw tray, so its temperature was lower (around 280°C), and it was first routed to the stabilizer reboiler and then to the preheat train. The stabilizer reboiler does not need a hot stream at such a high temperature (the stabilizer bottom temperature is only around 190°C). Furthermore, the large temperature difference between the hot stream and the cold stabilizer bottom can give rise to film boiling in the reboiler. The gas oil PA draw should be lowered as shown and the routed first to the preheat train and subsequently to the stabilizer reboiler.

The reboiler surface area was examined and found acceptable. The TCC was redrawn using the revised heat table and gave an attainable crude preheat of 280°C.

Some reorientation of the heat exchangers was carried out with surface area added in order to capture the high-temperature heat at a higher temperature range of crude. A preheat temperature of around 280°C was achieved with minimum modifications. Lowering the location of the gas oil PA draw as proposed was likely to affect the separation between kerosene and gas oil, but a high separation was not desired.

Case Study 2 (Preheat improvement in a crude distillation unit)

The preheat in another light crude processing unit was very low (around 235°C). The unit had a pre-topper with a furnace as the reboiler. The pre-topper bottom was directly routed to the charge heater. The pre-topper inlet temperature was identified as the preheat of the unit. It was found that the PA flows were low and operating with a high temperature differential. This downgraded the quality of the pump-around reflux heat (the return temperature was becoming too low). Further, the unit had crude preheat exchangers up to the pre-topper. The pre-topper bottom had a reboiler furnace. The bottom of the pre-topper was fed to the charge heater before being fed to the topper (Fig. 3.2.15).

It was found that if the pre-topper inlet temperature was increased by increasing the PA flows, a very large amount of heat would still be rejected to the pre-topper condenser due to the increased reflux ratio of the pre-topper at higher inlet temperature. The pre-topper inlet temperature was maintained at around 180°C. A third heat-exchanger train was added downstream of the pre-topper (Fig. 3.2.15). Gas oil PA and RCO were used in the third preheat train to heat the pre-topped crude. Additional surface area was provided, and a preheat of 270°C was achieved.

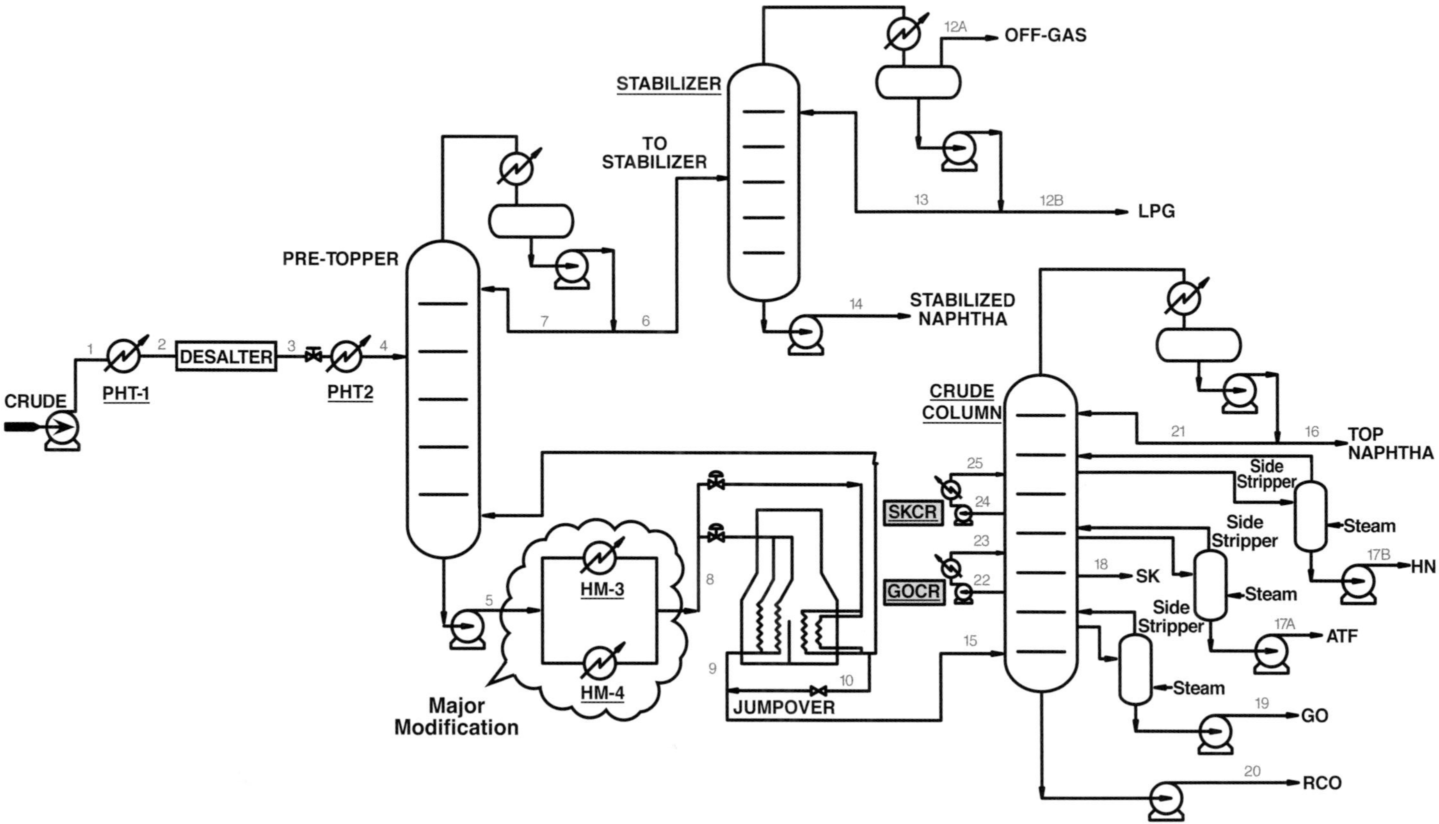

Fig. 3.2.15 Modification with the addition of a heat exchanger downstream of the pre-topper

Thus it can be seen that the pre-topper is to be operated with a low reflux ratio (preferably 0.3 to 0.5) in order to conserve heat, and the inlet temperature is to be fixed accordingly. After increasing the SK and gas oil PA flow rates, the heat source table is prepared, and from the heat source table, the TCC and GCC (grand composite curve) are drawn as below. The TCC shows that a crude preheat of around 275°C can be achieved (Fig. 3.2.16).

The GCC in Fig. 3.2.17 shows that some heat is available (around 2800 kW or 2.4 Gcal/hr) at around 200°C for steam generation. Accordingly, a steam generator was provided to generate steam from the SK + SK circulating reflux (CR) stream.

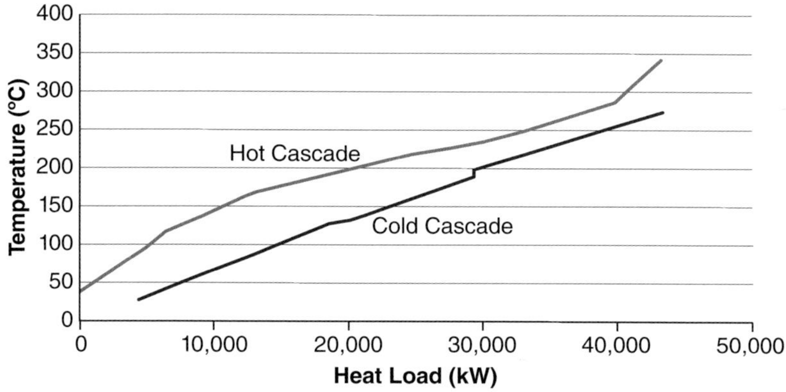

Fig. 3.2.16 TCC in revised configuration

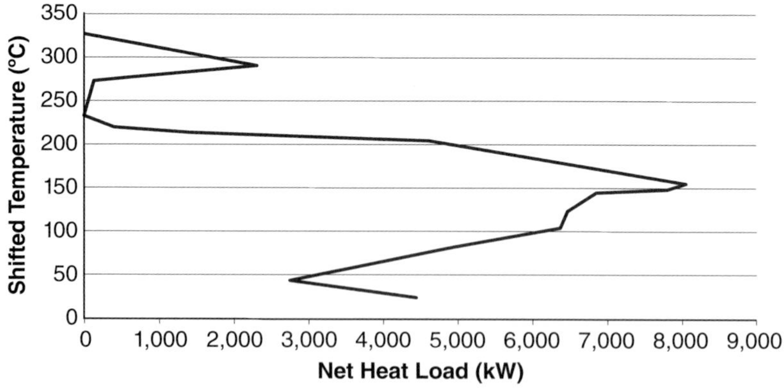

Fig. 3.2.17 GCC in revised configuration

The upper open jaw of the GCC is not shown and drawn only up to attainable preheat temperature.

Case Study 3 (Preheat improvement in a crude distillation unit)

Another crude distillation unit had only a topper, and no pre-topper. The topper had three pump-around (PA) refluxes: top PA, kerosene PA (KERO-1), and LGO PA (KERO-2). It also had an HGO or SRGO draw without a PA. The top PA duty was high. The crude preheat of the unit was low (205°C). The refinery wanted to look into the possibility of increasing crude preheat to increase the overall efficiency of the unit. In the unit hot RCO was first going to reboiler of naphtha splitter then being routed to the crude preheat train. Thus, the very high-temperature heat of the RCO was not wisely used. A simulation of the unit was carried out and a heat source table prepared. Further, it was found that very significant amount of heat was being transferred to the top PA and overhead condenser. Thus, there was a shortage of high-quality heat to increase the crude preheat.

A fourth PA, an SRGO CR or HGO PA was added (see Fig. 3.2.18). The heat was coupled to the heat-exchanger train, and surface area was added in the KERO-1 and in the heat exchangers in HGO PA service. Hot RCO was first routed to the last heat exchanger of the train, then routed to the naphtha splitter reboiler. The amount of heat rejected to the condenser and through the top PA was reduced by the incorporation of the HGO PA. A TCC and GCC were developed to estimate the attainable preheat and possibility of steam generation. Preheat increased from 205°C to around 275°C, even though the unit was only designed to deliver preheat temperatures of up to 220°C.

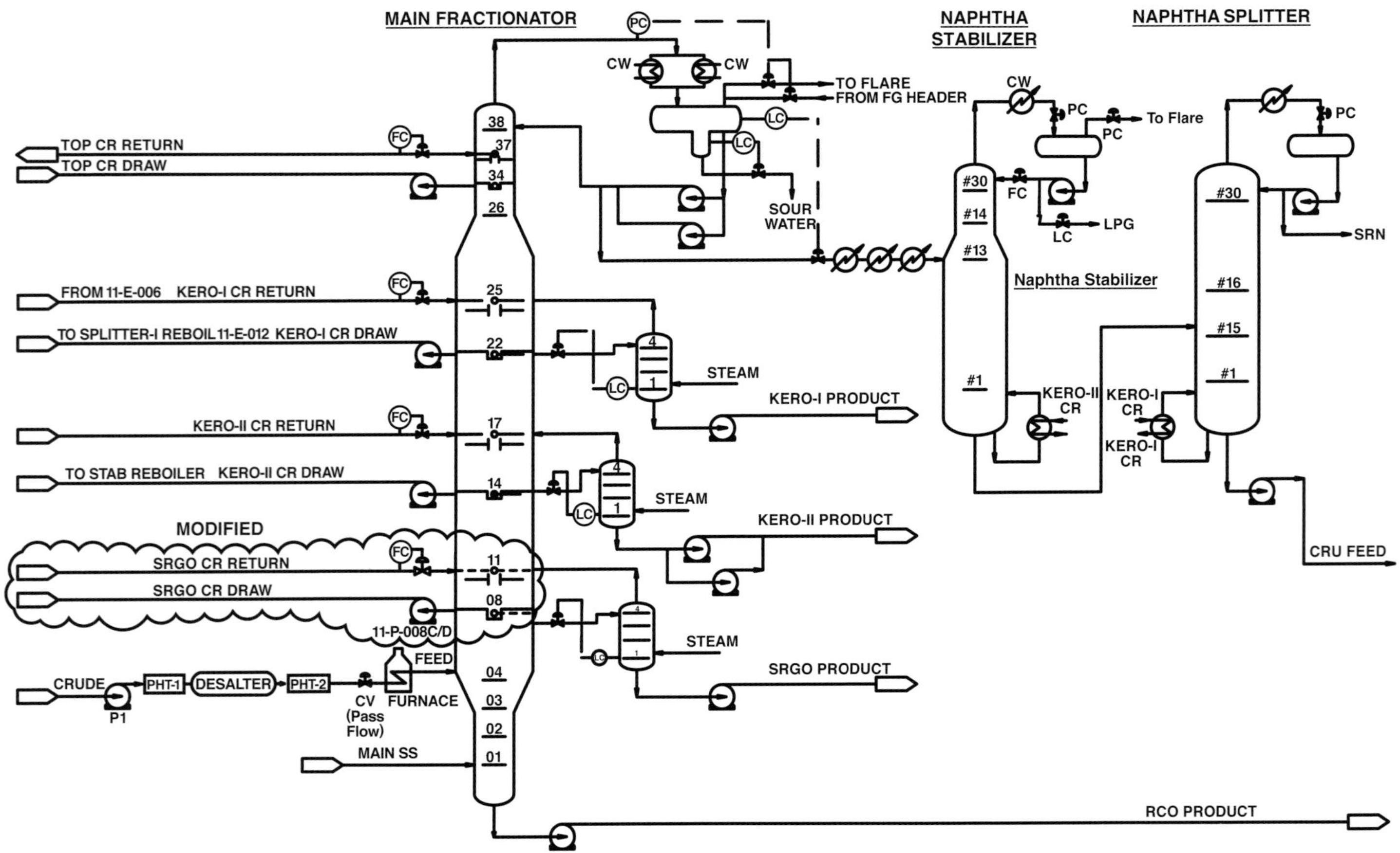

Fig. 3.2.18 Addition of an SRGO or HGO pump-around to increase preheat

3.3 Condensate Distillation Unit

3.3.1 Description of condensate distillation unit

Gas condensates are recovered during exploration and the production of sales-quality natural gas. Gas condensate typically contains a high proportion of naphtha with a very high proportion of distillate content and very low reduced crude oil (RCO). The condensate distillation units are similar to crude distillation units, except for the special ability to handle very low RCO content (as low as 0.5% wt) and very high naphtha content. This low RCO content necessitates a special approach in the design of gas condensate units.

The typical design has the unit pumping the condensate through a mercury trap, through the pre-desalter heat-exchanger train, and into the desalter, which is similar to the design of crude units. The desalted condensate is then further heated and fed to a pre-flash drum or pre-topper.

A pre-flash drum or pre-topper is a necessity as the bulk of the naphtha is flashed and removed from the condensate prior to being fed through the subsequent heat-exchanger train and furnace. In a crude unit design, a pre-flash drum or pre-topper is optional, but in condensate distillation units, the pre-flash drum or pre-topper cannot be avoided.

Adequate pressure is required at the upstream of the pre-flash drum or pre-topper to ensure that the condensate is in its liquid phase prior to entering the pre-flash drum or pre-topper upstream of the level control valve at the inlet temperature.

Parallel trains are used to contain the pressure drop in the heat-exchanger trains so that it requires minimum pump pressures and consequently minimum design pressures from the heat exchangers.

Pre-flash or pre-topper inlet temperatures cannot be increased much as the pressure required to maintain liquid phase will also increase, demanding higher design pressures of the heat exchangers and consequently increasing costs. However, the inlet control valve can be shifted upstream to the last one or two heat exchangers so that the feed is found in liquid phase at the upstream of the pre-topper/pre-flash drum inlet valve at the specified temperature and pressure. Normally, in condensate distillation units, the very high naphtha content does not require as high an inlet temperature for the pre-topper/pre-flash drum.

The major problem is encountered in the charge heater, as the charge is very light and RCO content is very low. To resolve this problem, a part of the RCO from the column bottom is pumped, cooled through heat exchange with the pre-flashed/pre-topped condensate, and introduced along with the feed to the furnace (Fig. 3.3.5).

This will contain the vaporization inside the furnace tubes, and help control the tube metal temperature (TMT) of the furnace tubes and high velocities in the tubes. This is an effective means of protecting the charge furnace from over-heating and flow-induced vibration. Except for this feature of RCO recycle, condensate units are similar in design to crude distillation units. However, mercury traps are most often installed upsteam of the unit to protect the downstream equipment.

Other aspects of the unit are essentially the same as a crude unit processing light crudes with a pre-flash drum or a pre-topper. The flow scheme of a condensate distillation unit with a pre-flash drum is given in Fig 3.3.1. The pre-desalter section, the post-desalter section, and the main column are shown in Figs. 3.3.2, 3.3.3, and 3.3.4, respectively. Fig. 3.3.5 shows the installation of a new pre-topper. The new pre-topper bottom is routed to the existing pre-flash drum to avoid too many modifications.

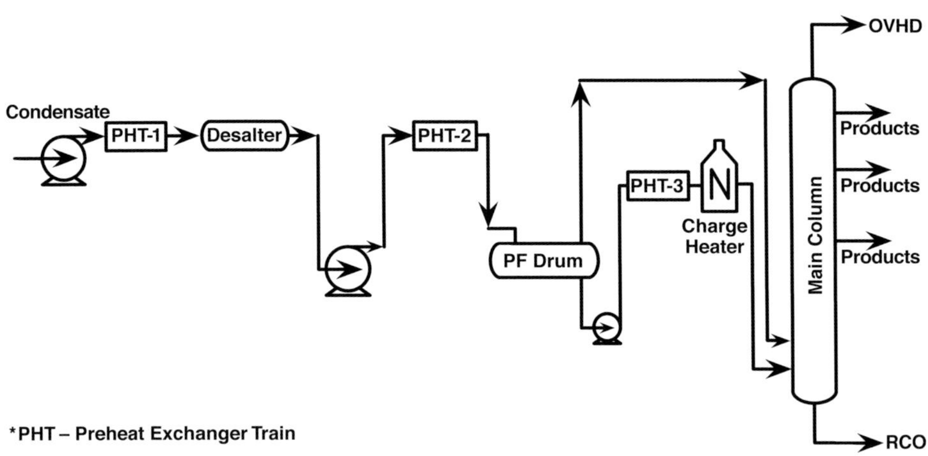

Fig. 3.3.1 Overall scheme

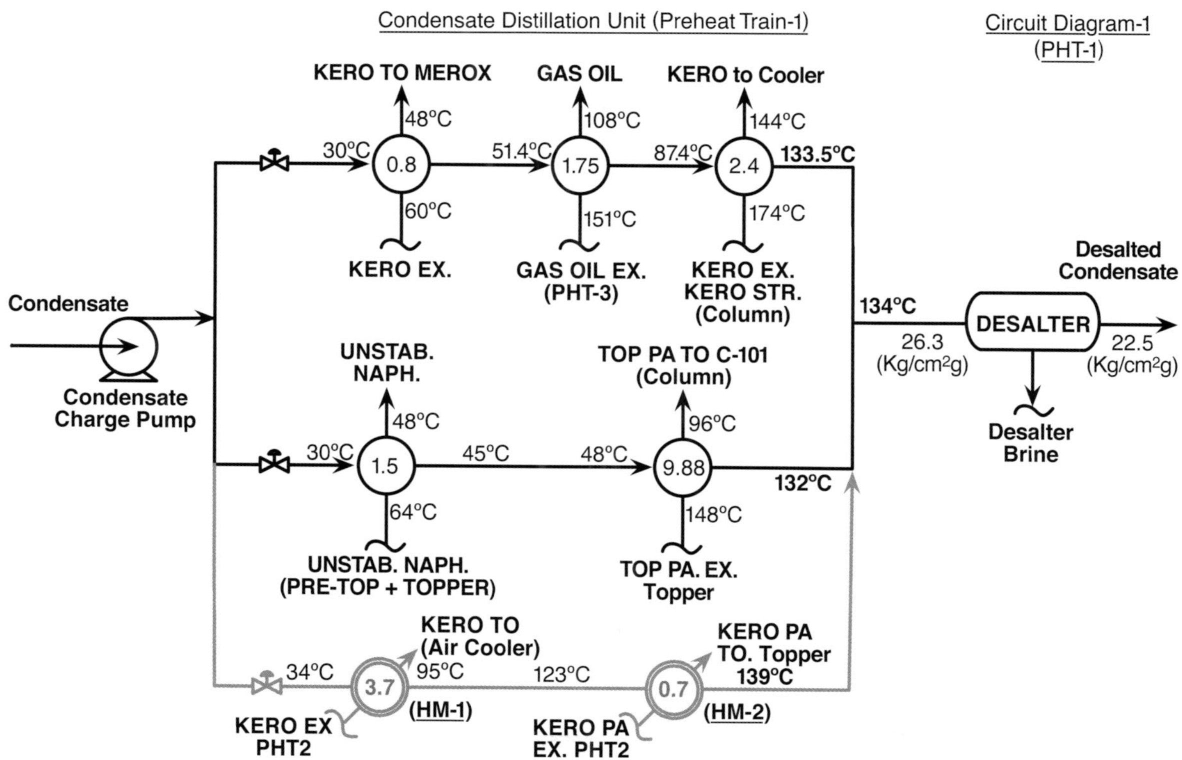

Fig. 3.3.2 Existing preheat train 1 with addition of a parallel train of exchangers

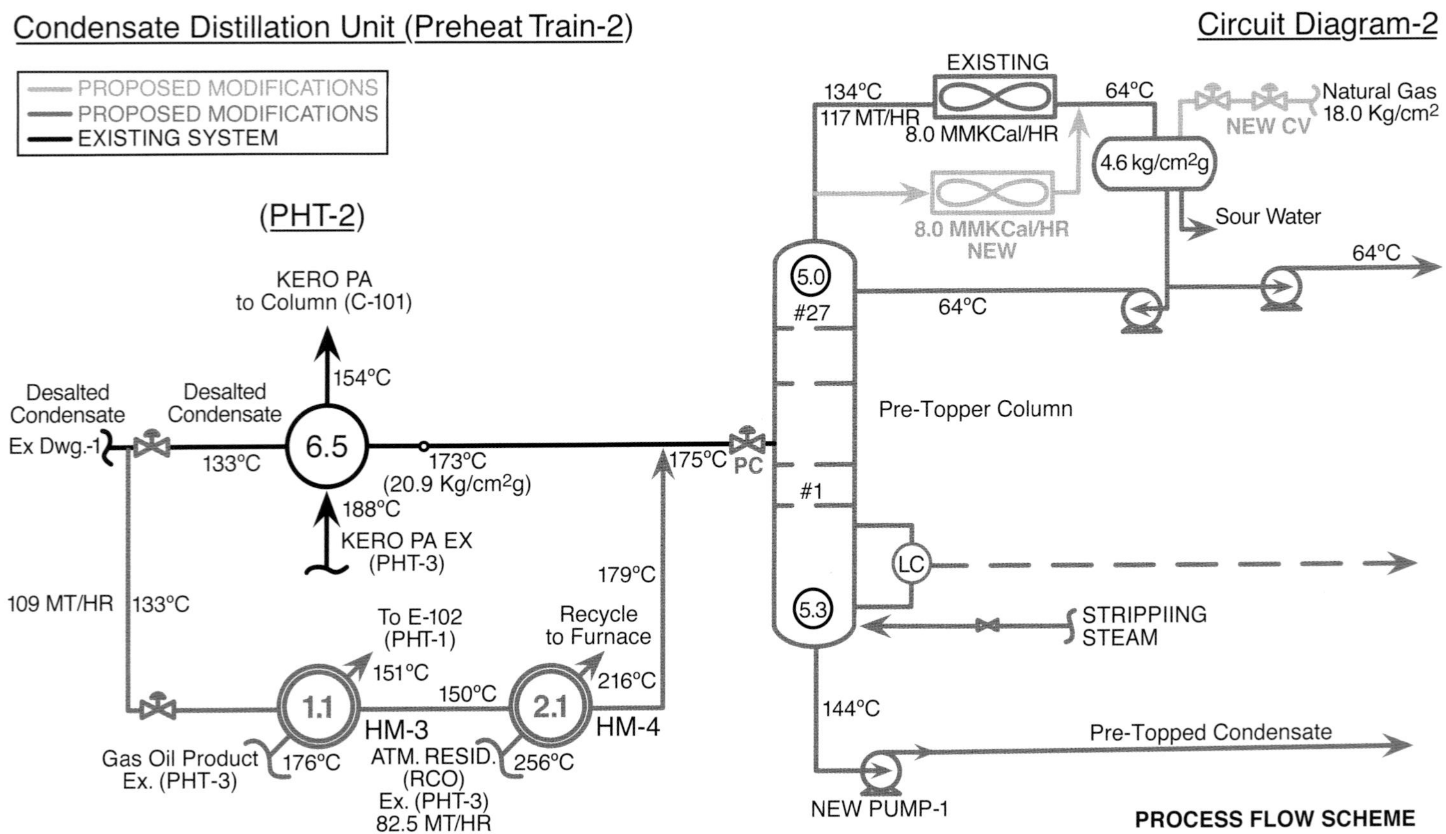

Fig. 3.3.3 Preheat train 2 with addition of a parallel train of exchangers, pre-topper

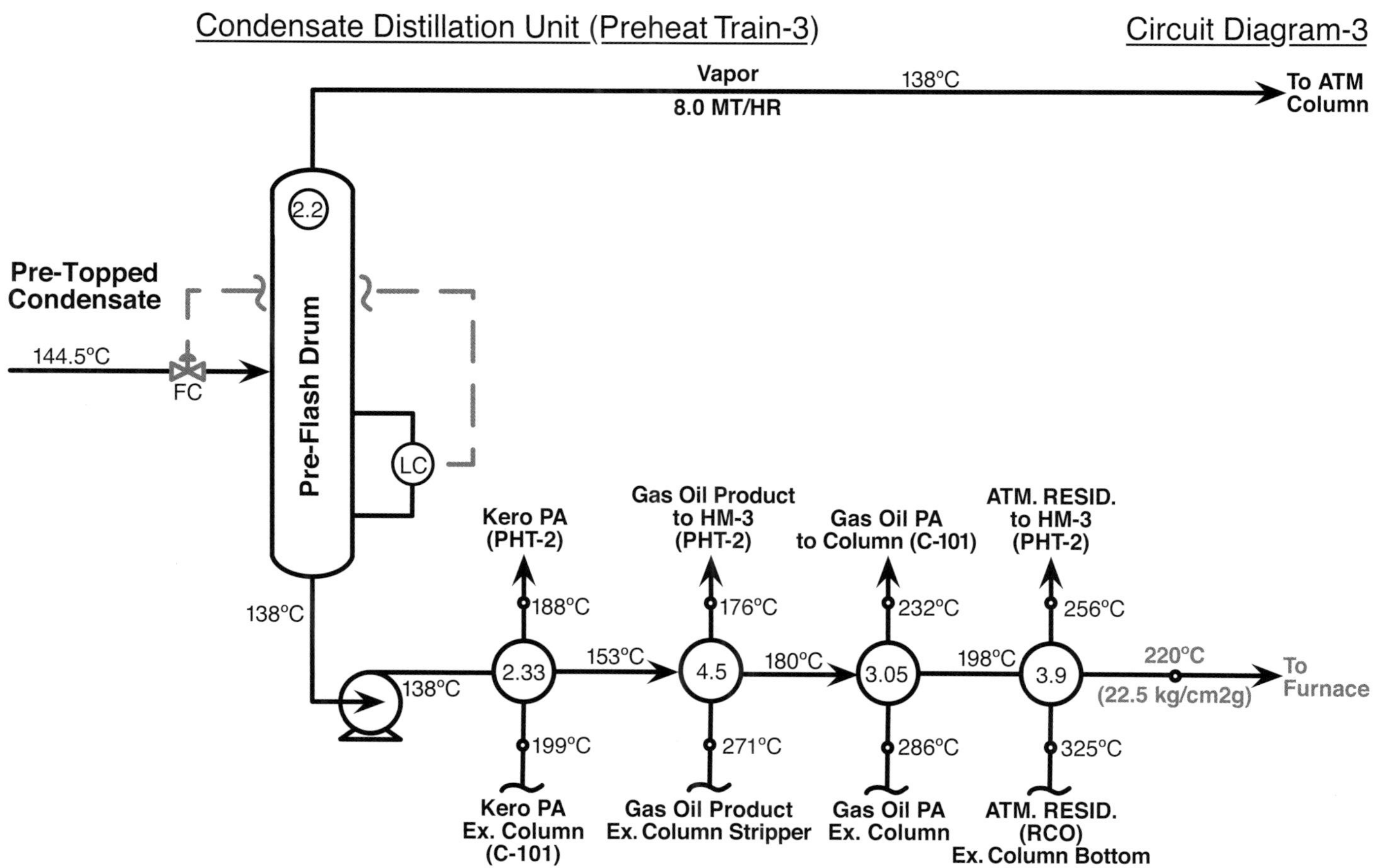

Fig. 3.3.4 Preheat train 3 with no modification

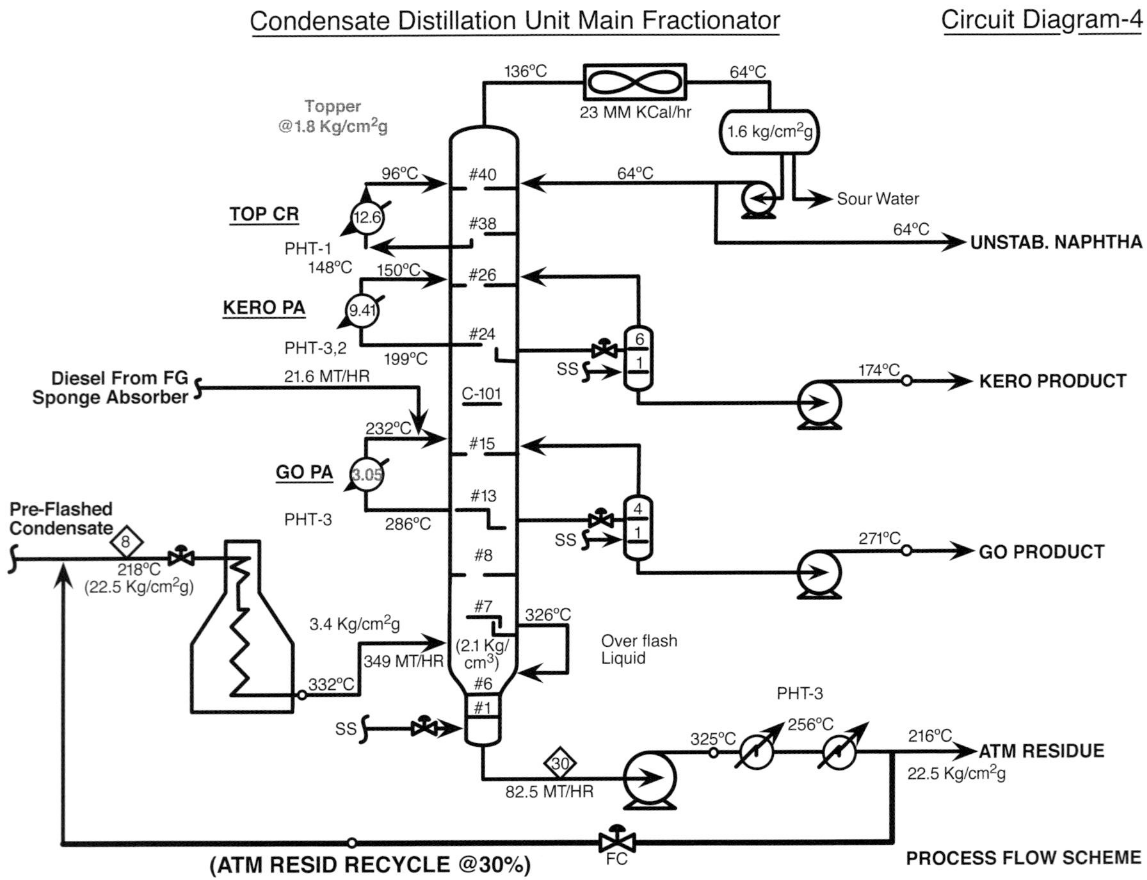

Fig. 3.3.5 Atmospheric column with furnace recycle

As the capacity of the unit was to be revamped to process 25% more, the pre-topper was introduced to remove higher quantities of naphtha from the pre-topper top product. This reduced the quantity of pre-topped condensate and decreased the existing charge furnace mass velocity, pressure drop, and main fractionator diameter limitation.

A parallel heat-exchanger chains (Figs. 3.3.2, 3.3.3) are introduced upstream and downstream from the desalter to contain the pressure drop in the overall heat-exchanger train. The third preheat train (PHT 3) remained the same (see Fig. 3.3.4), as a higher quantity of naphtha is removed from the top of the pre-topper column.

For simple execution, the pre-topped condensate is routed to the existing pre-flash drum and pumped through the charge heater then to the fractionating column (Fig. 3.3.5).

The main column has pump-around refluxes for kerosene and gas oil and a top pump-around as usual with side strippers for products (Fig. 3.3.5).

The unit has attained the desired capacity and improvement of preheat to around 285°C.

3.4 Crude Vacuum Units in Crude Oil Upgraders

Introduction and description

Crudes with a °API of around 18–22 are usually called heavy crudes. Crudes with a °API of 8–10 are called extra heavy crude oils (EHCO). Extra heavy crudes are available in the Orinoco Belt in Venezuela and the Athabasca tar sands region of Northern Alberta.

Latin American heavy crudes

Crude Name	°API (Approx.)
Mexican Maya	17.0
Merey	8.0–10.0
Zuata	8.0

Canadian heavy crudes

Crude Name	°API (Approx.)
Cold Lake	8.0–10.0
Lloyd Minster	8.0–10.0
Tar Sand Blends	8.0–10.0

Typical heavy crude characteristics

1. Low API
2. High total acid number (TAN)
3. High sulfur
4. High viscosity
5. High RCO and VR yield
6. VR has a high TAN and CCR, and contains large quantities of sulfur, nickel, and vanadium
7. High VGO yield
8. VGO has a high TAN and CCR, and contains large quantities of sulfur, nickel, and vanadium
9. Lower distillate yield

The TAN of some of the heavy crudes is given in Table 3.4.1.

Table 3.4.1 Total acid number (TAN) of some heavy crudes

Crude	TAN (mg KOH/g)
Kuwait	0.11
Arab Heavy	0.22
North Gujarat (INDIA)	2.06
Maya	0.4
BCF	2.5
Merey	1.2
Zuata	2.4
Cold Lake	1.0
Lloydminster	0.7
Canadian Blend	2.3

Usual method of processing

Heavy crudes are normally blended with lighter crudes before processing in refineries. During blending, compatibility in respect to precipitation must be addressed. Heavy crudes are normally blended with

1. Lighter crudes (compatible ones that do not result in asphaltene precipitation)
2. Hydrotreated lighter products from bitumen upgraders
3. Upgraded crude oil (UCO) or synthetic crude oil (SCO)

Heavy crudes (8°–10°API) are upgraded to synthetic crude oils of around 32°–42°API or blended with light crudes. Upgraders have a typical refinery-like configuration in that the essential features are carbon rejecting, hydrogen injecting, or a combination of both.

Two types of units used in carbon-rejecting facilities are VR coker units and VR de-asphalting units. In these units, VR is usually fed to a coking unit or a de-asphalting unit, and carbon is rejected as coke or asphalt.

Two typical kinds of units used in hydrogen-injecting facilities are RCO or VR hydrotreaters, VGO, hydrocrackers, or slurry hydrocrackers.

Some of the upgrader configurations have the following:

1. Delayed coker
2. RCO or VR hydrocracker or hydrotreater
3. Solvent de-asphalting

The liquid yield of synthetic crude oil (SCO) is around 88%–95% by volume, depending upon the upgrader configuration. Typically, the SCO is back-blended with heavy crude (unconverted crude) to produce crude oils of different grades based on their °API.

Description of a crude upgrader

An upgrader has essentially one crude unit and one vacuum distillation unit as the mother units. In a configuration with a coker unit, the coker unit is used to produce coke, the carbon is rejected, and all liquid products are blended back to the crude with or without hydrotreatment. The LVGO can be blended back with or without hydrotreatment. The HVGO cuts are typically hydrocracked at the desired conversion level and blended back into the crude. The extent of conversion in the hydrocracker is dependent on the desired °API of the UCO. A typical upgrader scheme with a coker is given in Fig. 3.4.1.

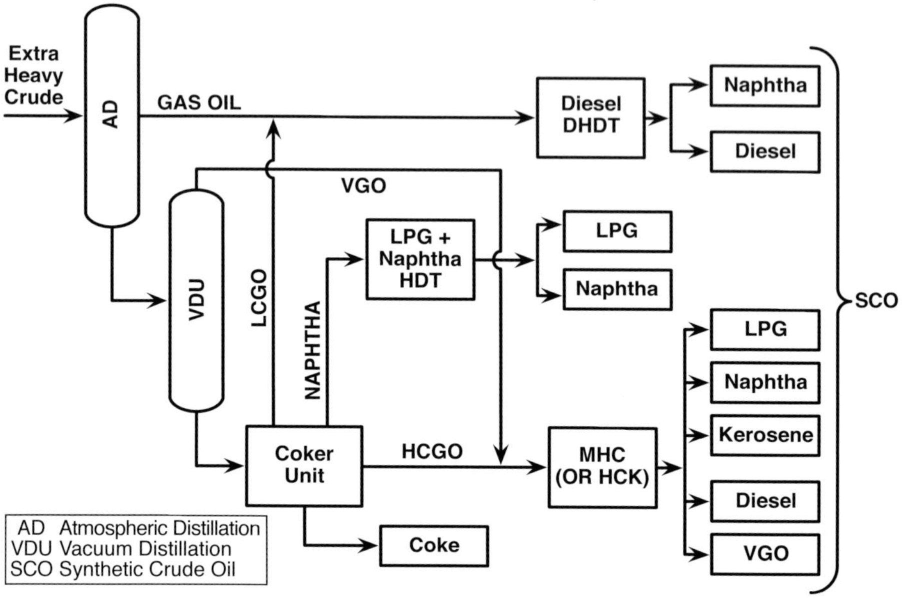

Fig. 3.4.1 Typical scheme for an upgrader with a coker

The vacuum residue (VR) is fed to the coking unit and the coke produced is rejected. The coker naphtha is treated in the hydrotreater to destroy all the diolefins and often to reduce monoolefins. The light coker gas oil and heavy coker gas oil can be blended with the crude with or without hydrotreatment based on the requirement of API of the UCO.

A solvent de-asphalting unit can be also used in place of a coker unit if the asphalt can be disposed of. In a hydrotreater-type configuration, the RCO, VR, or VGO is hydrotreated or hydrocracked and the products are blended back to crude.

The scheme of an upgrader with a coker and a DHDT is given in Fig. 3.4.2.

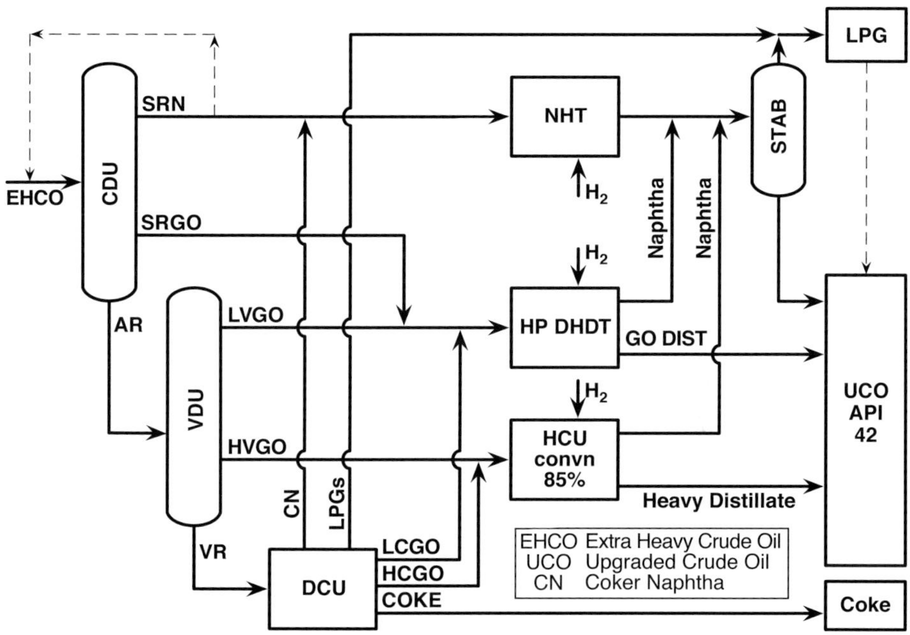

Fig. 3.4.2 Typical scheme for an upgrader with a coker and a DHDT

Features in crude units used for heavy oil processing

1. Use of efficient filtering media at the crude booster pump suction
2. Higher desalter temperature (in the range of 140°C–150°C)
3. A parallel train of heat exchangers to reduce pressure drop on account of higher crude viscosity
4. Additional area in the preheat exchangers to provide the desired preheat and desalter temperature
5. Better desalting facilities to handle higher viscosities and high salt content in the crude. (Tar sand crudes have high sediment and clay contents and some blends also have high viscosity, creating more difficulties in desalting.)
6. Maintaining igh velocity in heat exchangers to avoid asphaltene precipitation
7. Recycling of naphtha and heavy naphtha along with heavy crude
8. Special metallurgy in the sections of the column based on naphthenic acid distribution in the crude

9. Special metallurgy in furnace tubes and/or transfer lines (SS316L, SS317, etc.)
10. Increasing atmospheric tower bottoms (ATB) product cut points to stay within vacuum column diameter limits
11. Splitting of RCO and VR streams due to their high viscosities, particularly in revamp applications to contain the pressure drop

The problem of low preheat and low desalter temperatures is contributed to by the following:

- Low heat transfer coefficient (U_D) caused by high viscosity of streams
- Low pump-around duty in the crude column because of low distillate content
- Less heat is recovered from the VR as it cannot be cooled much due to its high viscosity

Features in vacuum units used for heavy oil processing

1. A vapor horn is installed at the column inlet to reduce entrainment of heavier ends, to limit metal content of the VGO stream.
2. The metallurgy of selected sections of column can be upgraded depending on the naphthenic acid distribution in the crude.
3. Vacuum slop pump-around may be provided.
4. XVGO or HHVGO withdrawal facilities are provided to draw the extra heavy VGO fraction and divert it to the FCCU feed. The lighter VGO is sent to the hydrocracker to avoid nitrogen, asphaltene, nickel, and vanadium in the HCU feed.
5. A very strong wash zone is required (e.g., a higher irrigation in wash beds to prevent entrainment and wash zone bed coking).

Crude TBPs of some heavy crudes are presented in Figs. 3.4.3 and 3.4.4.

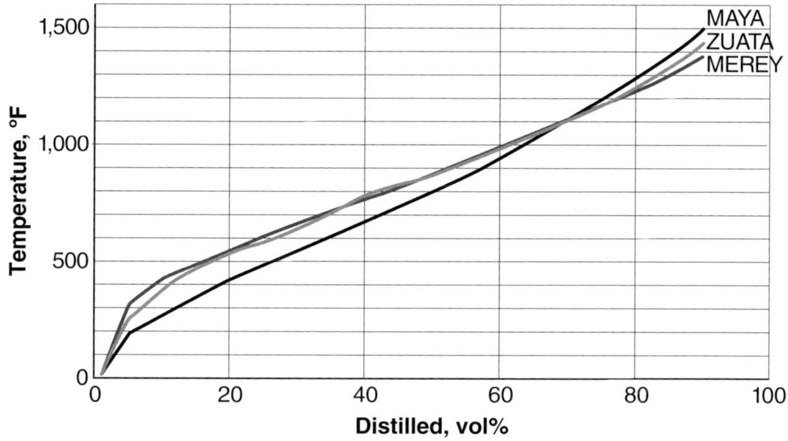

Fig. 3.4.3 Crude TBP of some heavy crudes of Latin American origin

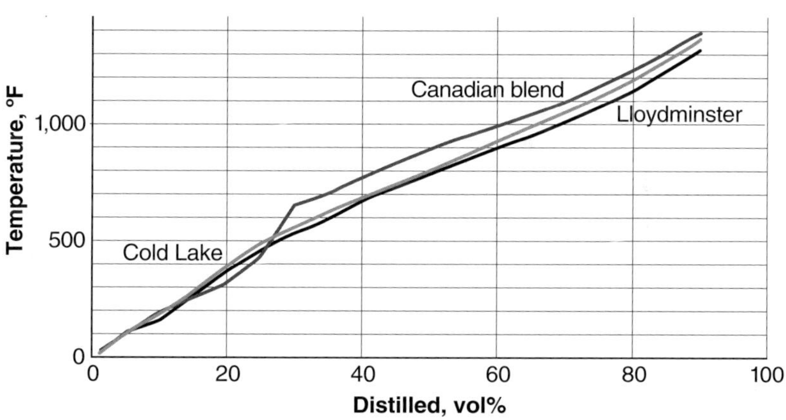

Fig. 3.4.4 Crude TBP of some heavy crudes of Canadian origin

3.4.2 Design methodology of crude or vacuum units used for extra heavy crude oils in upgraders

These crude units can be designed normally, but as the crude contains a very low proportion of atmospheric distillates, a two-column configuration can also be used. The first column would separate the RCO from the distillates with 6–7 trays at the top of the flash zone to achieve a good separation between the RCO and the distillates. Ammonia or a neutralizing amine is injected at the top of the first column at the upstream of the air cooler to neutralize the HCl. The top temperature of the first column will be higher than the top temperature with a normal configuration; thus, the chance of water condensation in the first column top is less likely, which is likely to reduce HCl corrosion at the top.

The distillate yield is low so the first column's overhead circuit will not be large. It can be seen from the TBP of some of the heavy crudes with different origins that the atmospheric distillates (those that boil below 360°C) will be at the most around 30%–33% by volume (Figs. 3.4.3 and 3.4.4).

Tar sand crudes from Canada have high sediment and clay contents and some blends also have a high viscosity, creating difficulties in the desalting process. Typically for such crudes, a desalter is not installed. Some have a low salt content. If no desalter is provided and the salt content is low, a two-column configuration may be preferred. The overhead liquid product of the first column, with much lower chloride, is fed to the second column, reducing the likelihood of HCL corrosion in the second column. This configuration also may not require much higher metallurgy/Monel lining in the column.

In this scheme, the overhead of the first column will preheat the crude. The distillate product can be back-blended or fed to the second column (after preheating with overhead) to separate the naphtha from the middle distillates with a provision of around 10–12 trays (Fig. 3.4.5). The naphtha can then be stabilized in a stabilizer for recovering the LPG.

However, if the distillate content is higher, a normal crude column with pump-around refluxes will be the better choice (discussed earlier in section 3.1). The vacuum distillation unit is typically configured as a normal fuel-type vacuum unit (discussed earlier in section 3.2). These heavy crudes typically contain a high proportion of VGO. The vacuum column is provided with pump-around refluxes as usual. The refluxes are cooled while preheating the crude.

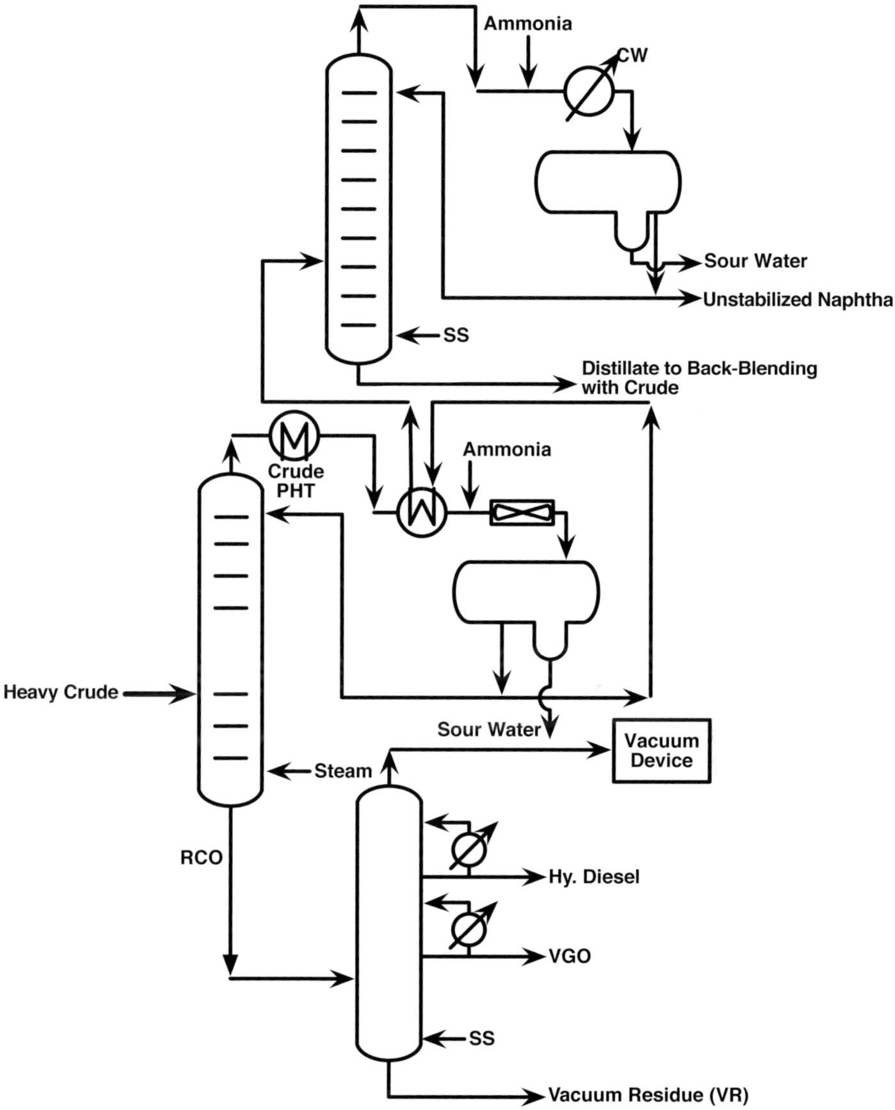

Fig. 3.4.5 Two-column crude distillation

3.4.2 Case study of a capacity revamp with the introduction of heavy crude as an alternative

A crude unit was originally designed for Kuwait crude and had a capacity of 3.0 MMTPA.

The unit originally had only a topper (no pre-topper or pre-flash drum) and was described earlier in Fig. 3.1.22 (reproduced in Fig. 3.4.6).

It was decided to increase the capacity of the unit to around 7.0 MMTPA. Accordingly, design was carried out with the addition of a pre-topper, the introduction of the split of gas oil cut (LGO, HGO) draws and pump-around refluxes, and an additional kerosene cut of linear alkylbenzene feedstock (LABFS) with its own pump-around reflux. Two additional strippers were added (Fig. 3.4.7), as well as a parallel furnace with a parallel heat-exchanger pass in the exchanger train and a method for splitting the RCO stream into two streams (Figs. 3.4.7, 3.4.8, 3.4.9, and 3.4.10).

After the package was finalized and some of the heat exchangers and pumps had already been purchased, the refinery wanted to introduce flexibility to consider a heavy crude of Indian origin (as an alternative crude) to be processed in a blocked out operation along with the Kuwait crude.

The heavy crude had a °API of around 28, and an RCO yield of around 77% by weight, but had a low TAN. It was decided to continue with the pre-topper and the pre-topper overhead coupled with the crude in order to achieve further energy improvement in the Kuwait operation, as the availability of the heavy crude was not certain.

In the Kuwait operation, the pre-topper would be working as regular pre-topper, producing pre-topped naphtha as an overhead product. During heavy crude operation the pre-topper was used as a pre-flash vessel (without stripping steam) and no overhead product was withdrawn.

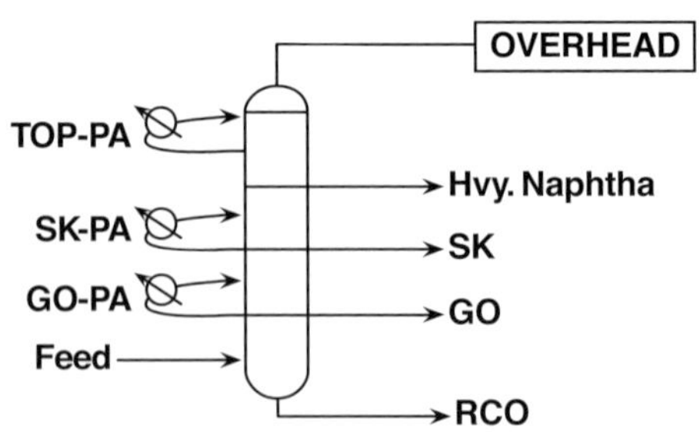

Fig. 3.4.6 Original configuration (from Fig. 3.1.22)

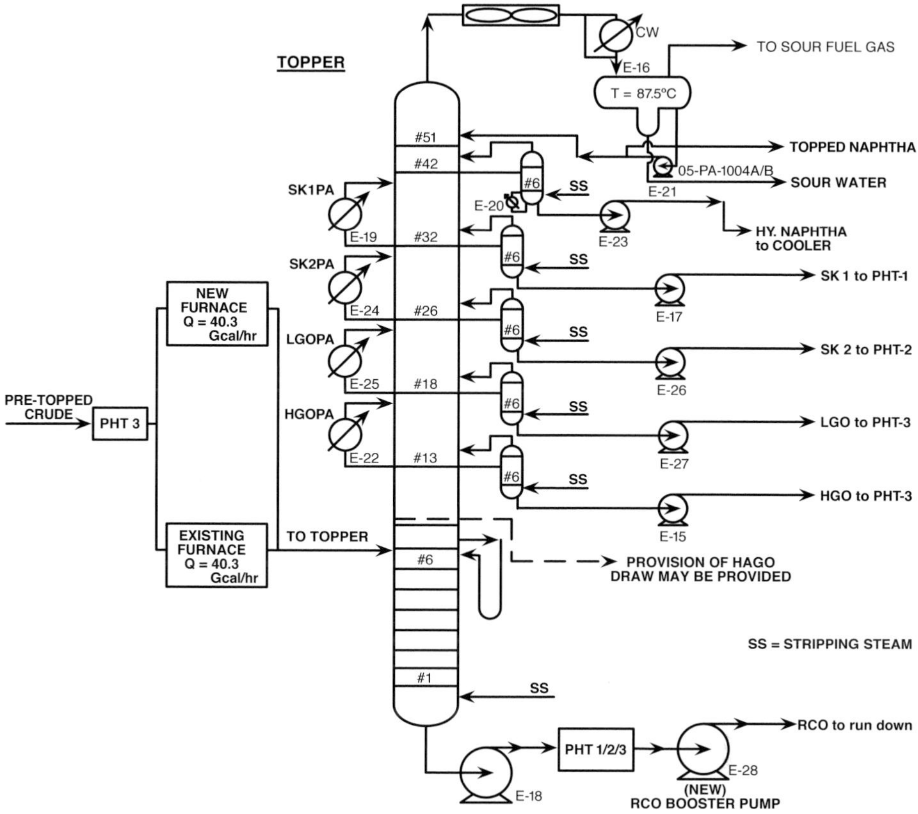

Fig. 3.4.7 Modified scheme for revamp

During the heavy crude operation, if the pre-topper had been operated to recover pre-topped naphtha as overhead, there would not have been any load in the top section of the topper due to a very low naphtha content of the heavy crude. Thus, the top section trays of the topper would have gone below turn-down (as the top section internals of the topper were designed for Kuwait crude and had a significant load even with the pre-topper upstream). In the simulation, the pre-topper has been configured as a vessel for the heavy crude. At the pre-topper pressure and inlet temperature, no vaporization of the heavy crude was expected to occur without the stripping steam. The difficulty was to keep the section of the pre-topper above the feed hot during the heavy crude operation, as no stripping steam was introduced and no overhead product was produced from the pre-topper during heavy crude operation. To resolve the problem, a purge steam connection (with a restricted orifice) was provided above the flash zone of the pre-topper (Fig. 3.4.8) to keep the upper section and overhead circuit of the pre-topper hot during heavy crude operation if required to avoid thermal shock during the changeover to the Kuwait crude operation. During heavy crude operation, there would be no pre-topped naphtha production; only a small amount of unstabilized naphtha would be recovered from the topper overhead and that would be routed to the stabilizer to maintain the vapor pressure of the stabilized naphtha. The pre-topper overhead or crude heat exchanger would not deliver any heat to the crude during operation with the heavy crude.

Provision of a heavy naphtha recycle was also introduced at the pre-desalter heat-exchanger train (PHT1) to cut back the viscosity of the crude.

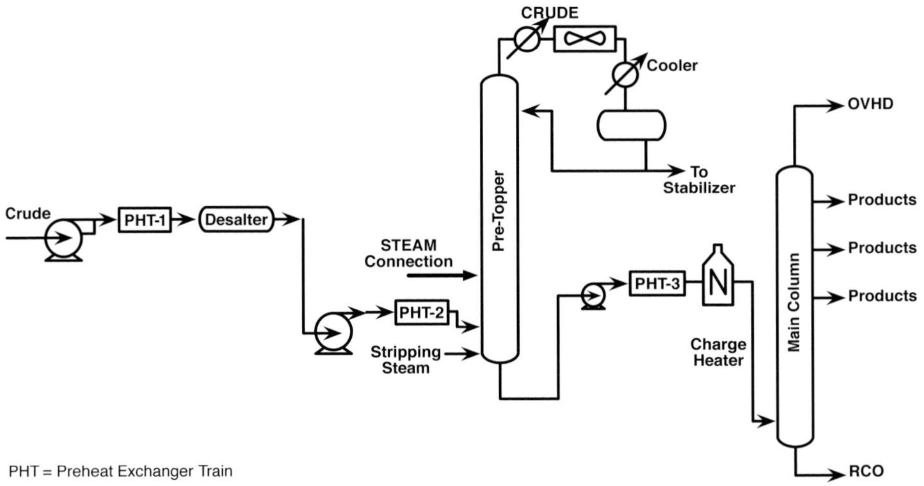

Fig. 3.4.8 Addition of a pre-topper unit

A RCO heat exchanger was added at the downstreams of PHT1, PHT2, and PHT3 to recover the RCO heat from the heavy crude and cool the RCO.

There was a greater pressure drop in the RCO circuit due to the addition of the heat exchanger necessary for handling the high flow and viscosity of the RCO from the heavy crude. A pump was added at the downstream of the RCO cooler to supply the battery limit pressure required to reach the RCO tank farm. The modifications in the heat-exchanger trains are shown in Figs. 3.4.9, 3.4.10, and 3.4.11.

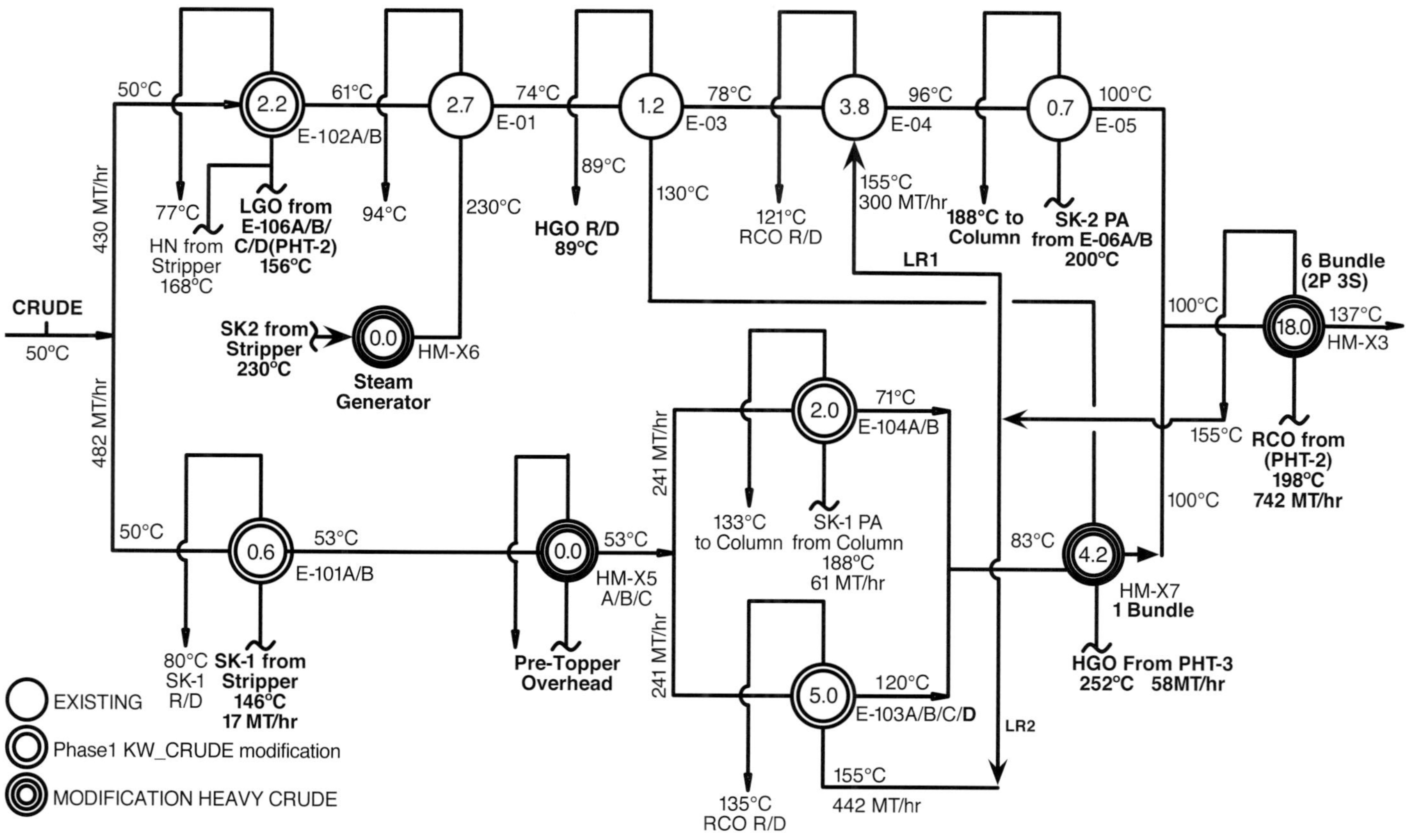

Fig. 3.4.9 Preheat Train 1 (PHT1) showing modifications

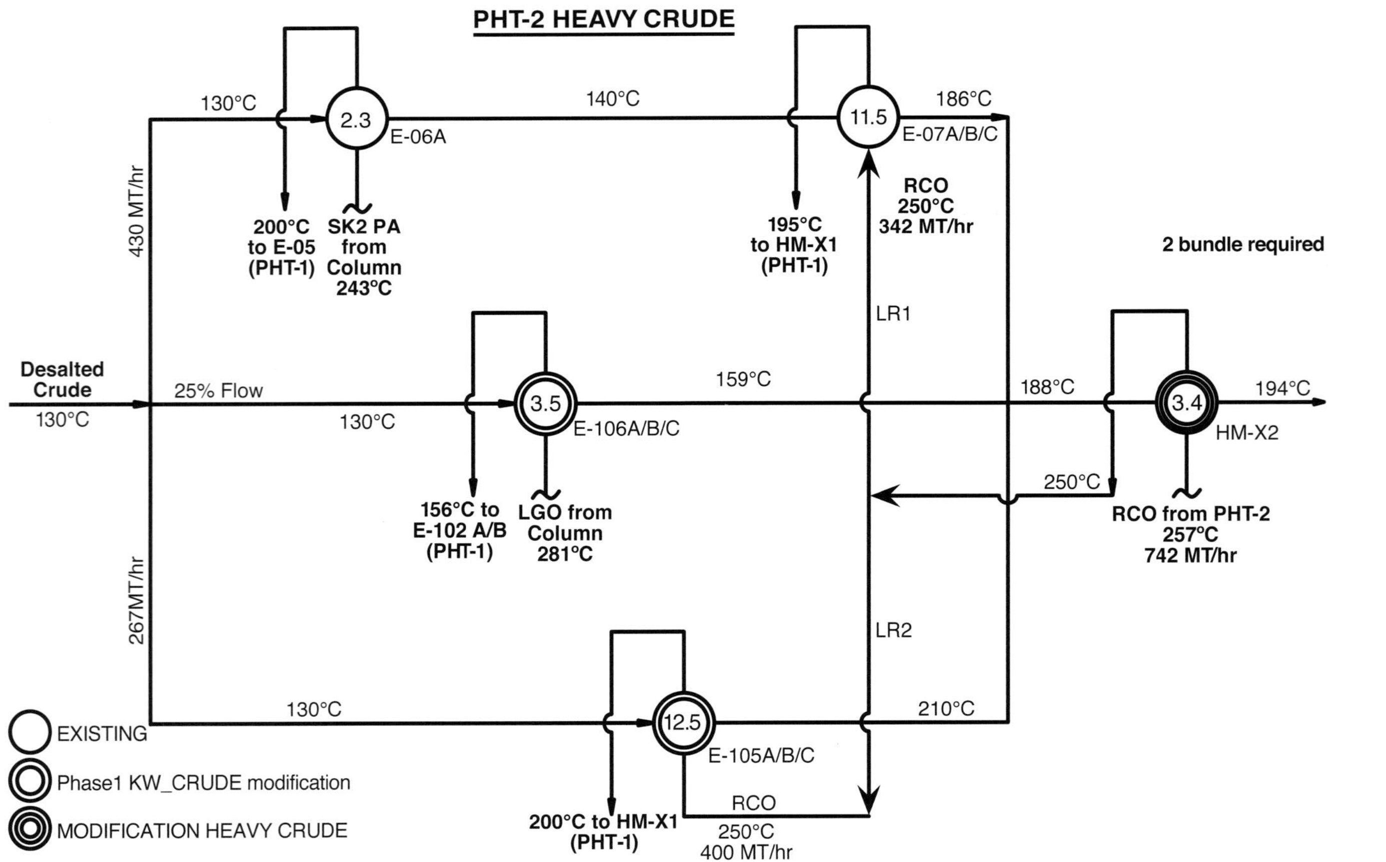

Fig. 3.4.10 Preheat Train 2 (PHT2) showing modifications

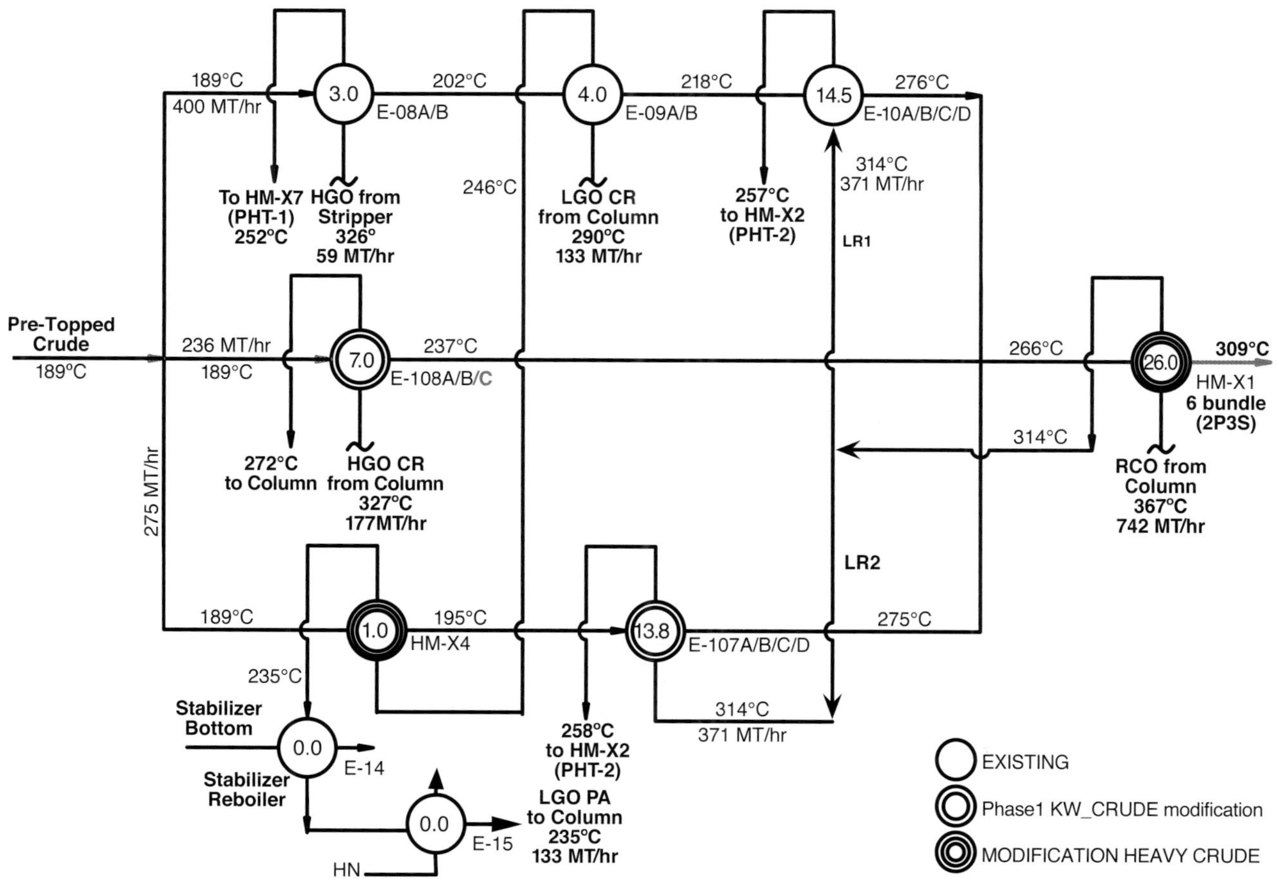

Fig. 3.4.11 Preheat Train 3 (PHT3) showing modifications

Abbreviations

°API: American Petroleum Institute gravity, (141.5 – (131.5/sp. gr.)), where sp. gr. is the specific gravity

ASTM gap: a positive difference of ASTM 95% point temperature of upper product and ASTM 5% point temperature of adjacent lower product

ASTM overlap: a negative difference of ASTM 95% point temperature of upper product and ASTM 5% point temperature of adjacent lower product

ASTM: American Society for Testing and Materials

ATF: aircraft turbine fuel

AVU: atmospheric-vacuum unit

COT: coil outlet temperature

CR: circulating refluxes

CRU: catalytic reformer unit

CCRU: continuous catalyst regeneration unit

DHDS: diesel hydrodesulfurization

DHDT: diesel hydrotreating/hydrotreater

EHCO: extra heavy crude oil

EHGO: extra heavy gas oil

FBP: final boiling point

FCCU: fluid catalytic cracking unit

GCC: grand composite curve

HAGO: heavy atmospheric gas oil

HCU: hydrocracker unit

HGO: heavy gas oil

HO: heavy oil

HHVGO: extra heavy vacuum gas oil

HVGO: heavy atmospheric gas oil

IO: inter oil

IR: internal reflux

K: vapor–liquid equilibrium constant

LABFS: linear alkylbenzene feedstock

LDO: light diesel oil

LGO: light gas oil

LO: light oil
LPG: liquefied petroleum gas
LVGO: light vacuum gas oil
MMTPA: million metric tonnes per annum
NG crude: North Gujarat crude (heavy crude of Indian origin)
PA: pump-around
PHT: preheat train
RCO: reduced crude oil
RON: research octane number
SCO: synthetic crude
SO: spindle oil
SRGO: straight-run gas oil
TAN: total acid number
TCC: true composite curve
TBP: true boiling point
TMT: tube metal temperature
UCO: upgraded crude oil
VGO: vacuum gas oil
VR: vacuum residue
XVGO: extra vacuum gas oil

Note

1. The design was jointly developed with Mr. M. M. Khan and Mr. Mukesh Sharma, and a paper was jointly authored about this unique design to debottleneck a vacuum unit. The paper was selected and presented at IRPC Singapore in July 2011.

Hydrotreating and Hydrocracking

4

4.1 Introduction

A hydrotreater is typically used to treat the following streams:

1. Vacuum gas oil (VGO)
2. Heavy coker gas oil (HCGO)
3. Reduced crude oil (RCO), vacuum residue (VR), or dewaxed oil (DWO)
4. Straight-run gas oil (SRGO), light coker gas oil (LCGO), or light cycle oil (LCO) from the fluid catalytic cracking unit (FCCU)
5. Straight-run kerosene
6. Lube oils

The VGO, HCGO, RCO, and VR hydrotreaters are used to desulfurize and denitrify the feed, as well as reduce its CCR and metal content. Some amount of conversion is also achieved during the process of hydrogenation in hydrotreating.

The gas oil components SRGO, LCGO, and LCO are typically treated in a hydrotreater to lower their sulfur contents, improve their cetane numbers, etc. Kerosene cuts are also treated in a hydrotreater to lower their sulfur contents, as well as to improve their smoke points. Lube oils are treated in a hydrotreater to improve certain qualities like color, stability, etc.

A hydrocracker is typically used to hydrogenate and crack the VGO or HCGO feeds to upgrade them to lighter distillates. Conversions of up to 100% are often attained. For a partial conversion unit, the unconverted residue is typically fed to the FCCU to upgrade it to a distillate, or used to produce feedstock for the production of lube oil cuts.

4.2 Description

Hydrocrackers and hydrotreaters both essentially consist of the following:

1. Reactor section
2. Effluent cooling section
3. Feed and recycle gas heating section
4. Cooled effluent heating section prior to fractionation section
5. Product fractionation or stripping
6. Makeup compressor section
7. High-pressure water injection system
8. Feed filtration section
9. High-pressure recycle gas amine wash system as optional system

4.2.1 Processing scheme

The processing schemes in hydrotreaters and hydrocrackers are essentially the same. The only difference lies in the fractionation section. In a hydrocracker, a fractionation facility is provided for separating the distillate from the unconverted oil (Fig. 4.2.1, 4.2.2), whereas in a hydrotreater, a product stripper strips off the light ends (Fig. 4.2.3).

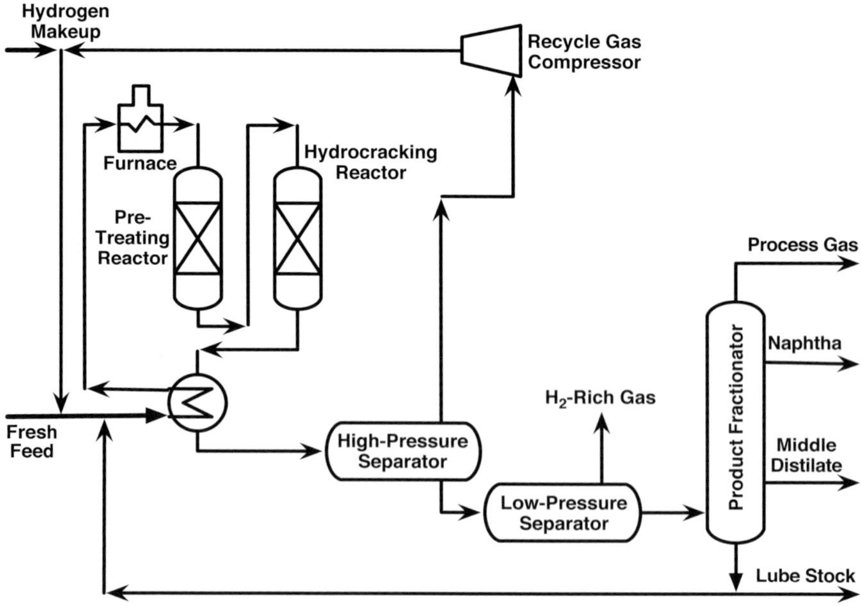

Fig. 4.2.1 Simplified scheme of a hydrocracker

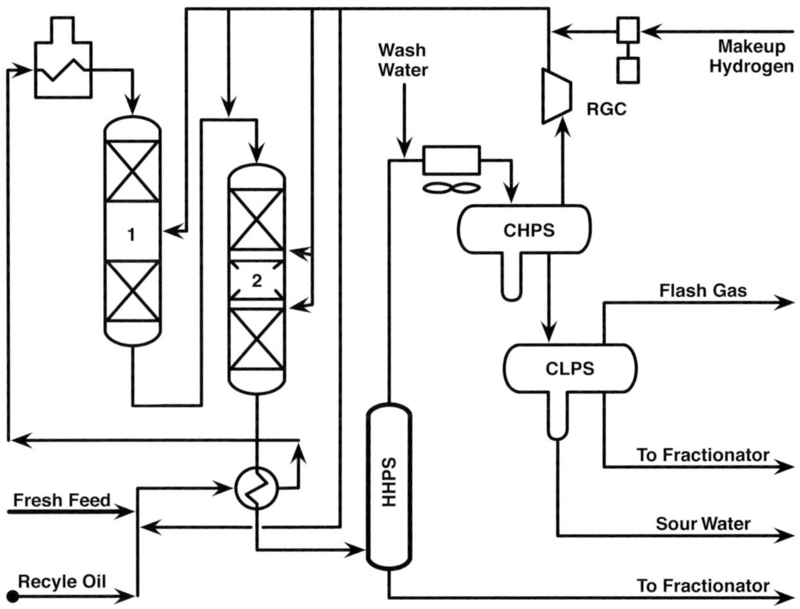

Fig. 4.2.2 Typical scheme of a hydrotreater

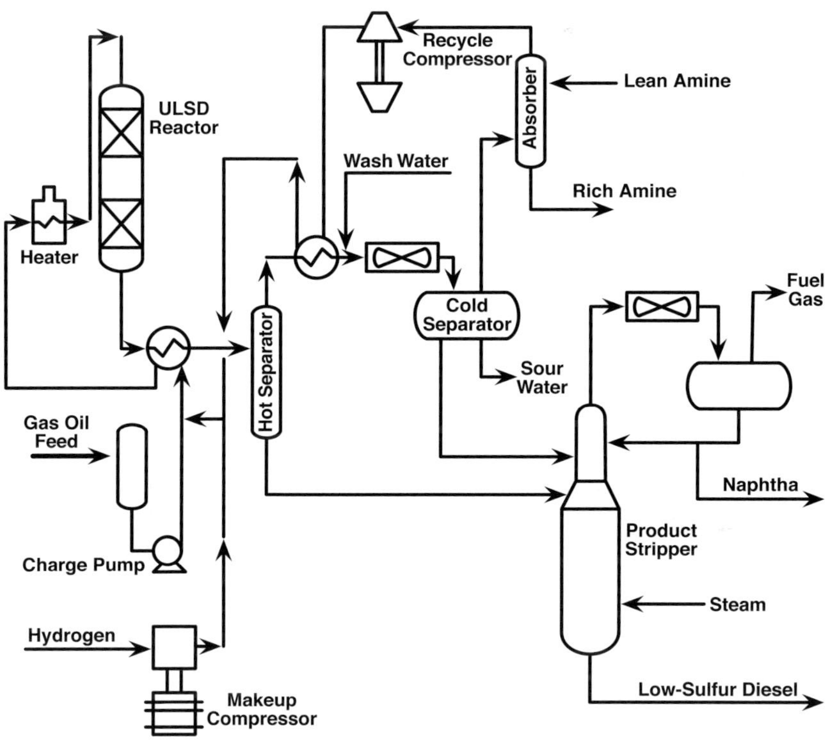

Fig. 4.2.3 Typical hydrotreater scheme

In a hydrocracker, the raw feed is first heated for filtration and, after filtering, normally collected in a filtered feed surge drum. From there it is picked up by a high-pressure pump, pumped and mixed with recycle gas, routed to the feed vs. effluent heat exchangers, then sent through a high-pressure reactor furnace to the reactors.

Often, recycle gas is heated separately in recycle gas vs. reactor effluent heat exchangers, then in a recycle gas heater (furnace for high-pressure recycle gas) before joining the heated feed at the reactor inlet.

Feed alone is never heated in a furnace in this configuration, as this can cause coking in the furnace. The mixture of hot feed and recycle gas is introduced in the reactor. The effluent from the reactors is cooled by heating the filtered feed, recycle gas, and feed to the fractionator or stripper, by steam generation (optional), and lastly by a high-pressure air cooler (HPAC), which will cool it to approximately 55°C–65°C before routing it to the cold high-pressure separator (CHPS). High-pressure water is injected at the upstream of the HPAC for washing and removing NH_4HS, and NH_4Cl. In the CHPS, the gas, liquid, and water phases are separated. The calculation for the water injection quantity is given later. The liquid from the CHPS is routed to the cold low-pressure separator (CLPS) at around 25–30 kg/cm^2 g. The liquid from the CLPS is fed to the stripper or fractionation section after preheating. CHPS or hot high-pressure separator (HHPS) liquid is often routed to the CLPS or hot low-pressure separator (HLPS) through a power recovery turbine (PRT) installed parallel to the pressure letdown control valve (Fig. 4.2.4).

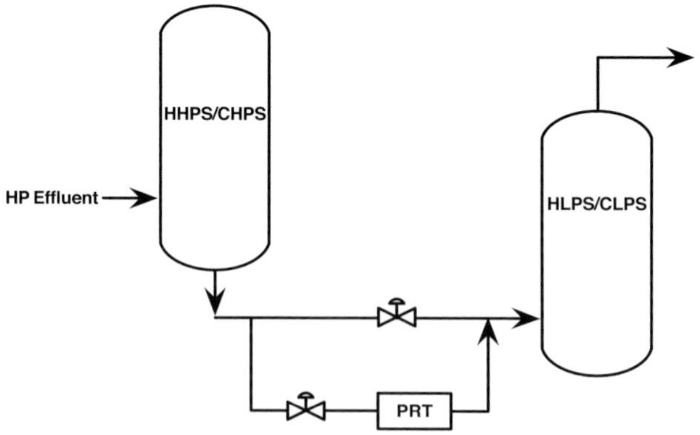

Fig. 4.2.4 Pressure recovery turbine (PRT) for pressure letdown from the HHPS or CHPS

The separated gas from the CHPS is picked up by the recycle gas compressor (RGC), circulated with the feed, and used as a quench at reactor bed outlets or reactor outlets going to the subsequent beds or reactors. Hydrogen is added at the RGC suction or discharge to make up for the hydrogen loss during the reaction, leak, and dissolution.

Reactors

The reactor contains several beds of catalyst to keep the temperature rise across each bed (exothermic reaction) within the limits of the reactor vessel, catalyst, and internal design temperatures. The catalyst in each bed is placed on catalyst supports.

The hydrogenation reaction is strongly exothermic. At the outlet of the beds, a recycle gas quench is injected to reduce the temperature of the mixture before it enters the next bed. Special vapor–liquid contacting facilities are provided to achieve the desired mixing and cooling of the effluent by the cooled recycle gas. Normally a guard catalyst or demetallization catalyst is placed at the top of the bed. Guard catalysts capture the extraneous muck to limit the pressure drop across the bed after continued use. Demetallization catalysts are mild hydrogenation catalysts that capture metals in their large pores. Typically, these are housed in a separate bed (treating catalysts are often housed in a separate reactor preceding the conversion reactors). The guard bed and demetallization catalyst bed are followed by hydrodesulfurization (HDS) and hydrodenitrification (HDN) catalysts. The HDS and HDN catalysts have relatively high hydrogenation capabilities. Typically, denitrification requires higher temperatures than desulfurization. The subsequent beds of catalysts can be for hydrocracking and for aromatic saturation. Aromatic saturation is favored at lower temperatures, as the reaction is exothermic.

Once-through hydrocrackers mostly use the single-stage configuration. Full-conversion hydrocrackers are also designed with the single-stage configuration, often with a recycle of unconverted oil from the fractionation section feeding into the reactor inlet along with fresh feed.

In a two-stage hydrocracker, the first stage purifies and partly converts the VGO to lighter distillates, while the unconverted VGO is fed to the second stage reactor, where the majority of aromatic saturation is done. The second stage inlet temperature in such a configuration is normally kept lower to facilitate aromatic saturation. The cracking activity is higher in the second stage catalyst, as it operates at a lower catalyst average temperature (CAT).

At the inlet of subsequent catalyst beds, the required partial pressure of hydrogen is maintained by recycle gas quench injections between beds, which are primarily used to cool the bed outlets. The reactors are also provided with a means of depressurizing using special control valves (dump valves) located at the top of the CHPS. Depressurization is required in order to arrest temperature excursions, arising from emergencies such as RGC failure, feed failure, etc. The maximum allowable pressure drop during depressurization is 10–14 kg/cm^2 in most of the supplied technologies which usually have a stipulated depressurizing rate specified by the reactor designer or process licensors. This stipulated rate is related to the reactor dimensions (diameter) and the design of the catalyst supports, as well as catalyst bed porosity, catalyst particle size distribution, catalyst bed height, etc. If the pressure drop goes beyond this value, there is a danger of damage to the catalyst supports and other reactor internals.

Addition of makeup hydrogen

Makeup H_2 can be added in the RGC suction or RGC discharge. Adding makeup H_2 at the RGC discharge requires that the makeup gas have a higher pressure and consequently needs a higher compression ratio for the makeup gas compressor (MUG compressor).This, however, provides extra recycle gas flow. Makeup H_2 is often added to the downstream of amine wash at the RGC suction.

For the units with a hot high-pressure separator (HHPS), the HHPS vapors are routed to the CHPS after being cooled and water-washed. The HHPS liquid is often routed to the HLPS through the pressure recovery turbine, while the CHPS vapors are routed to the RGC suction with or without an amine wash. The CLPS and HLPS gases are typically routed either through a membrane or through a pressure swing adsorption (PSA) after cooling for H_2 recovery. When a PSA is used, the gases must go through an amine wash prior to entry to PSA.

Air cooler inlet temperature

Air cooler inlet temperature is an important parameter. If the reactor effluent is cooled too much in the preceding heat exchangers, there is the likelihood that NH_4HS and NH_4Cl will be deposited upstream. The likelihood of NH_4HS and NH_4Cl deposition increases when processing nitrogen- and chlorine-rich feed stocks like coker HCGO. NH_4Cl deposition occurs at higher temperatures than NH_4HS and is dependent upon the NH_4Cl dissociation constant KP_{NH_4Cl}, while NH_4HS deposition is dependent on KP_{NH_4HS} as given below:

$$KP_{NH_4Cl} = [p_{HCl}] \times [p_{NH_3}] \text{ (psia)}^2$$

$$KP_{NH_4HS} = [p_{H_2S}] \times [p_{NH_3}] \text{ (psia)}^2$$

When only straight-run vacuum gas oil is processed (with less chloride), the typical air cooler inlet temperature is kept at around 120°C–125°C, again, depending on the K_p value of the NH_4Cl or NH_4HS formation (see Figs. 4.2.5, 4.2.6 and 4.2.7).

The typical air cooler inlet temperature with part coker product in the feed is 160°C–170°C, which is governed in most cases by the NH_4Cl deposition temperature (which is higher). A cushion of around 5°C–15°C over and above the deposition temperature (either NH_4Cl or NH_4HS, whichever is higher) is typically taken as the air cooler inlet temperature.

The NH_4HS deposition chart is given in Fig. 4.2.7.

The same NH_4Cl and NH_4HS deposition charts are also available in API RP 932B.[4]

NH_4Cl deposition charts are given in Figs. 4.2.5 and 4.2.6.

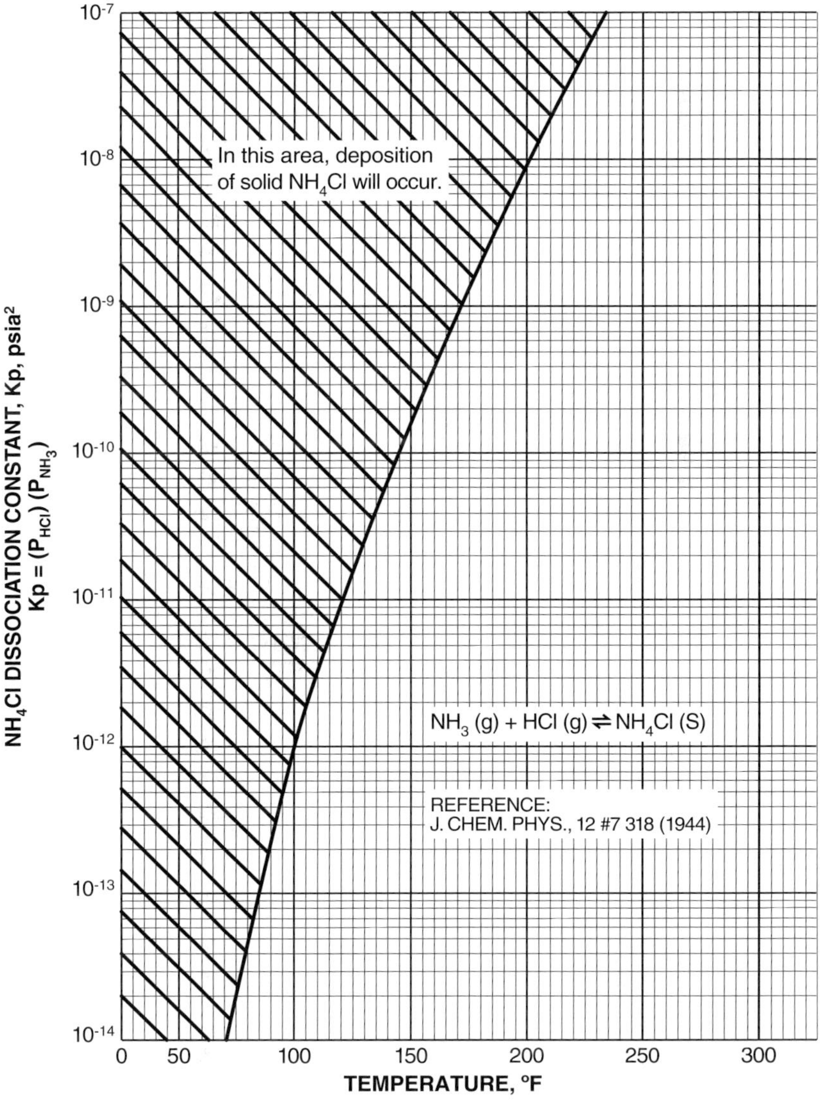

Fig. 4.2.5 NH_4Cl deposition chart low temperature range[1]

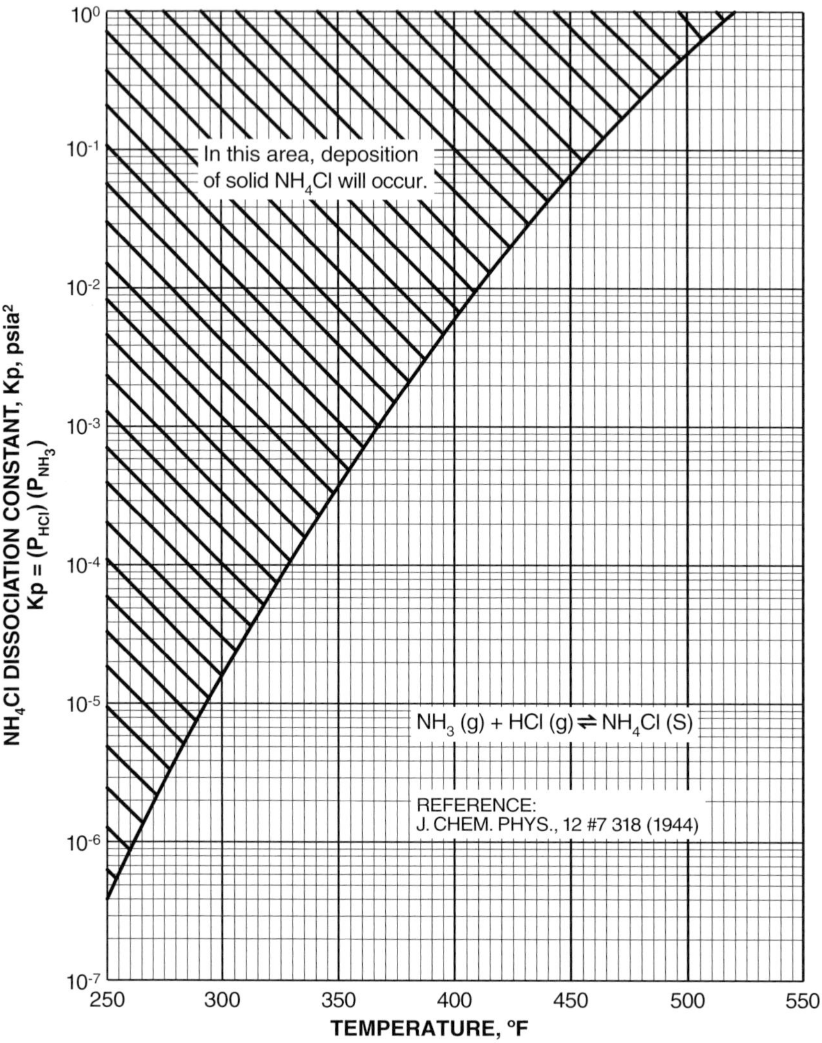

Fig. 4.2.6 NH_4Cl deposition chart high temperature range[2]

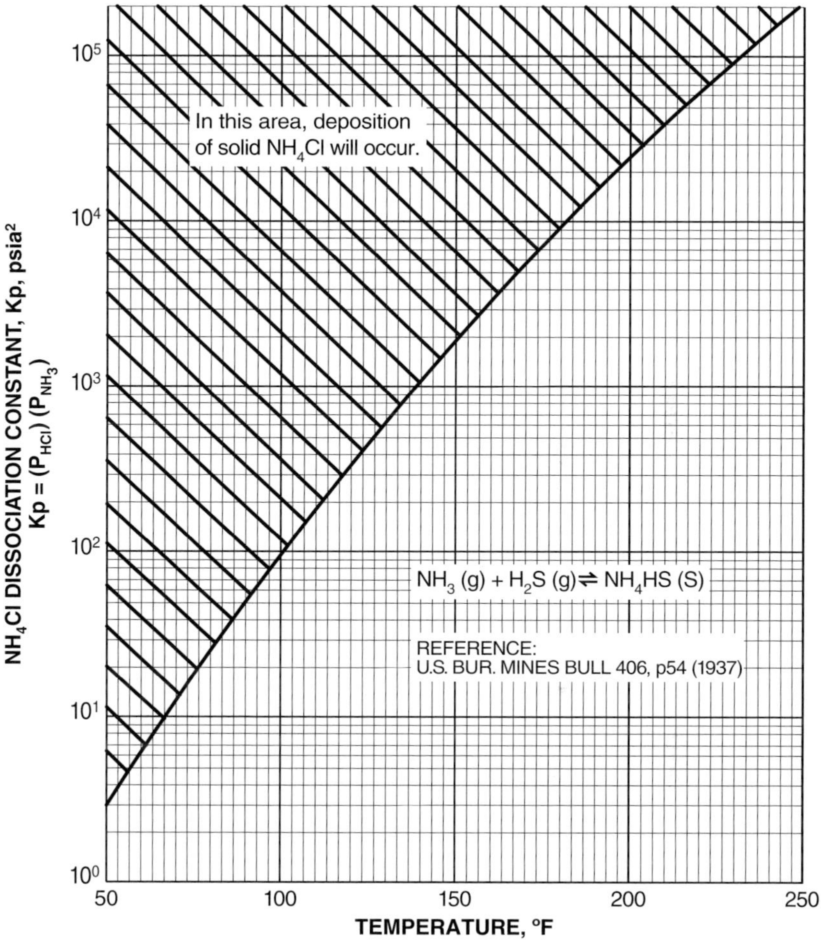

Fig. 4.2.7 NH_4HS deposition chart[3]

Water injection arrangement

Two arrangements are typically followed for water injection upstream of air cooler:

1. Single-point water injection
2. Multiple-point water injection

In single-point water injection, water is added slightly before the air cooler so that it can mix with the hot vapor, vaporize, and saturate the effluent vapor with water vapor at that temperature. The effluent vapor saturated with water then enters the air cooler, condenses, and produces water to wash and dissolve

NH_4HS and NH_4Cl. However, 25% extra water is typically added that is likely to distribute in the liquid phase only.

In a multipoint injection scheme, a measured quantity of water (provided with flow orifice and restricted orifice) is admitted with each inlet of effluent vapor to the HPAC. Both arrangements usually deploy symmetrical inlet and outlet pipings of HPAC.

Velocity limitation in HPAC

Typically, when a carbon steel air cooler is used, the velocity through the air cooler is within 6 M/sec. Incoloy 825 metallurgy is used for higher velocities (up to 9 M/sec). Sometimes 1.25% Cr, 0.5% Mo metallurgy is used to mitigate the risk of high-temperature hydrogen attack (HTHA), but when this metallurgy is used, it becomes necessary to heat-treat the welding between the air cooler tubes and tube sheet. Thus, people prefer to use a carbon steel air cooler if permitted by HTHA, velocity, and the NH_3 and H_2S contents of the effluent vapor at the inlet of air cooler.

Typically, the HPAC should remove 30%–40% of its design heat duty through natural convection during fan failure. This is achieved through the chimney effect. For specifying the HPAC, the heat removal requirement by the air cooler during fan failure must be included in the data sheet. Alternatively, some technology suppliers provide a steam generator in the effluent cooling circuit. The two features act as a heat sink while depressurizing the HP section in case of emergencies like feed failure, recycle gas compressor failure, etc.

The air cooler outlet is normally routed to the CHPS. The CHPS outlet vapor goes to the RGC with or without an amine wash. Makeup H_2 is added mostly in the RGC suction, although some designs add it at the RGC outlet.

Some amount of H_2S is maintained in the recycle gas stream even when an amine wash is provided at the RGC suction. This is done so that the catalyst is not desulfurized and converted to its metallic state, which would result in loss of activity.

Catalysts

The catalysts deployed in hydrocracking reactors have two functions: hydrogenation and cracking. The hydrogenation capability is contributed by earth metals, for example, nickel, molybdenum, cobalt, etc. Normally nickel is used in the hydrocracker catalyst to impart the metal function. The cracking function is contributed by the acid sites of silica and alumina catalysts. Metals are dispersed on alumina or silica catalysts (zeolites). The guard or demetalization catalysts have lower hydrogenation capabilities and having bigger pores for metal trapping.

4.3 Different Hydrocracker Configurations

Reactor section

There are various configurations of VGO hydrocrackers that are popular and in current operation (Figs. 4.3.1, 4.3.2, 4.3.3, and 4.3.4) and those are presented here.

In the scheme in Fig. 4.3.1, the final reactor effluent is cooled while heating the filtered feed, recycle gas, and CLPS liquid. The recycle gas quench is injected at each reactor and bed outlet to control the inlet temperature to the subsequent catalyst bed or reactor.

In the scheme in Fig. 4.3.2, interreactor exchangers are provided to heat the CLPS liquid. In this kind of scheme, the quench hydrogen required to cool the interreactor effluents is minimized, thus relieving the RGC. This method is often practiced for debottlenecking the recycle gas capacities in revamp applications. However, the quench flow also supplies the hydrogen to maintain minimum hydrogen partial pressure at the inlet of the next bed or reactor.

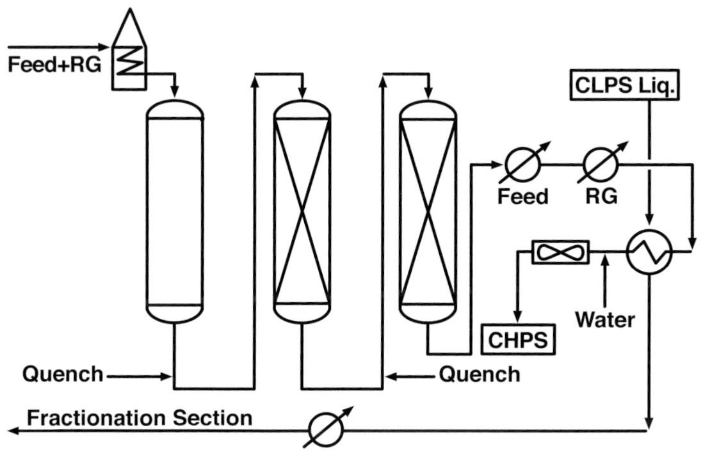

Fig. 4.3.1 Typical reactor section scheme with three reactors

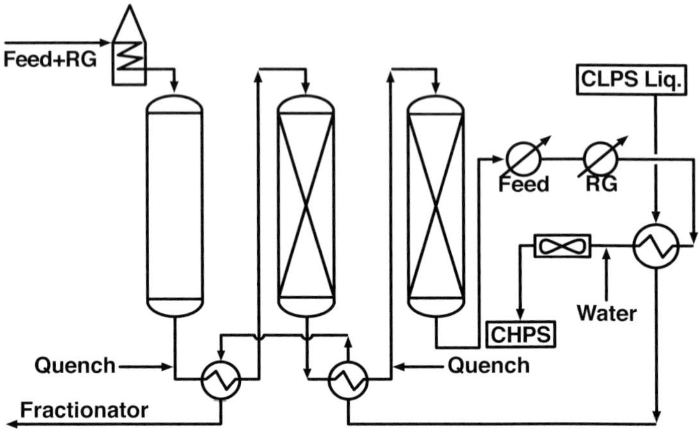

Fig. 4.3.2 Reactor section scheme with heat exchangers in between

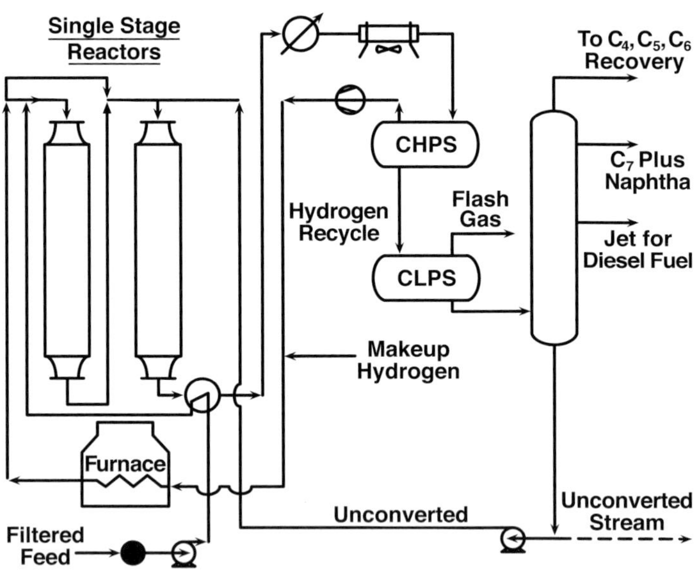

Fig. 4.3.3 Scheme with unconverted recycle to reactor

In this scheme (Figs. 4.3.3 and 4.3.4), unconverted oil from the fractionator bottom is recycled for further conversion. In Fig. 4.3.3, the unconverted oil is recycled along with the fresh feed in the second reactor, while in Fig. 4.3.4 the unconverted is recycled to a separate reactor designed to handle it. Often the cooled unconverted oil is recycled to the filtered feed surge drum and pumped along with fresh feed to avoid an additional high-pressure pump for recycling.

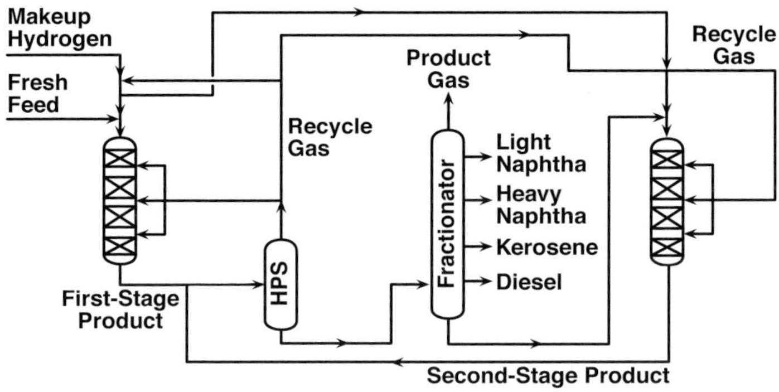

Fig. 4.3.4 Different scheme of unconverted recycle to reactor

Effluent cooling section

Two popular effluent circuit designs are available.

The two-drum system has one CHPS and one CLPS (Figs. 4.2.1 and 4.3.5). The four-drum system has one HHPS and one HLPS, plus a CHPS in combination with a CLPS (Fig. 4.3.6).

A two-drum system is shown in Fig. 4.3.5.

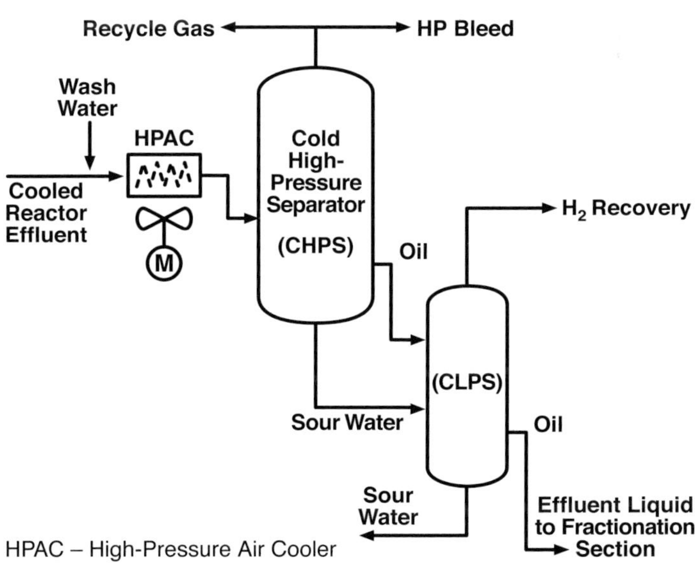

Fig. 4.3.5 Two-drum system

In the scheme in Fig 4.3.5, the effluent is routed to the CHPS after a water-wash and air cooling. The hydrocarbon from the CHPS is routed to the CLPS (at lower pressure), and then to the fractionation section. The CHPS liquid is routed to CLPS through the PRT installed parallel to the level control valve. Typically, the power recovered through the PRT is used to share part of the load of the high-pressure feed pump drive. The PRT typically delivers around 25%–30% of the power requirement of high-pressure feed pump motor drive.

A four-drum system is shown in Fig. 4.3.6.

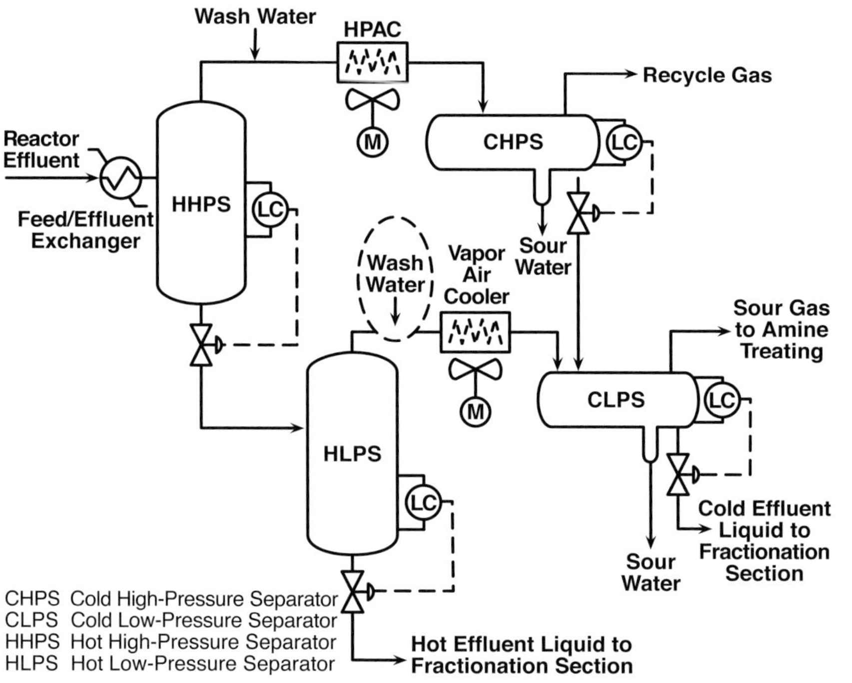

Fig. 4.3.6 Four-drum system

In the scheme in Fig. 4.3.6, the effluent is first routed to the HHPS. The liquid from the HHPS is then routed to the HLPS through the PRT and the vapor from the HHPS is routed through heat exchangers before being water-washed, air cooled, and routed to the CHPS. The hydrocarbon from the CHPS is routed to the CLPS.

In a four-drum system, the majority of the effluent liquid is typically obtained from the HHPS bottom as compared to the CHPS bottom. Thus, in order to recover maximum power, the PRT is often installed in the HHPS liquid line going to the HLPS parallel to the HHPS level control valve.

The benefits of a four-drum configuration

The biggest benefit of a four-drum configuration is that a large quantity of heat can be recovered by installing an HHPS and HLPS instead of rejecting the heat in the HPAC. The HLPS bottom can be fed to the fractionation section of a hydrocracker, VGO hydrotreater, or a stripper in the diesel hydrotreater (DHDT) at a higher temperature. The heat load and velocity in the downstream effluent heat exchangers and HPAC can be reduced. This is particularly useful in capacity revamp applications. This system will reduce the pressure drop in the downstream effluent exchangers and decrease the likelihood of a metallurgy upgrade in the air cooler.

Thus, the four-drum system helps in the following ways:

1. Conserves heat required for the fractionation section
2. Lowers the pressure drop in the effluent circuit
3. Reduces HPAC velocity and HPAC heat duty requirement
4. Reduces cooling surface area requirement in the effluent train
5. Allows feed splitting of effluent and the heating of cold streams through both HHPS liquid and vapor

Disadvantages of a four-drum configuration

A four-drum system has a higher H_2 dissolution in the HHPS liquid, as H_2 solubility in hydrocarbon increases with temperature and is released along with HLPS overhead vapor. H_2 dissolution loss will increase, demanding additional H_2 makeup. However, a part of dissolution hydrogen in HHPS can be recovered through H_2 recovery from HLPS vapor.

A part of the hydrogen finds a way to stripper or fractionators (along with the liquid from the HLPS) and may load the column. This arrangement of four drums, shown in Fig. 4.3.6, can reduce the H_2 purity of the CHPS vapor or recycle gas as the quantity of CHPS liquid reduces. Some amount of C1 or C2 may be liberated from the CHPS liquid, which can lower the purity (H_2 mole %) of the recycle gas (RG). In order to maintain the recycle gas purity, the CHPS vapor can be cooled from the CHPS temperature of 55°C–60°C (with the water cooler) to 40°C–45°C (Fig. 4.3.7).

In the scheme, the vapor condensed in the cooling is separated through a separator located at the downstream of the water cooler and returned to the CLPS outlet liquid stream. The actual arrangement is shown in Fig. 4.3.7. Often, a high-pressure amine wash is introduced at the CHPS outlet vapor line with or without a water cooler (Fig. 4.3.8), which increases the purity of the RG.

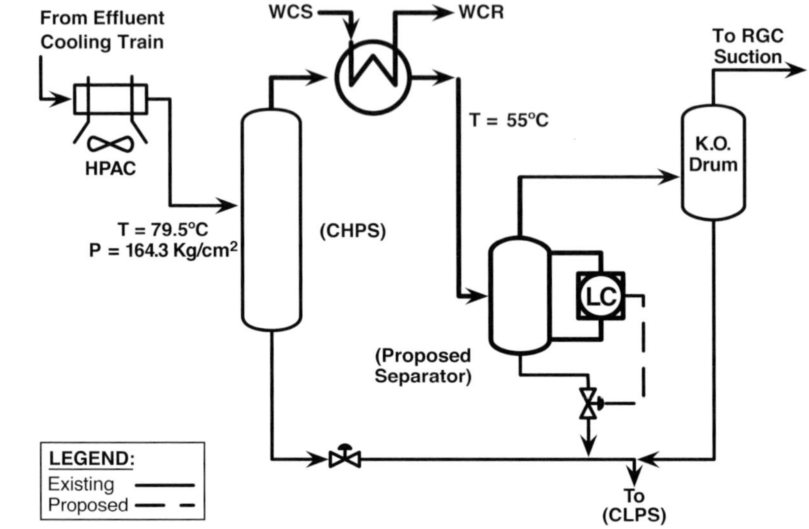

Fig. 4.3.7 CHPS outlet gas water cooler

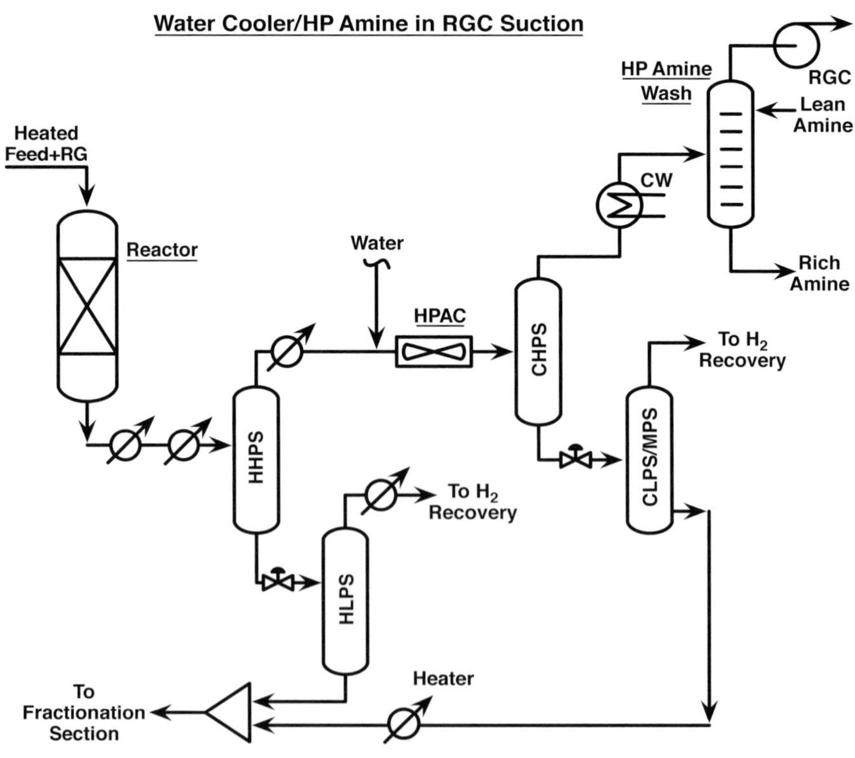

Fig. 4.3.8 Scheme with CHPS gas amine absorber

Alternatively, the liquid from the CLPS can be recirculated to the HPAC inlet to increase the RG purity (see Fig. 4.3.9). This may be mostly practiced in diesel hydrotreaters with an HHPS. This is also required in low conversion VGO hydrotreaters with an HHPS. The quantity of HHPS vapor decreases due to low conversion and consequently the CHPS liquid also decreases. The quantity of liquid in the CHPS becomes inadequate to keep the noncondensable hydrocarbons in a dissolved condition at the pressure and temperature of the CHPS. Thus, the recycle gas purity tends to decrease with the ingress of noncondensable hydrocarbons in the RG. The CLPS liquid recirculation helps to increase the hydrogen purity of the recycle gas. A high-pressure pump is used to pump the CLPS liquid to a higher pressure in order to introduce it to the HPAC inlet. The biggest problem is encountered when the HHPS is installed at a low temperature at the inlet of the HPAC. At low HHPS temperatures, the effluent vapor quantity from the HHPS reduces, and the partial pressures of NH_3, H_2S, and HCl increase as the vapor-phase mole fraction of the constituents increases with the condensation of hydrocarbon vapor. This may lead to the deposition of NH_4Cl in particular (as NH_4Cl is likely to deposit at higher temperatures than NH_4HS) in the heat exchangers located at the upstream of the HHPS.

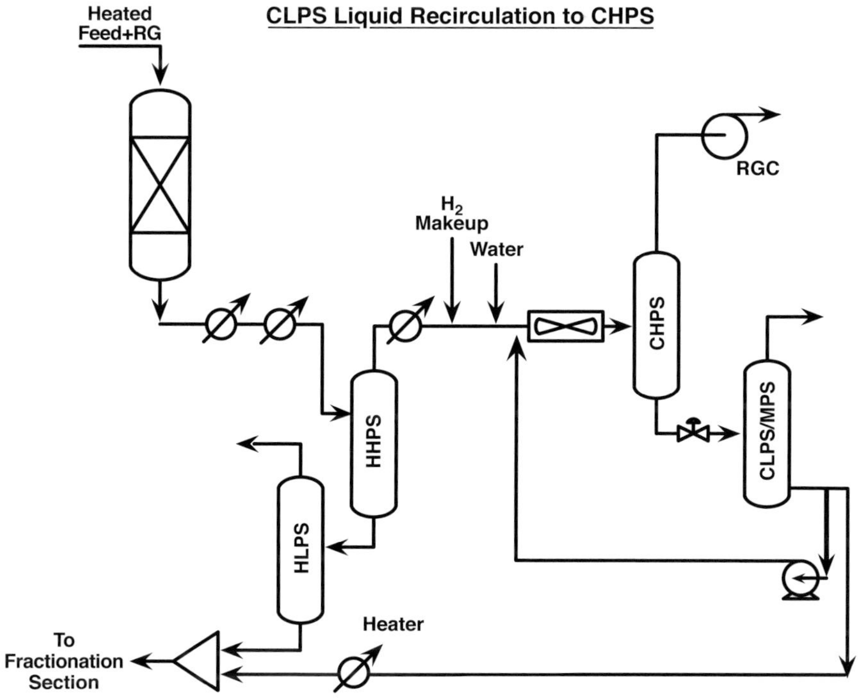

Fig. 4.3.9 CLPS/MPS bottom recycle to HPAC inlet

If an HHPS is provided in a diesel hydrotreater at the immediate upstream of the HPAC, the solid NH_4Cl may deposit and mix with the HHPS liquid (hydrocarbon). This may introduce deposited NH_4Cl into the stripper feed circuit, and in turn into the stripper, which will result in severe corrosion. This problem will be more pronounced in a DHDT, which the author has experienced. Even with a very low chloride content (around 1 ppm wt) in the feed, NH_4Cl deposition may occur at temperatures around 180°C–200°C, particularly in diesel hydrotreaters operating at pressures around 100–110 kg/cm^2 (g). Thus it is prudent to install an HHPS in a DHDT only when the temperature is around 250°C and above. This can be decided only after referring to the NH_4Cl deposition chart (see Fig 4.2.5). In the last couple of heat exchangers upstream of HPAC, the effluent is normally routed through the tube side and a water connection is provided upstream to wash the deposits periodically. The absence of periodic washing upstream can lead to sharp increase in pressure drop in the effluent circuit and in a hydrotreater that can lead to the opening of the anti surge valve of the RGC.

Different configurations of the fractionation section

A few common configurations are typically used.

In one configuration, the CLPS and HLPS liquid is fed directly to the fractionator after preheating and heating in the charge furnace to recover the products (see Fig. 4.3.10). In another, the CLPS and HLPS liquid is fed to an H_2S stripper after preheating, then the stripped product is further preheated and heated in the charge heater and fed to the fractionator (see Fig. 4.3.11).

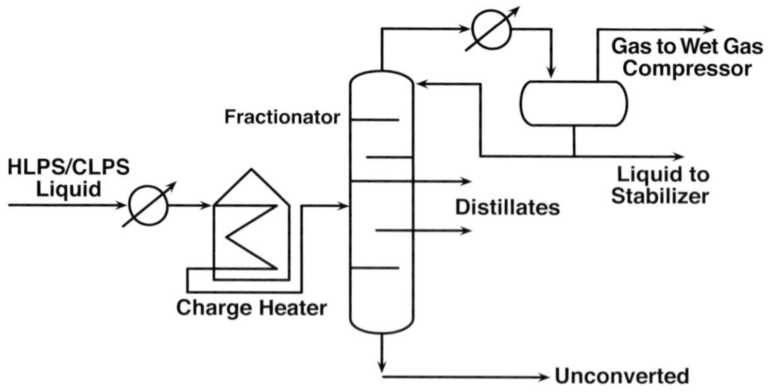

Fig. 4.3.10 CLPS/HLPS liquid feeding to fractionator

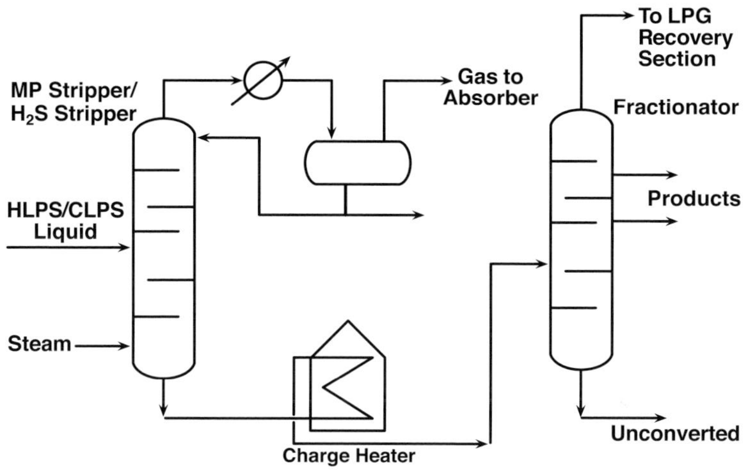

Fig. 4.3.11 CLPS/HLPS liquid feeding to the H_2S stripper

Configuration with CLPS/HLPS liquid feeding to the fractionator after preheating

In the configuration in Fig 4.3.10, the wet gas compressor is required. The compressor compresses the light ends from the reflux drum to the required pressure of the recovery section, typically 15 kg/cm^2. Further, the HLPS or CLPS liquid is usually fed to the heat exchangers without a pump. Thus, the charge will be at bubble point and can be partly vaporized after heating at the inlet of the charge heater. Thus, achieving equal distribution of flow in the passes of the heater may be difficult, as no control valve can be used at the heater inlet and distribution must be achieved through similar geometry of the passes. In one hydrocracker the author operated with a similar configuration, the charge heater pass became coked up during startup due to low flow in some of the passes, caused by a misdistribution of feed flow in the furnace passes. The problem was partly resolved by recirculating a substantial quantity of unconverted residue from the atmospheric column and vacuum column bottom along with the furnace feed.

Stripper preceding fractionator

In this configuration (Fig. 4.3.11), the stripper is placed upstream of the main fractionator. The light naphtha, along with the liquefied petroleum gas (LPG) components from the stripper, is fed to the light end recovery section (stabilizer). One advantage of this configuration is that it does not need a wet gas compressor, but it does need a reboiler furnace, or stripping steam in the stripper bottom. The stripper bottom is pumped through heat exchangers then routed to the charge heater and finally to the fractionator in order to obtain the desired product fractions. The second very important advantage is that part of the lighter product is removed as top product from the stripper and the stripper bottom is pumped as feed to the furnace, thereby avoiding the likelihood of vaporization at the charge heater inlet (liquid is subcooled at the pressure and temperature). A better pass flow distribution can be achieved in the charge heater with pass flow control valves, thus avoiding the likelihood of coking resulting from misdistribution.

A vacuum column also can be appended at the downstream of the fractionators for full recovery of the diesel components. With this scheme, part of the gas oil will be drawn from the atmospheric fractionator and part will be drawn from the vacuum column. This will reduce the load on the atmospheric column and the charge heater located upstream of the atmospheric column, as well as debottleneck the charge heater and the atmospheric column. The vacuum column requires a reboiler to provide the heat for recovering the high-speed diesel (HSD) components. The vacuum bottom product would in turn deliver its heat to the fractionator feed or the HP stripper feed to increase preheat and reduce the atmospheric charge heater/stripper reboiler heat duty requirement. If the charge heater mass velocity is found to be limiting, a pre-flash drum can be used upstream as in a crude distillation unit. The pre-flash drum bottom will go to the charge heater (through the pump) and then to the atmospheric fractionator, and the flash vapor will go to the flash zone of the fractionators. This will also remove the lighter hydrocarbons from the feed to the furnace and avoid a flow distribution problem in the furnace.

Simulation guidelines

A simulation model can be developed for an HDT or hydrocracker unit rating (excluding reactor) using the Grayson–Streed (GS) correlation for the K-value (VLE), the API correlation for liquid density, and the Soave–Redlich–Kwong (SRK) equation of state for vapor density. The Lee–Kesler method can be used to find vapor–liquid enthalpies, and petroleum correlations can be used to evaluate other transport properties like the viscosities of vapor or liquid, surface tension, etc.

Typically, the simulation begins with the reactor effluent obtained from the pilot plant data or proprietary mathematical models. The effluent flow and composition is often available with light ends along with narrow cuts that have a normal boiling point (NBP), molecular weight, and density of cuts, all of which are entered in the simulation as input.

Alternatively, the effluent can be generated (synthesized) by blending all the products—gas, LPG, naphtha, kerosene, diesel, and the unconverted—with the requisite amount of hydrogen. Catalyst experts also supply the effluent composition. The simulation flow scheme is developed with reactor effluent and the effluent coolers (heat exchangers) HHPS, HPAC, and CHPS, and the recycle gas flow is obtained after adding the makeup hydrogen. Subsequently, the feed, recycle gas, and fractionator feed-heating sections can be configured separately.

4.4 Methodologies for Debottlenecking and Revamping

Capacity augmentation in an HDT or hydrocracker can be limited by the following:

1. Catalyst volume
2. Recycle compressor capacity
3. Makeup compressor capacity
4. Feed pump capacity
5. High-pressure heat exchangers or HPAC in the effluent section
6. High-pressure furnace or reactor furnace duties
7. Feed preheat capability or recycle gas heating
8. Fractionators or gas concentration (gas-con) section capability

Catalyst volume

For processing higher feed rates, a higher quantity of catalyst is required for the same space velocity. However, with the advent of a new-generation catalyst that can handle higher space velocity, the requirement for processing additional feed can be fulfilled without adding more catalyst.

Often, if the catalyst is originally sock loaded, dense loading can be done to pack a higher quantity of catalyst into the reactors. Before dense loading, it is required that the licensor's advice be sought, as this type of loading may increase the pressure drop across the reactors during both normal and emergency dumping operation.

Addition of reactors in series

Additional reactors can also be added, either at the beginning or the end. When a reactor is added in front of the existing reactors, it should act as a guard or demetallization reactor. When added in the back, the reactor may only have the conversion catalyst, as the feed is already treated in the preceding reactors.

If a small volume of additional catalyst is required, a small guard reactor can be added at the beginning of the reactor train and the guard catalyst can be loaded. The original guard catalyst in the first reactor can be changed to an active catalyst to introduce the additional active catalyst. If a large volume of additional catalyst is required, it may be prudent to add it at the back of the last reactor in the series or in parallel. There are many configurations for adding an extra reactor. A new reactor can be added in a series with its own effluent cooling section terminating in a new HHPS (HHPS-2) with new effluent heat exchangers to heat the additional feed quantity and recycle gas. A new recycle gas preheat furnace can be added for the extra heat duty requirement. The same RGC can be used by the rotor augmentation if there is no cushion available in the capacity augmentation.

The new HHPS top will join the existing HHPS top and go to the existing air cooler (HPAC) after a water-wash. The new HHPS bottom will join the existing HLPS bottom and go to the common fractionation section.

The catalyst in the new reactor in such a configuration may take on functions like guarding, desulfurization, denitrification, or conversion. This concept may provide flexibility to add a reactor in series without disturbing the existing effluent cooling and feed heating sections. Only a few hookups will be needed to connect the new HP train to the existing one, which cuts the downtime of the unit for the capacity revamp.

RGC capacity debottlenecking in hydrocrackers and hydrotreaters

The RGC capacity requirement can be calculated as

$$Q_R + SQ_{\text{quench}}$$

where Q_R is the RG flow along with feed and SQ_{quench} is the total quench flow or sum of all the quench flows (Q_q).

The ratio of recycle gas flow to fresh feed flow is defined as recycle ratio and is a very critical parameter. The recycle ratio requirement often reduces with an increase in feed capacity, as the higher hydrocarbon flow requires less RG flow for uniform dispersion through the reactor bed, distributors, etc. The reduction in recycle ratio is limited by the minimum partial pressure requirement of hydrogen at the reactor and bed inlets.

The RGC capacity limitation can also be handled by the following:

1. Heat exchangers between reactors

The reactor outlet (for a series reactor system) can have heat exchangers between reactors to heat filtered feed, heat stripper, and fractionator feed (see Fig 4.4.2), and generate steam, reducing the quench flow requirement. The reduction or discontinuation of quench flow between reactors is limited by the recommended hydrogen partial pressure at the inlet of the succeeding reactor. The quench requirement is normally lower in this arrangement than in the reactor outlets with no heat exchangers (see Fig. 4.4.1).

The quench gas requirement can be reduced by installing heat exchangers or coolers between reactors (e.g., steam generators). The duties of the coolers are finalized based on the temperature requirement at the inlet to the next reactor in the series, which is limited by the partial pressure of the hydrogen in the next reactor. If the hydrogen partial pressure at the inlet of the next reactor is likely to fall below the prescribed value in a system using interreactor coolers, then a combination of part cooling through quench injection and part cooling through an interreactor heat exchanger is practiced.

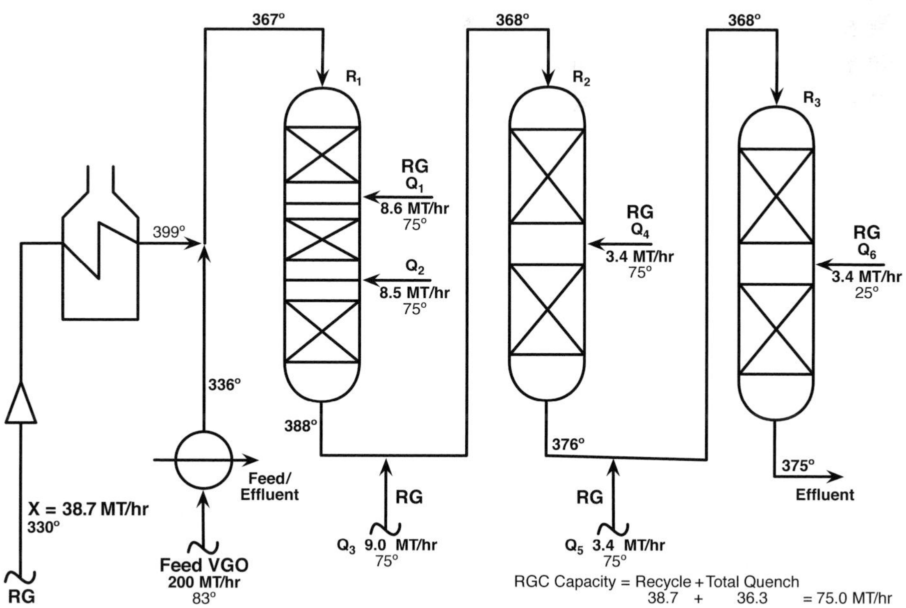

Fig. 4.4.1 RG quench at reactor outlets

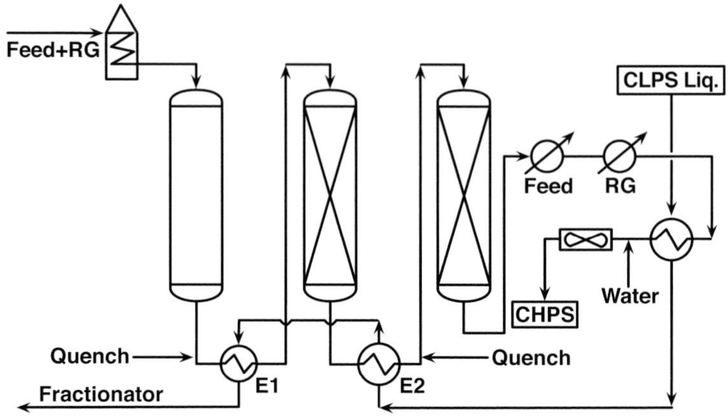

Fig. 4.4.2 Heat exchangers at reactor outlets

However, a quench to the inlet of the second reactor should be continued only if the hydrogen partial pressure reduces; otherwise, the quench facility may remain as a provision with normally no-flow (NNF).

Total quench flow (Fig. 4.4.1) is defined as $Q_q = Q_1 + Q_2 + Q_3 + Q_4 + Q_5 + Q_6$. With the introduction of interreactor heat exchangers E–1 and E–2 in Fig. 4.4.2, the quench flows Q_3 and Q_5 in Fig. 4.4.1 will reduce or become zero during normal run, thus reducing total quench flow Q_q and consequently reducing the total recycle gas flow requirement.

2. Using quench gas coolers

In this arrangement, the RGC discharge that acts as a quench to the reactors is cooled with water coolers to reduce its temperature when the CHPS temperature is 55°C–65°C. This consequently reduces the quench gas requirement, which in turn reduces the recycle gas flow requirement. This recycle gas is typically cooled to 5°C–10°C above the saturation temperature of the recycle gas at the discharge pressure of the recycle gas compressor. The recycle gas compressor discharge temperature typically goes above the saturation temperature at the discharge pressure due to the temperature increase caused by compression. Thus, some cooling is possible without causing the temperature to reach the saturation temperature of the gas. Typically in a low-pressure (CHPS pressure of 35 kg/cm^{2}pressure) hydrotreater, the RGC operates at a higher compression ratio: $(P_2/P_1) \approx (1.5–1.8)$. Thus, the RGC discharge temperature is typically greater than 100°C with a suction temperature of around 55°C–60°C. The cooling of quench gas is more effective in low-pressure hydrotreaters. The typical calculation compression ratios for a low-pressure hydrotreater and high-pressure hydrocracker or hydrotreater are given below.

For low-pressure hydrotreaters, the RGC discharge pressure is typically 20–25 kg/cm^2 higher than the suction pressure (the pressure required to overcome the pressure drop in the circuit).

Suction pressure P_1 = 35 kg/cm^2 (g) (assumed)

[$P_2 = P_1 + D_P$ = 35 + 25 = 60 kg/cm^2 (g): compression ratio (P_2/P_1) = 61/36 = 1.74]

Discharge temperature $T_2 = T_1 \times (P_2/P_1)^{(n-1)/n}$

Suction temperature T_1= 55°C (cooled by air cooler)

$T_2 = (55 + 273) \times (1.74)^{(1.3-1)/1.3}$ or T_2 = 373°K = 100°C

For high-pressure hydrocrackers and hydrotreaters, the RGC discharge pressure is also typically 20–25 kg/cm^2 higher than the suction pressure.

The suction pressure: 160 kg/cm^2 (g)

RGC discharge pressure: P_2 = 160 + 25 = 185 kg/cm^2 (g)

Compression ratio: $P_2/P_1 = 186/161 \approx 1.16$

Discharge temperature $T_2 = T_1 \times (P_2/P_1)^{(n-1)/n}$

Suction temperature T_1 = 55°C (cooled by air cooler) assumed

$T_2 = (55 + 273) \times (1.16)^{(1.3-1)/1.3}$ or T_2 = 339.4 K = 67°C

Thus, it can be see that the compressor discharge temperature increase is lower in the case of high-pressure hydrotreaters and hydrocrackers, and that recycle gas cooling is not an effective tool to reduce the quench gas quantity.

3. Recycle gas amine wash

A recycle gas high-pressure amine wash is often installed in a capacity revamp to increase the H_2 purity of the recycle gas, requiring less recycle gas flow along with feed and providing a capacity cushion in the RGC (Fig. 4.3.8, Fig. 4.4.3). The recycle gas flow requirement may reduce with the feed due to the improved H_2 purity of the recycle gas. Typically in an amine wash, the recycle gas temperature increases by around 15°C–20°C. So unless a water cooler or water-wash succeeds the amine wash, the inlet temperature to the recycle gas compressor increases. A water-wash is useful to knock off entrained amine particles, lower the temperature at the RGC suction, and provide some marginal capacity gain in the RGC. Typically, the water-wash is installed at the top of the amine wash column in a separate compartment at the top (Fig. 4.4.3). The sweet gas is admitted after the amine wash in the section with around 4–6 trays for contacting the gas with water. The gases enter the section through the chimney. The water collects on the pan of the chimney tray and is circulated by a pump at the top of the section for washing the gas. Initially, demineralized water is made up in the chimney tray through the high-pressure water pump.

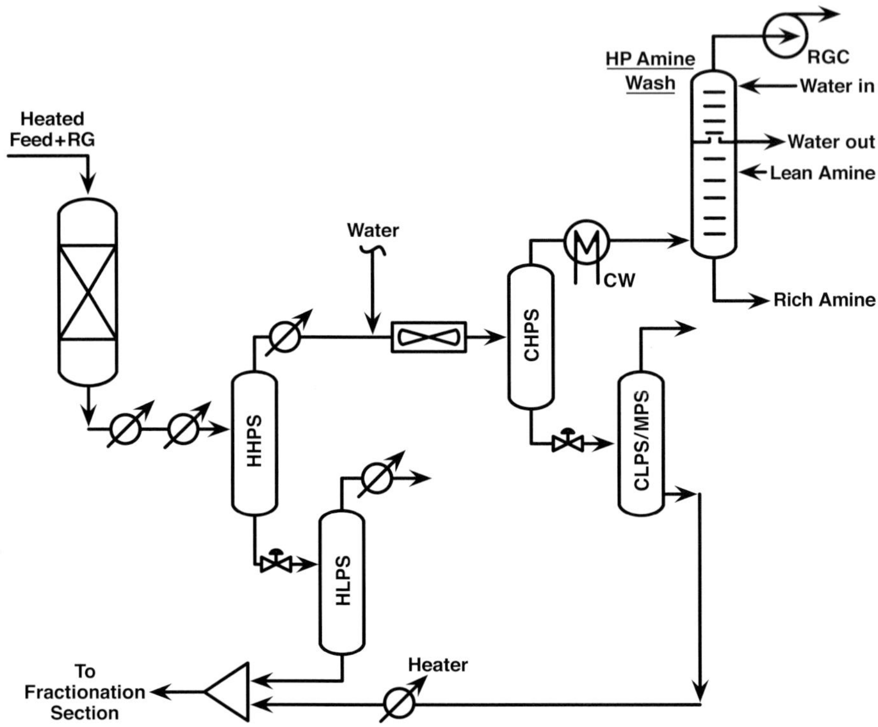

Fig. 4.4.3 Scheme of recycle gas amine wash followed by water-wash

A water-wash is not normally provided after an amine wash in high-pressure hydrotreaters/hydrocrackers, as the suction pressure of the water circulation pump may be higher than recommended by the API. Similarly, the recirculation of the liquid to the HPAC inlet as in Fig. 4.3.9 is performed from the lower-pressure CLPS and not from the CHPS, as this wold require a very high suction pressure of the recirculation pump.

4. Lowering the CHPS outlet gas temperature

The CHPS temperature in a hydrocracker ranges from between 55°C–65°C to facilitate water separation and maintain flow ability of the liquid hydrocarbon stock in the CHPS. Further reduction may not be advisable. The CHPS off-gas temperature can be further reduced by adding a water cooler downstream from the CHPS at the CHPS outlet vapor line (see Fig. 4.3.7), followed by a knockout drum to knock off the condensate produced. This added cooler will improve the H_2 purity of the recycle gas, which will in turn lower the recycle gas flow requirement along with the feed. This would also reduce the actual suction volume due to the decreased recycle gas suction temperature, which would help increase suction mass flow. In a revamp application, this is a very good alternative for improving the recycle gas H_2 purity and increasing H_2 mass flow.

The scheme shown in Fig. 4.3.7 has been implemented in a hydrocracker where the CHPS temperature has gone up due to an increase in conversion and marginal capacity increase.

Recycling unconverted oil or treated hydrocarbon liquid as quench

It is very difficult to install interreactor heat exchangers that save in quench gas requirement and reduce the recycle gas flow requirement during a unit revamp. This is due to the fact that high-pressure section welding takes a long time, leading to longer unit outage. Thus, using unconverted oil recycle as a quench can be an option for reducing the recycle gas flow requirement. Potential use of liquid quench between beds with unconverted oil or VGO is shown in Fig. 4.4.4. Normally, the flow requirement for liquid quench is more than the recycle gas quench and would need a differently designed mixing device to mix the unconverted oil and effluent to achieve the cooling. A high-pressure pump for the injection of liquid quench is also required.

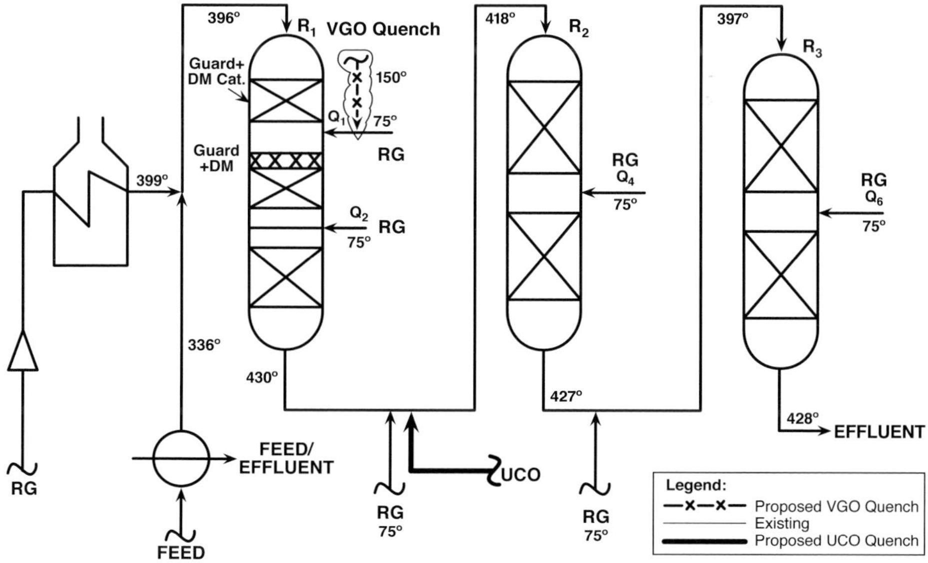

Fig. 4.4.4 Use of liquid quench at bed and reactor outlets

In some units, part of the unconverted oil from the fractionator is recycled along with the fresh feed in a single stage configuration (Fig. 4.3.3), or to a separate reactor in a two stages configuration (Fig. 4.3.4) to achieve the desired conversion. Part of this unconverted liquid, after cooling, can be used as liquid quench. This will not result in additional load on the effluent cooling circuit in a single-stage unit. Similarly, part of unconverted liquid after cooling can be used as quench in the second stage without additionally loading the effluent cooling circuit in a two-stage unit. The liquid unconverted oil can be injected between two reactors (Fig. 4.4.4), as there will be a good length of piping for mixing.

As an alternative to the same cooled unconverted oil, treated hydrocarbon liquid from other sources can be used as a quench medium, but this has the following two disadvantages:

1. The unit would need a high-pressure pump to introduce the treated hydrocarbon liquid into the high-pressure circuit, as with the use of unconverted oil as a quench.
2. Additionally, use of treated hydrocarbon liquid as a quench will result in its recirculation through the effluent exchangers and air cooler, the high-pressure and low-pressure drums, fractionator, and the low-pressure furnace, leading to an increase in the load on the downstream sections and equipments.

Raw feed can also be injected between the reactors to absorb the heat of the reaction. The same can be done in place of the quench gas flow, which in turn conserves the recycle gas flow requirement.

When raw feed is used as quench, the downstream reactor catalyst bed should have a guard and a demetallization and desulfurization catalyst at the top. This may be best applied in the desulfurization reactor, where the first bed is a guard catalyst with a low hydrogenation catalyst and metal traps. Raw feed quench may be possible at the downstream of the first bed. This will require a guard catalyst or a tailored catalyst above the second bed to handle the metals, sulfur, and the N_2 in the raw feed at the expense of some hydrocracking catalyst. The liquid quench quantity will be small, thus the unit may not require a large volume of tailored catalyst. This can provide additional feed to the unit in the form of quench oil as well as conserve the RGC capacity.

The addition of makeup gas at the discharge of the RGC instead of the suction can be a way to resolve the limitation in RGC capacity, but it is a difficult proposition, as the makeup compressor compression ratio will increase. This process is not normally used unless installed during design originally.

Makeup compressor capacity debottlenecking

1. The first option is to add another makeup compressor.

2. Cascade H_2-rich gas at high pressure (without use of the makeup compressor) in the recycle loop at the upstream of the amine column from higher pressure sources.

In one low-pressure diesel hydrodesulfurization (DHDS) unit, H_2-rich gas (95% purity) was introduced ahead of the high-pressure amine wash from the catalytic dewaxing (CDW) unit and the lube oil hydrofinishing unit (from a higher pressure source). It made up the additional required makeup gas quantity required, around 0.8 MT/hr, which is ≈ 4.5 KNM^3/hr or 4,500 normal m^3/hr (NM^3/Hr). Thus, the makeup hydrogen was added in the high-pressure loop without the addition of a makeup gas compressor. However, this method can only be used in low- to moderate-pressure hydrotreaters, and may not be applicable for high-pressure hydrocrackers.

Feed flow augmentation

Feed flow can be increased by the following:

1. Addition of a feed pump in parallel with the individual spill back facility for pump low-flow protection
2. Changing the feed pump or pump motor
3. Increasing the surge drum pressure to increase the feed pump suction pressure and thereby create additional flow. Typically, filtered feed surge drums are blanketed with nitrogen and operate at a pressure of 4–6 kg/cm^2 (g)

The typical drum design pressures are higher (around 10 kg/cm^2 g). They can be increased to around 9 kg/cm^2 by changing the blanketing agent to H_2. This can help to increase flow through the pump by 10–15 m^3/hr. Some amount of H_2 is dissolved in the feed instead of N_2, which helps increase the recycle gas purity to some extent.

In one low-pressure DHDS unit, the marginal shortfall was compensated by pressurizing a feed surge drum with pure H_2 instead of N_2 and fuel gas (FG). This was primarily done to increase suction pressure of the high-pressure feed pump by 1–2.0 kg/cm^2 to obtain a higher flow from the pump. Incidentally, blanketing the feed surge drum with H_2 also created some makeup H_2, introducing around 200 kg/hr of H_2 through dissolution with feed. It also increased the RG purity by eliminating the dissolution of N_2 or hydrocarbons in the feed from the blanketing gas. However, the primary achievement was a throughput increase of around 10%, with an increase in surge drum pressure.

Effluent cooling and feed, recycle gas, and stripper feed heating circuit

Increasing the number of separators (for example, introducing an HHPS) is an excellent way to

1. Lower the HPAC heat duty requirement
2. Conserve the heat for the fractionator section
3. Lower the surface area requirement for effluent cooling
4. Reduce likelihood of the formation of condensation products (polynuclear hydrocarbons) in the effluent HHPS.

During a capacity revamp, the existing heat transfer area in the effluent circuit is likely to be inadequate and cause the following:

1. Lower feed and recycle gas preheat temperature, which will increase the load on the high-pressure furnaces for heating the feed to inlet temperature of reactors
2. Higher CHPS temperature, which in turn will lower the recycle gas compressor capacity and the recycle gas purity

These issues can be solved using the following alternatives:

Alternative 1

Replace the existing elements with new elements to handle the increased flow with an increased surface area. Pressure drop is typically maintained the same way. The inlet and outlet line sizes will also need to change in order to handle increased flow on both sides. Thus, the entire heat-exchanger train gets revamped. This is very difficult to implement due to the longer outage period required for a job in the high-pressure circuit.

It is important to mention that the hydrocracker reactor effluent is a mixture of vapor and liquid. Typically, the vapor fraction is nearly equal to the fractional conversion. As the effluent is a mixture of vapor and liquid, it cannot be distributed into two parallel streams with the same vapor and liquid fraction in both; thus, a parallel effluent heat exchanger train cannot be installed in a hydrocracker to handle the additional effluent generated by capacity augmentation.

Alternative 2

If the unit has a two-drum system, the only option available is conversion to a four-drum system.

In a four-drum system, an HHPS is introduced in the effluent circuit at a suitable temperature. The flashed vapor from the HHPS is routed through the original high-pressure effluent train to cool in heat exchangers while heating the feed, recycle gas, and CLPS liquid. It is finally cooled in the HPAC before continuing to the CHPS.

The HHPS liquid is typically fed to the HLPS at a pressure of around 25–30 kg/cm^2, while the HLPS liquid is fed to the fractionation section along with the heated CLPS liquid.

In the scheme in Fig. 4.4.5, an HHPS is installed to debottleneck the effluent circuit flow of hydrocarbon vapors downstream of the HHPS, so the duty requirement will reduce and the required heat removal for reactor effluent cooling will also reduce. This scheme will also lower the velocity in the HPAC. However, the upstream heat exchangers will continue to have a higher flow.

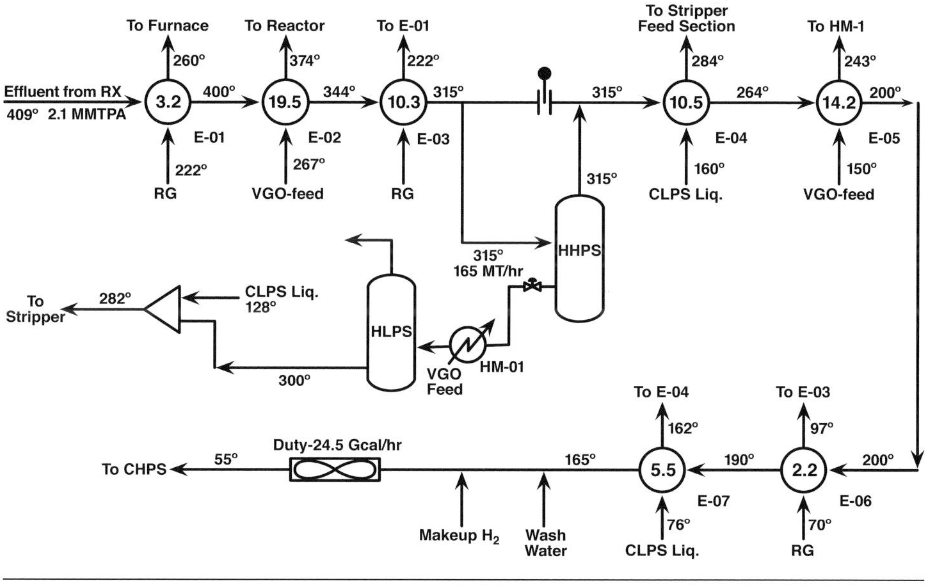

Fig. 4.4.5 Installation of an HHPS in an effluent cooling train

An unconverted oil separator (UCOS) can be used to separate the effluent into liquid and vapor right at the outlet of the last rector (Fig. 4.4.6). The vapor can be routed through the existing effluent train and the liquid can be cooled by heat exchangers with cold streams, usually the fresh feed, recycle gas, CLPS liquid, etc. The vapor stream is cooled through normal circuit and gets water-washed upstream of the HPAC.

Installing a UCOS is difficult, as the operating temperature will be high and would demand metallurgy similar to that of the reactor itself. It is not common to install a UCOS in revamp applications. The vapor from the UCOS can be routed through the existing effluent train and finally to the HPAC and CHPS.

The liquid from the UCOS can be routed through a separate new circuit, cooled through heat exchange with the feed, RG, CLPS liquid, etc., then fed to the HLPS at a pressure of around 30 kg/cm^2. This HLPS liquid will contain mostly unconverted oil and can be fed to a separate stripper or an existing stripper at a lower feed point before being fed to the main fractionator after heating, if heating is required. A HLPS can be used directly to receive the liquid from the UCOS or an additional HHPS can be used to receive the liquid from the UCOS after cooling by cold feed, CLPS liquid, RG, etc. Only with this kind of UCOS arrangement can the reactor effluent be divided into two streams, where one is total vapor and the other is total liquid. This particular installation of a UCOS at the last reactor outlet (Fig. 4.4.6) is not uncommon when the UCOS vapor is fed to a separate reactor for hydrogenation, saturating the aromatics in the diesel and lighter produced through hydrotreating in order to improve the cetane number (CN) of the diesel and smoke point of the kerosene.

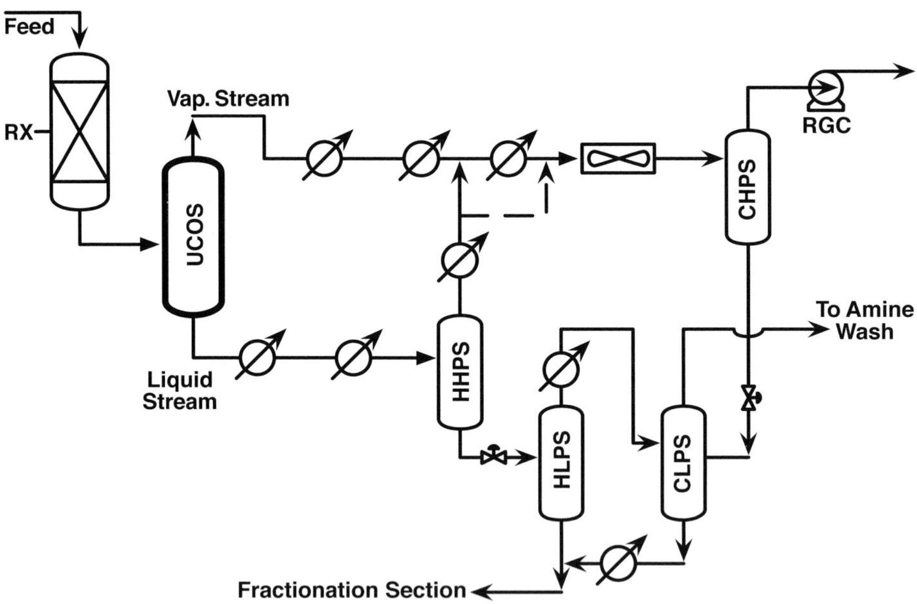

Fig. 4.4.6 Installation of a UCOS in an effluent cooling train

In contrast to a hydrocracker, in a low-pressure DHDT, the reactor effluent is 100% vapor. Thus, the effluent can be distributed into a parallel chain and cooled in the additional parallel train while heating either the reactor feed or the stripper feed. For this reason it is much easier to revamp a DHDS or low-pressure DHDT with a parallel effluent circuit.

However, in high-pressure DHDT units (those that operate at a pressure of 120–125 kg/cm^2), the reactor outlet is a mixed phase, so an additional parallel effluent circuit cannot be introduced to debottleneck the existing effluent circuit. Other options include installing an HHPS in the effluent circuit, or augmenting the effluent heat exchangers.

Debottlenecking the fractionator and gas-con section

The essential differences between the two popular configurations are given below:

1. In one configuration, the H_2S stripper precedes the fractionator. Some amount of light naphtha is also recovered from the top of the stripper. The stripper is provided with either a stripping steam for stripping or a fired reboiler. The stripper bottom is further heated in a charge furnace, then fed to the fractionator for product recovery.
2. If the hydrotreater or hydrocracker has CLPS or HLPS liquid going directly to the fractionators (with no H_2S stripper), the capacity of the low-pressure section can be augmented by adding an H_2S stripper upstream of the fractionators, similar to the pre-topper in a crude unit. The stripper pressure must be sufficient to contain the LPG fractions. If there is any gas bleed, it is to be washed by kerosene or gas oil in a sponge absorber. The high-pressure stripper naphtha will be routed to the deethanizer and then to the stabilizer. This debottlenecks the existing main fractionator, the deethanizer, and the stabilizer, and also eliminates the gas compressor requirement for compressing the gas from the low-pressure fractionators overhead to the deethanizer.

Calculation of required water injection rate in a hydrotreater

Water required for injection is calculated using the following information:

1. How much water is required for saturating the gas at the pressure at the HPAC inlet temperature.
2. Typical water has a temperature 17°C–55°C less than the process temperature before injection is considered.
3. A water injection of around 25% in excess of the amount required to saturate the gas is used. Thus, 25% of the water must be in a liquid state.

Let the water quantity be W in kmoles/hr to saturate the vapor.

Let A be the kmoles of H_2 in the gas (vapor phase) at the inlet of the HPAC.

Let B be the kmoles of HC in the gas (vapor phase) at the inlet of the HPAC.

Let 148°C be the vapor temperature at the HPAC inlet.

$P_{w@148\,°C} \sim 5.0\ \text{kg/cm}^2\ (\text{a})$[5] (from the standard steam table by Keeman and Keyes).

Total moles in the vapor = $W + A + B$ kmoles.

Partial pressure of water in the vapor phase = $(W/(W + A + B)) \times P_{system}$ kg/cm^2. At saturation, partial pressure of water vapor in the HPAC inlet vapor stream = saturation pressure or vapor pressure of water. So $(W/(W + A + B)) \times P_{system} = 5.0\ \text{kg/cm}^2$.

Calculate moles of water required to be injected: W kmoles/hr from above equation.

Over and above this quantity of water, a factor of 1.4 (as per API 932B, appendix A) is typically used in high-pressure hydrocrackers, primarily to make up for imperfections and water dissolution in the hydrocarbon liquid phase.

Wash-water molar flow = $1.4 \times W$ kg-mole/hr to saturate the vapor in the air cooler.

Actual water flow is calculated 25% higher than the above.

Actual wash water rate = $1.25 \times (1.4 \times W) \times 18$ kg/hr.

Calculation in accordance with API

In API RP 932 B Appendix A, a similar method is recommended to calculate the water quantity required for injection upstream of the HPAC.[4] Make sure that the sour water produced contains less than 8% NH_4HS to reduce the corrosion in the sour water lines. A simulation can be done to calculate the water quantity using a stream representing the vapor stream entering the air cooler (Fig. 4.4.7).

A flash drum is used as shown in the above figure and a controller is used to vary the water input rate so that 25% water appears in the bottom stream of the flash drum. The input water rate obtained gives the water rate required (Fig. 4.4.7).

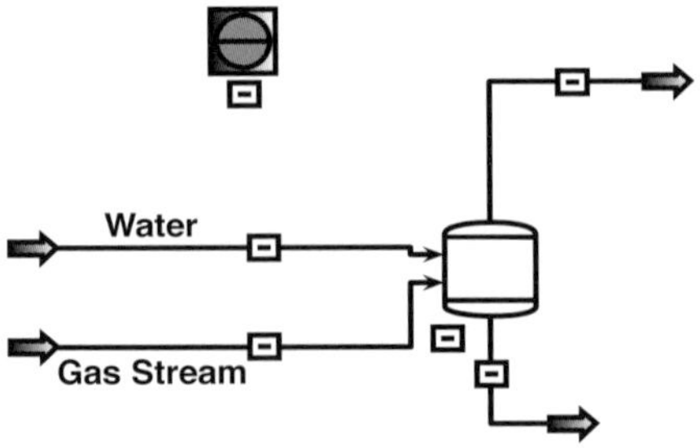

Fig. 4.4.7 Simulation for water rate calculation ahead of HPAC

The injected water must meet the stipulated quality. It is typically pumped at the inlet of HPAC through high-pressure reciprocating pumps. Typically, water is either stored in a closed vessel with vessel pressure control and safety valves or stored in an open vessel.

4.5 Case Studies

Case Study 1 (Frequent choking in the H_2S stripper in a VGO hydrotreater)

The issue was the frequent choking experienced in the H_2S pressure stripper of a moderate-pressure VGO hydrotreater. The unit was primarily used to treat the VGO and produce feed stock for the FCCU. The unit was designed for a conversion of around 15%–20%. It had an H_2S stripper preceding the fractionator, and four drums: an HHPS, an HLPS, a CHPS, and a CLPS. It was reported that the stripper experienced a high pressure drop across the column after 2–3 months of operation, necessitating a shutdown. During shutdown, the stripper column was washed with water and it was reported that the pressure drop reduced after the restart.

It was understood that the choking may have been caused by a deposit of NH_4Cl and NH_4HS in the dispersers of the stripper column trays. These salts are water soluble and likely to be produced in a hydrocracker or hydrotreater and to be removed almost completely by a water-wash provided at the inlet of the HPAC.

The water injection quantity at the effluent HPAC was checked and found adequate. The unit had an HHPS and the vapor from the HHPS was fed to the high-pressure heat exchangers and the HPAC after water injection, then to the CHPS (Fig. 4.5.1).

The HHPS liquid was routed to the HLPS after cooling in a heat exchanger, and from there it was routed to the stripper after further heating in the heat exchanger and furnace. The CHPS liquids were fed to the CLPS. The HLPS vapors were routed to the CLPS after cooling in a heat exchange and air cooler, along with the CHPS liquid feed to the CLPS.

The CLPS liquid was also routed to the high-pressure stripper at a colder location above the heated HLPS liquid feed. It was found that the HHPS and HLPS temperatures were low, around 150°C–160°C. It was suspected that the lower temperature of HHPS may dissolve some NH_3, H_2S, and HCl in the HHPS liquid and lead to the liberation of NH_3, H_2S, and HCl at the lower pressure of the

HLPS. Thus, some of the NH_3, H_2S, and HCl may get into the CLPS liquid and go to the stripper, and some will go with the CLPS vapor to stripper overhead. The gases may combine and deposit as NH_4CL or NH_4HS in the overhead air cooler of the stripper. Further there is always the likelihood that the HLPS liquid may contain some NH_3 and HCl and get into stripper and deposit in the colder section of the stripper.

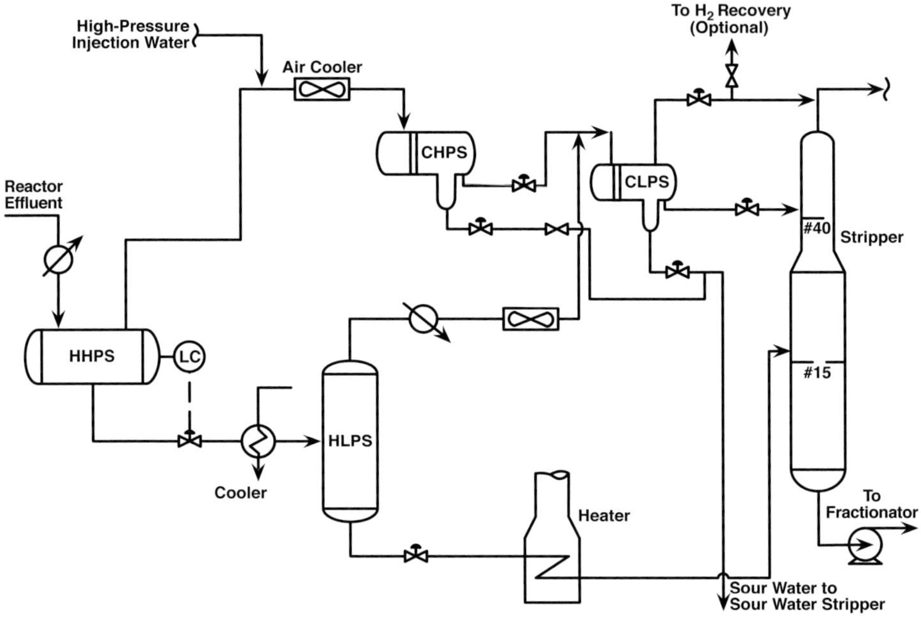

Fig. 4.5.1. Original scheme of VGO hydrotreater

It was learned that the refinery had started injecting visbroken naphtha into the VGO HDT feed as a means of disposing of it. This most certainly triggered the deposition of NH_4Cl.

Solution Suggested

An additional water injection was suggested for the HLPS vapors upstream of the HLPS overhead air cooler (Fig. 4.5.2). All the injected water was to be drained from the CLPS. It was advised that the engineering contractor engaged for detail engineering check the adequacy of the CLPS and its water boot to handle the additional water injected downstream of the HLPS overhead air cooler. A water injection upstream of the HLPS overhead was also recommended by API RP 932B, published in 2004.

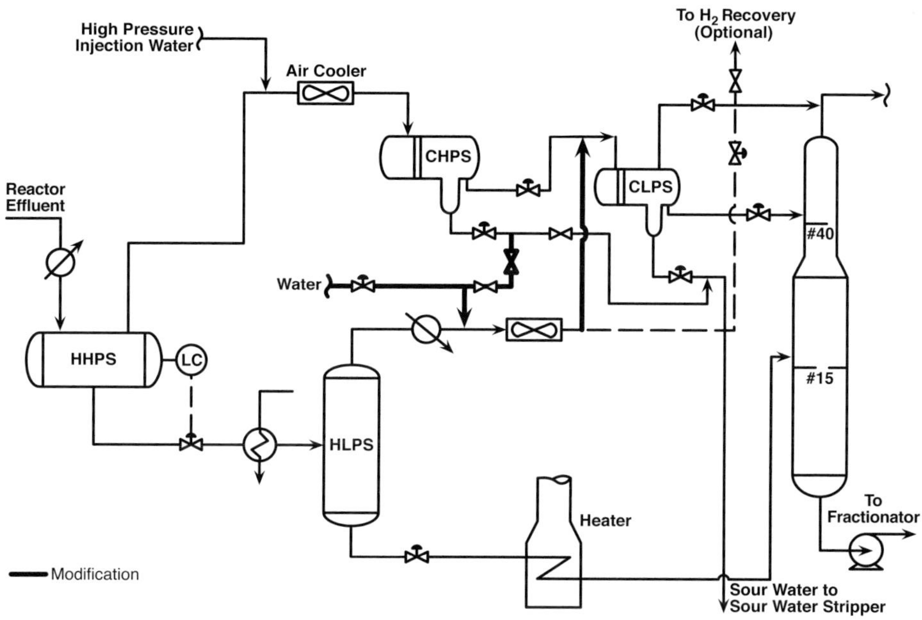

Fig. 4.5.2 Modified scheme with water injection

Further, it was suggested that the HLPS temperature be increased until it was nearly the same as that of the HHPS in order to liberate all the NH_3 and HCl, wash it with water, then route it to the CLPS to completely eliminate the likelihood of stripper choking. Thus, water injection was added at the upstream of the air cooler at the HLPS overhead to wash any NH_3, H_2S, and HCl to avoid NH_4Cl and NH_4HS deposition. This suggested modification was implemented and resolved the problem.

Case Study 2 (Debottlenecking a once-through hydrocracker [OHCU] consequent to conversion increase)

One partial conversion hydrocracker operated with conversion of 60%. Conversion capability was increased to 65% with a new catalyst. The refinery was operating at 65% conversion in a bid to increase distillate.

There were many problems in the OHCU, such as low LPG yield, foaming in recycle gas amine absorber, high CHPS temperature, low recycle gas purity, and high fractionator reflux drum temperature leading to fractionator pressurization.

The entire unit was simulated with higher conversion. After a thorough rating of the unit and a thermal rating of the high-pressure effluent heat exchangers and HPAC, it was found that the air cooler (HPAC) duty was inadequate, leading to the increase in the CHPS temperature (up to 85°C at 65% conversion). Due to the high temperature of the CHPS, the recycle gas purity dropped below design level. Thus, the unit was operating at lower recycle gas purity than envisaged in the design. The fractionator overhead condenser duty was found to be limiting and causing the high reflux drum temperature. The wet gas compressor was reaching its maximum capacity, leading to the pressurization of the fractionator.

In the actual run, a CHPS temperature of 80°C–85°C was observed, as opposed to the design temperature of 55°C. The high temperature of inlet gas to the high-pressure amine absorber was likely to result in foaming. Other observations made from the rating of column were as follows:

1. In the main fractionator, the zone above the kerosene draw was heavily loaded at a higher (65%) conversion.
2. The sponge absorber was very heavily loaded.
3. The stabilizer stripping section (below the feed) was flooded.

Solutions offered:

High-pressure section

It was suggested that a water cooler (Fig. 4.5.3) with a duty of around 2.5 Gcal/hr be placed downstream of the CHPS to cool the vapor to around 55°C. It was also suggested that a knockout drum with a continuous draining facility to the CLPS be implemented to improve the recycle gas purity and decrease the temperature of the gas going to the high-pressure amine absorber.

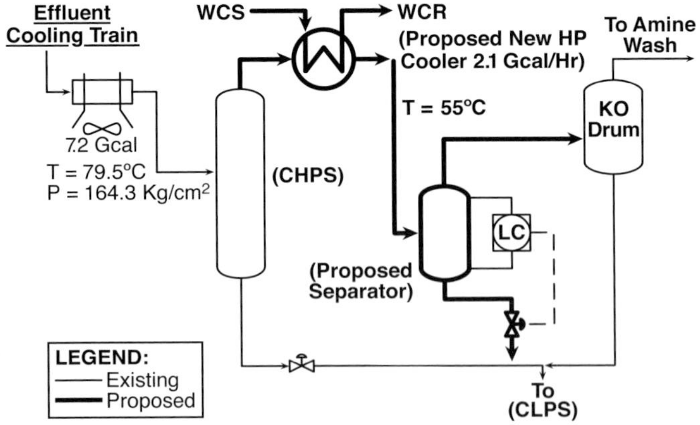

Fig. 4.5.3 Installation of a cooler and separator to cool CHPS vapor

Low-pressure section

In the fractionation section, it was suggested to add a condenser with a duty of around 3 Gcal/hr at the main fractionator overhead. Alternatively, it was also suggested to install one kerosene pump-around reflux at the kerosene draw tray. There were already 10 trays between top and kerosene draw.

Even after using three trays for the additional pump-around (the kerosene PA), there would still be enough trays for mass transfer, and the separation of naphtha and kerosene would not be greatly affected. This would also reduce the required duty in the fractionator overhead condenser and help maintain the reflux drum temperature.

The stabilizer feed temperature was increased by 20°C to 120°C in an attempt to relieve the stabilizer reboiler duty. This reduced flooding below the feed section (stripping section).

The refinery implemented the suggested modifications and the problems were resolved.

Case Study 3 (Hydrocracker capacity revamp)

A revamp was planned for a hydrocracker unit with three reactors in series and a medium-pressure (MP) stripper preceding the fractionator. The unit had two drums, a CHPS and a CLPS (Fig. 4.5.4). The CLPS liquid was heated and fed to the MP stripper to remove the H_2S. The stripper had a fired heater as a reboiler and the stripper bottom was fed to the main fractionator through the charge heater. The configuration is shown below in Fig. 4.5.4. The original effluent cooling circuit and reactor section is presented in Figs. 4.5.5 and 4.4.1.

For this hydrocracker capacity revamp, the original plan was to replace all the effluent exchangers from EE-02 to EE-07 (Fig. 4.5.6), and use two interreactor heat exchangers (Fig. 4.4.2) to reduce the quench flow requirement, as well as heat the stripper feed. The air cooler velocity was high, so it was recommended that the metallurgy of the air cooler be upgraded to Monel. One additional makeup gas compressor was also planned and installed.

However, installing additional interreactor heat exchangers and changing all the effluent circuit heat exchangers would have caused a very long unit outage. So the first phase (phase 1) of the unit revamp was carried out without additional interreactor heat exchangers and effluent circuit heat exchangers and around 15% capacity increase was achieved.

Subsequently, an HHPS was included in the design along with an HLPS to debottleneck the effluent circuit instead of replacing the heat exchangers. This was likely to relieve the load on the downstream effluent heat exchangers and HPAC.

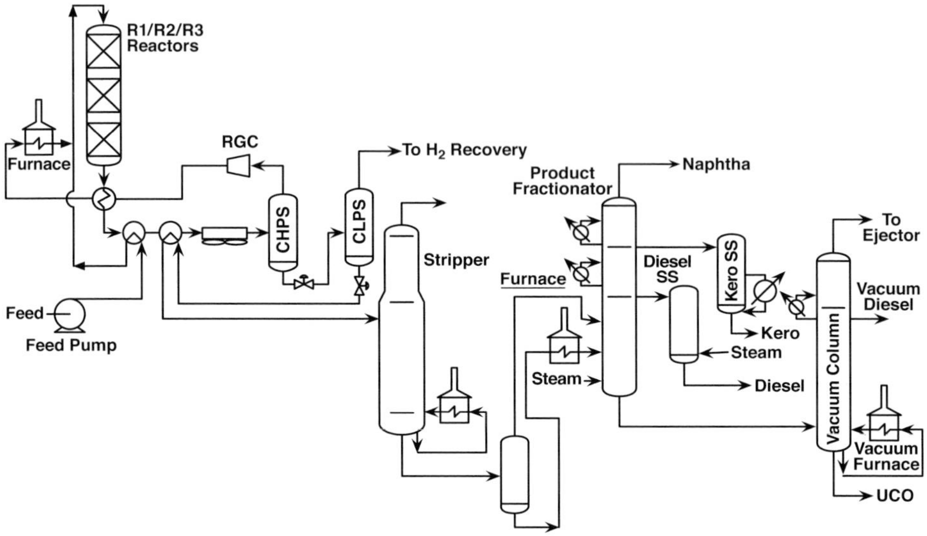

Fig. 4.5.4 Original configuration of the unit

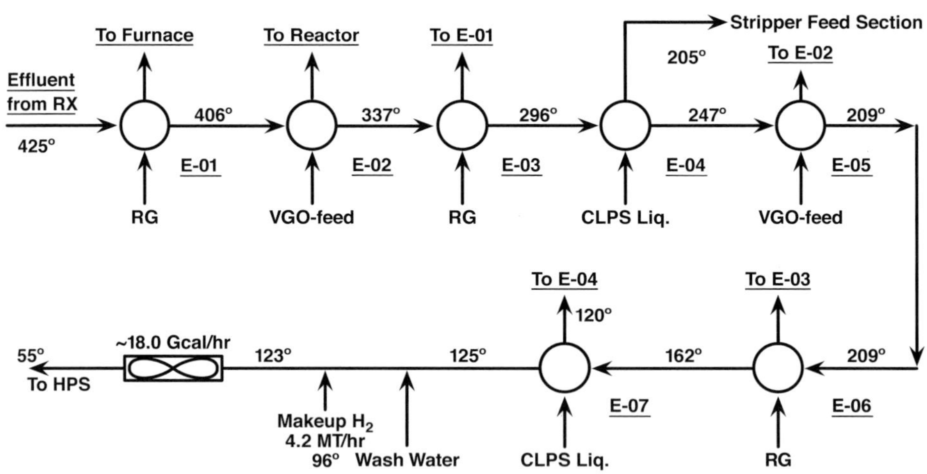

Fig. 4.5.5 Original effluent heat exchanger train

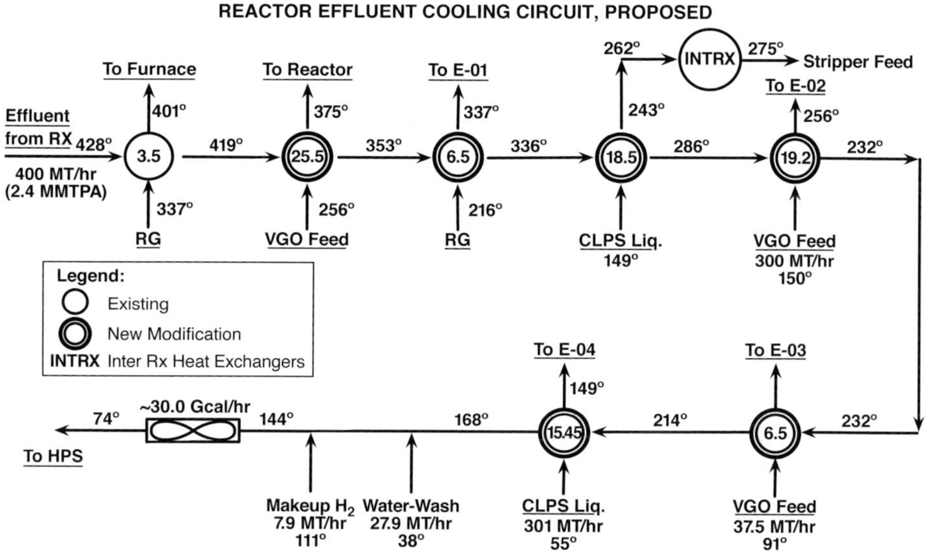

Fig. 4.5.6 Heat exchanger needed replacement in effluent service

The heated CLPS feed joined the HLPS bottom and was fed to the stripper. A pre-flash drum was also introduced at the inlet of the charge furnace. The modifications are indicated in Fig. 4.4.5, which shows the revised effluent cooling train, and Fig. 4.5.7, which shows the revised overall scheme. The installation of the HHPS and HLPS was likely to save around 8–10 Gcal/hr of heat in the stripper reboiler, in addition to increasing the capacity.

In the second phase (phase 2), the additional drums were installed to save heat and increase the capacity by 10%. However, the RGC capacity reached its limit at this point. Thus, further capacity increase was not possible without the installation of interreactor heat exchangers.

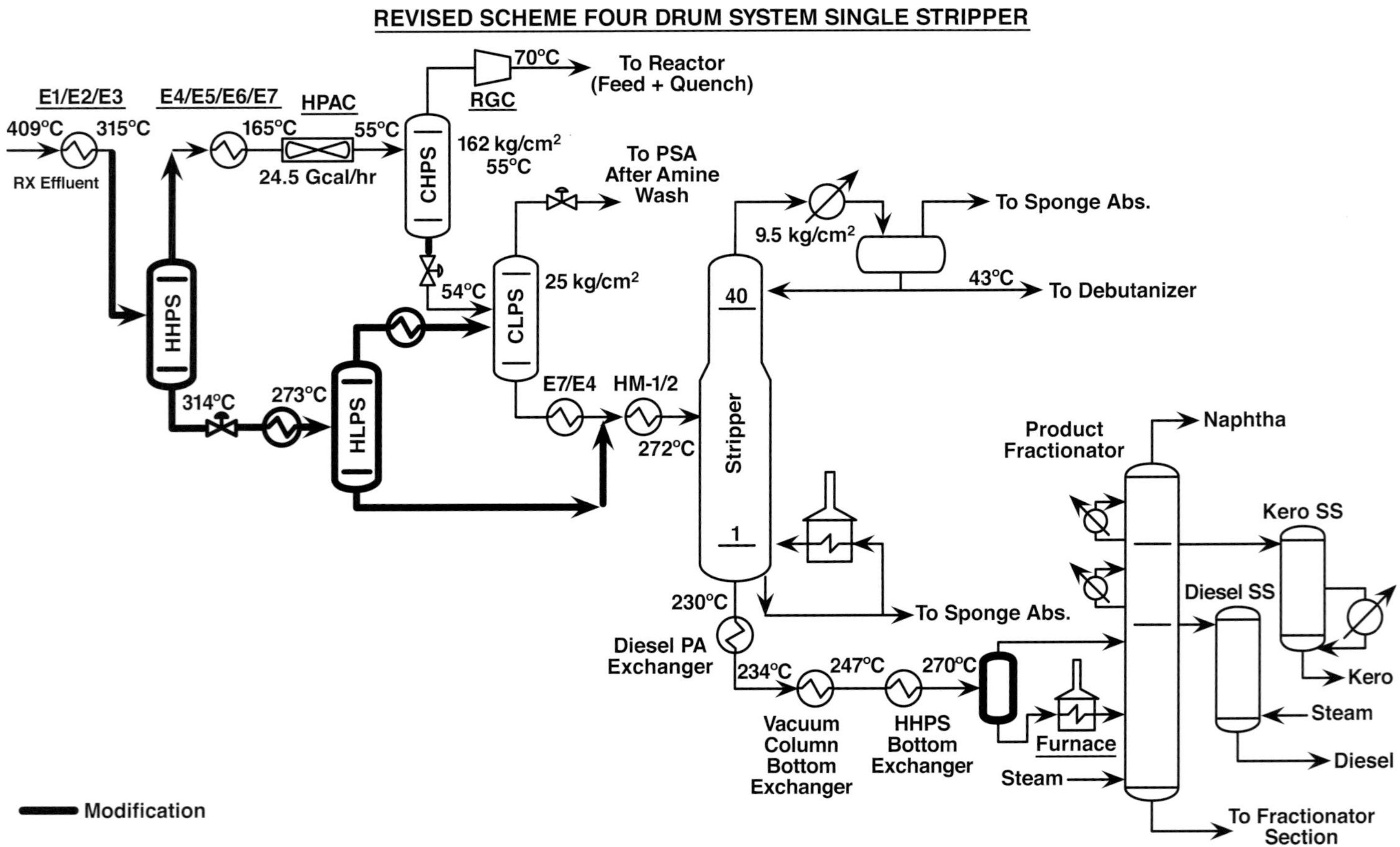

Fig. 4.5.7 Proposed overall scheme

Benefits are summarized below:

1. Total direct fuel savings in terms of the furnace absorbed duty, which was approximately 8.0 Gcal/hr in the stripper reboiler.
2. An additional 2.0 MT/hr MP steam through the existing vacuum column bottom steam generator, as the vacuum bottom was no longer required to heat the stripper feed in a four-drum configuration.
3. The reactor effluent air cooler (HPAC) duty was reduced to 25 Gcal/hr from an earlier duty of 29 Gcal/hr in the two-drum system.
4. The CHPS and CLPS drum temperatures can be easily maintained at 55°C, as opposed to the 74°C of the two-drum system.
5. The H_2 purity of the recycle gas is the same in both the cases (90 mole %).
6. The low suction drum temperature (55°C) of the RGC will provide additional margin in the RGC capacity.
7. RGC discharge temperature will decrease to 70°C (as opposed to the original 91°C), which will reduce the H_2 quench requirement. Thus, it may lower the quench flow requirement and provide additional margin in the RGC capacity.
8. Pressure drop in heat exchangers in the effluent exchanger train will be reduced to almost the same value as the original. This will lower the RGC discharge head requirement and may provide more flow.

All the above will provide the flexibility required to enhance the capacity of the unit.

Recent trends in hydrocracking and hydrotreating operations

1. The conversions in VGO and HCGO hydrotreaters are being increased through the use of more active catalysts with small revamps in the cooling, fractionation, and treating sections.
2. The unconverted oil from the hydrocrackers is being processed for the production of Group II and Group III lube-oil-based stocks.
3. Incorporation of raffinate from the solvent de-asphalting unit (SDA) along with feed of VGO hydrocrackers.
4. Lube oil feed stocks are treated in CDW units for conversion of wax components to increase the yield of lube oil.

4.6 Metallurgy

A hydrocracker reactor is typically made of 2¼% chromium (Cr) and 1% molybdenum (Mo) steel and is lined inside with a 3 mm layer of SS 347 stainless steel. The 2¼% Cr and 1% Mo steel has a minimum pressurization temperature (MPT) of 93°C due to the material's transition from ductile to brittle behavior. Thus, during operation, the reactors are never subjected to pressures greater than ¼ the design pressure until all the points of the reactor are heated beyond 93°C. Hydrogen can degrade the mechanical properties of steel, and there are two main ways in which it does so:

1. At high temperatures, hydrogen can attack and combine with the carbon in steel, producing methane, and causing temperature embrittlement or decarburization.
2. Hydrogen can permeate through steel and cause hydrogen embrittlement.

Nelson curves indicate that the temperature limit for 2¼% Cr—1% Mo steel at the hydrogen partial pressures typically encountered in a hydrocracker (averaging at around 146 kg/cm^2) should be 440°C to avoid HTHA. Nelson charts or curves are normally used as a guide to specify the operating limits for steels in hydrogen service (these limits are also presented in API RP 941).[5] The guidelines for all the material used in constructing equipment are also based on these operating limits. Hydrogen also has permeability in steel, therefore reactor metal is susceptible to hydrogen embrittlement. During operation, the heating and cooling rates are rigorously maintained to decrease the likelihood of embrittlement and fissuring. During the cooling operation, the temperature is not decreased below a stipulated limit until the pressure is also decreased to the prescribed level. This is also to avoid hydrogen embrittlement.

For higher operating temperatures (temperatures from 440°C–500°C), API suggests the use of 3% Cr–1% Mo steel or 2¼% Cr–1% Mo steel with 0.25% to 0.35% vanadium (V).[5] The modern reactors are made with 2¼% Cr–1% Mo steel with 0.25%–0.35% V in order to maintain better mechanical properties at higher temperatures. The addition of vanadium also increases the tensile strength of the reactor material.

Abbreviations

CAT: catalyst average temperature
CHPS: cold high-pressure separator
CLPS: cold low-pressure separator
CN: cetane number (of diesel)
DAO: de-asphalted oil
DWO: dewaxed oil
DHDS: diesel hydrodesulfurization
HCGO: heavy coker gas oil
HDN: hydrodenitrification
HDS: hydrodesulfurization
HHPS: hot high-pressure separator
HLPS: hot low-pressure separator
MPT: minimum pressurization temperature
HPAC: high-pressure air cooler
LCGO: light coker gas oil
LCO: light cycle oil
MUG: makeup gas
NH_4Cl: ammonium chloride
NH_4HS: ammonium hydrosulfide or ammonium bisulfide
PRT: power recovery turbine
RGC: recycle gas compressor
SRGO: straight-run gas oil
UCOS: unconverted oil separator

Notes

1. Reprinted with permission from AIP Publishing LLC. The NH_4Cl dissociation constant (Kp, NH4Cl) chart. *Journal of Chemical Physics* 12, no. 7: 318.
2. Ibid.

3. US Bureau of Mines, "NH_4HS Deposition Chart," in *Bulletin 406* (1937).
4. American Petroleum Institute, *API RP 932-B: Design, Materials, Fabrication, Operation, and Inspection Guidelines for Corrosion Control in Hydroprocessing Reactor Effluent Air Cooler (REAC) Systems* (Washington, DC: API Publishing, 2012).
5. American Petroleum Institute, *API RP 941: Steels for Hydrogen Service at Elevated Temperatures and Pressures in Petroleum Refineries and Petrochemical Plants* (Washington, DC: API Publishing, 2008).

5 Gas Amine Absorbers

5.1 Introduction

Gas amine absorbers use aqueous solutions of various types of alkyl amines (commonly referred as amines) to remove acid gases (H_2S and CO_2) from sour gases. It is used in refineries, petrochemicals and natural gas processing plants, and in many other industries. In this process, acid gases react with the basic alkanolamine solution through an exothermic reversible reaction in a gas/liquid contactor (absorber). The reverse reaction occurs in the amine regenerator (stripper). Amine absorbers are primarily used to

1. Scrub sour gas to remove H_2S and CO_2
2. Wash liquefied petroleum gas (LPG) to remove H_2S
3. Scrub gases to remove CO_2

Different amines that are used in gas treating include:

1. Monoethanolamine (MEA)
2. Diethanolamine (DEA)
3. Methyldiethanolamine (MDEA)
4. Diisopropanolamine (DIPA)
5. Diglycolamine (DGA)

Other amines that are used for removal of acid gases are 2-amino-2-methyl-1-propanol (AMP) and tri-ethanol amine (TEA or TEOA).

Using alkanolamines for removal of H_2S and CO_2 from natural gas has been in practice for decades. Natural gas (NG) can have a wide range of acid gas concentrations, from parts per million to around 50 volume % or higher. Due to the corrosive nature of moist H_2S and CO_2 and due to no heating value of

CO_2 in NG, the gas needs treatment to sweeten to reduce corrosion as well as to remove CO_2 in order to increase the heating value of sales gas to typically around 950–980 BTU/SCF in line with the agreement. The typical pipeline-grade sales gas has 2% CO_2 by volume and 3 ppm wt of H_2S.

In refineries, there is a wide range of applications for amine gas scrubbers for removal of H_2S. Gases with H_2S and CO_2 are washed with amine to convert the sour gases to sweet gases. The H_2S is picked up by the amine and the amine is regenerated in the amine regenerator to liberate the H_2S and feed it to the sulfur unit to recover the sulfur.

An amine wash in a refinery is provided for the recycle gas in the hydrotreaters and hydrocrackers in order to decrease the H_2S level and increase the H_2 purity of recycle gas. This is normally done in high-pressure (HP) systems.

Any low-pressure sour gas is also washed with amine to decrease the H_2S content. Low- to moderate-pressure sour gases include dry gas from the fluid catalytic cracking unit (FCCU) and coker units, and very sour gases from naphtha hydrotreaters and diesel hydrotreaters, residue hydrotreaters, and hydrocrackers.

Tail gas in the sulfur recovery unit with a lower concentration of H_2S is also amine-washed to remove the low level of H_2S in the gas stream. This requires a lean amine with very low H_2S loading to ensure the difference in concentration gradient between the H_2S-bearing sour gas and the lean amine at near atmospheric pressure.

Monoethanolamine (MEA) is used to wash compressed charge gas in steam crackers upstream of the caustic scrubber to free it of H_2S. It is also used to scrub CO_2 from the gases. The sour LPG is also scrubbed with amine in the LPG amine contactors to remove the dissolved H_2S and lower the H_2S content before going through a caustic wash to help it pass the copper corrosion test. These are liquid–liquid contactors.

The process of separation and capture of CO_2 from flue gases from power plants by absorption with amine is also gaining importance with respect to reducing greenhouse gas emissions. The amines that may find application are MEA, DEA, AMP, and TEA. The heats of absorptions of MEA, AMP, and TEA are 88.91, 44.71, and 63.95 kJ/mol CO_2 respectively. The heat of absorption (of CO_2) reduces from MEA to DEA to AMP to TEA. Diglycolamine (DGA) also has a higher affinity for CO_2 than H_2S. The advantage of DGA is that it can be used at a concentration of 50%–65% weight in solution, maximum acid gas loading is around 0.35 moles of acid gas/mole of DGA, and it results in a lower circulation rate, and is also lower in freezing point but has higher heat of reaction for both H_2S (around 675 BTU/lb) and CO_2 (around 850 BTU/lb). Experts observe that the net energy requirement for DGA may still be low. The amine is not likely to react with COS, CS_2, SO_2, or SO_3 irreversibly.

5.2 Description

The sour gases are admitted to the amine absorber bottom with a knockout drum and filter in order to remove the liquid hydrocarbon and dirt, as they can cause foaming (Fig. 5.2.1).

Lean amine is introduced from the top, preferably with a filter. A semi-lean amine from the regenerator or partly loaded amine from another absorber can also be fed at a lower location of the column. If the absorber is at higher pressure (such as in a high-pressure hydrocracker or a hydrotreater recycle gas scrubber), an amine booster pump is used to elevate the pressure and help the amine enter the high-pressure system. Otherwise, no separate pump for the lean amine is required.

The amine is introduced at a temperature 6°C–8°C higher than the gas entering the bottom of the absorber so that it does not condense the hydrocarbons from the gas, which would result in foaming. An amine steam heater with a bypass provision is used to keep the temperature of the lean amine higher than that of the sour gas. The rich amine exits to the amine regenerator under the level control of the absorber (Fig. 5.2.1).

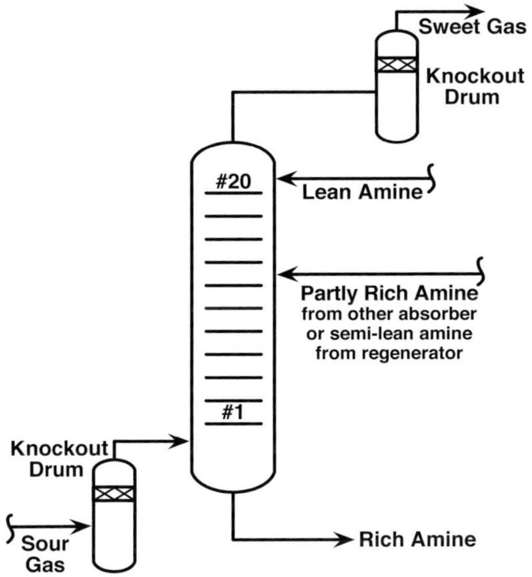

Fig. 5.2.1 Typical scheme of an amine gas absorber

Conventionally, three types of amines are used: monoethanolamine (MEA), diethanolamine (DEA), and methyl diethanolamine (MDEA). The inherent corrosiveness of amine in the water solution reduces from MEA to DEA to MDEA.

MDEA-based mixed amines having MEA and DEA are used to enhance CO_2 pickup by MDEA. The MEA and DEA proportion in MDEA is usually limited to 15%–20% weight to avoid the requirement of higher metallurgy.

DGA/DIPA can also be used mostly in admixture with other amines for selective absorption.

General features for different amines are presented below for selection of a particular amine for a specific service.

1. **Monoethanolamine (MEA)** is a primary amine with a molecular weight of 61 and an atmospheric boiling point of 171°C. The strength of an MEA solution in water is limited to 15%–20% in carbon steel. Reaction stoichiometry indicates that two moles of the amine can react with one mole of CO_2. Thus, maximum theoretical loading is 0.5 mole of CO_2/mole of amine. Typical rich amine acid gas loading for MEA is 0.15–0.2 moles of acid gas (H_2S or CO_2)/mole of MEA in carbon steel to minimize corrosion. MEA is a primary amine and has a high pH and can meet pipeline grade specification of quarter grain H_2S per 100 SCF even when contacted with gas at a very low pressure. The heat of reaction of MEA with CO_2 is around 825 BTU/lb of CO_2 (very high) and with H_2S about 550 BTU/lb of H_2S. (The heat of reaction is dependent on amine loading, temperature, etc., but usually varies only by around 45 to 60 BTU/lb with acid gas loading.) Thus, a sour gas containing a high concentration of CO_2 will need an extremely high reboiler duty in the regenerator for full regeneration of the amine due to its high heat of reaction. MEA reacts almost irreversibly with CO_2, CS_2, SO_2, SO_3, and oxygen, and products of reaction remain in solution and poses a serious corrosion problem and degrades the amine and lowers the effective strength. A higher regenerator temperature increases the deactivation. The amine can be reclaimed using caustic or soda ash followed by distillation to recover activated amine off the top.

2. **Diethanolamine (DEA)** is a secondary amine with a molecular weight of 105.6 and an atmospheric boiling point of 268°C. The strength of DEA solution in water is generally limited to 25%–30% in carbon steel. Typical rich amine acid gas loading for DEA is 0.3 to 0.35 moles/mole of DEA in carbon steel to minimize corrosion. As a secondary amine,

it has lower affinity for H_2S and CO_2 and may not be in a position to meet pipeline grade from low-pressure gas streams. The gas may need compression to enable removal of the acid gas components by the amine. The heat of reaction of DEA with CO_2 is around 650 BTU/lb (much lower than MEA), and with H_2S around 510 BTU/lb. The amine also irreversibly reacts with COS, CS_2, SO_2, SO_3, and oxygen and degrades, but the products of degradation are less corrosive than MEA. Higher regenerator temperature increases degradation. DEA cannot be reclaimed like MEA due to its higher boiling point, but it reportedly can be done under vacuum distillation to prevent thermal degradation. The degradation rate is slow and mostly does not need conventional reclaiming.

3. **Methyl diethanolamine (MDEA)** is a tertiary amine with a molecular weight of 119.5 and an atmospheric boiling point of 247°C. The strength of MDEA solution in water is limited to 40%–50% wt in carbon steel. Typical rich amine H_2S loading for MDEA is 0.4–0.5 moles of acid gas/mole of MDEA in carbon steel to minimize corrosion. Due to the high affinity for H_2S, it can be used in the tail gas cleanup section in a Claus unit at low pressure to capture H_2S and achieve the desired CO_2 slip. The heat of reaction of MDEA with CO_2 is around 600 BTU/lb (much lower than MEA), and with H_2S around 520 BTU/lb. It has higher resistance to degradation and selectivity toward H_2S in the presence of CO_2 but reacts irreversibly with COS, CS_2, SO_2, SO_3, and oxygen. Due to the lower heat of reaction with CO_2 and a weak bond with CO_2, the bulk removal of CO_2 from MDEA can be achieved by letdown of pressure to near atmospheric (pressure swing).

The following are acid gas absorption reaction mechanisms:

- H_2S is known to react with amine by proton transfer. MDEA is also known to react with H_2S through proton transfer.
- CO_2 is known to react with primary and secondary amines to form a carbamate.
- MDEA does not have hydrogen attached with nitrogen. Thus, it cannot react with CO_2 to form a carbamate. Reaction can only occur once CO_2 dissolves in water to form a bicarbonate ion and performs an acid-base reaction with amine.

5.3 Design Concerns

A high CO_2 concentration in sour gas when treated with amine to remove H_2S and CO_2 will release a substantial quantity of CO_2 along with H_2S during regeneration. Consequently, the concentration of CO_2 in the acid gas from the amine regenerator (stripper) will increase and in turn will lower the concentration of H_2S in the gas to the Claus unit, may result in lowering the temperature of the main combustion chamber (MCC) of the Claus unit, and can result in flame instabilities.

Further in the tail gas treatment section of a sulfur unit, the gas primarily containing N_2, CO_2, SO_2, and H_2S reacts with hydrogen in the reactor to convert SO_2 to H_2S, and then the gas is contacted with amine and the amine is regenerated. If the pickup of CO_2 in amine is high, this may increase CO_2 concentration at the Claus inlet as the gases produced after regeneration are recycled to the front of the Claus unit and may reduce the H_2S concentration in the gas. Thus, CO_2 slip may be desired in such a scenario. Many proprietary solvents are in use to selectively absorb the H_2S from high CO_2 content in a mixture at the low pressure of the tail gas stream.

A different scenario may emerge in natural gas (NG) processing plants. CO_2, having no calorific value, may need to be removed in substantial quantities by amine to achieve the agreed calorific value of sales gas (typically 950 to 980 BTU/lb) along with the removal of H_2S to meet the pipeline grade specification of quarter grain of H_2S per 100 SCF (or 4 wt ppm). In this case, the amine regenerator is likely to produce a considerable amount of CO_2 in the acid gas and may pose a problem for the sulfur unit if regeneration is done in a single stage. Pressure swing regeneration may be relevant where the rich amine pressure is reduced to liberate the majority of CO_2 at near atmospheric pressure (only applicable in case of tertiary amines like MDEA). Other options observed by experts are installing absorbers in series, in parallel, and using a static mixer (precontactor) for partial CO_2 removal.

5.3.1 Rich amine flow scheme

The rich amine should not be routed directly from the high-pressure amine wash system (the recycle gas treatment of hydrocrackers and hydrotreaters) to the amine regenerators. It should be brought down to an intermediate lower pressure before being sent to the amine regeneration unit (ARU). Otherwise, the low-pressure amine regenerators can be endangered by any gassing that may occur through the rich amine line at the higher pressures of amine absorbers.

Very-high-pressure systems should not be connected to low-pressure ones without sufficient caution and safety measures.

A typical scheme (Fig. 5.3.1) for a high-pressure hydrotreater recycle gas and sour gas amine wash is given below. The spent amine from the high-pressure amine wash column operating at 115 kg/cm^2 is introduced to the bottom sump of the low-pressure amine absorber at 7 kg/cm^2 (for sour fuel gas) and from there, sent to the amine regenerator. The problem with this setup is that the H_2 dissolved in the amine solution at a high pressure is released in the low-pressure amine absorber. Typically, the low-pressure amine absorber operates at around 7–10 kg/cm^2. Thus, the H_2 released in the low-pressure absorber from amine will get into the fuel gas and will be difficult to recover.

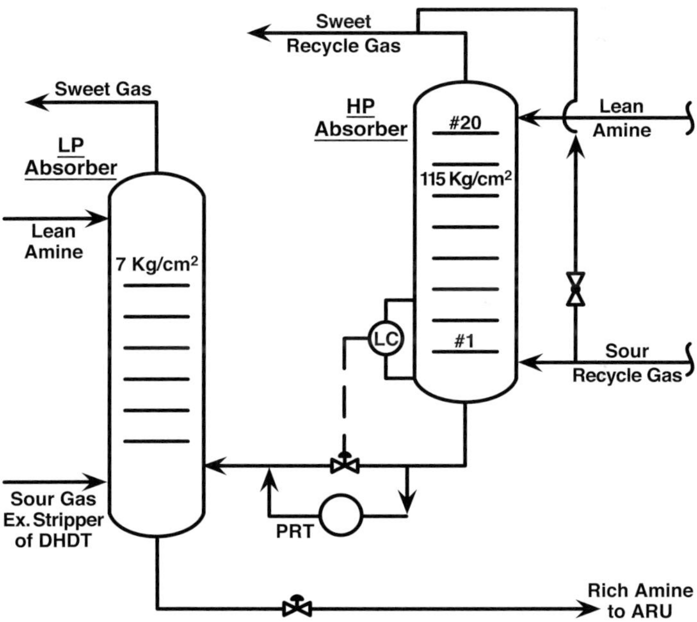

Fig. 5.3.1 High-pressure amine routing to the low-pressure amine column spout

It may be advisable to install a flash vessel in the spent HP amine line to flash the hydrogen at a medium-pressure level of around 25–27 kg/cm^2. The the amine can be further routed to the low-pressure amine absorber bottom spout. Using this setup, the dissolved H_2 can be recovered at medium-pressure and reused (Fig. 5.3.2), which will also reduce the low-pressure absorber load to some extent and can be used to debottleneck the LP absorber.

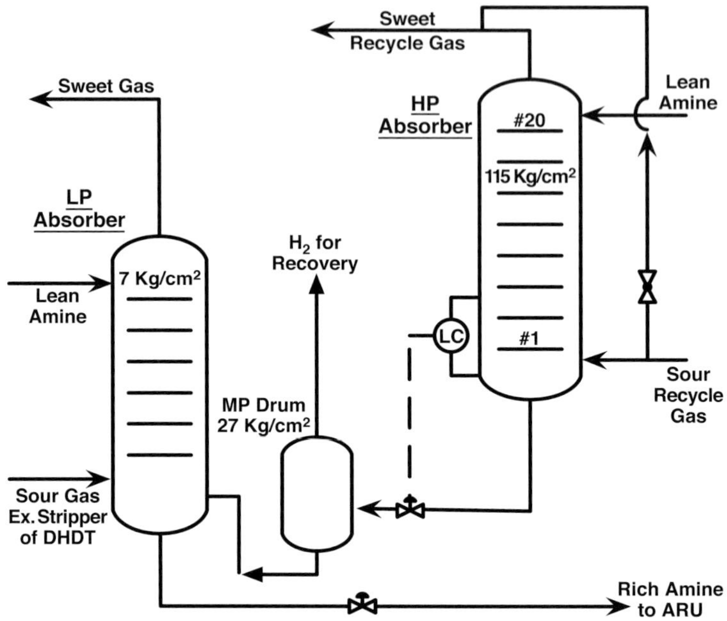

Fig. 5.3.2 Spent high-pressure amine routing to the medium-pressure flash drum, then to the low-pressure amine absorber spout

A power recovery turbine (PRT) is often installed (Fig. 5.3.1) in the high-pressure amine line going to the low-pressure amine column bottom sump to recover power if possible.

Typically, a bypass across the recycle gas scrubber is provided with a control valve to partially route sour gas to sweet recycle gas, so that a minimum H_2S concentration is maintained in the recycle gas to avoid the loss of sulfur from the catalyst.

Amine absorption is an exothermic process. The rich amine is heated in the process of absorbing the H_2S. The rich amine can be pumped if the pressure of the absorber is not adequate to route it to the desired destination. Both the lean amine and rich amine lines need to be heat insulated. A partly loaded amine stream can be introduced at a lower location (see Fig. 5.2.1), if one is available to reduce the fresh lean amine requirement at the topmost tray of the absorber.

Amine used for LPG treating preferably should not be mixed with amine used for gas scrubbing, particularly with the amine used for scrubbing the FCCU dry gas or sulfur unit tail gas, as both the gases contain oxygen and produce oxygenated product in the amine, which may cause the LPG to fail in the copper corrosion test. The author has experience with such failures, where the amine used to treat FCCU dry gas (sponge gas) or sulfur unit tail gas was mixed and used for LPG treatment after regeneration. Normally, the amine system for tail gas treatment is separate (having its own amine regenerator).

5.3.2 Design parameters

A gas amine absorber is typically provided with 20–22 trays with an efficiency of approximately 25%–30%, and a system factor of 0.7–0.8 considered for design. The typical sweet gas H_2S ppm desired is less than 80–100 ppm (wt) and often even less in refinery operations. To lower the sweet gas H_2S content, the lean amine used as an absorbent should have a low H_2S loading, around 0.001 to 0.01 moles H_2S/moles of MDEA. For the VLE data generation Kent and Eisenberg correlation, an amine package or NRTL electrolyte can be used. Some software has a kinetic model that takes into account residence time, temperature, and solution concentration on the rate of H_2S and CO_2 absorption.

The process variables that are likely to influence the absorption are lean amine temperature and acid gas loading, amine circulation rate, liquid residence time on the tray, and pressure of the absorber. A higher temperature (above 60°C) would help in selective rejection of CO_2. CO_2 reaction with amine is slow, and a high residence time increases CO_2 pickup. Low lean amine loading and higher absorber pressure increase absorption.

5.4 Case Studies

5.4.1 Capacity augmentation of an amine wash system

A gas condensate refinery amine system was revamped (the original system is shown in Fig. 5.4.1). The unit had two amine systems with two amine regenerators. During revamp, it was found that the regenerators would require reboiler and condenser duty augmentation due to an increased amine circulation requirement.

In the original processing scheme in the old plant (Fig. 5.4.1), there was only one LPG amine contactor and the rich amine was routed to the old regenerator for amine regeneration. In the new plant, there were three amine absorbers (Fig. 5.4.1):

1. LPG amine wash
2. Fuel gas amine wash
3. Sulfur unit tail gas amine wash

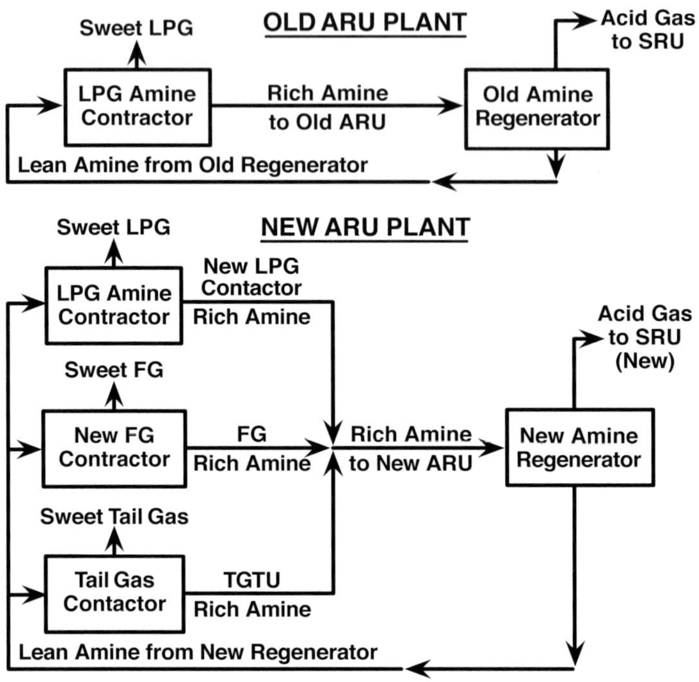

Fig. 5.4.1 Original processing schemes of the old and new plants

In this system, the spent amine from the three absorbers was routed to the new amine regenerator in the new block. The rich amine from the tail gas treating unit was mixed with the total amine and the regenerated amine was used for treating the LPG, apparently causing the LPG to frequently fail during the copper corrosion test.

During the condensate refinery's capacity augmentation, the amine circulation requirement increased, pushing the equipment in the regenerators to its capacity limits. A new configuration was evolved, which can be seen in Fig. 5.4.2. In the new configuration, the LPG amine was segregated from the gas amine, which was expected to solve the repeated failure of LPG during the copper corrosion test.

It was found that the H_2S loading of spent amine from tail gas treating was low, and thus had a residual absorption capacity. This partly loaded amine was introduced in the sour fuel gas amine-absorption column at a lower location, while freshly regenerated amine with very low H_2S loading was introduced at the top of the fuel gas amine absorber. Thus, the amine leaving from the bottom of the fuel gas absorber was adequately loaded before returning to the amine regenerator. This reduced the amine circulation requirement compared to that of the old configuration at the revamped capacity.

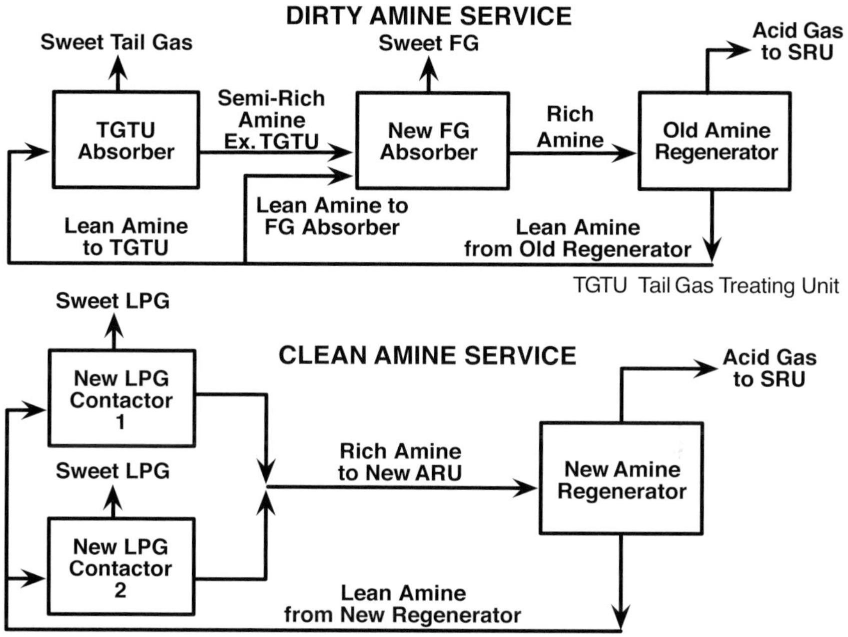

Fig. 5.4.2 Modified scheme with segregation of dirty and clean service

Benefits of the proposed configuration (Fig. 5.4.2) over the original configuration (Fig. 5.4.1) at a higher capacity are given in Table 5.4.1.

Table 5.4.1 Benefits of the proposed scheme over the existing one

S.N.	New equipment	Existing config.	Proposed config.
1	Total amine circulation rate	~86 MT/hr	~71 MT/hr
2	Total regenerator reboiler duty	~8.7 Gcal/hr	~7.3 MT/hr
3	Total regenerator condenser duty	~5.4 Gcal/hr	~3.9 Gcal/hr
4	Regenerator reboiler steam consumption	~17.0 MT/hr	~14.0 MT/hr
5	Pump power (kW)	Reduction in power by ~20%	

5.4.2 Case study of the low flow of rich amine from the gas absorber

A diesel hydrotreater's (DHDT) recycle gas amine scrubber (operating at 120 kg/cm^2) was malfunctioning: the rich amine was not flowing to the amine regenerator when it was commissioned during winter. In the plant, the rich amine at higher pressures was brought down to around 7–8 kg/cm^2 at the sump of the sour gas amine absorber, operating at around 7 kg/cm^2 (see Fig. 5.3.1) before routing the spent amine to the amine regenerator.

It was suspected initially that there was some obstruction in the line. A pressure survey was conducted in the long line to identify the point of choking. The survey was not conclusive. There was also no additional gas bleed observed from the rich amine surge drum of the amine regeneration unit, confirming that no two-phase flow or gassing was occurring in the rich amine line and causing an increased pressure drop.

Eventually, it was found that the amine heater was not in line and that the rich amine temperature in the absorber was too low. The line to the regenerator was very long (around 1 km), and part of the line was cold due to low flow and heat loss. It was suspected that hydrates might have formed at the low ambient temperature and consequent low line temperature.

The amine heater was put in line and the lean amine temperature increased to around 60°C, which caused the rich amine temperature to increase. The sour gas amine system pressure increased, limited by the safety setting. Gradually, a low flow of amine started. After around 12–18 hours, the line was cleared and regular flow began. A large volume of gas entered the rich amine flash drum once the line was cleared, then became normal after a while. It is important to remember that when H_2S is mixed with propane or butane gas, hydrates can form even at 4–5°C.

Metallurgy

1. Absorber column is made of killed carbon steel (KCS) with a corrosion allowance (CA) of 3mm.
2. Rich amine pump is made of KCS casing and a 410S impeller with no CA.
3. Lean amine pump is made of KCS casing and a 410S impeller with no CA.
4. Sour gas line is made of KCS with a CA of 3mm.
5. Lean amine water cooler shell is made of KCS with a CA of 3mm and the tube is made of KCS with a CA of 3mm.
6. Feed line after the feed control valve (where vaporization is likely) is made of stainless steel (SS) 304.
7. Reboiler shell is made of KCS with a CA of 6mm and the tube is made of SS 304L with no CA.
8. Rich amine lines and lean amine lines are made of KCS with a CA of 3mm.

9. Line from the downstream of the flash drum inlet control valve to the flash drum is made of SS 304 with no CA.
10. Rich amine pump is made of SS 316.
11. Lean amine pump is made of SS 316.

5.5 LPG Amine Contactors

The LPG is washed with amine in an LPG amine contactor to reduce the H_2S content to around 5–10 ppm. Typically this uses 2–4 stages, with the LPG introduced from the bottom and the amine from the top.

The LPG outlet from the absorber top typically has a knockout drum with or without a demister to arrest the amine particles in the LPG. A caustic wash followed by a water-wash is necessary at the downstream of the LPG amine wash to reduce the H_2S content below 2–3 ppm, which will allow the LPG to pass the copper corrosion test. The caustic wash also removes the mercaptans to ensure that the LPG passes the copper corrosion test.

Design guidelines:

1. The normal operating pressure is decided: a minimum 3kg/cm^2 higher than the LPG vapor pressure at the maximum temperature in the absorber column. This is to avoid the formation of LPG vapor bubbles in the absorber. It is preferred that the typical operating pressure for a grassroots unit be 7 kg/cm^2 higher than the maximum vapor pressure of LPG in the absorber.
2. Column diameter is typically estimated as the diameter needed to contain the total volumetric flow of LPG + amine to 10 gpm/ft^2 of the column cross-sectional area for grassroots designs and to 12 gpm/ft^2 for revamp applications. Flow of amine is estimated with the maximum loading of H_2S in amine (e.g., 0.4–0.5 moles of H_2S/moles of MDEA).
3. Flooding velocity is often calculated with the Crawford–Wilke chart (correlation curve).
4. Maximum flooding allowed is around 20% by the above correlation.
5. Height of the column is estimated by the height of the packing, the height for distributors/redistributors, and the height for top and bottom disengagement sections.

The column with trays can use ten sieve trays with a static mixer upstream for mixing part of the amine with the LPG. Around 80%–90% of the amine is routed to the static mixer and around 10%–20% is admitted from the top of the absorber.

A column with a packed bed arrangement uses three to four beds, each around 3 meters in height. Random packing is normally used. Redistributors are provided between the packed beds.

A typical LPG amine absorber with a water-wash is shown in Fig. 5.5.1. Here, no caustic wash (downstream to amine wash) is shown.

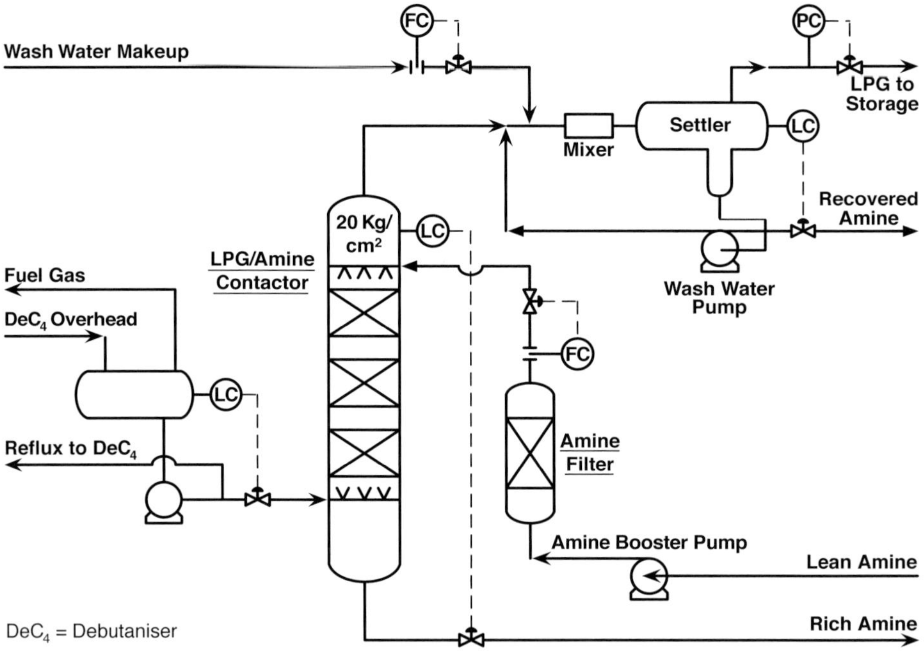

Fig. 5.5.1 A typical LPG amine absorber with a water-wash

Normally, in an LPG amine contactor the LPG is the dispersed phase and the amine is the continuous phase. Thus, the LPG–amine interphase is maintained at the absorber top.

For packed columns, the design of the LPG distributor at the bottom and the amine distributor at the top is critical. The LPG velocity is to be limited to 1.25 ft/sec through the nozzles to avoid the formation of emulsion and the distribution nozzles should point downwards. The amine distributor needs to be designed to minimize turbulence. The amine velocity should not be greater than 2.8 ft/sec through the nozzles. The temperature of amine at the entrance is to be adequate to maintain the amine viscosity at less than 2.0 centipoise (cP). Normally, it is desirable that amine be admitted into the column or distributor through a filter.

The residence time of LPG at the top is critical. Eight to ten minutes of LPG residence time is normally provided at the top disengaging space. Sometimes the diameter of the disengaging section is increased to provide the desired residence time and restrict the height. Amine residence time at the bottom of the column is maintained to avoid LPG slippage in the spent amine. A residence time of around 10 minutes is typical for rich amine; otherwise, a minimum disengaging height of 2.5 ft needs to be provided.

Abbreviations

ARU: amine regeneration unit

CA: corrosion allowance

cP: viscosity in centipoise

DEA: diethanolamine

KCS: killed carbon steel

LPG: liquefied petroleum gas

MDEA: methyl diethanolamine

MEA: monoethanolamine

PRT: power recovery turbine

Amine Regeneration Units 6

6.1 Description

Rich amines from all absorbers are routed to the amine regeneration unit (ARU) under the pressure control of the spent amine line to avoid vaporization and consequent two-phase generation and corrosion in the line. The control valve is located at the inlet of the amine flash drum in the ARU. The amine flash drum eliminates hydrocarbons with an amine residence time of about 30 minutes or more in the drum. Rich amine from the flash drum is pumped through a mechanical filter to the rich amine vs. lean amine heat exchangers, where it is heated to around 70°C–110°C (the maximum temperature to avoid vaporization at the upstream of the control valve in the rich amine line going to the regenerator column) and routed to the regenerator column under the level control of the flash drum. The level control valve is located close to the regenerator. Vaporization takes place at the downstream of the control valve and the line is typically expanded. Bends are normally avoided in the line; however, if bends are unavoidable, long radius elbows are used instead of short radius elbows. The feed tray location is normally the fourth or fifth tray from the top. The regenerator column overhead gas is cooled in an air cooler, followed by a water cooler to lower the reflux drum temperature to between 40°C and 45°C. The overhead gas is let out under the pressure control of the column to the sulfur recovery unit. Overhead liquid is totally refluxed to the column.

The column bottom is heated by a closed reboiler; normally, kettle-type (once-through) reboilers are used. Low-pressure steam at saturation temperature is used as the reboiler heating medium. Amine is degraded by temperature; therefore, a bottom temperature of 135°C is seldom exceeded. The bottom temperature is a function of the column bottom pressure and is almost equal to the boiling point of water at the column bottom pressure. Thus, if the column pressure is high in order to supply acid to the sulfur unit, the bottom pressure and temperature will also be high. Therefore, the ARU should be as close as possible to the sulfur unit. Lean amine from the column bottom is pumped under the level control of the column through the lean vs. rich amine heat exchangers to a water cooler, then through mechanical filters before finally being routed to the lean amine tank. From the lean amine tank, it is pumped to different units (gas amine absorbers, LPG amine treatment units, etc.).

A slip stream of amine is also routed to the absorber mounted on the flash drum, used to capture H_2S from the flash drum vapors. A typical amine regenerator flow scheme is shown in Fig. 6.1.1. The acid gas concentration of H_2S is typically 92%–96% by mole.

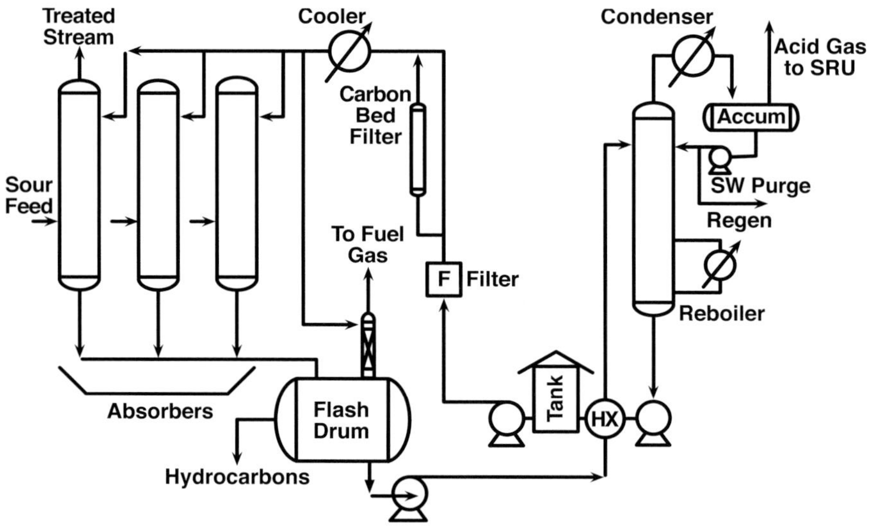

Fig. 6.1.1 Flow scheme of an ARU

Amine cleanup circuit

The total cooled lean amine flow is taken through a mechanical filter. Subsequently, a slip stream of amine (around 20%–30%) is taken through a charcoal filter and a cartridge filter. The charcoal filter removes hydrocarbons from the amine and the mechanical filter arrests particulates. The cartridge filter succeeding the charcoal filter captures any charcoal slipping from the charcoal filter in case of damage.

The lean amine tank is a floating roof tank. This keeps the amine from coming in contact with or being contaminated by air, which will oxidize and degrade the amine. The degraded products cause foaming. In fixed roof tanks, a kerosene layer is often used above the amine to keep it from coming in contact with air.

ARUs are very prone to foaming, triggered by particulates and hydrocarbons; thus the amine flash drum is a very important component in an ARU to separate oil from amine.

The reboiler duty of an ARU is a sum of the following:

1. Heat required for dissociation of the weak H_2S–amine bond.
2. Heat required for dissociation of the weak CO_2–amine bond.
3. Heat required for dissociation of the NH_4OH if there is NH_3 ingress.
4. Heat required to vaporize water from the amine solution and to generate steam to strip the H_2S and CO_2 from the amine and generate reflux.
5. Heat required to heat the feed amine solution from the feed temperature to the bottom temperature of the column.

The heat required for dissociation of the H_2S–amine bond is lowest in tertiary amines (MDEA) and becomes progressively greater in secondary amines (DEA) and primary amines (MEA). Further, the CO_2 capture capability is lowest in MDEA, becoming progressively greater for DEA and MEA as the polarity increases from tertiary to primary amines. Thus, MDEA consumes the least heat in the regenerator column reboilers, followed by DEA, then MEA. However, MDEA is seldom used for CO_2 absorption, even though if a higher residence time in the amine absorber trays is used, MDEA can also absorb CO_2.

1. The typical H_2S loading in MDEA is 0.4–0.5 moles of H_2S/mole of MDEA to minimize corrosion in carbon steel.
2. The typical H_2S loading in DEA is 0.33 moles of H_2S/mole of DEA to minimize corrosion in carbon steel.
3. Typical H_2S loading in MEA is 0.15 moles of H_2S/mole of MEA to minimize corrosion in carbon steel.

Beyond the above loading values, the amine will be very corrosive in carbon steel.

6.2 Degradation Products of Amine

Normally strong anions like formates, acetates, thiosulphates, thiocyanates, and chlorides can tie up a molecule to form a salt that cannot be regenerated by the addition of heat, making it a heat-stable salt (HSS). HSSs are produced as a degraded product of amine and reduce the strength of the amine, resulting in foaming. Ion exchange resins are often deployed for HSS removal and keep the HSS level below the stipulated value of around 1% wt.

6.3 Design Parameters

Amine regenerator

1. System factor 0.85 for regenerator
2. System factor 0.8 for absorber
3. Number of trays: 24–26 (typical)
4. Number of trays above feed: 3–4
5. Tray η: 50%
6. Typical reflux ratio: DEA = 1.3

   MDEA = 1.1 (The reflux ratio is expressed as a ratio of mass flow of reflux to mass flow of acid gas) for typical feed preheat of 90°C and normal lean amine loading of 0.006 mole H_2S/mole amine.
7. Feed tray location: Fifth tray from top
8. Reboiler duty cushion in design: 20%
9. Condenser duty cushion in design: 30%
10. Reflux drum temperature is 45°C (max.) and 55°C to reduce buildup of NH_3 in case of ammonia ingress
11. Rich amine line velocity: 1 m/sec (max.)
12. Overhead line velocity: around 10 m/sec for grassroots designs
13. Lean amine line velocity: 2 m/sec (max.)
14. Amine circulation rate: This calculation is shown in 6.4.

A typical amine regenerator column is shown in Fig. 6.8.1.

6.4 Calculation of Amine Circulation Rate

Let 10 Mt/hr of sour gas with 1.5% H_2S (wt%) be scrubbed with a lean amine that has an H_2S loading of 0.01 mole H_2S/mole MDEA to produce 100 ppm of H_2S in sweet gas.

H_2S pick up from sour gas = 10 × (1.5 – 100/10,000)/100 = 0.15 Mt/hr = (0.15 × 1,000)/34 kmole/hr = 4.41 kmole/hr

Amine circulation = 'X' kmole/hr (assumed). Rich MDEA loading desired = 0.4 moles H_2S/mole MDEA

(0.4 – 0.01) × X = 4.41 or X = 11.3 kmole/hr of MDEA = 11.3 × 119.5 = 1,350 kg/hr of neat MDEA. Mole wt of MDEA = 119.5

Lean amine MDEA strength in demineralized (DM) water assumed to be 40% by wt.

Circulation of MDEA solution in kg/hr and Mt/hr = 1,350/0.4 = 3,376 kg/hr = 3.4 Mt/hr

This quantity of rich amine, along with rich amine produced from other sources, will be the feed to the ARU. If the strength of the amine solution is lower than 40%, a proportionately higher lean amine flow will be required. For example, if a 30% amine strength (wt%) is maintained in the amine solution, then the flow of MDEA will be 1,350/0.3 or 4.5 Mt/hr.

The additional heat requirement in the ARU reboiler will be approximately equal to the additional sensible heat required to heat the extra feed quantity from the feed temperature to the bottom temperature, which is the product of (flow increase due to lower strength of the amine solution), (bottom temperature – feed temperature) and (specific heat of the amine solution).

In the case above, additional sensible heat in the ARU reboiler would be (4.5 – 3.4) × (130 – 110) × (0.95 × 1,000) = 20,900 kcal/ hr. This additional heat will be required on account of the higher circulation rate attributable to a lower amine strength of 30% compared to 40%.

Assumed feed temperature = 110°C

Assumed bottom temperature = 130°C

Average specific heat for amine solution assumed = 0.95 kcal/kg°C

6.5 Fundamentals in Amine Regeneration

The reverse reaction of amine absorption, an endothermic reaction, occurs in the amine regenerator. The amine–H_2S or amine–CO_2 bond is broken with heat delivered by steam in the closed reboiler. The steam generated through the closed reboiler reduces the vapor phase partial pressure of H_2S and CO_2, thus H_2S and CO_2 strip from their aqueous phase into their vapor phase with steam.

Typically, some residual steam flow (generated from the bottom reboiler) at the flash zone is ensured in design to achieve good performance. The residual steam is condensed in the condenser, then flows to the drum from where it is refluxed. The typical reflux ratio (mass rate of reflux/mass rate of acid gas) is 1.1 for MDEA and 1.3 for DEA.

Out of 24 total trays, 20 are used for stripping, while 4 trays are used as wash trays above the feed zone to arrest any amine carryover to the reflux drum.

An anti-foaming agent is injected at the top of the ARU to arrest foaming; however, excess or continuous injection of an anti-foaming agent may often cause foaming in the ARU. The regenerator reflux drum is equipped with DM water as a makeup provision.

6.6 Calculation of Water Loss in an Amine Absorption Unit (AAU)

When scrubbing coker or FCCU dry gas, water is taken from the amine to saturate the gas, which results in a loss of amine solution.

- FCCU gas being handled = 12 Mt/hr, mol wt = 25
- Moles of FCCU gas = (12 × 1,000)/25 = 480 kmoles/hr
- Temperature of sour FCCU gas at inlet to amine absorber = 45°C
- Temperature of sour FCCU gas at outlet to amine absorber = 53°C
- Vapor pressure of water at 53°C = 0.15 kg/cm^2 (a)
- Pressure of amine absorber = 3.5 kg/cm^2 (g) = 4.5 kg/cm^2 (a)

At saturation, the partial pressure of water = vapor pressure of water at the same temperature. So 0.15 = $(Xw/(Xw + 480)) \times 4.5$, where Xw = moles of H_2O required to saturate the dry FCCU gas. Thus, Xw = 16.55 kmoles = 16.55 × 18 kg/hr = 298 kg/hr.

The above shows the additional water requirement at the ARU reflux drum. In similar way, the loss of water along with the ARU bleed gas is estimated at 65 kg/hr, making the total loss of water 363 kg/hr.

6.7 Calculation of Reboiler Duty in the ARU

A good rule of thumb is to remember that

- Reboiler duty with MDEA = 100 kg/m^3 of circulation of lean amine solution

- Reboiler duty with DEA = 120 kg/m^3 of circulation of lean amine solution

6.8 Common Problems

6.8.1 Lean amine loading 0.01 mole H_2S/mole MDEA not achievable

Possible reasons:

1. Higher H_2S loading in rich amine than design.
2. Higher circulation rate than design due to lower strength of amine solution demanding more heat and causing the reboiler to reach its limit.
3. Not enough circulation through the ARU thermosiphon reboiler due to the following:
    - Unavailability of required head across the thermosiphon reboiler (kettle) due to lower height difference between chimney tray and steam reboiler provided in design
    - Inadequate size of thermosiphon inlet and outlet lines, resulting in higher pressure drop (ΔP) than available for the recirculation line
    - Many bends or short radius elbows and longer lines at the inlet and outlet of the reboiler
    - Leak from the chimney tray feeding the reboiler
4. Heavy leak through trays below feed zone resulting in fewer stages.
5. Weeping of column trays and unregenerated amine reaching bottom.
6. Foaming in the column.
7. Flooding of the regenerator, usually reflected by a higher ΔP across the column (>0.3 kg/cm^2).
8. NH_3 ingress along with rich amine demanding higher reboiler and condenser duty. The reaction, $NH_3 + H_2O \rightarrow NH_4OH$, with heat of reaction (ΔH) = 87 kcal/g-mole of NH_4OH (exothermic), will take place in the condenser, and the condenser duty requirement will be higher and lead to increased drum temperature and moisture in acid gas. In the reboiler, the reverse reaction will occur as the reaction of dissociation will be endothermic and demand greater heat in the reboiler. There may be a shortfall in the reboiler duty during NH_3 buildups in the system. In case of ammonia buildup, often the reflux drum temperature is increased to

around 55°C during operation and maintained for some time to liberate NH_3 from reflux drum along with the acid gas. The drum temperature is later normalized. Alternatively, a water stream is drawn from the reflux drum and routed to the sour water stripper unit for some time then stopped.

9. Amine loss from the system is often compensated by DM water makeup in the drum, causing amine strength to decrease and more amine to be circulated through the amine absorber in order to maintain the desired sweet gas H_2S ppm.
10. Amine degradation due to the formation of HSS.
11. Inadequate filtration and foaming.
12. Lower preheat than design (typical temperature is 90°C and maximum is 100°C) due to fouling of the lean or rich amine exchangers.
13. Rich or lean amine heat-exchanger leakage. Usually rich amine side pressure is higher than lean amine in the heat exchanger. Thus, rich amine can leak into the lean amine side and increase the acid gas loading of lean amine.
14. Higher feed temperatures will demand less reboil and consequently reboiler duty would reduce, which would adversely affect the stripping efficiency. This has been experienced by the author in a number of units where preheat was increased to improve the energy efficiency and lower the reboiler duty, which resulted in higher lean amine loading due to higher H_2S partial pressure in the bottom stripping steam. When reboil duty is again increased to maintain the lean amine loading, additional heat is rejected in the condenser.

6.8.2 High reflux drum temperature

High reflux drum temperature will reduce the H_2S concentration in the acid gas as more moisture gets into the acid gas due to the higher drum temperature. This can result from

1. Increased reboiler duty
2. Deposition of salts in the condenser and overhead air cooler
3. NH_3 buildup in the system
4. Design deficiency
5. Operation at higher capacity
6. Flooding in the column

6.8.3 Case Study

Amine carryover to the reflux drum

An amine concentration of <1 wt% in the reflux liquid is not unlikely. Amine carryover may result from problems in the wash trays. Typically three to four trays are provided above the feed zone to arrest amines carried over to the reflux drum. In the wash trays, the liquid load is very low so the trays are likely to lose level and result in no wash or poor wash.

The wash trays normally have a very low liquid load (ranging from 0.2 to 1.0 gpm/inch of weir length), and are shown in Fig. 6.8.1.

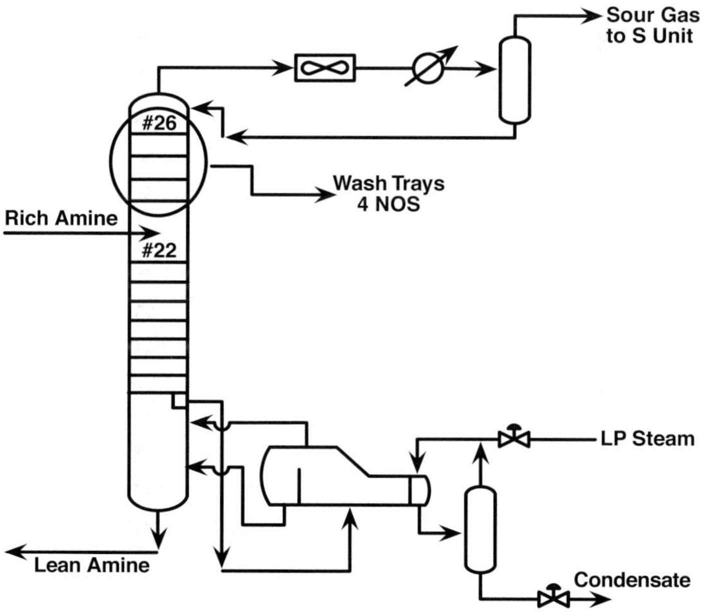

Fig. 6.8.1 Wash section of amine regenerator column

These wash trays operate in a spray regime with a very low liquid load. Often splash baffles may be required to arrest the liquid droplets from being flung and reaching the downcomer and direct them to the tray to maintain liquid level and downcomer seal. A small leak from the tray will aggravate the problem. Thus, there is a likelihood of liquid level loss in the trays, which will lead to no wash or a poor wash. This may often expose the downcomer seal and allow vapor to blow through the downcomer instead of passing through the tray dispersers. It is a good practice to gasket the trays to arrest leakage.

Normally, the diameter of the column above the feed zone is reduced to resolve the problem of low liquid loading, particularly considering the turndown. In such cases, it is not required to use gaskets or seal welding in wash trays.

The wash trays and feed distributor play a vital role in containing the amine carryover to the reflux drum. A large carryover was observed in an ARU. The amine strength in the drum was almost the same as the amine circulated. Gaskets were installed with partial seal welding in the wash sections between tray segments. The problem was resolved. Foaming can also lead to a carryover of amine to the reflux drum. This phenomenon usually occurs with an increase in pressure drop across the column.

6.9 Methods Used for Revamp of Amine Absorber and Regenerator

Methodologies broadly consider the use of a precontactor, semi-lean amine in amine absorber (multiple amine inlet points), pressure swing regeneration, ultra-lean amine (use of two regenerators), and multiple absorbers in series or in parallel with or without compression. These changes in configuration can be adopted in addition to the normal actions such as change of amine strength, use of different type of amines, increase in pressure of the absorber, addition of a reboiler in the regenerator, etc. A configuration may be adopted based on the specific requirement. A precontactor (static mixture) can be deployed to treat a slipstream of acid gas in parallel or can be used in a series at the upstream of the absorber. The author advised a refinery to use a redundant gas/naphtha recontactor (originally with two stages) as a gas amine precontactor (upstream) to resolve the limitation of the diameter of the amine absorber. Multiple amine inlets can be adopted for achieving CO_2 slip. Fig. 6.9.1 shows the different configurations typically deployed in a single figure.

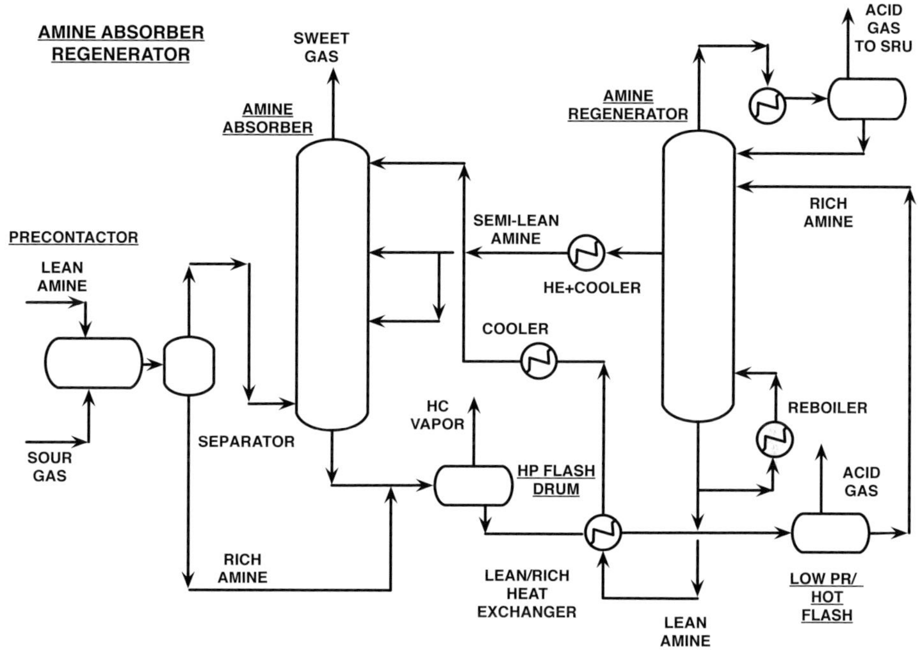

Fig. 6.9.1 Amine absorber and regenerator configurations

Metallurgy

Amine is inherently corrosive, and in the presence of H_2S, CO_2, and moisture, the system is very prone to corrosion. The unit needs special metallurgy in its equipment and special design consideration. A typical metallurgy is given below.

1. Metallurgy used for amine flash drum is killed carbon steel (KCS) with a corrosion allowance (CA) of 6 mm.
2. Metallurgy used for an ARU column is KCS with a CA of 3 mm.
3. Overhead air cooler is made of SS 304 (stainless steel) with no corrosion allowance.
4. Overhead water cooler shell is made of KCS with a CA of 3 mm. The tube is made of SS 304L with no corrosion allowance.
5. Overhead reflux drum is made of KCS with a CA of 6 mm.
6. Regenerator trays and feed distributors are made of SS 304L.
7. Lean amine air cooler shell is made of KCS with a CA of 3mm. The tube is made of KCS with a CA of 3mm.

The author has experienced the severe thinning and failure of a regenerator overhead air cooler that used KCS. Later the air cooler was changed to SS 304, and the problem resolved.

Abbreviations

AAU: amine absorption unit
ARU: amine regeneration unit
DM: demineralized
gpm: gallons per minute
HSS: heat-stable salt

Gas Concentration Units

7

7.1 Description

A gas concentration unit (GCU) is typically appended with a FCCU, hydrocracker, and a coker to recover the LPG and stabilized naphtha in a refinery. An overall GCU scheme is shown in Fig. 7.1.1.

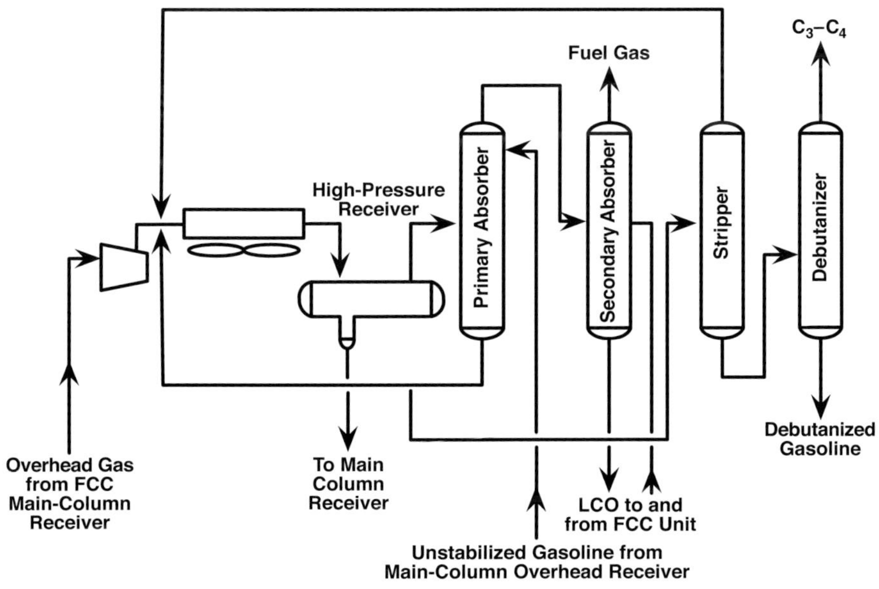

Fig. 7.1.1 Overall GCU scheme for an FCCU

The overhead gas from the fractionator overhead receiver is compressed and mixed with the stripper overhead gases and primary absorber bottom liquid and fed to the air cooler and then to the high-pressure receiver. The compressor used is typically a two-stage and pressure is increased from a low range to around 15–16 kg/cm^2 (g).

The HP receiver liquid is fed to the stripper and the stripper bottom in turn is fed to the stabilizer to produce LPG and stabilize naphtha. The stripper overhead gases are recirculated back to the inlet of the high-pressure air cooler (HPAC) along with the wet gas compressor (WGC) discharge and the primary absorber bottoms to be cooled with air cooling and, in some cases, water cooling before entering the high-pressure receiver.

The fractionator overhead drum liquid is fed to the primary absorber to absorb the LPG components from the high-pressure receiver gases at a higher pressure. The unabsorbed gases from the primary absorber are fed to the secondary absorber in the FCCU, where either the heavy naphtha cut or light cycle oil (LCO) is used as an absorbent. The unabsorbed gases (sponge gases) from the secondary absorber are bled under pressure control to maintain the GCU system pressure at around 15 kg/cm^2 (g). Often an amine absorber is used to scrub the sponge gases. Pressure control for the whole system is located downstream of the amine absorption unit (AAU).

Fundamental working principle

The fundamental principle of the GCU section is that a majority of the LPG components must be captured (absorbed) at the system pressure and temperature with the liquid stream available as naphtha. To increase the quantity of naphtha, a part of stabilized (cooled) naphtha is recirculated at the primary absorber top.

In the coker unit, naptha production is low compared to the FCCU. Though the LPG production from the coker is also low, an appreciable quantity of cooled, stabilized naphtha still needs to be recirculated from the stabilizer to absorb the LPG components at system pressure.

A hydrocracker GCU is similar. It typically produces more naptha than a coker, but produces similar quantities of LPG based upon the conversion level.

The primary absorber uses two configurations: with a pre-saturator and without a pre-saturator. The scheme in Fig. 7.1.2 shows a pre-saturator. In this scheme, stabilized (cooled) naphtha is introduced in the upstream of the overhead cooler of the primary absorber. The off-gas from the absorber column, along with the stabilized naphtha, is cooled and goes to the reflux drum with a water boot. The overhead liquid from the reflux drum is introduced at the topmost tray as an absorbent. The off-gas from the drum is routed to the sponge absorber. The main fractionator (MF) drum's naphtha is introduced at a location 9–10 trays below the top.

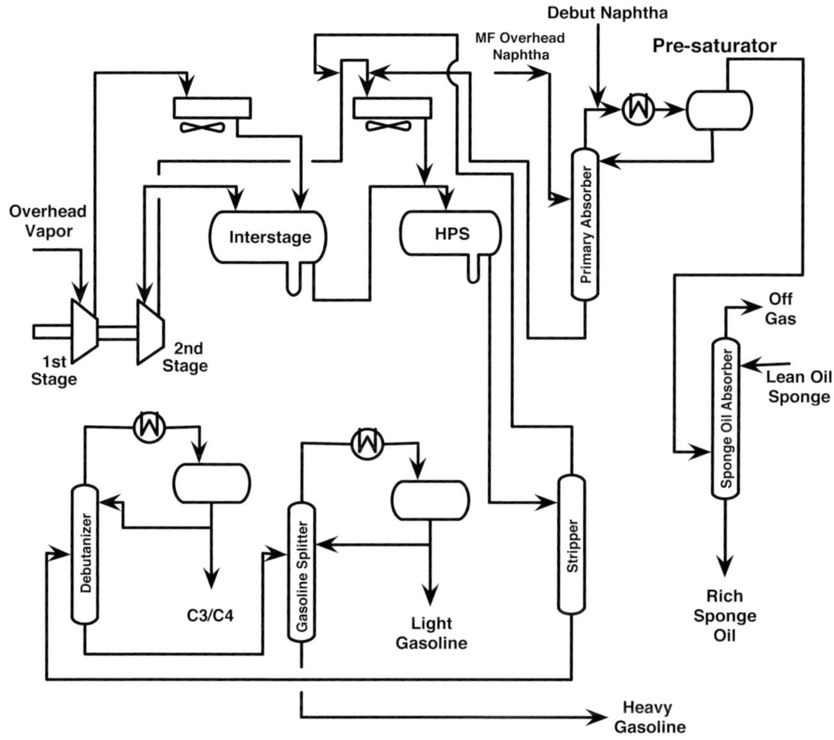

Fig. 7.1.2 GCU scheme with a pre-saturator

The scheme without a pre-saturator does not have an overhead cooler and drum in the primary absorber. The stabilized naphtha is introduced at the uppermost tray and unstabilized naphtha from the fractionator reflux drum is introduced at a location 9–10 trays below the top. In this arrangement, intercoolers are often provided in a primary absorber, along with a water draining facility. The cooling in the intercooler in the absorber increases LPG yield. In the presence of water and iron, the alkyl cyanides can produce nascent hydrogen, which can give rise to hydrogen-induced stress corrosion cracking in steel. Thus, water draining is preferred. The typical number of trays used in GCU columns is shown in Table 7.1.1.

Table 7.1.1 Typical number of trays in GCU columns

Columns	No. of trays
Primary absorber	36–38
Stripper	30–35
Stabilizer	36–44
Secondary absorber	20

The primary absorber is a moderate foaming system. Typically, a system factor of 0.65 to 0.8 is considered during the sizing of a primary absorber. The absorption of gas in the naphtha stream is moderately exothermic. If a pre-saturator is used, intercoolers are not required, otherwise intercoolers are used to cool and increase LPG recovery.

Typically, a GCU section is designed for the recovery of 90%–95% of the LPG content of the feed. However, if the naphtha quantity recirculation is decreased, the LPG recovery will be also lower. The cooled lean sponge oil is introduced as an absorbent in the sponge absorber and rich oil is returned to the MF after heating.

In FCCUs, the propylene yield is quite high in the LPG stream. Often the propylene-rich LPG is fed to propylene recovery units (PRUs) where petrochemical-grade propylene is produced. Thus, maximum propylene recovery occurs in GCUs clubbed with FCCUs. The secondary absorber (sponge absorber) plays a big role in capturing the propylene.

In actual practice, it has been observed that in the majority of the units, the primary absorber is limiting due to its high foaming nature. The secondary absorber or sponge absorber is also often a problem. In actual operation, the recirculation of stabilized naphtha and sponge oil is reduced to overcome the limitations. All these actions adversely affect LPG and propylene recovery.

An alternative scheme developed to resolve the problem of the primary absorber, stripper, and stabilizer limitations in the GCU of an FCCU is given below.

7.2 Case Studies

GCU section modification with a trim absorber

The original configuration of the unit is shown in Fig. 7.2.1.

During the revamp of the FCCU unit, the GCU section was found to be limiting. The major limitations were observed in the following equipments:

1. The primary absorber was very heavily loaded.
2. Loads were observed to be on the higher side in the stripper and stabilizer. The stabilizer bottom section was limiting.

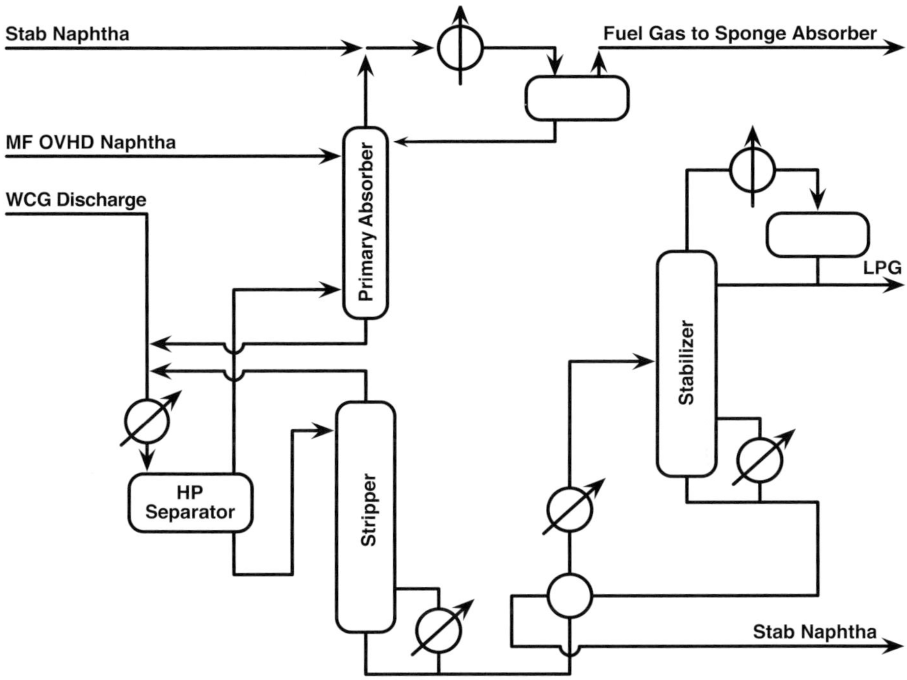

Fig. 7.2.1 Original flow scheme of the unit

The only solution was to replace the primary absorber column and stripper column. Any reduction in the stabilized gasoline circulation would have reduced the LPG and propylene component recovery. To resolve the problem, an additional trim absorber was introduced (with around 5–7 stages). The primary absorber top gas was fed to the trim absorber and stabilized naphtha was used as an absorbent in the trim absorber. The following is a summary of the resolutions proposed:

1. Trim absorber (small column with 10 trays) to be installed to relieve the primary absorber (Fig. 7.2.2)
2. Two-point feed in C_2 stripper to reduce reboil duty and in turn relieve the stripper load. The trim absorber bottom is fed at the top, and the HP receiver liquid is fed at a lower location.
3. Increase stabilizer feed temperature from 126°C to 137°C to relieve the stripping section of the stabilizer

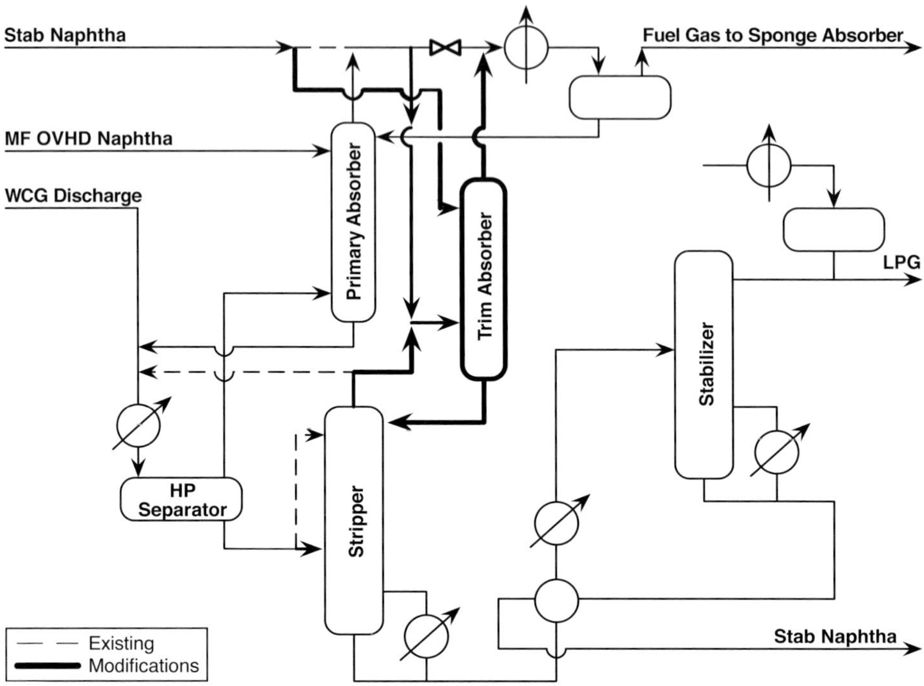

Fig. 7.2.2 Modification with trim absorber

Main features of the modification:

1. In the revised configuration (Fig. 7.2.2), the stripper outlet gases (cooled) will be routed to the trim absorber bottom instead of going to the WGC discharge, thereby relieving the primary absorber through a reduction in recirculation.
2. The LPG components will be absorbed by feeding stabilized naphtha at the top of the trim absorber.
3. Some quantity of unstabilized naphtha from the MF overhead reflux drum and pre-saturator, which cannot be fed to the primary absorber due to its limitation, shall also be fed to the trim absorber along with the stabilized naphtha.
4. Primary absorber outlet gas will also be routed to the trim absorber. The trim absorber top gases will then go to the sponge absorber via the pre-saturator.
5. Trim absorber bottom liquid will be routed to the stripper top and the primary absorber bottom liquid will continue to be routed to the high-pressure receiver at the air cooler upstream.

6. The high-pressure receiver liquid will be fed to the stripper at a lower feed location and the liquid from the trim absorber will be fed at the top of the stripper column.
7. Feeding the two naphtha streams with different compositions at two points will reduce the reboiler duty and partly reduce the load of the stripper.
8. The revised configuration will reduce the load on the primary absorber considerably and the load of the stripper to some extent without sacrificing the LPG yield.
9. Low ΔP has been specified in the trim absorber design so that the total ΔP in the trim plus the primary absorber remains the same as in the original configuration.
10. An inter-stage cooler has been provided in the trim absorber to further enhance LPG recovery and minimize propylene slippage along with sponge gas.

This resolved the very big limitation of the primary absorber, as liquid flowing down the primary was reduced at least to the extent of the stabilized naphtha recirculation. The feed gas quantity was also reduced, which relieved the absorber. The two-point feed in the stripper reduced the reboiler duty to some extent and relieved the stripper. In the stabilizer, the higher feed preheat reduced the reboiler duty and the section below the feed was debottlenecked. This method has been implemented in an FCCU, and the LPG and propylene recovery increased marginally. The trim absorber height and diameter was small and served uniquely.

The primary absorber flooding with and without a trim absorber is presented in Fig. 7.2.3. Stripper flooding with single- and two-point feeds is given in Fig. 7.2.4. Stabilizer flooding with low and high feed temperatures is shown in Fig. 7.2.5.

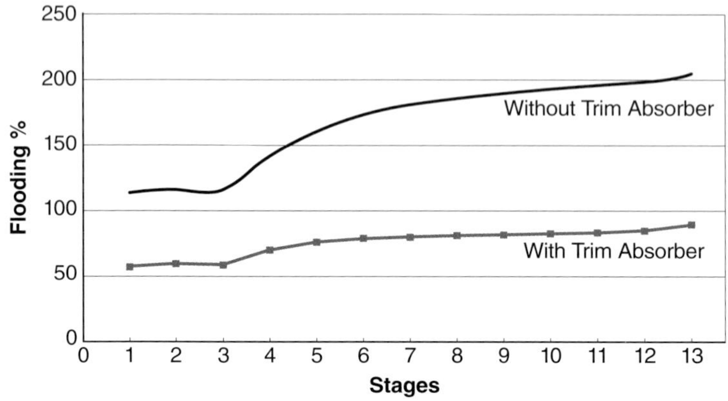

Fig. 7.2.3 Primary absorber flooding with and without trim absorber

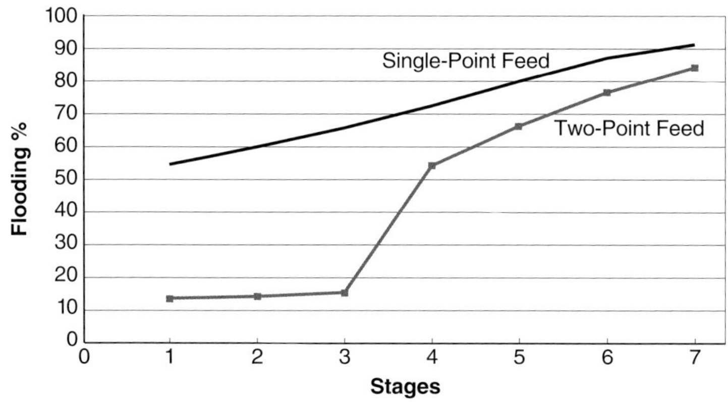

Fig. 7.2.4 Stripper flooding with single- and two-point feeds

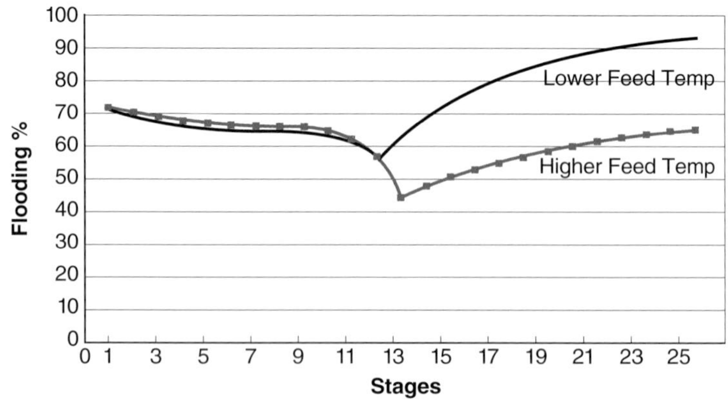

Fig. 7.2.5 Stabilizer flooding with lower and higher feed temperature

Capacity augmentation in GCUs often begins with the installation of a parallel compressor for handling the increased quantity of wet gas from the fractionator receiver if the existing capacity cannot be increased to the required level by a wet gas compressor rotor change, speed increase, etc.

While revamping a coker unit, the design pressure of the existing sponge absorber was found to be 8 kg/cm^2 only. Thus, the operating pressure was not adequate for absorbing the LPG components bled from the primary absorber and increasing the LPG yield. The typical pressure requirement in a sponge absorber is around 15 kg/cm^2. A scheme was developed to cool the lean sponge oil to 2°C–5°C by refrigeration and to absorb LPG components in the sponge. The hot lean sponge oil was cooled by delivering its heat to the cold rich oil from the sponge absorber, then by refrigeration before it was fed to the sponge absorber. The cold rich oil was heated by hot lean oil and returned to main fractionator. The LPG yield was increased by using the existing sponge absorber with a lower design pressure.

Abbreviations

GCU: gas concentration unit

PRU: propylene recovery unit

WGC: wet gas compressor

Sour Water Strippers 8

8.1 Description

A sour water stripper (SWS) unit is deployed to strip the H_2S and NH_3 from sour water in order to bring down the H_2S and NH_3 levels in the sour water (typically less than 50 ppmw of each or lower). A typical scheme is shown below in Fig. 8.1.1.

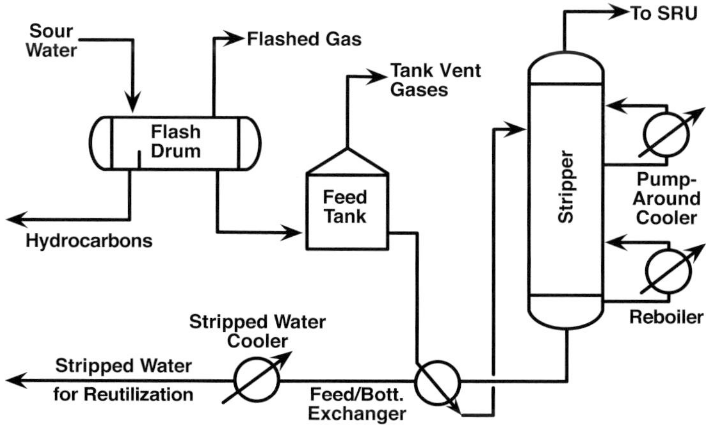

Fig. 8.1.1 Typical SWS unit

The sour water is first fed to the feed surge drum (flash drum) with pressure control with gas makeup and bleed facility. The bled gas is routed to the H_2S flare or acid gas flare. This feed surge drum separates hydrocarbon in order to avoid hydrocarbons getting into the sour water. The sour water is routed to an equalizing or harmonizing tank, then fed to the sour-water vs. stripped-water heat exchanger and then to the tower. The feed control valve is located downstream of the heat exchanger. The feed pump pressure must be sufficient to keep all the ammonia and H_2S in the aqueous phase at the feed inlet temperature to avoid corrosion.

The sources of sour water are usually organized in two groups:

1. Hydrotreater and hydrocracker sour water
2. Sour water from other refinery units, such as FCCUs, cokers, visbreakers, crude distillation units (CDUs), and vacuum distillation units (VDUs)

The sour water from hydrotreaters and hydrocrackers contains very high levels of H_2S and NH_3. Typically, the H_2S concentration (around 4%–6% wt) is double the NH_3 concentration (around 2%–3% wt content) in refinery applications; however, when crudes with very high nitrogen content are processed, the ammonia concentration in sour water may be as high as H_2S content or sometimes higher. Ammonia and H_2S can coexist in sour water in much higher quantities than estimated from the solubility of either component alone in water. No major contaminants are there. Thus, this can be injected at the air cooler in the hydrotreaters after stripping.

The sour water from other refinery units like coker units, FCCUs, etc., usually contain around 9,000–12,000 ppmw of H_2S and around 4,000–6,000 ppmw of NH_3. The sour water from the cracking units would have plenty of contaminants like phenols, cyanides, thiocyanates, polysulfides, and other metal complex compounds. This is not preferred as injection water in hydrotreaters and hydrocrackers, although it may be reused as injection water in crude. The phenols and heavier cyanides will partly dissolve in the large volume of crude and get distributed in different hydrocarbon products. The heavier alkyl cyanides present in sour water remain in the stripped sour water, but the majority of the hydrogen cyanide (HCN) goes along with the sour gases.

The typical arrangement of a sour water stripper column is shown in Fig. 8.1.2. The stripper column is usually provided with 40 trays. The reboiler can be open type (stripping steam) or closed type (low-pressure steam reboiler). The column normally deploys a once-through type of reboiler. Bigger units need multiple reboilers to provide the required duty and are normally provided with isolation valves to facilitate isolation and cleaning of reboilers during operation. Recirculation type of reboilers are often used in such applications. The column top is provided with a pump-around reflux for maintaining the column top temperature.

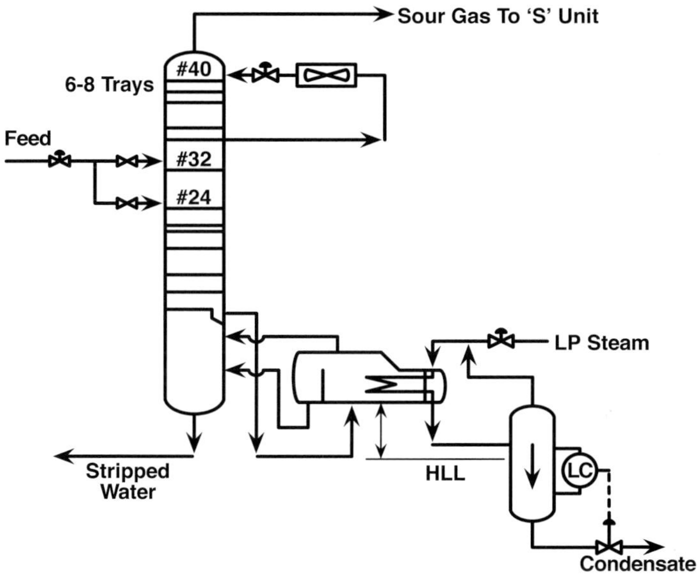

Fig. 8.1.2 Sour water stripper column

8.2 Feed Impurities

The sour water from the catalytic cracker and coker contains many impurities:

1. **Phenols.** Phenols are weak acids that can lower the pH and may make ammonia liberation difficult.
2. **Oil.** Oil may cause the ingress of hydrocarbon to the sulfur unit along with the sour gases.
3. **Silica.** Due to its low solubility, silica can produce deposits caused by its reaction with magnesium salts.
4. **Carbonates and bicarbonates.** These are present in catalytic cracking unit sour water. If cations like Mg++ and Ca++ are present, they can produce deposits of $Mg(OH)_2$ and $CaCO_3$ and block the dispersers in the column trays.
5. **Amines.** MEA and DEA end up in stripped water. They do not inhibit NH_3 removal.
6. **Cyanides.** Cyanides can cause hydrogen blistering and corrosion. They can be partially stripped (overall 30%–40%). HCN is removed almost entirely, while a considerable amount of the CH_3CN and part of the

ethyl cyanide is removed. Other cyanides convert to thiocyanates and form metal complexes, which cannot be stripped.

7. **Thiocyanates.** When polysulfides are present, part-cyanides convert to thiocyanates. To achieve this conversion, polysulfides are often added to FCCU and coker sour water before it's routed to the sour water strippers in order to reduce corrosion. The thiocyanates remain in the stripped water.
8. **Polysulfides.** Often found in sour water from FCCUs and coker units. Polysulfides can form from organic sulfide reactions with air. They inhibit NH_3 stripping. Polysulfides remain in stripped water.

8.3 Process Chemistry and Operating Conditions

Ammonia and hydrogen sulfide are known to dissolve in water and ionize to the equilibrium extent according to the pH and temperature and can only be stripped if present in the parent form (i.e., the components exerting partial pressures). The ionized components do not strip. Ions that lower pH tie up ammonia and aid stripping of H_2S, and ions that raise pH tie up H_2S and help in the stripping of ammonia.

Typical feed inlet temperature to the column is 70°C–95°C. The bottom is heated by a reboiler using low-pressure steam. A higher feed temperature will increase the pump-around (PA) duty and increase the column diameter in the location and would practically result in no reduction in reboiler duty. Normally, 24 trays are used in the stripping section.

For a sour water stripper for a hydrotreater, around 32 trays are used in the stripping section for stripping the NH_3. Often fewer trays (as few as 16) are used in stripping. Typically 6–8 trays are provided above the feed tray. A pump-around reflux is provided above the feed tray and the pump-around is returned at 6–8 trays above that. The pump-around stream is cooled by an air cooler and returned to the column at about 65°C–70°C.

An alternative feed nozzle for the sour water feed to the column with the inlet device is provided 2–4 trays above to handle the large quantities of NH_3 in the feed.

The top temperature in the sour water stripper is maintained above the deposition temperature of NH_4Cl and NH_4HS at the partial pressures of the NH_3, HCl, and H_2S at the column top. Typical top temperature is 90°C. The concentration of H_2S in the acid gas typically ranges from 48% to 55% by mole, ammonia concentration ranges from 25% to 35% by mole, and the balance is water vapor (20%–25% by mole). Stripping of ammonia is more difficult from sour water due to its higher solubility in water. Typically, stripping of H_2S takes place in the upper section of the stripper, while the stripping of ammonia takes place in the lower section. Often, facility for injection of caustic solution (6% wt) is provided in the bottom 6 to 8 trays, typically at three locations, to increase the pH of water in the section to facilitate stripping of ammonia.

Some hydrocracker/hydrotreater sour water stripper designs deploy two columns for stripping H_2S and ammonia. The first column operates at a higher pressure (around 2.5–3.0 kg/cm^2(g)) and is primarily used for stripping H_2S (ammonia is contained in the bottom liquid). The liquid from first column bottom is fed to the second column and the column is used for stripping ammonia. This column operates at a lower pressure of around 0.5kg/cm^2 (g). Both the columns are provided with individual pump-around refluxes and reboilers heated by low-pressure steam. The H_2S liberated from the first column goes to the main combustion chamber of the sulfur unit, and the ammonia liberated from the second column is routed to the incinerator of the sulfur unit, or alternatively to the incinerator stack. This two-column configuration is practiced where the design temperature of the primary combustion chamber of the sulfur unit is lower, as a higher temperature (1,350°C–1,400°C) is required for the destruction of ammonia in MCC. The two-column design is no longer used for grassroots units, but is often adopted for capacity revamp applications where a second column is added to create the two-column configuration described. A two-column configuration may be deployed when feed NH_3 ppm is very high. A 26° Baume ammoniacal solution can be produced by the condensation of overhead of the second column (after trace H_2S removal). This provides an opportunity to recover some nitrogen in crude as ammoniacal solution and may result in value addition.

8.4 Unique Scheme

In hydrotreater SWS units, the sour water is almost saturated with H_2S. So when the water is routed to the equalizing tank at a lower pressure (an API tank at almost atmospheric pressure), some of the H_2S is inadvertently bled from the tank to the stack of the sulfur unit incinerator (Fig. 8.4.1). This results in a 7%–10% wt loss of H_2S, and may also have an environmental implication (Table 8.4.1).

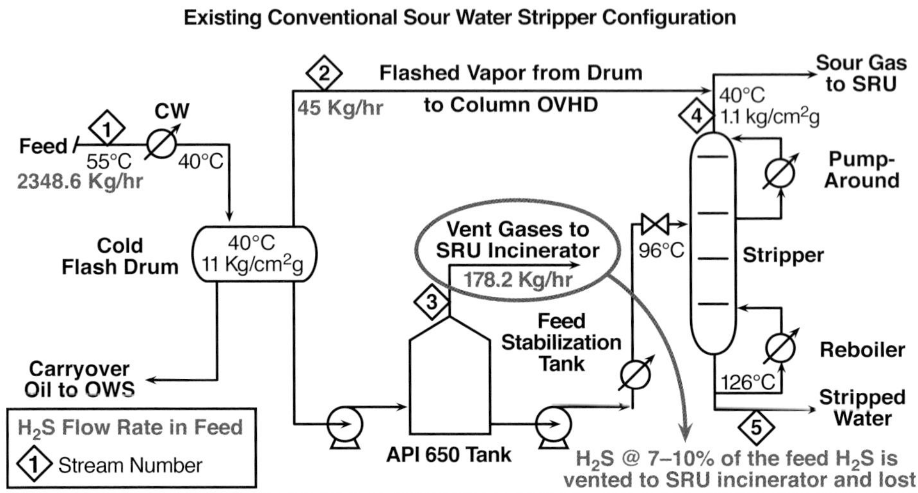

Fig. 8.4.1 Conventional scheme

Table 8.4.1 Stream summary of the original configuration (Fig. 8.4.1)

Stream No.		1	2	3	4	5
		Feed	Flashed Gas from Cold Flash Drum	Vent Gas from Tank	Sour Gas from Stripper to SRU	Stripped Water from Sour Water Unit
Temp. °C	55.0	40.0	40.0	90.0	126.0	
Pressure	$Kg/cm^2(g)$	5.0	1.1	0.015	1.1	1.4
Composition						
NH_3	Kg/hr	1,058.10	0.122	0.19	1,056.79	<50 ppmw
H_2S	Kg/hr	2,348.60	45.0	178.20	2,124.4	<50 ppmw
H_2O	Kg/hr	53,995.3	0.854	6.80	1,300	52,687.65
Total		57,402.0	45.976	185.19	4,481.19	52,689.65

A scheme has been developed to avoid the loss of H_2S (jointly authored with Mukesh Sharma, published in the *Oil Gas Journal* in May 2009).[1] The scheme has a hot surge drum with a temperature of around 55°C–60°C that precedes the regular cold surge drum (Fig. 8.4.2). Some H_2S is liberated at the hot surge drum temperature and is routed to a small adjacent amine absorber with two to three stages. The liberated H_2S is absorbed and the rich amine is sent to the amine regenerators. The gas is then routed to the incinerator stack or incinerator of the sulfur unit. The sour water is cooled in a water cooler and sent to the cold sour-water flash drum with a hydrocarbon separation facility.

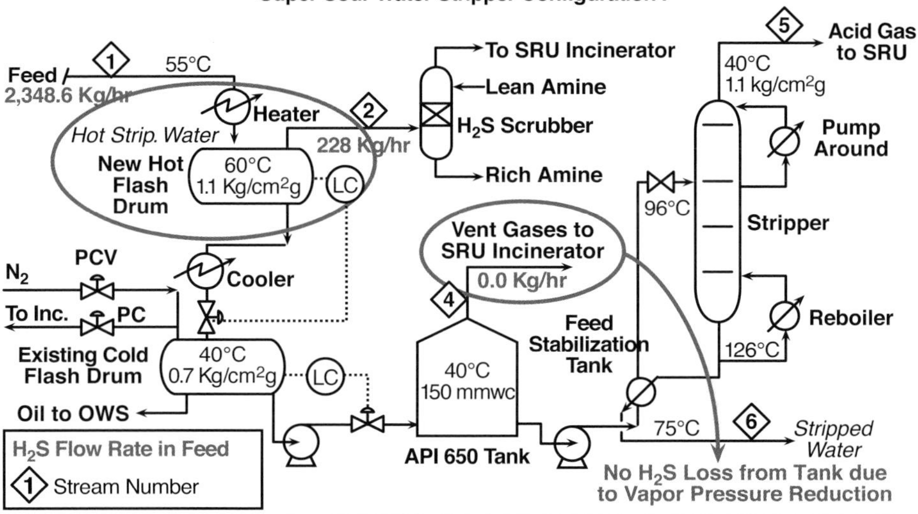

Fig. 8.4.2 Modified scheme with a hot drum with an amine scrubber

An alternative configuration (Fig. 8.4.3) routes the hot surge drum gases to the column overhead. The sour water is then routed to the equalizing tank with an N_2 blanket. No H_2S bleed takes place from the tank (Table 8.4.2). If the sour water feed is cold, a heat exchanger can be placed ahead of the hot surge drum to heat it using hot stripped sour water before it is routed to the stripped sour water cooler. Thus, no additional heat is required.

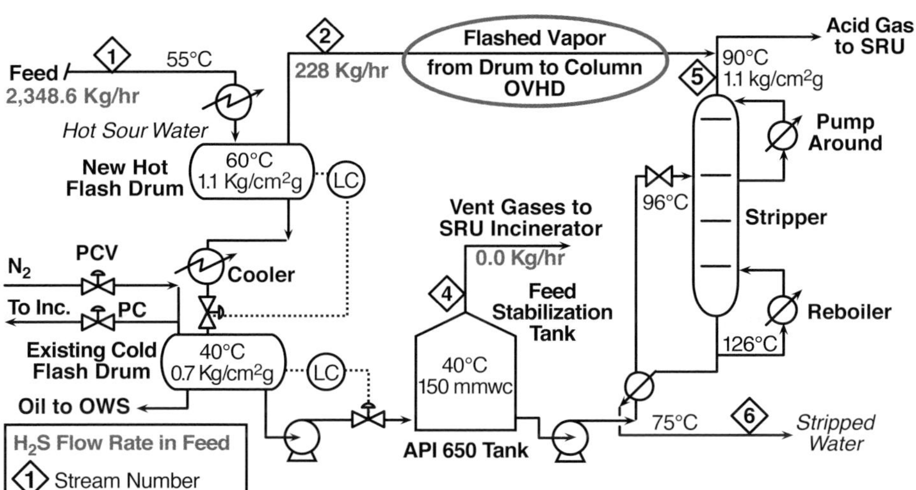

Fig. 8.4.3 Modified scheme with a hot drum and gas to the column

Table 8.4.2 Stream summary with super sour scheme

Stream No.		1	2	3	4	5	6
		Feed	Flashed Gas from Hot Drum	Flashed Gas from Cold Drum	Vent Gas from Feed Tank	Sour Gas from Stripper to SRU	Stripped Water from Sour Water Unit
Temp.	°C	55.0	59.1	40.0	40.0	90.0	75.0
Pressure	$Kg/cm^2(g)$	5.0	1.1	1.1	0.015	1.1	8.0
Composition							
NH_3	Kg/hr	1,058.10	0.380	0	0	1056.72	<50 ppmw
H_2S	Kg/hr	2,348.60	228.0	0	0	2119.6	<50 ppmw
H_2O	Kg/hr	53,995.3	12.0	0	0	1300	52,683.3
Total		57,402.0	240.38	0	0	4,476.32	52,685.3

8.5 Simulation and Design Guidelines

1. Tray efficiency: 25%–30%
2. System factor: 70% for tower size
3. Total column pressure drop: 0.3–0.4 kg/cm^2
4. Column pressure: as low as possible; depends on the proximity to the sulfur unit, as the sour gas is being routed to the sulfur unit with the column pressure
5. Reboiler cushion: 50%
6. Pump-around duty cushion: 30%
7. Sour water special system package to be considered for K values (VLE) in simulation, or an NRTL electrolyte correlation can also be used
8. Redlich–Kwong (R–K) equation of state may be used to calculate liquid phase properties
9. Column overhead line vapor velocity: maximum 6 m/sec
10. Normally towers with trays are used for turndown
11. Pump-around reflux return temperature: 65°C–70°C

8.6 Case Studies

Case Study 1 (High NH_3 and H_2S in stripped sour water)

A SWS unit designed for treating sour water from cokers, FCCUs, and other refinery units was put into operation. However, it was found that the H_2S in the sour water feed was around 690–700 ppmw and the NH_3 was around 500 ppmw, rather than the expected 3000 ppmw H_2S and 1500 ppmw NH_3.

Additionally, the pH of the stripped sour water was low: 5–5.5. It was suspected by the client that due to the low pH, the NH_3 was not being released, and that the unit might need a caustic injection at the bottom to raise the pH of the stripper and remove the NH_3.

All the relevant data were analyzed. It was found that the pump-around flow and duty were too low to maintain the top temperature, which created a focus on the heat input in the reboiler and the exact reboiler arrangement. It was also found that there was no condensate level in the condensate pot and that the level control valve for the condensate pot was wide open (Fig. 8.1.2). It was likely that the steam would not condense and the heat input to the reboiler would be low. Thus, the low heat input would not have produced enough steam at the bottom of the column to strip off the NH_3 and H_2S.

The operation group explained that whenever they attempted to increase the level in the condensate pot column, the bottom temperature was dropping, confirming that the reboiler tubes were being submerged by the condensate. It was found that the elevation difference between the bottom of the tubes in the reboiler and maximum water level in the condensate pot was inadequate. This was creating the problem of condensate draining from the reboiler tubes and heat input through the reboiler. It may be noted that the difference in elevation between the bottom of the kettle reboiler and the high-level alarm point in the condensate pot (if the pressure of the pot is equalized with inlet) overcomes the steam side pressure drop (DP_E) through the tubes of the kettle reboiler and allows the smooth flow of condensate to the condensate pot. Further, the condensate outlet line from the reboiler had too many sharp bends before entering the condensate pot, which was causing a higher pressure drop and submergence of the tubes of the reboiler, and the condensate outlet line from the pot also had many bends, causing extra pressure drop. The height between the lowest level of condensate in the pot and the condensate outlet control valve was found to be inadequate, making the condensate vaporize at the inlet of the control valve. This resulted in a full opening of the valve and loss of level in the condensate pot. It may be worth mentioning that the condensate is always likely to vaporize at the inlet of level control valve, as it is a saturated liquid and liquid head is required to arrest vaporization at the inlet to overcome the pressure drop due to flow in the line. The elevation difference between the lowest level (LL) in the condensate pot and the condensate outlet control valve provides the head and arrests vaporization at the condensate control valve inlet. To remedy the situation, the elevation of the condensate pot was lowered to provide necessary head for the smooth flow of condensate out of the reboiler tubes. Additionally, the elevation of the condensate line control valve lowered and was rerouted to eliminate the sharp bends to reduce pressure drop and ensure availability of adequate pressure at the inlet of the control valve to arrest vaporization. Further, it was apparent that the low reboiler heat input was causing the HCN to remain in the stripped water and decrease its pH. The elevation differences mentioned are very important and require attention to ensure smooth operation of the reboilers and the unit (Fig. 8.1.2).

Case Study 2 (High ammonia in stripped water)

In one unit, the H_2S content in the stripped sour water was less than 50 ppm wt, but the ammonia was reported to be high (100–150 ppm wt) and was not reduced with increased reboiler heat input. An analysis of the entire unit was carried out, including the anions and cations. It was found that many anions (sulfates, sulfites, thiocyanates, chlorides, and acetates) were present in the feed (around 300 ppm wt). These were suspected to cause an inhibition in NH_3 stripping by binding with ammonium ions.

The unit was simulated with the sour water impurities (anions and cations) and specific composition of H_2S and NH_3. It was found that the anions present in the feed were inhibiting the release of NH_3. The findings were forwarded to the client and sour water system experts. A caustic injection in the bottom stripping section was suggested by experts. It should be noted that H_2S is stripped in the upper stripping section while ammonia is stripped at the bottom stripping section.

The author has experienced a problem in a reboiler of another sour water stripper with a once-through vertical reboiler with a partition plate in the tower bottom that was not delivering design heat duty. It was diagnosed that the large height of the fall of internal reflux in the reboiler side was causing froth (waterfall pool effect) that was entering the reboiler and lowering the heat transfer, resulting in higher ammonia levels in the stripped water.

Metallurgy

The system handles H_2S and needs special emphasis on the metallurgy. Accordingly, the metallurgies normally used and considered during design are described below.

1. Metallurgy used for the SWS column is killed carbon steel (KCS) with a corrosion allowance (CA) of 6mm.
2. Metallurgy used for the column internals and trays is SS 316L.
3. Metallurgy used for the reboiler shell is KCS with a CA of 6mm; reboiler tube is SS 316.
4. Metallurgy used for the pump-around air cooler is KCS.
5. Sour water pumps use duplex stainless steel.
6. Stripped sour water pumps use duplex stainless steel.
7. Sour water lines use KCS with a CA of 3 mm.
8. Stripped sour water lines use KCS with a CA of 3 mm.

Abbreviations

CDU: crude distillation unit

SWS: sour water strippers

VDU: vacuum distillation unit

Notes

1. Mukesh Sharma and Ashis Nag, "Process Developed for Enhanced H_2S Recovery from Sour-Water Strippers," *Oil and Gas Journal,* 107: 18 (2009).

Natural Gas Processing 9

9.1 Introduction

The major constituent of natural gas is methane. Heavier hydrocarbons like ethane, propane, butane, pentane, and hexane are also present but in relatively smaller proportions. Some impurities like CO_2, N_2, helium, H_2S, R-SH, and water are also present in natural gas (NG).

Some basics of NG reservoir

A natural gas with a composition of around 75 mole % of methane is given in Table 9.1.1. The gas, with a mass flow rate of 1,000 kg/hr at 140 kg/cm^2 pressure and a temperature of 20°C, is expanded to the lower pressure of 60 kg/cm^2 in a flash vessel isothermally. Condensation takes place despite decreasing pressure at a constant temperature (around 161 kg/hr condensate produced) (see Fig. 9.1.2). This phenomenon is known as *retrograde condensation*. If the gas above critical point (total vapor) is heated isobarically, the condensate will appear (Fig. 9.1.1). The phase envelope generated through the simulator for the composition in Table 9.1.1 is given in Fig. 9.1.1.

The phenomenon is further explained with another NG of similar composition: around 75 mole % of methane and other normal components (Fig. 9.1.3). Now the NG follows a path from state 1 (15 MPa at 60°C) to state 2 (15 MPa at 25°C) to state 3 (9 MPa at 25°C) and lastly to state 4 (1 MPa at 25°C). The gas was full vapor at state 1. The path from state 1 to state 2 to state 3 to state 4 has been simulated. It has been found that no condensation takes place in state 2. Condensation takes place in state 3 when pressure is lowered isothermally and re-vaporization takes place at state 4 when the pressure is decreased at the same temperature. This is explained in Fig. 9.1.3.

Table 9.1.1 Composition of the NG from the well

Stream Name	S1
Temperature °C	40.000
Total Molar Comp. Percents	
N_2	1.0000
H_2S	0.1000
CO_2	2.5000
Methane	75.0000
Ethane	10.5000
Propane	4.3000
Ibutane	1.5000
Butane	1.8000
Pentane	1.3000
Hexane	2.0000
Heptane	0.0000

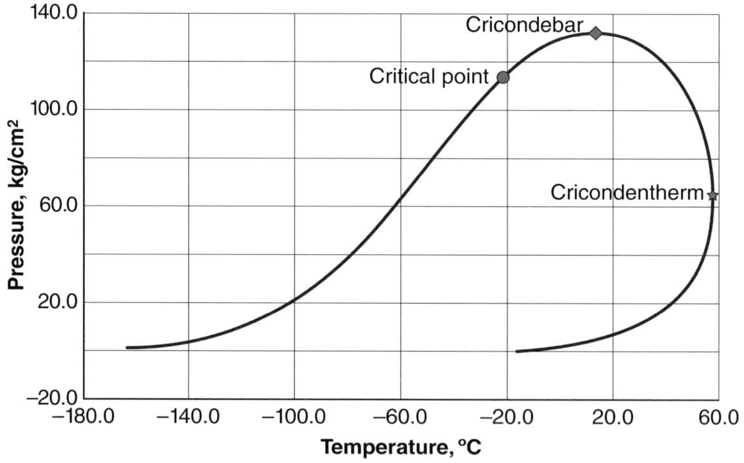

Fig. 9.1.1 Phase envelope of NG

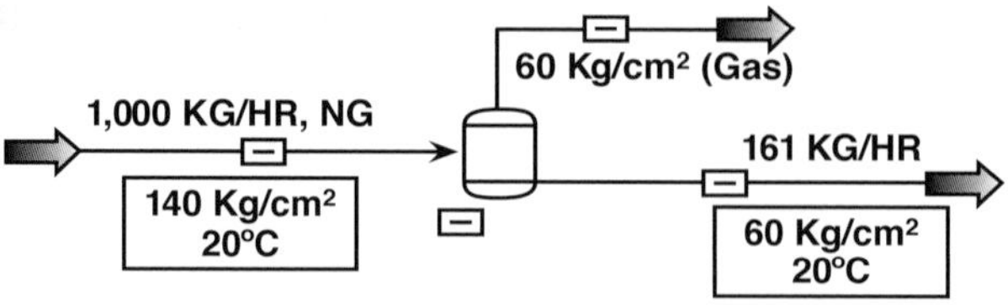

Fig. 9.1.2 Condensation of NG when pressure is decreased at a constant temperature

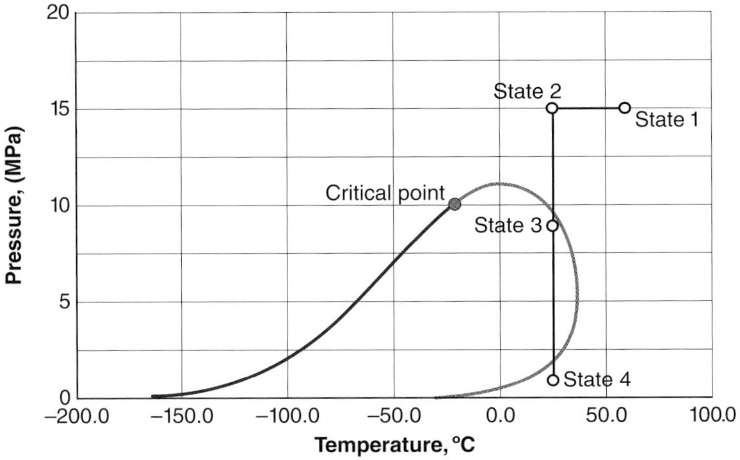

Fig. 9.1.3 Phase envelope of NG showing a hypothetical production path (1–4)

Retrograde phenomena

In retrograde condensation, condensation takes place when the pressure is reduced isothermally or when temperature is increased isobarically.

The composition of the gas determines the shape of the Pressure–Temperature (P–T) curve (phase envelope). The curve changes shape along with location of the critical point, the cricondenbar, and the cricondentherm. The left side of the critical point on and outside the envelope is the total liquid region, and the right side of the critical point on and outside the envelope is full vapor. The critical point, cricondenbar, and cricondentherm are shown on the curves.

The cricondenbar is the maximum pressure or the highest pressure at which two phases can coexist at a constant temperature, and the cricondentherm is the maximum temperature below which two phases can coexist at a constant pressure. The critical point can be between the cricondenbar and cricondentherm or to the left of the cricondenbar and cricondentherm.

In a well, the NG may be at state 1 and full gas. It may be cooled to state 2 and still be in full gas form. With the continued production of NG, the well pressure may fall at a constant temperature. This will condense the gas and the condensed liquid will remain in the well. The gas composition will also undergo change: when the pressure is decreased to point 4, the liquid may again vaporize.

Desired quality

The processing of NG involves removing impurities to acceptable levels, decreasing moisture content, lowering the hydrocarbon dew point, and meeting the heating value of the gas. The typical heating value desired is 950–1000 Btu/scf, the desired dew point is –10°C at 800 psi, the desired H_2S content is typically below 4 ppmw, the desired R-SH content is a maximum of 0.2 grains/100 scf, the desired water content is a maximum of 4 lbs/MMSCF, and the CO_2 spec is often kept at 2% by volume.

9.2 Recovery Process

The recovery process used for natural gas liquid (NGL) is greatly dependent on supply pressure, temperature, the contaminant levels and, to some extent, on the P–T envelope. Typically, the process involves the following:

1. Lowering the pressure (if required)
2. Inlet gas compression (if required)
3. Treatment for H_2S, R-SH, and moisture
4. Chilling and cooling (with or without expansion) for lowering the hydrocarbon dew point
5. Recompression for the delivery pressure requirement

The liquid stream obtained can be sold if the plant is large and there is local demand. Otherwise, the liquid obtained is treated in the NGL plant. The liquid is first routed to the demethanizer/deethanizer, then the depropanizer, debutanizer, or stabilizer for LPG recovery.

Before chilling the gas, extensive drying is required to avoid the formation of gas hydrates. A tentative hydrate analysis of the gas needs to be performed and can be done with the help of a simulator or available literature. Water removal is required to avoid hydrates at the likely minimum temperature attained during the process.

A typical processing scheme for an NG plant is shown in Fig. 9.2.1.

The natural gas from the well is received to the processing plant based on the pressure of the well. The gas is routed to a three-phase separator. The pressure may be decreased in the three-phase separator if the original condition of the gas from the well is beyond the P–T envelope to the left of the critical point. This is to promote retrograde condensation and knock off some condensate if it is found to be economic at this stage.

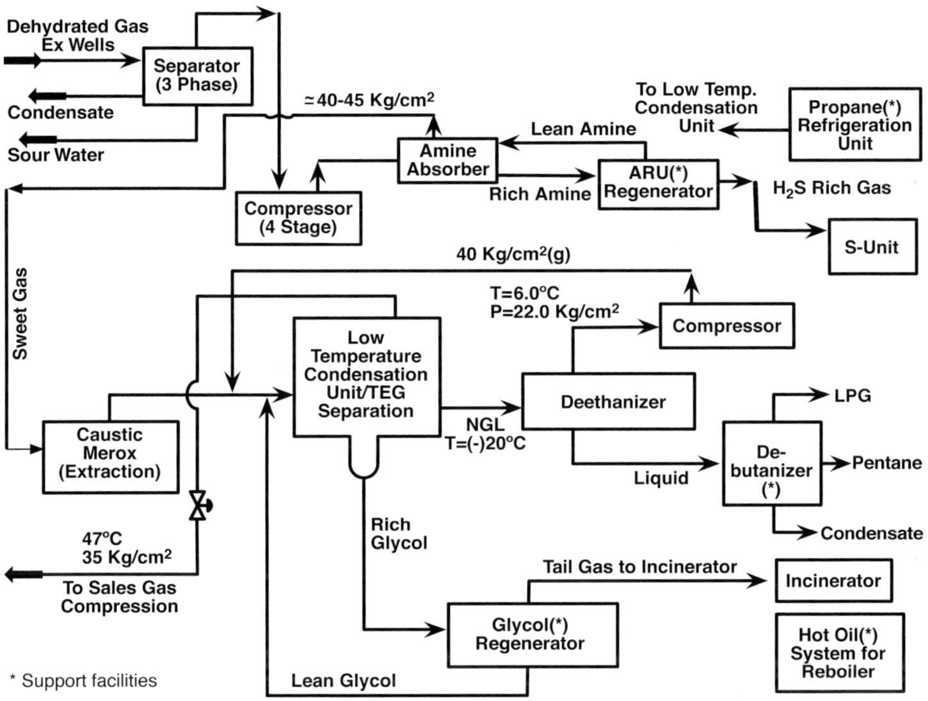

Fig. 9.2.1 Typical processing scheme for a NG plant

After the separation of the condensate and water, the gas is compressed to around 40–45 kg/cm^2 and absorbed with amine to remove the H_2S and CO_2. The gas is then routed to a caustic extractive Merox unit before being sent to a low-temperature condensation unit to condense heavier liquids from the gas to reduce the hydrocarbon dew point as desired.

If the gas is available at a higher pressure, no inlet gas compression is required. Prior to reducing the temperature, the gas is dehydrated with glycol to avoid hydrates. A hydrate inhibitor is also used. Often, solid bed dryers are also used for gas drying before the low-temperature condensation unit. The low temperature is typically achieved using a Joule–Thomson (JT) valve or mechanical refrigeration using propane refrigerant or a combination of a JT valve and mechanical refrigeration or expansion through the turbine. A temperature drop of up to –20°C to –25°C is typical.

The refrigeration is typically achieved through

1. **Mechanical or external refrigeration.** This is accomplished using a vapor compression refrigeration cycle that normally uses propane as a refrigerant.

2. **Self-refrigeration.** In this method, the inlet gas is pre-cooled with the cold-treated gas in a gas–gas exchanger and further cooled by isoenthalpic expansion (Joule–Thomson expansion) through a JT valve and chilled to condense the heavy hydrocarbons. To avoid hydrate formation, a hydrate inhibitor is usually added upstream of the gas–gas exchanger. If the water content is higher, a dryer is placed before the chilling unit.
3. **Cryogenic refrigeration.** This method is used in plants when gas pressure is not sufficient to attain the dew point with self-refrigeration. Here the JT valve is replaced by an expansion turbine. As the gas expands, work is extracted through the turbine shaft, reducing the gas enthalpy. The decrease in enthalpy produces a larger temperature drop than found when using the JT valve, which has a constant enthalpy. The turbine shaft can be connected to the compressor used for recompressing the gas. The cryogenic process can be applied if the gas pressure is high enough after expansion to condense the heavier components. A process scheme with a solid bed dryer and chilling section is shown in Fig. 9.2.2.

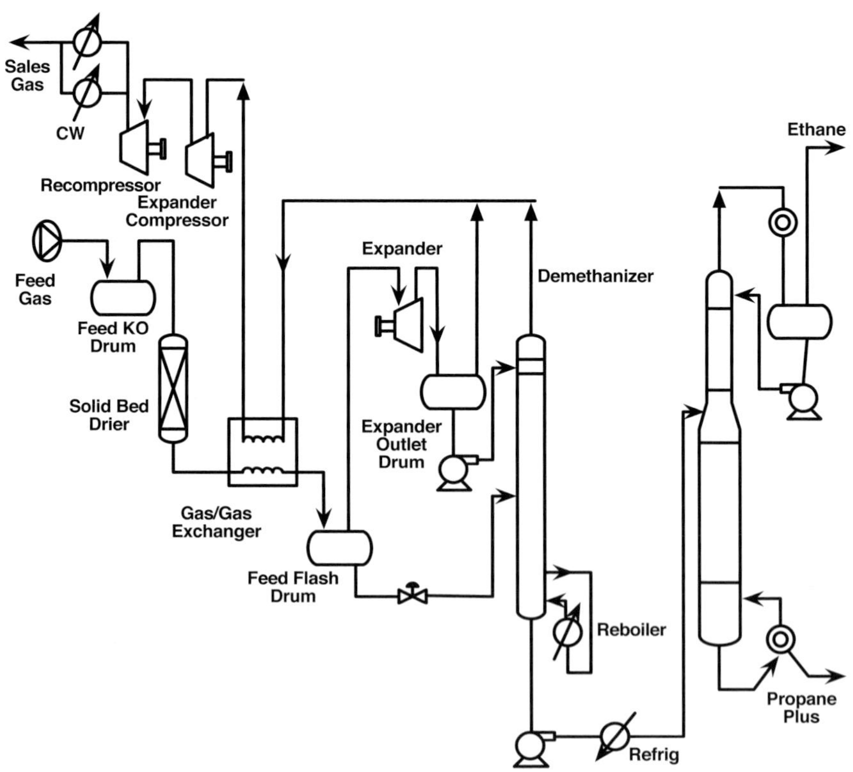

Fig. 9.2.2 Process scheme with solid bed dryer and chilling section

The NG composition for the problem simulated is given in Table 9.2.1.

Table 9.2.1 NG composition considered in simulation of Fig. 9.2.3

Stream name	FEED (S1)
Temperature °C	48.89
Pressure kg/cm^2	42.37
Rate kmol/hr	1,284.91
N_2	101.64
Methane	938.63
Ethane	98.68
Propane	73.11
Ibutane	12.72
Butane	31.35
Ipentane	8.87
Pentane	10.54
Hexane	5.40
Heptane	3.98
Total	1,284.91

The simulation is shown in Fig. 9.2.3 with a demethanizer and the results of the simulation are presented in Tables 9.2.2 and 9.2.3.

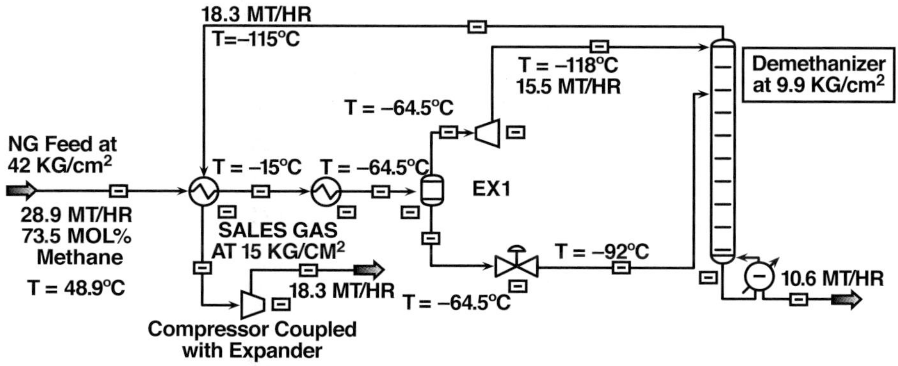

Fig. 9.2.3 Simulation of a typical NG chilling section

Table 9.2.2 Results of the reboiler duty from the simulation of Fig. 9.2.2

Column description	Demethanizer	
Condenser duty	Mkcal/hr	NIL
Reboiler duty	Mkal/hr	0.689
Column reflux ratio		0.228

Table 9.2.3 Results of the expander parameters from the simulation of Fig. 9.2.3

Expander name		EX1
Inlet pressure	kg/cm^2	42.17
Outlet pressure	kg/cm^2	9.82
Outlet temperature	°C	–118.38
Actual work	kW	384.93

Abbreviations

Btu: British thermal unit (of heat)

JT: Joule–Thomson (valve, effect)

MMSCF: million standard cubic feet

MPa: megapascal (10.2 kg/cm^2)

NG: natural gas

NGL: natural gas liquid

P–T: pressure and temperature

R-SH: thiols or mercaptans

scf: standard cubic feet

Delayed Coker Units 10

10.1 Introduction

A delayed coker unit is a low-cost versatile unit for upgradation of residues. Earlier coker units were designed to produce coke and products lighter than the fractions boiling at 370°C. They operated on high recycle and at relatively higher pressures, with mostly reduced crude oil (RCO) as feed. There were not many conversion units for the 370°C–540°C fraction produced from a coker unit. Thus, the heavy coker gas oil (HCGO) produced in the coking process was recycled to the furnace along with the fresh feed to further crack them to distillates lighter than 370°C.

Presently, with the advent of HCGO hydrotreaters and hydrocrackers, coker units are designed for producing HCGO (370°C–540°C fraction) and lighter fractions as products from the fractionator. Therefore, they can operate with a very low recycle rate (typically around 5%–10%) and a chamber pressure of 0.7 kg/cm^2 (g) or lower. The 370°C–540°C HCGO cut is fed to hydrocrackers or HCGO hydrotreaters to partly or fully crack the heavy fractions to lighter distillates (lighter than 370°C).

In a coker unit, the fractionator pressure governs the coke drum pressure. Lower drum pressure reduces the production of coke but causes a significant proportion (around 25%–35% wt) of HCGO to be produced.

The light coker gas oil (LCGO) is typically treated in a diesel hydrotreater (DHDT) to lower the sulfur (S) content, improve the cetane number (CN), saturate olefins, etc. The HCGO needs to be treated in a hydrotreater with part conversion so that the unconverted product can be further converted to distillates in an FCCU. Sometimes it is fed in conjunction with straight-run vacuum gas oil (SRVGO) and processed in hydrotreaters or hydrocrackers. Often the combination of LCGO and HCGO is treated in a single CGO hydrotreater and fractionated into a diesel fraction and a heavy fraction. The quality of the LCGO fraction improves and is recovered as diesel. The heavier fraction (hydrotreated HCGO) is further processed in an FCCU for the production of distillates.

The feed to the coker unit is typically RCO or vacuum residue (VR). A coker unit can be preceded by a residue (RCO or VR) hydrodesulfurization (HDS) unit. The hydrotreated residue can also be a feed stock for the coker unit. The hydrotreated feed stock will have a lower Conradson carbon residue (CCR) and sulfur content. The lower CCR content in the feed stock would decrease the coke yield and increase the valuable distillate production. The clarified oil (CLO) or decant oil (DO) from the FCCU is often blended with the coker feed. A delayed coker is normally used as a conversion unit where coke is a byproduct. The key properties of typical fuel-grade coke are given in Table 10.1.1.

At times, coke also becomes the desired product for anode-grade or needle-grade coke production. The properties of typical calcined anode-grade and calcined needle-grade coke are given in Table 10.1.2. The production of needle coke requires a feedstock with high aromatic content, low sulfur, and low metal, and usually operates with a high recycle ratio.

Table 10.1.1 Typical key properties of fuel-grade coke

Properties	Fuel grade coke
Carbon, wt%	94.55
Hydrogen, wt%	3.6
VCM, wt%	8–12
Gross heating value, kcal/kg	8,936
Net heating value, kcal/kg	8,749
Ash content, wt %	0.35
Sulfur, wt%	Based on feed

Table 10.1.2 Calcined anode-grade and calcined needle-grade qualities

Properties	Calcined anode-grade coke	Calcined needle-grade coke
Bulk density, kh/m^3	720–800	620–720
Sulfur, wt%	1.0–3.5	
Nitrogen, ppmw	—	50
Nickel, ppmw	200	5–7
Vanadium, ppmw	350	—
VCM, wt%	0.5	0.5
ASH content, wt%	0.4	0.1
Moisture, wt%	0.3	0.1
HGI (Hardgrove Grind ability Index)	60–100	—
Coefficient of thermal expansion, × 10^{-7}/°C		1–5

10.2 Description of the Unit

A coker unit can be divided in several different sections (Fig. 10.2.1):

1. Fresh feed preheat circuit
2. Coker furnace (2)
3. Coke drum (3, 3)
4. Fractionators/strippers (1) and gas concentration unit (4)
5. Blowdown system
6. Coke handling system
7. Water handling system

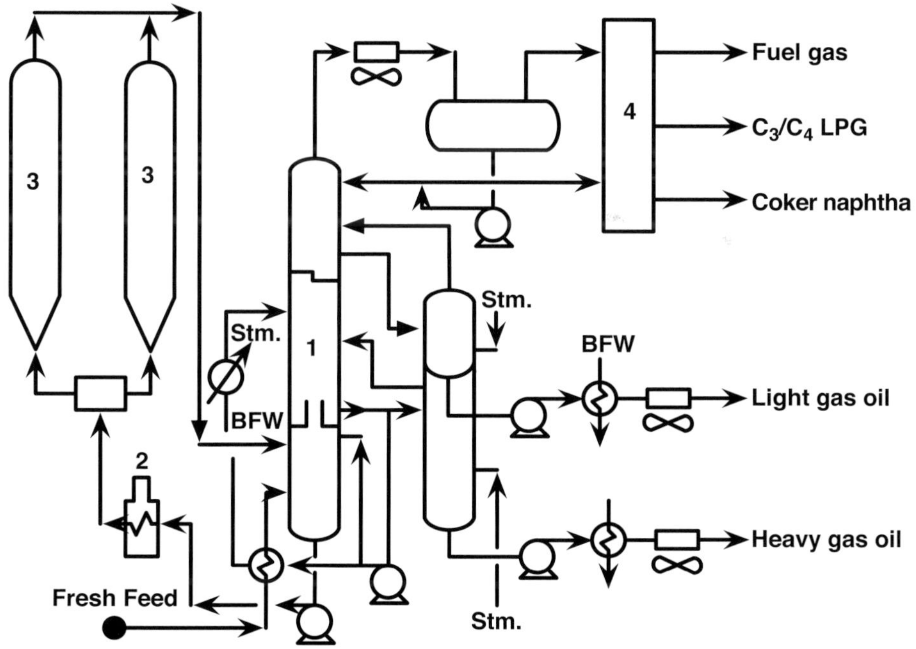

Fig. 10.2.1 Overall process flow scheme of a coker unit

The feed stock (either RCO or VR) is heated through heat exchange with hot products and fed to the main fractionator. The fractionator bottom is introduced in the coker furnace, where it is heated to around 500°C and introduced in the coke drum from the bottom. The drum is allowed to fill with hot liquid over a period of time (called the coking time) until it reaches a safe level in the drum. The hot liquid fills in the drum cracks and produces lighter distillates and coke. The furnace outlet is then introduced (switched) to the other coke drum (through switch valves). The drum filled with coke is steamed (wax tailing) and cooled with water (deluge cooling), then the water is drained (draining) and the drum covers opened (deheading), and coke is cut by a high-pressure water jet (coke drilling and cutting) and removed. After removal of the coke, the drum covers are boxed up (heading), pressure tested, heated (vapor heating), and taken on line to again receive the furnace effluent for coking. The above operations are called decoking operations and constitute the decoking cycle. Typically, the time of decoking cycle matches the time taken for coking or the coking cycle.

Clearance of the highest coke level from the top of the drum is maintained to avoid carryover. A minimum clearance of around 10–12 feet was practiced earlier, with a quench tower preceding the coker fractionator. The current trend is to provide a single fractionator column and maintain higher clearance above the coke (a height of 7–10 meters) to minimize coke carryover to the fractionator. Anti-foam agent is injected to the top of drum to reduce foaming and carryover. The overall process flow scheme of a coker unit is given in Fig.10.2.1.

In coker units, the fresh feed is introduced at the main fractionator bottom where it joins the recycle from the upper section of column and is fed to the coking furnace through a high-pressure secondary feed pump. The recycle ratio is defined as the ratio of the secondary feed flow to the fresh feed flow.

A coker unit flow scheme with one side stripper for the LCGO is shown in Fig. 10.2.2. A typical coker unit flow scheme with two side strippers is shown in Fig. 10.2.3.

The following are typical products from the coker unit:

1. Gas + liquefied petroleum gas (LPG)
2. Coker naphtha
3. LCGO
4. HCGO and extra heavy coker gas oil (XHCGO)
5. Coker fuel oil (CFO) and residual fuel oil (RFO)
6. Coke

The typical yields of products of a low-recycle VR coker unit are presented in Table 10.2.1.

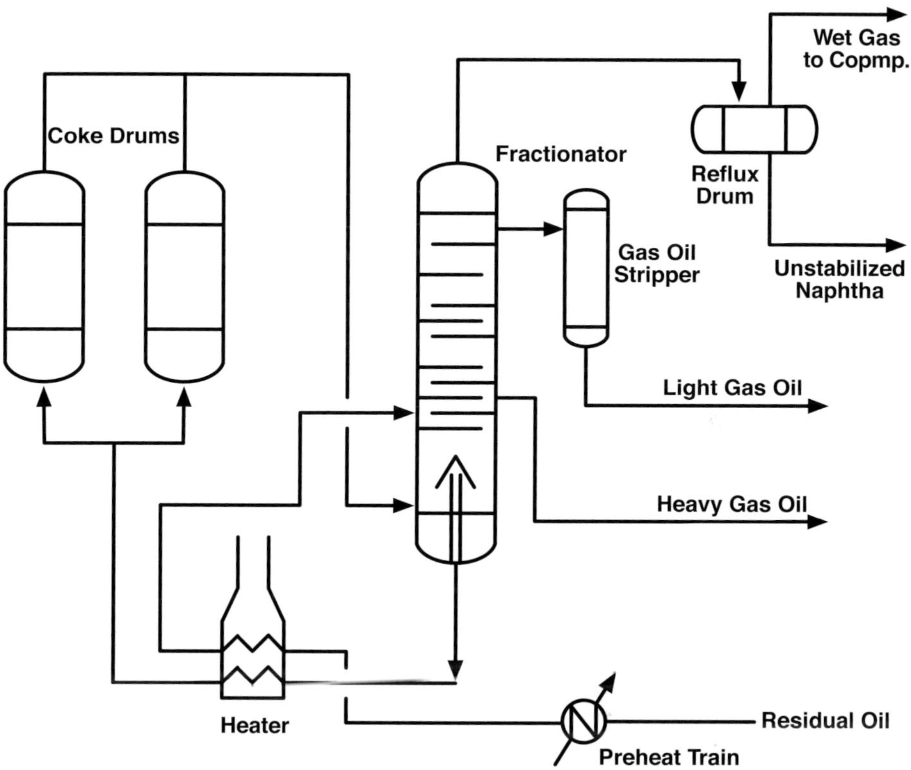

Fig. 10.2.2 Typical coker unit flow scheme with one side stripper

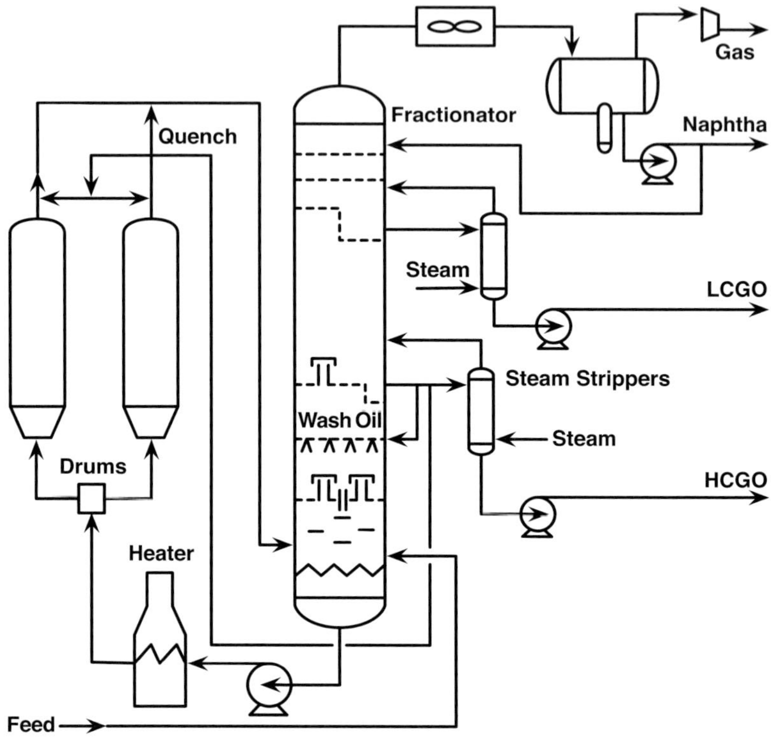

Fig. 10.2.3 Typical coker unit flow scheme with two side strippers

Table 10.2.1 Yield of products of a low-recycle VR coker unit

Products	Yield Pattern (Typical) WT%	
	Design	Actual
Fuel Gas	3.71	3.72
LPG	2.8	3.09
Lt Naphtha	6.48	6.56
Hvy Naphtha	1.94	5.36
LCGO	28.78	23.34
HCGO	24.76	25.43
CFO	1.42	1.22
Coke	28.82	31.28
H_2S + Misc.	1.29	–

Coker fractionator

The preheated fresh feed is routed to the fractionator spout. The spout liquid is picked up by the pump and fed to the furnace, then to the coke drum. The coker fractionator receives hot vapors from the coke drums. The vapor is superheated, so it is first cooled to bring it to the saturation temperature for fractionation. In coker units, the fractionator has baffle trays or disc and doughnut trays (counterflow trays) for this cooling process (desuperheating).

The fractionator is typically provided with 22–24 trays.

The cracked vapors are routed to the fractionators and one of two configurations are used:

1. Cracked vapors contacting part of the fresh feed for high-recycle units.
2. Cracked vapors not contacting the fresh feed for low-recycle units. The HCGO wash liquid from HCGO tray provides the recycle and cools the hot vapors to nearly saturation.

Low-recycle units

In low-recycle units, the feed preheat is maximized to increase the bottom temperature of the fractionator and reduce the coker furnace duty. Normally, the fractionator has two to three side cuts with side strippers. Typically, two pump-around (PA) refluxes are provided: one is the HCGO PA for cooling and condensing the HCGO and other is the LCGO PA for cooling and condensing the LCGO. A CFO cut with a side stripper is often provided below the wash section to draw the liquid from the wash section as CFO and regulate the recycle. The recycle is supplied by the HCGO internal reflux reaching the bottom and flowing down the wash section below the HCGO draw tray. The typical recycle ratio is 1.1, although units are also being designed for the very low recycle of 5% (recycle ratio of 1.05). A heat insulation tray is used in the fractionator bottom so that the hot vapors are not condensed by the colder fresh feed. The typical fractionator bottom temperature for such a low-recycle coker (recycle less than 10%) unit is around 270°C–330°C (approximately 10°C–20°C above the fresh feed preheat temperature). The column bottom is also provided with a coke clean-up system with a pump of low discharge pressure to circulate the bottom liquid through the duplex filters to keep coke fines from accumulating in the column bottom. The pumps take suction from the column bottom through the duplex filters and discharge it to the column bottom. The furnace heats the combined feeds from the fractionator bottom temperature to around 500°C–505°C.

High-recycle units

High-recycle units are normally provided two or three gas oil side cuts with side strippers. The fractionator is typically provided with two or three pump-around refluxes. In high-recycle units, the preheated fresh feed is fed into the fractionator at approximately 220°C–250°C and contacted with hot vapors from the coke drums. In a high-recycle coker, the fresh feed enters the fractionator at two locations. Around 80% of the fresh feed goes to the upper location (above the heat transfer trays and above the hot vapor feed from the coke drums). The fresh feed flows down the heat transfer trays, coming in contact with rising hot vapors from the drum, and condenses part of the heavy fraction from the drum vapor to generate recycle. The remaining 20% of the feed is routed to the bottom spout in normal operation. In such coker units, the fractionator bottom temperature is normally 350°C–360°C. The recycle ratio is typically 1.5–1.6. The feed is preheated only to around 220°C–260°C. It is heated to around 360°C by coming in contact with hot vapors in the heat transfer trays of the fractionator and condensing the heavy fractions from effluent vapors. Often these units are provided with two towers (a quench column preceding the main fractionator) (see Fig. 10.2.4).

The vapor from the coke drums (chambers) is cooled to around 430°C and is admitted to the quench tower preceding the main coker fractionator. The vapor in the quench tower is further washed with the CGO stream and the top vapor from the quench tower is introduced in the main fractionator succeeding the quench tower. This arrangement produces some heavy fraction from the quench tower bottom (Fig. 10.1.4), a product called RFO, which contains the coke fines carried over from the coke drums and may contain some amount of the CGO used as a wash in the quench tower. The bottom of the fractionator is fed to the coker furnace as usual. Thus, the feed to the furnace is reconstituted with the fresh feed and recycle heavy stock generated by the cooling and condensation of hot vapors.

Processing scheme with quench tower

In this arrangement, the coke drum vapors first enter the quench tower (Fig. 10.2.4) with counter flow trays, where they are washed free of heavy products and coke fines. Typically, a coker gas oil wash is provided. This mostly revaporizes, condensing the heavier fractions from the coke drum vapors. The hot vapor from the quench tower is routed to the main fractionator. The liquid from the quench tower bottom is drawn from the column as RFO, cooled, and typically blended with fuel oil. These units with quench towers do not need a coke clean-up system in the main fractionator bottom, as the majority of the entrained coke is removed along with the RFO. The RFO product may contain some gas oil fraction and there may be a loss of gas oil component with the RFO.

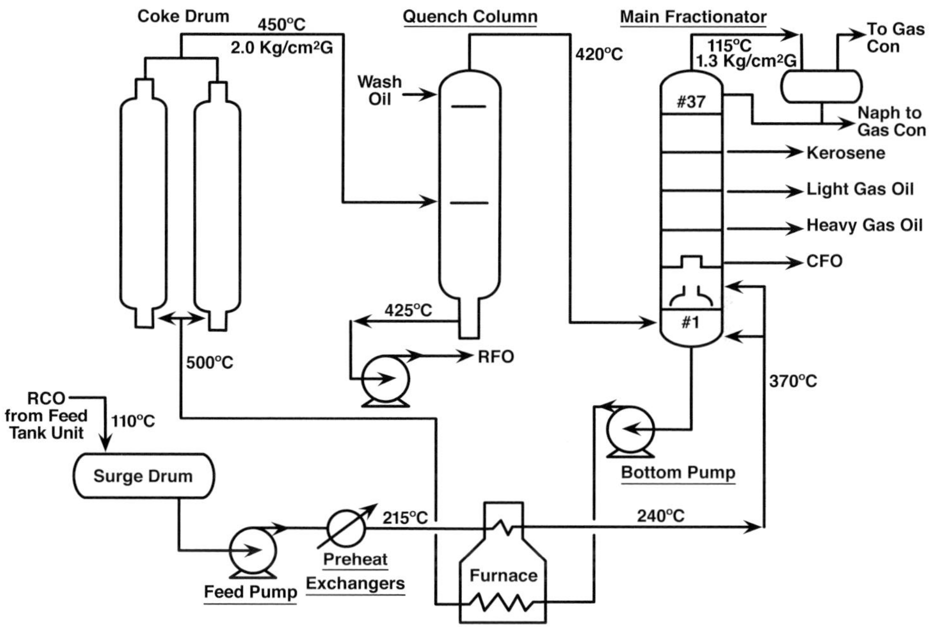

Fig. 10.2.4 Configuration with a quench column

Blowdown system

The blowdown system is comprised of the blowdown scrubber (BDS) with its accessories, the overhead air cooler, blowdown drum (BDS overhead drum), and chamber condensate pot (drum).

Blowdown tower or scrubber

The blowdown system is comprised of a BDS with six to ten baffle trays with an overhead condenser (air cooler) and a drum as an accumulator for overhead. The tower has a bottom pump to circulate the liquid collected in the bottom through an air cooler to the top of the tower in order to cool the vapors and maintain the desired top temperature and a provision of bottom liquid pump out (Fig. 10.2.5). An LCGO makeup connection is used in the tower to build up the initial level for bottom pump start up. The tower typically uses a low-pressure steam coil at the bottom to avoid congealing of the heavy hydrocarbons in the bottom.

In the delayed coker units operating at relatively higher pressures, the blowdown tower overhead drum is often provided with a vent gas compressor (low capacity) with a spillback facility to deliver the vapors from the BDS overhead drum to the wet gas compressor suction to minimize vapor loss from the blowdown tower top to the flare. A scheme of a BDS with a vent gas compressor is presented in Fig. 10.2.5.

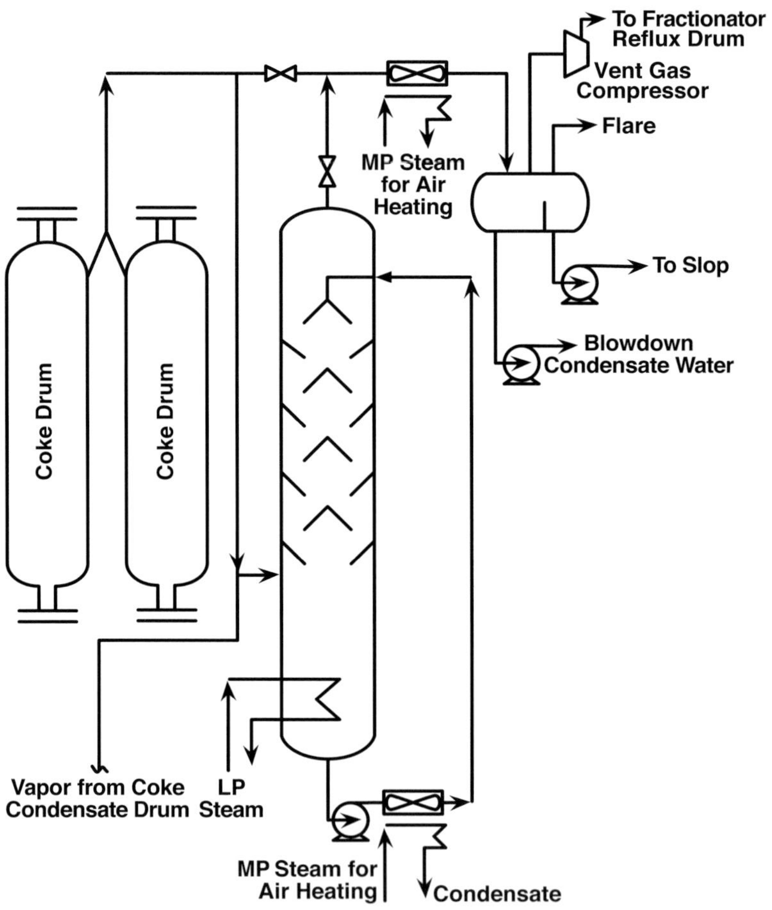

Fig. 10.2.5 Scheme of a BDS with a vent gas compressor

Chamber condensate pot (drum)

A chamber condensate pot is located below the coke chambers to collect the condensates during vapor heating operation. The vapor from the chamber condensate drum is routed up to the blowdown tower during the initial period of heating (up to 150°C) and finally to the main fractionator. In some plants it continues in a BDS.

Water handling system

The water for cooling of the coke drum is supplied by a low-discharge pressure pump called the deluge water pump or cooling water pump. The water drained from the chamber goes to coke dump area and flows to a maze of iron grids for catching the coke lumps and is then sent to a clear water sump and pumped to the decoking water storage tank, often through hydrocyclones.

The water for the cutting and boring of coke is supplied at a high pressure (150–300 kg/cm^2) through a coke-cutting water pump. The coke-cutting water follows the same route and is recirculated to the water tank after the removal of any solids or fines. The water handling system of a coker unit is presented below in Fig. 10.2.6.

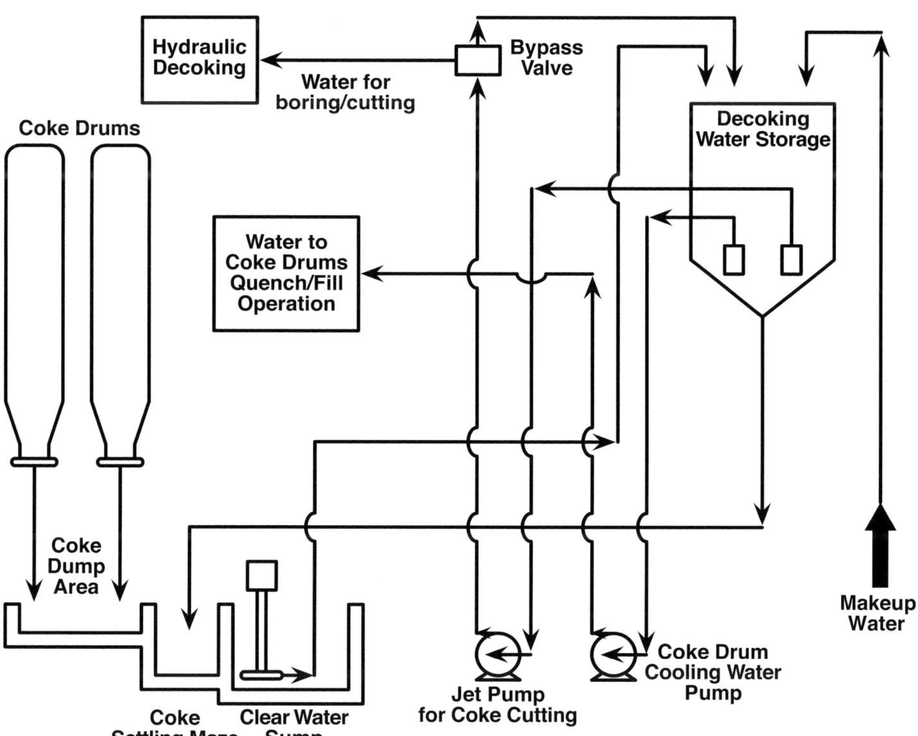

Fig. 10.2.6 Water handling system

Reactions

The coking reaction is an endothermic reaction. The heat input to the furnace supplies the necessary heat for the reaction.

Operating variables

1. Coil outlet temperature

Temperature is used to control the volatile combustible material (VCM) content of the coke. The VCM content ranges from 8%–12% wt, although recent trends are for a VCM content of 6%–8% wt. At constant pressure and recycle ratio, the coke yield decreases as the drum temperature is increased. If the temperature is too low, pitch or soft coke formation occurs. If temperature is too high, the coke formed is very hard, which may render it difficult to cut and remove with the hydraulic jet equipment used for coke cutting.

2. Drum pressure

At constant temperature and recycle ratio, the increase in pressure will lead to the retention of higher quantities of heavy hydrocarbon in the coke drum. This would eventually crack and increase the coke yield and gas yield. Lower pressure operation increases liquid product yield.

3. Recycle ratio

An increase in the recycle ratio increases the coke and gas yield. The recycle ratio is maintained to retain the cut point of gas oil or HCGO. An increase in the recycle ratio at the constant secondary feed reduces the fresh feed rate. The effects of lowering the recycle ratio are given below:

1. Generates more HCGO
2. Decreases coke yield
3. Increases fresh feed at constant combined feed rate
4. Can produce shot coke
5. Reduces the production of gas and naphtha

Higher recycle ratio is preferred for the production of needle coke. The refineries that do not have an HCGO conversion unit (such as HCGO or VGO hydrotreaters or hydrocrackers) are required to operate with a high recycle ratio to recover distillates lighter than 370°C.

10.3 Description of Decoking Cycle

Steaming and water cooling

After switchover, the coke in the chamber is steamed and cooled with medium-pressure steam (wax tailing) and the volatile content is initially sent to the main fractionator column to recover the hydrocarbons before being sent to the blowdown tower. The chamber is then cooled with a mixture of steam and water to avoid thermal shock. The temperature of the chamber at the beginning of this process is typically around 448°C. It is then cooled using only water, called deluge water. Typically, the initial deluge water cooling starts when the temperature in the chamber is around 400°C, then the middle deluge starts at 358°C and cools the drum to around 120°C. Later, the final deluge is done and the drum is filled to the maximum coke level. At the end of the cooling process, the chamber has been cooled to around 100°C–110°C.

During the deluge, the water used for cooling vaporizes and is converted to saturated steam and later to superheated steam. The steam produced is typically routed to the blowdown tower (often directly to the air cooler of blowdown tower overhead) and condensed by the overhead air cooler. Thus, the blowdown tower experiences different feeds and loads during the cooling cycle.

During steaming (wax tailing), a mixture of steam and hydrocarbon vapors is routed to the blowdown tower. During deluge, a large quantity of superheated steam (resulting from the vaporization of the deluge water) is routed to the blowdown tower.

The subsequent activities in the decoking cycle are: depressurization of the coke drum to atmosphere through the silencer, draining the water, unheading, drilling, and coke cutting. After coke cutting, the chamber is headed, pressure tested, and heated with steam, followed by heating with vapor from the chamber on line. Typically, a coke drum is heated to around 250°C–320°C before receiving the effluent from the furnace. During the initial period of chamber heating, vapor from the drum is admitted to the BDS (through the coke condensate drum), and this results in a different kind of load to the BDS.

Coker unit cycle time

Typical coking cycle time is 48 hours, 24 hours for coking and 24 hours for decoking. The decoking cycle time is comprised of activities like wax tailing and cooling, draining, unheading and decoking, heading and pressure testing, vapor heating, and switch over.

Often the coker cycle time is reduced to increase the fresh feed capacity of the unit. This means that there is less decoking time available for chamber steaming, water filling and cooling, water draining, unheading, coke cutting, heading, steaming and pressure testing, chamber heating, and feed introduction.

This decoking cycle is shortened typically by using a modern device for chamber heading, unheading, coke cutting, etc. The typical activities involved in a coker unit and the time required for each activity in the cycle times of 48 hours, 36 hours and 28 hours are presented below (Table 10.3.1).

Table 10.3.1 Cycle time with times for different activities

SL. no.	Activity	48-hr cycle	36-hr cycle	28-hr cycle
1	Coking	24.0	18.0	14.0
2	Switch + steam to fractionator	1.0	1.0	0.5
3	Steam to blowdown	1.0	1.0	1.0
4	Water quenching	5.0	4.0	4.0
5	Draining	3.0	2.0	2.0
6	Cover removal	1.0	0.5	0.5
7	Boring/cutting	4.0	4.0	2.5
8	Reheading/pressure testing	2.0	1.0	0.5
9	Drum warmup	7.0	4.5	3.0
10	**Total**	**48.0**	**36.0**	**28.0**

It can be seen from the above table (Table 10.3.1) that the main reduction in decoking time is seen in water cooling and quenching, cover removal, boring and cutting, reheading, and drum warm-up.

The reduction of time for cover removal and reheading is accomplished by using an improved arrangement and improved devices. The time reduction in the boring and cutting of coke is possible with improved tools. The reduction in cooling time and warm-up time produces rapid changes in the skin temperature of coke drums. The time required for cooling and heating is normally limited by the rate of change of skin temperatures. This is likely to have an effect on the service life of the coke drums. The coke drum metal has skin temperature thermocouples to monitor the skin temperature changes of the metal during the cooling and heating cycles.

The cooling rate and heating rate of the drums affects the design of the BDS and its accessories. The steam produced during coke drum cooling with water gets routed to the BDS. Thus, faster cooling produces steam that enters the BDS at a higher rate and is cooled and condensed by the BDS overhead air cooler.

Similarly, faster drum heating can result in a higher load of hydrocarbon going to the BDS, as the BDS tower is lined up through the chamber condensate pot and experiences a higher rate of drum vapors during the initial heating. The vapor feed to the BDS is also cooled and condensed by a pump-around of the BDS and its air cooler.

In practice, cooling in the decoking cycle can be monitored from the following:

1. Total quench and cooling water flow to the drum
2. Total sour water production from the BDS overhead
3. Coke drum overhead pressure (under decoking)
4. Coke drum outlet temperature
5. Change in the skin temperatures of the coke drum

During cooling, the deluge water input rate and the sour water generation needs to be balanced. A balance of the deluge water input and sour water generation indicates that the water is vaporizing and cooling the coke in the drum. This may provide the opportunity to increase the water input rate and achieve faster cooling. If the sour water production is less than the water input, an increase in the water rate is not likely to cause faster cooling. Very fast quenching and cooling has an adverse impact on the coke drum life. The rate of quenching should be gradually increased through a ramping program. A high rate of deluge may pressurize the drum under cooling due to a restriction in the overhead line to the BDS, and may also lead to high scrubber overhead receiver temperature, etc.

Modern coker units are designed with 18 hours each for coking and decoking, making the total cycle time 36 hours. This is achieved with automated tools for deheading and heading, improved tools for coke boring and cutting, and a reduction in the time required for cooling and heating the coke drums.

Number of coke drums or chambers

For a high-capacity coker unit, the number of coke chambers increases. The maximum diameter of a chamber is limited by fabrication (the maximum diameter seen by the author is 32 feet). Beyond this limiting diameter, the number of chambers is increased to contain the volume of coke produced.

For a four-chamber (drum) coker unit, two coking furnaces are used, and for a six-chamber coker unit, three coking furnaces are deployed. Typically, two coke chambers are fed by one coking furnace. Another advantage of having more than two chambers is that during vapor heating, a smaller quantity of feed is abstracted from the fractionator. Thus, the fractionator feed does not decrease much, which ensures its better operation. For a four-chamber coker unit, there are always two drums at different phases of the coking cycle to meet the capacity of the unit. Six-chamber designs are also common for large coking units. In a six-chamber configuration, the BDS remains almost fully engaged during operation. Any increase in the number of coke drums in a configuration (beyond six) may need an additional BDS with its accessories and preferably with an additional coke condensate drum.

10.4 Design Approaches

The coke drum effluent composition is normally derived from pilot plant trials. Typically, in a pilot plant, the fresh feed is initially introduced in the furnace and the outlet of the furnace is fed to the coke drum. The coke drum vapors are cooled and condensed and analyzed for the gas, distillates, and heavier products that boil above the desired products (recycle).

In another run, the heavier quantity or recycle is mixed with the fresh feed, fed through the furnace and drum, and vapor-analyzed. The coke produced and the recycle quantity and quality is determined. Several runs are carried out with different proportions of recycle estimated from the previous run. When the recycle quantity in the feed almost matches the recycle quantity produced during coking, the trial is terminated. The coke yield and recycle quantity and distillate produced are considered as final inputs for carrying out the design.

Often, the drum overhead condensed liquid from the pilot plant is fractioned in a separate distillation facility or analyzed with gas chromatography and the quantities of each fraction are determined. The drum effluent is generated and normally supplied to the process designer as a combination of pure components (light ends) and normal boiling point (NBP) cuts of effluent with their quantities.

Alternatively, the effluent data can be supplied with the composition and quantity of the gases along with the yield and distillation of the liquid fractions obtained in the pilot distillation facility. The coke yield and recycle quantities are also supplied by the pilot plant studies. Often the effluent, coke, and recycle quantities are generated by the mathematical model of the unit by experts.

The combined drum effluent is the starting point for the design of a fractionation facility. The fractionator feed is generated by blending all the fractions and the drum outlet quench (injected to drum overhead to lower the temperature of the drum effluent to around 420°C). If the combined effluent is supplied in NBP format, then only the drum outlet quench quantity is combined to constitute the feed to the fractionator. Otherwise, gas and liquid products are blended along with the drum overhead quench to constitute the fractionator feed. The furnace coil turbulizing water is also added at a rate of 0.5 m^3/hr per pass.

Sizing the coke drum

A coke drum is used to contain the coke produced, leaving a clearance of around 7–10 meters at the maximum coke height for a grassroots design. This fixes the height of the coke drum. The vapor velocity at the drum overhead is maintained at around 0.5 ft/sec for grassroots designs. This determines the diameter of the coke chambers.

Sizing the BDS

The blowdown tower is lined out at different operations in the decoking cycle. Thus, it experiences different loads at different operations it is subjected to as given below.

1. Steaming and cooling

During steam-out, the vapors from the chamber are routed to the blowdown tower for approximately 1 hour as indicated in the list of activities in the decoking cycle (Table 10.3.1). Later, during the cooling operation, the BDS is lined out to receive the steam produced during the chamber cooling with water as in the initial deluge, middle deluge, and final deluge (Table 10.4.1). Often the steam generated during the deluge operation is directly routed to the air cooler of the blowdown tower overhead, thereby reducing the steam load to the BDS.

2. Vapor heating

During vapor heating, the chamber condensate drum vapor outlet initially remains lined up to the BDS and the BDS experiences a different load. The top temperature of the blowdown tower is maintained at around 150°C to avoid water condensing in the column. The return temperature of the pump-around is maintained at around 120°C. The maximum hydrocarbon vapor routed to the tower during vapor heating is typically 20% of the vapor outlet of the chamber in-line.

After attaining a coke drum skin temperature of 150°C–200°C during vapor heating, the vapors are usually routed to the main fractionator instead of the blowdown tower to reduce the loss of hydrocarbon products. Table 10.4.1 shows typical chamber cooling during different activities.

Table 10.4.1 Different steps in coke drum cooling with estimated deluge water rates

Cooling steps	Time in hours	Mode of cooling	Start temp. (°C)	End temp. (°C)	Coolant flow rate (kg/hr)
Step 1	1.0	Steam	450	448	1,315
Step 2	1.0	MP steam + water	448	403	3,260
Step 3	0.5	Initial deluge water	403	358	6,000
Step 4	2.0	Middle deluge water	358	120	9,350
Step 5	1.0	Final deluge water (up to top of chamber)	120	100	116,500

Coke-specific heat variations with temperatures are given in Fig. 10.4.1.

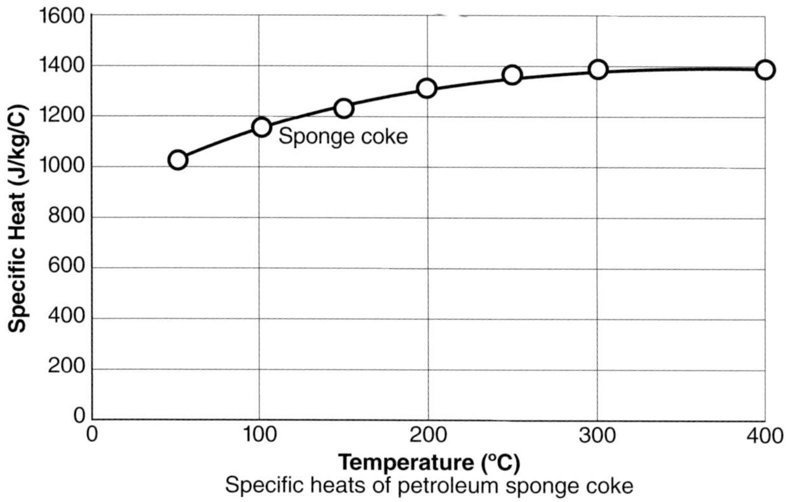

Fig. 10.4.1 Coke specific heat variations with temperatures

The temperature change and time required for coke drum cooling are derived from experience. Two different cases need to be considered for sizing the blowdown tower and its accessories as given below.

1. Steaming and cooling operations (i.e., initial deluge, middle deluge, and final deluge)
2. Vapor heating of the drum chamber

The coolant heat removal rate is calculated based on the ((coke quantity × the temperature change × specific heat at constant pressure (C_p) of the coke)/(the time of operation)). Typical coke C_p ranges from 0.24–0.29 kcal/kg°C (Fig. 10.4.1). This is divided by the latent heat of water vaporization in order to approximately provide the flow rate of the deluge water or the steam generated while cooling the coke.

Alternatively, the water required or steam generated can be calculated by dividing the heat removal calculated above by the difference in the enthalpy of the steam and that of the deluge water. The water flow required during the different steps of the deluge is presented along with the temperature changes in Table 10.4.1. Normally, the medium deluge water vaporizes and can be measured for sizing the BDS. Water may also vaporize during final deluge and can be considered for routing to the BDS overhead air cooler directly.

The maximum hydrocarbon vapor routed to the tower during vapor heating is 20% of the outlet of the chamber inline, which is also considered for column sizing and arriving at the controlling case. The drum outlet vapor composition needs to be considered for a temperature of 350°C–400°C. A higher inlet temperature to the BDS provides cushion in design.

The BDS column is separately simulated with all the cases mentioned (e.g., steam flow rate during water cooling and vapor flow during vapor heating). These simulation results deliver the maximum diameter of the tower and overhead air cooler duty, pump-around flow, and pump-around air cooler duty.

During simulation, a top temperature of 150°C and a pump-around return temperature of 120°C is considered for the BDS column. Typically, the steam generated during the deluge controls the BDS column diameter and overhead air cooler duty, while vapor heating controls the pump-around flow and pump-around air cooler duty.

This concept for sizing the BDS column with its accessories is required when the coking cycle is reduced by a process engineer to increase the fresh feed rate, or for a grassroots design.

Tentative design guidelines

The feed to the fractionator is generated by blending the products based on pilot plant data or empirical correlations. The fractionator typically has 22–24 trays. For low-recycle coker units, the column typically has an LCGO pump-around and HCGO pump-around for recovering the heat available in the column for feed preheat and steam generation.

The tray efficiencies above the HCGO can be conservatively taken as 65%. Below the HCGO wash zone, the tray efficiencies are 30%–40%, and heat transfer trays have an efficiency of around 20%–25%. A foaming factor of 0.85–0.9 is better for main fractionator sizing. Products are to be drawn through side strippers, preferably with a minimum of 3 stages in each.

Summary of design methodologies

1. In simulation, the Grayson–Streed (GS) can be used with API densities.
2. The chamber overhead line is to be sized based on the typical pressure drop of 0.2 kg/cm^2 with one-inch coke thickness inside the line.
3. A foaming factor of 0.85–0.9 is better for main fractionator sizing.
4. Coke drum height must contain the coke in the cycle and still have at least 7 meters of vapor space (conservative 10–12 meter minimum clearance).

   The coke production is typically taken as 1.35 times the feed CCR for low-recycle coker units and higher for high-recycle coker units. Some other correlations are also used. It may be wise to consider the coke yield from the pilot plant run (with some cushion on coke yield) for calculating the drum size.
5. Coke drum diameter should be based on the volume flow rate of vapor at actual pressure and temperature with a vapor velocity limited to 0.5 ft/sec for grassroots units and up to 0.6 ft/sec for revamp applications.
6. Coke-cutting water flow can be estimated as 40–50 gpm/ft diameter of the coke drum or with some other empirical correlations.
7. Coke-cutting water pressure may be estimated as 140–150 psi/ft diameter of the coke drum or with some other empirical correlations.

The above guidelines can be used by a process engineer to estimate the modifications required for capacity increase and may be endorsed by a technology supplier.

A gas concentration unit (GCU) or LPG recovery unit is normally installed at the fractionator downstream. The feed to the GCU is fractionator reflux drum gas and reflux drum naphtha. Facility for injection of polysulfide solution is provided at the WGC first stage discharge cooler and the second stage discharge cooler inlet. Discontinuation of polysulfide injection may cause leaks in the coolers.

Coker furnace

A coker furnace is used to heat the combined feed to the required temperature of 505°C–510°C. The design is performed with consideration of the following:

Average heat flux is around 24,400–30,000 Kcal/hr m^2°C.

Maximum radiant heat flux is around 35,000 Kcal/hr m^2°C.

End of run (EOR) pressure drop is around 30–35 kg/cm^2 (in VR coker units).

Coke thickness considered when calculating the maximum tube metal temperature (TMT) is typically 3mm.

The minimum cold oil velocity at the inlet is 1.5 m/sec.

Typical liquid mass velocity ranges from 1,500–2,200 kg/m^2 sec.

Typical vaporization at the outlet of a VR coker is around 30%. This vaporization can be 20% in the last tube and 10% in each of the tubes before the last tube. Often it is considered as 5%, 10%, and 15% in last three tubes (with progressive increase toward outlet tubes).

The furnaces of RCO coker units operate with a slightly higher pressure drop than VR low-recycle coker units. For RCO as feed, the vaporization is much higher at the outlet (around 60%–70%). The pressure drop at the EOR is calculated based on the same. Vaporization in the last five to six tubes is considered for pressure drop calculation. The EOR pressure drop in an RCO coker furnace is higher due to higher vaporization at the outlet. Typical pressure drop is 40–45 kg/cm^2.

For rating purposes, the pressure drop can be estimated the same way as atmospheric and vacuum heaters, which is described in Section 13.4 of Chapter 13. Only the vaporization in the last three tubes of the VR coker furnace and five tubes in the RCO coker furnace should be considered. A mixture of steam gas, naphtha, and diesel is the vaporized product produced by cracking in the furnace. In the simulation, part of the feed is withdrawn and vaporizable stock is introduced ahead of the tube in which vaporization is assumed to start.

A process engineer may carry out a tentative rating of the unit using the above method for identifying a bottleneck. The preliminary rating of the furnace by a process engineer may be checked and endorsed by a coker furnace expert. A new coker furnace should always be designed by a coker furnace expert only.

10.5 Capacity Augmentation through Revamp

Coke drum

Cushion is required in order to contain the volume of coke produced in coke drums operating at a higher capacity. A minimum clearance of 7–12 meters is required from the coke level to the top tangent length (TL) of the coke drum. This is to avoid any carryover of liquid coke to the column.

If the required clearance of 7–10 meters cannot be provided at a higher capacity due to an increased amount of coke production, the only feasible method is to reduce the cycle time from the typical 48 hours by reducing the time in the decoking cycle (mainly in the activities of cooling, draining, unheading, coke cutting, heading, vapor heating, etc.). Another issue that needs to be addressed is the superficial velocity of cracked vapors through the coke drums. Typical superficial velocity maintained in the coke drums of an industrial coker unit is 0.2–0.6 ft/sec. Higher superficial velocities can lead to the production of shot coke (particularly in low-recycle operations), and can increase the carryover. A slight increase in drum pressure may help reduce the superficial velocity in the coke drums at the cost of increased gas and coke production.

Furnace

A higher capacity demands a higher furnace heat duty and mass velocity. The combined feed temperature (column bottom spout temperature) for a high-recycle coker is typically 360°C. Furnace limitation can be addressed by the following alternatives:

1. Increasing the temperature at the column spout by increasing the fresh feed temperature and the temperature of the combined feed
2. Lowering the recycle ratio to reduce the combined flow through furnace and heating the additional fresh feed to the required temperature (preheat)

A preheat furnace can be used to heat the fresh feed to higher temperatures to resolve any limitations in the main coker furnace (used for heating the combined feed) for coker unit capacity augmentation. This methodology can only be applied in a low-recycle coker unit where the column bottom temperature is usually low (270°C–310°C). A higher fractionator bottom temperature (higher inlet temperature of the coking furnace) is likely to reduce the heat transfer in the convection section of the coking furnace due to lower differential temperature, and may need augmentation of heat transfer surface area, primarily in the convection, and possibly in the radiation section as well to maintain the original heat duty. In one coker unit capacity revamp, the fractionator bottom temperature is increased from around 290°C to around 325°C–330°C with the addition of a preheat furnace in the fresh feed at the upstream of the fractionator.

Fractionation and gas-con section

Fractionation and gas-con sections can be modified in the same way as normal petroleum fractionation and gas-con sections. An extra draw of HCGO may be necessary from a location above the wash tray and below the HCGO draw.

A very low recycle ratio may increase the cut point of HCGO, affecting its quality. A strong wash zone is required to maintain the quality of HCGO. An XHCGO cut below the HCGO may be advisable. The wash zone will be located below the XHCGO draw and this XHCGO can be routed to the furnace oil (FO) to maintain the HCGO quality and its acceptability in the hydrotreater or hydrocracker. The XHCGO can be also routed to the residue fluid catalytic cracker unit (RFCCU). However, the clarified oil yield from the FCCU may increase with the incorporation of XHCGO in the RFCCU feed.

In all coker fractionators, the drum vapors are first desuperheated and then fractionated in the wash section below the HCGO or XHCGO draw. Typically, the wash section consists of two to three desuperheating stages made of a combination of counterflow trays (deck, disc, doughnut) and cross-flow trays (bubble cap trays, proprietary fixed-opening valve trays). Alternatively, grid or packing combinations are used. Normal valve trays are never used in the wash zone. The HCGO and XHCGO are introduced in the wash section as internal reflux and supply the wash liquid.

The liquid introduced as wash liquid revaporizes while condensing the heavier portion from the feed vapor, and is typically reduced to 25%–30% before joining the fresh feed at the column bottom. This partly dried liquid from the wash zone generates the recycle in low-recycle coker units.

10.6 Case Studies

Case Study 1

A study was carried out on an RCO coker unit whose capacity was increased by around 20%. There was a marginal additional duty requirement in the heater, but the pressure drop in the existing heater (two-pass) was limiting. Thus, it was decided that a parallel coking heater (with an absorbed duty of around 3–4 Gcal/hr) with only one pass should be installed. The coke drum height was limited in order to contain the coke production for a 24-hour coking cycle at the increased capacity. The only option was to reduce the decoking cycle time, so the coking cycle and decoking cycle times were reduced from 24 hours to 20 hours (Table 10.6.1).

Table. 10.6.1 The original and revised decoking cycle for revamp

Activity	Initial operation time in hours	Revised operation time in hours
Coking	24	20
Switch drum	0.5	0.5
Steam out to coker fractionator	0.5	0.5
Steam out to blowdown system	1.0	1.0
Slow water cooling (low range)	1.0	1.0
Fast water cooling (deluge cooling)	4.0	3.5
Drain coke drum	3.0	2.0
Remove top and bottom head	1.0	1.0
Hydraulic coke boring and cutting	4.0	4.0
Reheading (box up)/pressure testing	2.0	2.0
Drum warmup	7.0	6.0
Idle hours	0.0	0.0
Total	48	40

In the decoking cycle, cooling time was reduced, necessitating a higher deluge water flow and a higher steam flow through the BDS. Chamber heating time was also reduced, which in turn resulted in an increase in hydrocarbon flow to the BDS for vapor heating.

The BDS is then rated with a revised hydrocarbon flow for chamber heating (20% of the vapor at the new capacity), and the steam flow for chamber cooling (Table 10.4.1). The rating study indicated that the unit required augmentation of the blowdown condensate water pump, a blowdown tower circulating reflux (PA) air cooler, and BDS overhead air coolers.

If the BDS column diameter is inadequate to handle the steam quantity from the cooling operation during revamp, a bypass line can be used, which will bypass the steam and admit it to the overhead air cooler of the BDS, allowing the BDS column to be retained. However, some technology suppliers do not recommend this, possibly apprehensive of coke fines choking the BDS overhead air cooler.

Case Study 2 (Case study for the reduction of gas oil components in CFO + RFO)[1]

Some old RCO coker units (high-recycle) are used in operations with two consecutive towers, a quench tower followed by the main fractionator (Fig. 10.1.4). The RFO content in a particular unit was high, around 10% wt (Table

10.6.2), and had around 15% gas oil component in it. The refinery wanted to recover the majority of the gas oil components from the RFO and reduce the RFO + CFO content.

For a high-recycle coker unit, a scheme was developed to recycle part of the RFO from the quench tower to the fractionator bottom after mixing it with the fresh feed going to the bottom spout of the main fractionators in order to cool it to approximately 330°C (for filtering at a lower temperature in a duplex filter). It was then returned to the bottom spout along with the fresh feed line going to the fractionator's bottom spout and also partly at the upper feed location. The unit is reportedly operating fine and achieved the objectives of the revamp (Fig. 10.6.1).

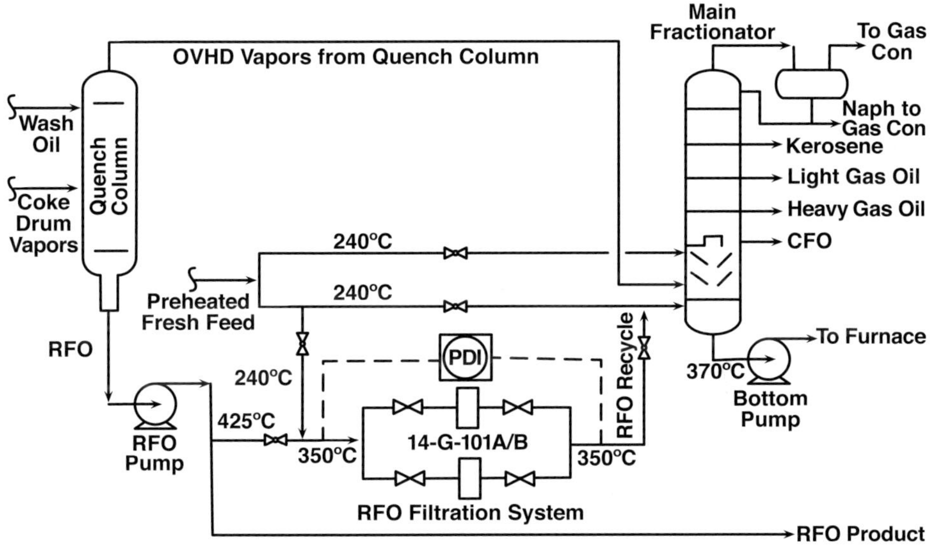

Fig.10.6.1 Modified configuration with quench column bottom fed to main fractionator

This system creates the effect of a combination tower to this two-tower setup, and makes it possible to operate with a high wash rate in the quench column to provide a longer run length of the quench tower. Whenever the RFO recirculation is discontinued, the fresh feed will keep flowing through the line and avoid congealing, coking, etc.

The RFO jump-over line to the fresh feed line should be the shortest and heat-traced, so that it can also be connected to the existing run-down line and maintain a small flow to the run-down to avoid coking and congealing due to idling of the line. A flash calculation for the coke drum reveals that the heavier fraction of the RFO does not vaporize in the condition of the drum and is likely to coke and crack.

This further provides the flexibility to recycle the heaviest stock to the coker furnace. The main fractionator in a single-tower operation usually has a coke filtration circuit at the bottom for the continuous removal of coke. The RFO recycle filtration and simultaneous routing of part of the recycle to the run-down will almost enable this function of coke clean-up. The yield of products before and after revamp is given in Table 10.6.2.

Table 10.6.2 Yield of products showing reduction of RFO

Stage no.	Products	Yield wt% (without RFO recycle)	Yield wt% (with RFO recycle)
1	Gas	5.0	5.5
2	LPG	3.3	3.4
3	Coker naphtha	6.2	6.7
4	Coker kero	22.0	23.0
5	Coker LGO	12.9	12.8
6	Coker HGO	12.5	12.4
7	CFO	13.4	13.9
8	RFO	10.7	7.8
9	Coke	14.0	14.5

Some other case studies of coker units are presented in the chapter on trouble-shooting (Chapter 15).

Metallurgy

The coke drum is subjected to rapid temperature changes and usually exposed to sulfur-bearing hydrocarbon at high temperatures. Thus special metallurgy is used in coke drums and associated lines and equipments.

Typical metallurgy of coke drum is 1.25 chromium (Cr) and 0.5 molybdenum (Mo) with 3 mm SS 410 lining inside. Special welding procedures are followed during fabrication, such as post-welding heat treatment, more longitudinal joints, etc. The coke drum inlet line metallurgy is usually 9 Cr–1 Mo with a corrosion allowance (CA) of 3 mm. The drum outlet line metallurgy is usually 5 Cr–0.5 Mo with a CA of 3 mm. The coker furnace metallurgy is usually 9 Cr–1 Mo with a CA of 3 mm for the radiation tubes and 11/4 (1.25) Cr–0.5 Mo with a CA of 3 mm for the convection tubes.

Abbreviations

BDS: blowdown scrubber
CCR: Conradson carbon residue
CFO: coker fuel oil
CLO: clarified oil
HCGO: heavy coker gas oil
LCGO: light coker gas oil
NBP: normal boiling point
RCO: reduced crude oil
RFCCU: residue fluid catalytic cracker unit
RFO: residual fuel oil
SRVGO: straight-run vacuum gas oil
TMT: tube metal temperature
VCM: volatile combustible material
VR: vacuum residue
XHCGO: extra heavy coker gas oil

References

1. M. M. Khan, M. K. Sharma, and Ashis Nag, "Unique Approach in Coker Design for Minimization of RFO" (paper presented at the International Refining and Petrochemical Conference in Rome, Italy, June 2010).

Fluid Catalytic Cracking Units and Residue Fluid Catalytic Cracking Units

11

11.1 Introduction

Fluid catalytic cracking units (FCCUs) and residue fluid catalytic cracking units (RFCCUs) are secondary units used by almost all refineries for converting residues to increase the bottom of the barrel. Distillate FCCUs use raw vacuum gas oil (VGO) or hydrotreated VGO, which is converted to LPG, gasoline, and diesel. Some unconverted clarified oil (CLO) or decanted oil (DO) is also produced along with coke on the catalyst. These units have feeds with a relatively low Conradson carbon residue (CCR): the typical CCR is maximum of around 0.5%.

The feed to the residue fluid catalytic cracking units (RFCCUs) usually has a high CCR (typically up to 6%) and may be constituted with

1. Reduced crude oil (RCO)
2. Hydrotreated RCO, hydrotreated vacuum residue (VR), or deasphalted oil (DAO), etc.
3. Mix of VGO and RCO, VR, or DAO

The FCCU typically consists of three different sections:

1. Reactor–regenerator section (R–R)
2. Fractionation section and feed preheat section
3. Gas concentration unit (GCU)

Petrochemical FCCU/RFCCU

In recent years, FCCU units have also been used for the production of propylene along with the other distillates as normally produced. These units use hydrotreated VGO or hydrotreated DAO as feed and operate with very high catalyst:oil ratios (8–10:1). The units produce higher quantities of LPG (around 30%–35% on feed) with higher propylene content than in conventional units. The propylene content in the LPG increases to around 35%–40% as compared to around 25% in conventional units. Typically these FCCUs operate with a very low CCR feed of around 0.1% (mostly with hydrotreated feed) containing very little metal to sustain high catalyst activity and operate with a very high reactor temperature of 550°C–570°C.

Process description

In the FCCU, the preheated feed is mixed with very hot flowing (fluidized) catalyst in the riser. The feed is heated to the reaction temperature of around 500–510°C by the hot catalyst. The reaction occurs in the riser with the stipulated residence time. The heavier feed is converted to lighter fractions by the effect of the catalyst at a high temperature. The catalyst particles are then separated from the hot product vapors in termination devices and cyclones. The vapor product is routed to the fractionator for further separation into various fractions.

The catalyst particles separated from the hydrocarbon vapors are routed to the stripper housed in the reactor to strip them free of hydrocarbons. The coke is deposited on the catalyst particles during the reaction and the activity of the catalyst declines. The fluidized catalyst particles are transferred to the regenerator through the spent catalyst standpipe (SCSP), normally by the head of the reactor stripper. A slide valve is provided in the spent catalyst standpipe to control the flow of the spent catalyst to the regenerator. Normally the stripper level controls the flow of spent catalyst by manipulating the spent catalyst slide valve (SCSV) opening. This also serves to isolate the oil side (reactor) and the air environment in the regenerator side.

In the regenerator, the coke on the hot spent catalyst is burned with air to CO and CO_2. Thus, the catalyst is regenerated. Air is supplied by an air blower through air distributors or grids located inside the regenerator. Heat is generated by burning the coke in the regenerator, which increases the regenerator catalyst temperature. The air rate is controlled to burn the carbon to the greatest extent possible without the excursion of regenerator dense bed and flue gas temperatures. Typically, a differential temperature is observed between the dense phase (catalyst bed) and the dilute phase in the regenerator. The differential temperature often controls the air rate to the regenerator.

The hot regenerated catalyst flows out of the regenerator through the regenerated catalyst standpipe (RCSP) to the wye or J bend (intersection of RCSP and the riser). The heat from the regenerated catalyst is used to heat the preheated feed to the reactor temperature and provide the endothermic heat of reaction for catalytic cracking. The regenerated catalyst slide valve (RCSV) is used to control the flow of regenerated catalyst to the riser and also to isolate the air side of the regenerator from the hydrocarbon side of the reactor. A differential pressure (typically 0.3 kg/cm^2) is always maintained across the slide valves to prevent the flow reversal.

The flue gas from the regenerator flows through the cyclones to separate catalyst particles from the flue gas. The separated catalyst particles are returned to the dense catalyst bed of the regenerator. The flue gas is let out of the regenerator typically through a double disc slide valve followed by an orifice chamber to lower the pressure. The flue gas containing CO and CO_2 is routed to the CO incinerator to convert the CO to CO_2. The very hot flue gas from the CO incinerator is cooled through the generation of high-pressure steam in the flue gas cooler and diverted to the stack with or without flue gas desulfurization (FGD).

The fractionation section consists of the main fractionator and side strippers used to fractionate the reactor effluent into different product fractions. The wet gas from the fractionator reflux drum is picked up by the wet gas compressor (WGC) and routed to the GCU along with the unstable naphtha from the fractionator in order to recover the LPG and stabilize the naphtha. The hot products and pump-around refluxes are used to preheat the feed and generate steam.

Reaction and catalyst in FCCU

The catalyst used is essentially silica and alumina. The catalyst used is typically a zeolite type. These have acidic functions that cause cracking reactions with the carbonium and carbeneium ion mechanism. Thus, the products have a higher distribution of isomeric hydrocarbons due to the reaction mechanism.

Coke and metals like nickel (Ni), vanadium (V), iron (Fe), sodium (Na), etc., deposit on the catalyst during operation. The coke is burned in the regenerator but the metals Ni, V, Fe, and Na remain in the regenerated catalyst (equilibrium catalyst).

A higher proportion of Ni in the catalyst promotes a dehydrogenation reaction and is reflected in the increased H_2:CH_4 ratio in the sponge gas. The typical H_2:CH_4 ratio mole:mole is 2.5:3.0 (0.7–0.85) in the sponge gas. The increased hydrogen content in the wet gas bled from the fractionator reflux drum demands a higher volumetric capacity of the WGC due to the low molecular weight of the wet gas. The lower molecular weight of the wet gas may result in lower discharge pressure of the WGC and may lead to the opening of the anti-surge valve and consequently result in flaring from the reflux drum. Ni passivators are used to reduce the dehydrogenation reaction, if required. The typical loading of Ni on the equilibrium catalyst is around 2,500–3,500 ppm. The total metal content normally found can be up to 10,000 ppm as (Ni + V) in the RFCCU equilibrium catalyst.

The V content ranges from 6,000–7,500 ppm. High V content in the catalyst may result in the production of vanadic acid at higher temperatures in the presence of moisture. A higher V content can be tolerated in a two-stage regenerator. The first stage operates at a relatively lower temperature, which decreases the likelihood of the formation of vanadic acid, even in the presence of water vapor. In the presence of air at higher temperatures, V converts to V_2O_5, which reacts with moisture (formed during the combustion of coke containing hydrogen) to result in the formation of vanadic acid. This destroys the matrix structure of the zeolite catalyst.

With an increased level of V in an equilibrium catalyst, vanadium traps are used as an additive in the catalyst. Typically they are used for V levels beyond 7,000 ppm. Equilibrium catalyst activity is estimated with a microactivity test (MAT) or tested by catalyst suppliers. A MAT activity of 65%–70% is typical of an equilibrium catalyst. The MAT of the equilibrium catalyst is increased by adding fresh catalyst. For this, a part of the equilibrium catalyst is normally withdrawn from the regenerator to the hopper and the fresh catalyst is introduced. This improves the MAT activity, controls the metal loading, and helps maintain the average particle size (APS). The proprietary additive ZSM-5 is mostly added to the catalyst to enhance LPG and propylene production and is maintained around 10%–15% weight on the catalyst.

Desired catalyst size distribution

The particle size distribution of the catalyst plays an important role in its performance. Catalyst particles less than 40 micron need to be around 8%–10% minimum in the equilibrium catalyst for good fluidization. The desired APS may be around 70–80 microns for better fluidization and FCCU operation. After continued operation, the minus-40 micron proportion reduces due to carryover from the reactor and regenerator. Fresh catalyst with a particle size distribution of around 20% less-than-40-micron particles is added to compensate for the loss.

Fines produced during attrition are lost through carryover from the cyclones. The fines normally escape from the system via reactor vapor and regenerator flue gas. The loss from the reactor ends up in decanted oil (DO) or clarified oil (CLO). The loss from the regenerator finds its way to the atmosphere, the flue gas cooler, or gets into the solid cakes formed in the flue gas desulfurization unit (FGDU).

Reactor

The reaction occurs in the riser pipe with a defined residence time. The reacted vapor along with the spent catalyst is fed to disengaging devices to separate the hydrocarbon vapors from the catalyst particles. The cracked hydrocarbon vapor is routed to the fractionator and catalyst particles are directed to the reactor stripper. The disengaging devices used are essentially cyclones mounted on the riser top or placed separately in the reactor enclosure. Often a riser termination device is also provided along with the cyclones. The spent catalyst is stripped of hydrocarbons at the stripper located at the bottom section of the reactor.

The feed is introduced through nozzles in the riser. The nozzles help in atomizing the feed to fine droplets for better contact with the catalyst. This in turn helps to reduce the amount of bigger liquid droplets coming in contact with the hot catalyst, which reduces the likelihood of coke formation during the reaction. The typical pressure drop desired for better feed nozzle performance is around 10 kg/cm^2. This requires a higher feed pump pressure compared to the older units that use a feed pipe as the riser feed device.

The atomization of the feed is also assisted by steam admitted into the feed nozzle along with feed. Feed nozzles are installed circumferentially, typically at an angle of 45° in the riser little (around 2 meters) above the wye (intersection of the RCSP and the riser). A minimum of four nozzles is common. Lift steam or lift gas vapor is sometimes also used at the bottom of the riser to pre-accelerate the catalyst before the hot catalyst meets the preheated feed at the wye. The pressure of the reactor is controlled by the fractionator overhead receiver or WGC suction pressure.

In the stripper section of the reactor, the oil from the catalyst is stripped with steam. Superheated steam is used in strippers as a stripping agent. A good stripping reduces the hydrocarbon (oil) content (which would add to the coke produced) in the catalyst going to the regenerator. Thus, better stripper performance reduces the regenerator combustion load. A higher ingress of oil from the stripper will be reflected in the higher H_2 percentage (calculated) in the coke. A typical hydrogen percentage is 4%–6% by weight in the coke.

The catalyst fines escaping from the reactor cyclones end up in the column bottom circuit and bleed through the DO. Catalyst carryover is reflected in the basic sediment and water (BS&W) of the DO. A typical BS&W value for DO is around 0.1–0.3 wt% before filter. In case of higher catalyst escape from reactor cyclones, the BS&W goes up. Filters are used on the DO to reduce the catalyst content in order to sell it as a product in isolation, in a blend with other products, or often for reprocessing in the coker unit.

Typically, reactor strippers are provided with 8–10 trays (counterflow devices, disc and doughnut trays, etc., are common). Improvement in the contacting device between catalyst and steam is a subject of continuous development. A good stripping of the catalysts decreases the hydrocarbon content of the spent catalyst going to the regenerator and results in less coke formation (strippable coke or occluded coke). Below are some of the different proprietary riser termination devices that are used in commercial units.

- Vortex separation system (VSS)
- Ram's horn
- Amoco liquid quench

Typically two stages of cyclones are used in a reactor with or without a riser termination device. A riser termination device followed by a single stage of cyclone is also common in a reactor. A minimum of three cyclones in each stage are common. More cyclones can be used in each stage for bigger units.

The volume of a reactor is essentially the volume required to house the cyclones and the stripper. An external cyclone in the reactor is not preferred as hot oil may spill outside in the event of a likely leak in the cyclones. Normally, the riser height governs the elevation of the reactor and also provides the static head for catalyst flow to the regenerator (normally at higher pressure) and the differential pressure across the SCSV. Typically the catalyst level and the elevation of the stripper provide the necessary pressure at the SCSV inlet.

Regenerators

Regenerator systems can have several different configurations of the regenerators.

1. Single regenerator
2. Two-stage regenerator
3. Regenerator with catalyst cooler
4. Two-stage regenerator with catalyst cooler

An air grid is used in the regenerator to blow air for combustion. The air grid has different configurations and designs (lateral grids, circular rings, etc.) in units supplied by different technology suppliers. The typical differential pressure maintained across the air grid is around 0.3 to 0.4 kg/cm^2. The spent catalyst enters the regenerator through a catalyst deflector to uniformly distribute the catalyst in the regenerator.

The air outlet nozzles in the distributor and proper distribution of air play a significant role in efficient coke burning. The air flows up through the bed of catalyst in the regenerator and ensures the burning of coke. A specific catalyst bed height must be maintained in the regenerator in order to ensure good contact between air and catalyst coke for efficient burning. Air channeling through the catalyst bed results in a higher oxygen percentage in the flue gas. This in turn results in afterburning, the excursion of the temperature differential between the dilute phase and the dense bed. A typical differential temperature is around 15°C–20°C. A higher differential temperature causes the operator to reduce air flow to the regenerator and as a consequence, the coke on regenerated catalyst (CRC) builds up and results in loss of catalyst activity.

A desired CRC is around 0.02 wt%. A CRC as low as 0.02 wt% is often difficult to achieve. Often CO promoters are used to control the afterburning by promoting the CO burning in the dense phase (bed) rather than in the dilute phase, which helps lower the CRC. The efficiency of carbon burning is often expressed as: *delta coke* = wt% carbon on spent catalyst – wt% CRC, or as (% coke yield)/(catalyst-to-oil ratio).

Normally two stages of cyclones are used in a regenerator, with a minimum of three cyclones in each stage. Higher numbers of cyclones are provided for bigger units to handle the higher volume of flue gas. The drip legs of the first-stage cyclones of the regenerator are normally dipped into the catalyst level so that air cannot flow up the drip legs and carry the catalyst fines to the vapor outlet. The drip legs of the second-stage cyclones are not dipped below the catalyst level and have trickle valves (as the second stage handles a relatively lower load of catalyst). The trickle valve opens after adequate catalyst accumulates in the drip leg and prevents air flow up through the drip legs. Often, the second-stage cyclones are placed outside the regenerator shell with drip legs extended to the regenerator catalyst bed, and the flue gas terminates in the plenum chamber before being routed out. External cyclones can be provided only in full-burn type regenerators as there will be no danger of a CO leak into the atmosphere, even if a crack/hole develops in the cyclones during operation.

The pressure of a regenerator is normally kept higher than that of the reactor. This is required for the pressure balance and circulation of the catalyst. The regenerator pressure is controlled by a double disc slide valve (DDSV) on the regenerator flue gas line. For single stage regenerators, the regenerator pressure is kept higher than reactor pressure by around 0.6–1.2 kg/cm^2. In case of the two-stage regenerator, the elevation of the second regenerator (RG2) becomes higher. Thus, it can provide adequate catalyst head at the upstream of the regenerated catalyst slide valve (RCSV). The pressure of the second regenerator can be lower than the reactor in this case, but the pressure of the first-stage regenerator (at lower elevation) is usually higher than the reactor. The flue gas from the regenerator passes through the DDSV and flows through the orifice chamber to reduce the pressure before it enters the CO incinerator or flue gas cooler.

For regenerators with two stages, two orifice chambers are common (one for each). Often for large FCCU units (more often with a single regenerator), pressure recovery turbines (PRTs) are used across the orifice chamber for decreasing the flue gas pressure. The PRT delivers part or full power requirement for the main air blower drive. A third-stage separator (TSS) is installed upstream to eliminate particles above 20 microns and reduce the concentration of particles to typically less than 50 mg/m^3 (with an average particle size of 10 microns) in flue gas going to the PRT.

Often an FGDU is placed in the regenerator flue gas line at the outlet of the flue gas cooler (steam generator) to arrest the catalyst fines and neutralize the SO_2 and CO_2 formed from the combustion of coke. This reduces particulates and SO_2 emissions as a step toward environment control.

A typical reactor–regenerator (R–R) system is shown in Fig. 11.1.1 with a single-stage regenerator.

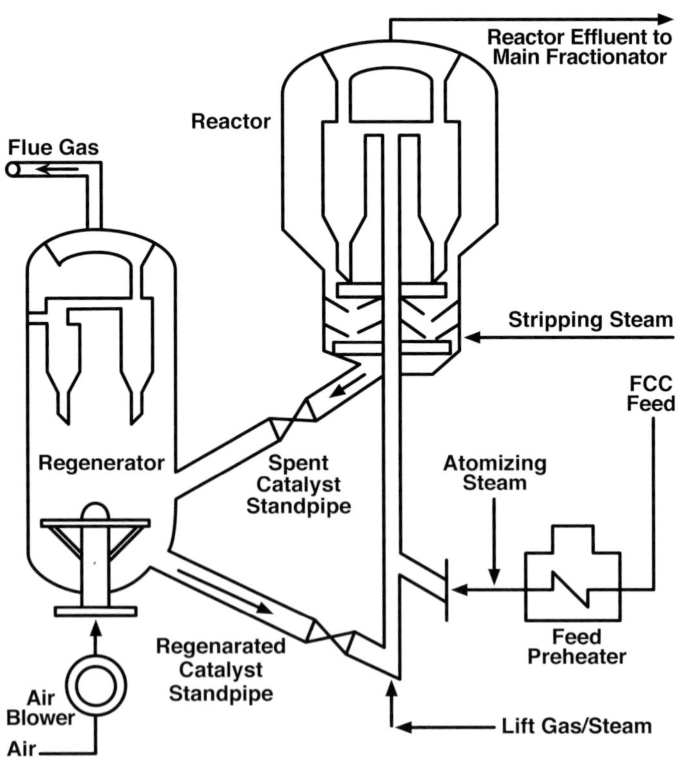

Fig. 11.1.1 Typical R–R system

Two-stage regeneration

In single-stage regenerators, the combustion air is controlled to minimize the afterburning to keep the dilute phase temperature from going beyond the allowable limits. Thus, decreasing the CRC content to a very low level may require CO promoters. Lower CRC helps maintain a higher equilibrium catalyst activity.

The coke on the spent catalyst is related to the CCR of the feed, the reaction severity in the reactor, and the metal content of the catalyst. If the feed's CCR or reaction severity is high, then the coke deposit on the spent catalyst is also increased.

The burning of the coke to CO and CO_2 releases a very large quantity of heat. The heat from the formation of CO_2 is 94.05 kcal/gmole and that of CO is 26.4 kcal/gmole at 25°C. The heat of combustion of C to CO (26.4 kcal/gmole) and heat of combustion of CO to CO_2 (67.63 kcal/gmole) is much higher (almost double). In a two-stage regenerator system, the coke is partially burned to CO by regulating the air flow to the first regenerator and then diverted to the CO incinerator to release the heat of CO combustion to CO_2 in the CO incinerator instead of the regenerator. The flue gas with around 7%–9% volume of CO flows out of the first regenerator through the DDSV and orifice chamber and is admitted to the CO incinerator furnace, where it is burned completely to CO_2 with the excess air supplied by a forced draft (FD) fan. The hot flue gases are routed to the flue gas coolers to generate normally high-pressure steam and cool the flue gas. The partially regenerated catalyst from the first-stage regenerator is transferred to the second regenerator by pressure difference through a lift line and plug valve. The plug valve, located at the bottom of the lift line, regulates the flow of catalyst from RG1 to RG2. Air is supplied in the RG2 to burn the balance of coke on the catalyst. The coke on the catalyst is burned completely to CO_2 in the second regenerator. As the catalyst in the second regenerator contains less coke, the heat liberation is manageable. Thus the overall heat generation in a two-stage regenerator system is lower. The heat balance of a two-stage regenerator in subsequent pages further clarifies this point. Typically, two-thirds of the coke on the catalyst is burned in the first regenerator and one-third in the second regenerator. The CRC content of the regenerated catalyst may often be lower than would be possible in a single-stage regenerator. Moisture in the regenerator comes from the combustion of the hydrogen from oil carried along with spent catalyst, burning of the hydrogen in coke to water vapor, and the carryover of steam from the reactor stripper, along with spent catalyst. In the presence of moisture and oxygen in a single-stage regeneration, the V can combine with oxygen to produce V_2O_5 and eventually react with moisture to produce vanadic acid. This in turn can attack the alumina of the zeolite structure and causes the collapse of the catalyst structure. This deactivates the catalyst. This phenomenon is more pronounced in an RFCCU where metal loading is high in the equilibrium catalyst. Thus, for an FCCU with a higher feed CCR, the two-stage regenerator system is often considered an advantage, as very little V_2O_5 is produced in the first stage, as it has an oxygen-deficient environment and operates at a relatively lower temperature (650°C–670°C). The partly regenerated catalyst from the first regenerator is further regenerated in full burn mode in the second stage, which has a lower moisture content, minimizing the likelihood of the formation of vanadic acid.

Regenerator with catalyst cooler

Often catalyst coolers are used to maintain the regenerator temperature. This system circulates hot catalyst through steam generator tubes that reject the excess heat for producing high-pressure steam. This helps keep the regenerator temperature within prescribed limits, particularly in cases where the feed has a CCR of 6% or higher.

11.2 Heat Balance across Regenerator

The heat in the R–R section is supplied by the burning of coke in the regenerator. The heat from the combustion of the coke primarily performs following:

1. Heats the spent catalyst from the reactor temperature to the regenerator temperature (after regeneration, the hot catalyst flows out to heat the preheated feed to the reactor temperature and provides heat for the endothermic cracking reaction).
2. Heats the air used for combustion to the flue gas temperature
3. Heats the coke from the reactor temperature to the regenerator temperature
4. Provides the heat loss to the surroundings

Inputs required

1. Regenerator flue gas compositions on dry basis (Orsat analysis)
2. Air flow rate to regenerators and air temperature
3. Catalyst temperature from reactor and first regenerator
4. Flue gas temperature leaving the regenerators

5. Heat of formation of CO_2, SO_2, H_2O, and CO
6. Specific heat, (cP in kcal/kg°C) of the catalyst, coke and flue gas at corresponding temperatures

Steps involved

1. Calculate average molecular weight of the dry flue gas
2. Calculate water formed during coke burning and calculate wet flue gas composition through stoichiometric calculations
3. Calculate the average molecular weight of the wet flue gas and wet flue gas quantity corresponding to 100 t-moles of dry flue gas (by adding the water formed with the dry flue gas quantity)
4. Calculate actual rate of wet flue gas from actual rate of air using the nitrogen balance
5. Calculate coke burned in the regenerator
6. Calculate heat liberation per mole of flue gas from the heat of formation of CO_2, CO, H_2O
7. Calculate total heat liberation based on actual wet flue gas flow
8. Perform heat balance around regenerator as below to calculate catalyst circulation rate

Envelopes have been considered for the heat balance for RG1 and RG2 (Fig. 11.2.1).

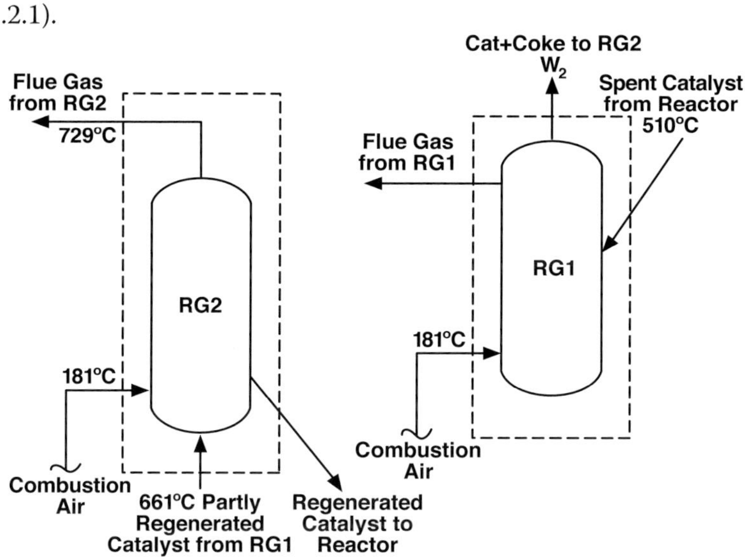

Fig. 11.2.1 Envelopes for heat balance in a two-stage regenerator

Heat balance across the envelope of RG2 is performed below:

Heat input = heat of spent catalyst + heat of air for combustion + sensible heat of coke along with catalyst + heat from combustion of coke (heat liberation)

Heat output = heat of regenerated catalyst flowing out + heat of flue gas leaving regenerator + heat loss to surroundings

A sample calculation has been presented for an FCCU with a fresh feed rate of 1.7 MMTPA (212.5 MT/hr). The unit has a two-stage regeneration facility. The heat balances of RG1 and RG2 are presented as samples.

Input data

RG2 flue gas analysis (dry basis) given in Table 11.2.1.

Table 11.2.1 RG2 dry flue gas composition

Components	Mole %	Mol. wt
Nitrogen	80.70	28
CO_2	16.60	44
CO	0.00	28
O_2	2.58	32
SO_2	0.12	64
Total	100.00	

Air flow to RG2 is 51.8 kNm^3/hr. Air temperature in RG2 is 181°C. Flue gas temperature at the outlet of RG2 is 729°C. Hot catalyst temperature from RG1 is 661°C. The average molecular weight of dry flue gas is calculated as ((80.7 × 28) + (16.6 × 44) + (0 × 28) + (2.58 × 32) + (0.12 × 64))/100, which is equal to 30.8.

Calculation of wet flue gas quantity, mol. weight, composition

Basis: 100 t-mole/hr (100,000 kg-moles/hr) of dry flue gas.

Oxygen supplied with air is (80.7 × 21)/79 =21.45 t-mole/hr.

Oxygen consumed for CO_2 is 16.6 t-mole/hr.

Oxygen consumed for CO is 0 t-mole/hr.

Oxygen consumed for SO_2 is 0.12 t-mole/hr.

Oxygen left out is 2.58 t-mole/hr.

Oxygen consumed for H_2O is 21.45 – (16.6 + 0 + 0.12 + 2.58) = 2.15 t-mole/hr.

H_2O formed is 2 × 2.15 = 4.3 t-mole/hr.

Wet flue gas quantity is 100 + 4.3= 104.3 t-mole/hr.

Wet flue gas composition calculated is given in Table 11.2.2.

The average molecular weight of wet flue gas is 30.27.

Actual air supplied is 51.8 kNm^3/hr, or (51.8 × 29)/22.4 = 67 MT/hr.

N_2 supplied along with air is 67 × 0.77 = 51.6 MT/hr.

N_2 percentage in air by weight is 77%.

Table 11.2.2 RG2 wet flue gas composition

Components	Mole %	Mol. wt
Nitrogen	(80.7/104.3) × 100 = **77.37**	28
CO_2	(16.6/104.3) × 100 = **15.92**	44
CO	0 = **0**	28
O_2	(2.58/104.3) × 100 = **2.47**	32
SO_2	(0.12/104.3) × 100 = **0.11**	64
H_2O	(4.3/104.3) × 100 = **4.12**	18
Total	**100**	

By a nitrogen balance: X × 0.7737 × 28 = 51.6, where X is the wet flue gas quantity in t-moles/hr.

Solving, X becomes equal to 51.6/(28 × 0.7737) t-moles/hr.

Thus, wet flue gas generated is (51.6 × 30.27)/(28 × 0.7737) = 72.1 MT/hr.

Carbon burned in RG2 is ((72.1/30.27) × (15.92 + 0))/100 = 0.379 t-moles/hr = 0.379 × 12 = 4.55 MT/hr.

H_2 content in the coke burned is ((72.1/30.27) × 4.12)/100 = 0.098 t-moles/hr, or 0.098 × 2 = 0.196 MT/hr.

Total coke burned (hydrogen of coke + carbon of coke) is 4.55 + 0.196 or 4.75 MT/hr. The heat of formation of the constituents in flue gas is given in Table 11.2.3.

Table 11.2.3 Heat of formation at 25°C

Components	Heat of formation, kcal/gmole
H_2O	57.8
CO_2	94.05
CO	26.4
SO_2	71

Heat liberation for 100 t-mole/hr of wet flue gas is (15.92 × 94.05) + (0 × 26.4) + (0.11 × 71) + (4.12 × 57.8) = 1,743.2 Gcal/hr.

Total heat liberation in RG2 is (1,743.2 × 72.1)/(100 × 30.27) = 41.52 Gcal/hr.

Heat balance across the RG2 is given in the following equation with 25°C as the datum temperature as the heat of formations of the components is considered at 25°C.

Equation 11.2.1

$$W2\ (661 - 25) \times 0.4 + (67 \times (0.3(181 - 25))) + (4.51(661 - 25)) \times (0.3 + 41{,}520)$$
$$= W2\ (729 - 25) \times 0.4 + (72.1(729 - 25) \times 0.3)$$

Solving for W2, W2 = 1,187.4 MT/hr, where W2 is the catalyst circulation rate in metric tons per hour (MT/hr) (neglecting the coke quantity on the catalyst).

Catalyst specific heat assumed is 0.4 kcal/kg°C at the average temperature of the catalyst. Flue gas specific heat assumed is 0.3 kcal/kg°C at the temperature of the flue gas.

Calculation for RG1

Inputs required for the calculation are presented in Table 11.2.4.

Table 11.2.4 RG1 dry flue gas composition

Components	Mole %	Molecular. wt
Nitrogen	79.70	28
CO_2	13.20	44
CO	7.08	28
O_2	0.02	32
SO_2	0.00	64
Total	100.00	

Wet flue gas composition of RG1 is given in Table 11.2.5

Table 11.2.5 RG1 wet flue gas composition

Wet fluegas	Mole %
SO_2	0.00
N_2	73.22
CO_2	12.13
CO	6.50
O_2	0.02
H_20	8.13
Total	100.00

Air flow to RG1 = 93.54 kNm^3/hr.

Air temperature to RG1 = 181°C.

Flue gas temperature at the outlet of RG1 = 661°C.

Hot catalyst temperature from reactor = 510°C.

A similar calculation is performed for RG1 where the spent catalyst enters at the reactor temperature of 510°C to estimate the catalyst circulation rate.

The average molecular weight of RG1 wet flue gas is 29.13.

The air supplied to RG1 is 93.54 kNm^3/hr.

Nitrogen supplied with RG1 air is (93.54/22.4) × 0.79 × 28 = 92.3707 MT.

From the nitrogen balance the wet flue gas quantity is determined:

Wet flue gas quantity is equal to (nitrogen supplied/28) × (average mol. wt of wet flue gas)/(percentage nitrogen in wet flue gas/100), or (92.37/28) × (29.13/0.7322) = 131.24 MT. The calculation does not consider the humidity of the air and the heat loss, which is estimated as 2% of the heat liberated in each regenerator.

A similar calculation provides the amount of carbon burned in RG1 as 10.3 MT/ hr, the hydrogen burned as 0.474 MT/hr, and the total coke burned as 10.3 + 0.474 = 10.77 or 10.8 MT/hr.

Some other items of heat, such as carbon desorption, are not considered for the sake of simplicity. The catalyst circulation rate calculation is very sensitive to the specific heat of catalyst and flue gas.

The summary of results is shown below in Table 11.2.6.

Table 11.2.6 Summary of results for coke burned and catalyst circulation

Regenerator	Coke burnt, MT/hr	Heat release in coke burning, gcal/hr	Catalyst circulation, MT/hr
RG1	10.8	80.3	1,156.0
RG2	4.72	41.52	1,187.4
Total	15.52		1,172 (avg.)

Total coke burned is (15.55/212.5) × 100 = 7.3 %wt of feed.

Carbon in coke is 10.30 + 4.55 = 14.85 MT/hr (95.7% by weight).

H_2 in coke is 0.67 MT/hr. % H_2 in coke is 4.3 wt%.

Calculated catalyst–oil ratio is 1,172/212.5 or 5.5.

Delta coke = wt% carbon on spent catalyst – wt% CRC, or as (coke yield wt% on feed)/(catalyst:oil ratio) = 7.3/5.5 = 1.33. The results obtained are summarized in Table 11.2.7.

Table 11.2.7 Summary of the results for coke composition and other important parameters

Parameter	Value
%wt of coke on feed (WT)	7.3%
%wt of carbon in coke (WT)	95.7%
%wt of hydrogen in coke (WT)	4.3%
Cat-oil ratio (WT/WT)	5.5
Delta coke % (WT)	1.33%

11.3 Concepts Used for Design

For a grassroots FCCU design, coke production in the reactor is typically estimated using a MAT, FST, or pilot plant studies with different catalyst-to-oil ratios and temperatures. The product yield along with the coke factor for that severity can be obtained from the MAT (ASTM D-3907). A MAT can deliver the reactor effluent, coke factor, hydrogen factor, etc. A pilot plant study conducted at likely cat-oil ratio and reactor temperature can estimate the product yields, properties, and coke production to a reliable extent to be used for commercial applications. A pilot plant study is required mostly for the estimation of properties of the products.

Alternatively, the coke factor obtained from MAT can be increased by 5%–6% on its weight, or coke production can be obtained from pilot plant analysis for commercial plant applications. Later, based on the regenerator configuration used, the quantity of coke, preferably estimated from pilot plant study, is burned either in a single-stage regenerator (partial combustion) or in a two-stage regenerator (full combustion). The air quantity required may be determined with the typical composition of flue gas from the regenerator required to burn 99.97% of the total coke produced on the spent catalyst (the CRC is assumed to be 0.03 wt%). In the case of a two-stage regenerator, the air quantity to each is calculated with the assumption that typically two-thirds of the total coke burning occurs in the first stage, and one-third in the second stage with the likely flue gas composition of the stages. When determining the actual air quantity, the air quantity to each regenerator is assumed to be in excess of the calculated requirement by around 1.5%–2% by volume.

The heat liberation in each regenerator is calculated from the flue gas quantity and composition and standard heat of formations as in section 11.2. For working out the heat balance for a two-stage regenerator, the amount of coke burned in each regenerator (distribution of coke burning) is assumed and the air quantities required in each regenerator are calculated. A heat balance is then performed to calculate the catalyst circulation rate and regenerator temperatures in a way similar to the one mentioned above in section 11.2.

The standpipes are sized based on the catalyst circulation rates and on the permitted catalyst flux in $kg/sec/m^2$ discussed later in this chapter.

11.4 Pressure Balance in the Catalyst Circulation System

A pressure balance in the R–R deals with the hydraulics of the catalyst circulation system which ensure the smooth flow of catalyst. The additional capacity of the unit comes from the increased catalyst circulation or from altering the pressure balance in the new set. The pressure measurement along the catalyst flow path provides a very good indication of the pressures required for a smooth catalyst circulation.

The objective of the pressure balance in the R–R is to identify problems in catalyst circulation, maximize and ensure steady catalyst circulation, and maximize the available pressure drop across slide valves. The reactor and regenerator pressures are often slightly altered to obtain adequate differential pressure across the slide valves. Optimum pressure in the reactor and regenerator also ensures optimum blower and WGC loads.

The regenerator pressure is typically maintained by manipulating the DDSV at the flue gas outlet. The RCSP height provides the driving force for transferring the catalyst from the regenerator to the reactor at the wye section, where the pressure is higher than the reactor top pressure. Longer standpipes may require external aeration to compensate for the volume lost during head buildup. Low pressure at the upstream of slide valves may be caused by under- or over-aeration. A pressure survey in the running unit for the catalyst standpipes needs to be performed by a process engineer to ensure optimum catalyst circulation.

This is important historical data that needs to be developed and used to ascertain the performance of the catalyst circulation system at any point in time. This involves the measurement of static pressure along the standpipes, particularly at the inlet and outlet of the SCSVs and RCSVs. Theoretically, the SCSV upstream pressure should be equal to the reactor overhead pressure + pressure due to the height of the dilute phase in the reactor + pressure due to the height of the catalyst in the reactor stripper – frictional losses in the standpipe due to the flow of catalyst.

Normally, the frictional loss due to catalyst flow is negligible at the design circulation rate. If the pressure at the upstream of a slide valve drops, it is usually caused by the following:

1. Restriction in the standpipe resulting in a higher valve opening and a decrease of differential pressure (DP) across the slide valve
2. Low catalyst bulk density
3. Fewer proportions of 0–40 micron catalyst particles leading to defluidization

The pressure survey in the R–R section is shown below in Fig. 11.4.1.

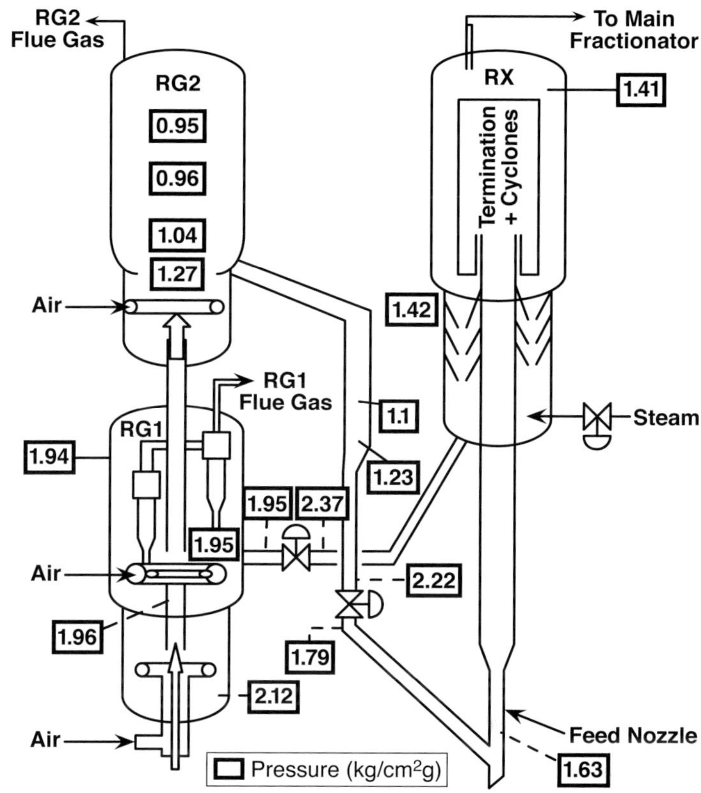

Fig. 11.4.1 RFCC unit typical R–R pressure survey

RFCC unit typical R–R pressure balance

Reactor side

Using the reactor dilute phase as the starting point, the pressure head is calculated as follows:

The reactor or stripper vessel dilute phase pressure is 1.41 kg/cm^2, and the pressure head of the dilute phase is 0.01 kg/cm^2 (calculated from density and elevation); therefore, pressure at the top of the stripper bed is 1.41+ 0.01, or 1.42 kg/cm^2.

Elevation from the reactor stripper bed to the head of the stripper outlet nozzle is 9.9 m (as measured), and the corresponding static head is 9.9 × 634 (assumed density), or 0.63 kg/cm^2; therefore, pressure above the SCSP is 1.42 + 0.63, or 2.05 kg/cm^2.

Pressure buildup in a spent catalyst standpipe with an elevation of 5.25 m is 5.25 × 634 = 0.32 kg/cm^2. Pressure above the SCSV should be 1.42 + 0.63 + 0.32 = 2.37 kg/cm^2. The actual pressure at the upstream of the SCSV can be measured to correlate with the pressure available using the following calculation: SCSV delta pressure is 0.42 kg/cm^2. So the pressure below the SCSV is 2.37 – 0.42 = 1.95 kg/cm^2.

Regenerator side

The first regenerator (RG1) dilute phase pressure measured is 1.94 kg/cm^2, the elevation from the RG1 bed to the withdrawal point (plug valve) is 3.703 m, and the catalyst density is assumed to be 459 kg/m^3; thus, the differential pressure available at the plug valve upstream (withdrawal point) is 3.703 × 459, or 0.17 kg/cm^2. The following are the calculations for evaluating pressures at other critical locations in the regenerator and regenerator standpipe:

Pressure outside catalyst plug valve = 1.94 + 0.17 = 2.12 kg/cm^2

Catalyst plug valve delta P = 0.43 kg/cm^2

Pressure inside lift line (RG1 to RG2) at bottom= 2.12 – 0.43 = 1.69 kg/cm^2

Lift line delta P including distributor = 0.42 kg/cm^2

Pressure in RG2 bed outside lift line slots = 1.69 – 0.42 = 1.27 kg/cm^2

Head corresponding to RG2 bed elevation of 6.11 m = 0.31 kg/cm^2

Pressure above RG2 dense bed = 1.27 – 0.31 = 0.96 kg/cm^2

The pressure above RG2 dilute phase = 0.96 – 0.01 = 0.95 kg/cm^2

RCSP head = 23 m × 545 kg/m^3 = 1.26 kg/cm^2

Pressure above RCSV is 0.96 + 1.26 = 2.22 kg/cm^2

RCSV delta P = 0.43 kg/cm^2

Pressure at the reverse seal to fresh feed injector head = 0.16 kg/cm^2

Available pressure at fresh feed injection point in riser
= 2.22 – 0.43 – 0.16 = 1.63 kg/cm^2

The actual measured pressure matches the calculated riser bottom pressure at the feed injection point, which shows that the pressure is balanced and concludes that there are no abnormalities in the catalyst circulation and that adequate pressure is available to overcome the head loss due to flow in the riser. A pressure survey is carried out at a calculated catalyst circulation rate. The survey results can be developed for different catalyst circulation rates and can be preserved as important historical data to assess the performance of catalyst circulation at a different point in time.

Typical catalyst densities obtained from online density meters:

For reactor or stripper bed: 634 kg/m^3

For SCSP: 609 kg/m^3

For firs-stage regenerator bed (upper/lower section): 459/550 kg/m^3

For second-stage regenerator bed (upper/lower section): 431/507 kg/m^3

11.5 Typical Design Data for the Catalyst Standpipes and Slide Valves

Regenerated catalyst standpipes and slide valves

1. Typical standpipe catalyst flux considered for design is 100–200 $lb/sec/ft^2$ (around 540 $kg/sec/m^2$).
2. Desired bulk density in the RCSP is around 40–45 lb/ft^3 (around 550–600 kg/m^3).
3. Typical RCSV pressure drop is 3–6 psi.
4. At design catalyst circulation rate, the RCSV opening is around 35%–55%.
5. Standpipe aeration points are typically provided every 5–8 ft.
6. Slide valve upstreams and downstreams are provided aeration points (instrument purge points) to measure upstream and downstream pressures across the slide valve. The upstream is typically purged with air and the downstream with dry steam with a restricted orifice in the purge lines.
7. The RCSP is normally longer and provided with supplemental aeration points.
8. The pressure drops in the catalyst standpipes are calculated using Particulate Solid Research, Inc. (PSRI) correlations.

Spent catalyst standpipe and slide valve

1. Typical standpipe catalyst flux design is 100–250 $lb/sec/ft^2$ or (540 $kg/sec/m^2$).
2. Typical bulk density in the SCSP is around 35–40 lb/ft^3.
3. The typical pressure drop in SCSV designs is 3–6 psi.

4. At design catalyst circulation rate, the SCSV opening is around 35%–55%.
5. Standpipe aeration points are typically provided every 5–8 ft. With short standpipes, supplemental aeration points are not required. Often a process engineer is required to calculate the density of the catalyst in the standpipes to adjust the aeration. Typically, they are not used in SCSPs.
6. Slide valve upstreams and downstreams are provided with aeration points (instrument purge points) to measure upstream and downstream pressure. The upstream is typically purged with dry steam and the downstream with air with a restricted orifice in the purge lines.

11.6 FCCU Fractionators

The feed to the FCCU fractionator is superheated reactor effluent vapor. The feed vapor is first cooled to saturation in the fractionator heat transfer trays located above the flash zone (Fig. 11.6.1). The column bottom spout has a coke trap. A steam connection is common at the bottom spout to reduce coke-fine accumulation at the bottom. The main column bottom (MCB) PA cools the superheated feed vapor going to the column fractionation section. The column bottom outlet to the bottom pump usually has duplex-type strainers. The MCB is cooled by preheating feed, generating steam, and heating the boiler feed water. Demineralized water (DMW) is injected at the upstream of the fractionator overhead air cooler to wash the deposition of salts and is again pumped from the fractionator reflux drum and injected into the WGC first-stage cooler inlet to the interstage drum. From the interstage drum bottom, it is pumped to the second stage at the inlet to the air cooler and led to the HP receiver and finally routed to the sour water stripper by the pressure of the reciever. This scheme of water injection is often referred as forward coupling and needs a pump for pumping the DMW to the overhead air cooler, from the fractionator reflux drum to the interstage drum, and from the interstage drum to the HP receiver. In another scheme (backward coupling), fresh boiler feed water (BFW) is injected through the pump at the HP receiver air cooler inlet, then recirculated back to the interstage cooler inlet by pressure difference. From the interstage receiver, it goes to the main fractionator overhead air cooler inlet (again by pressure difference) and is routed to the sour water strippers through the pump. Facility for injection of polysulfide solution is provided typically at the inlets of coolers of the WGC first-stage discharge and second-stage discharge.

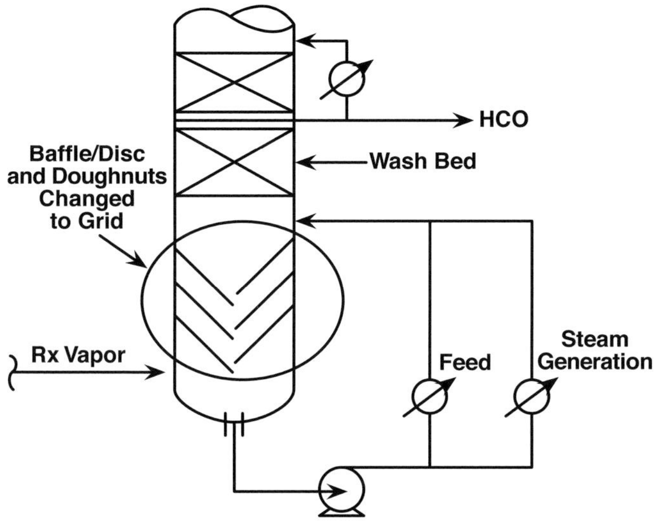

Fig. 11.6.1 Desuperheating zone in main fractionator

An FCCU typically produces the following products:

1. Gas
2. LPG
3. FCCU gasoline
4. Heavy naphtha (a heavy naphtha side cut with stripper)
5. Light cycle oil (LCO)
6. Heavy cycle oil (HCO), often recycled with feed
7. Clarified oil (CLO)
8. Coke on catalyst

For simulation purposes, the feed is constituted by blending the products and then fed into the fractionator for simulation after preheat (at reactor temperature).

A preheat furnace is often provided to further heat the preheated feed and can be also used to control the catalyst circulation. The units with heavier feeds (% CCRs of 4.0 and higher) generate more coke to provide adequate heat. These units need higher catalyst circulation to maintain the regenerator temperature for a specified reactor temperature. A preheat furnace is normally not required in these plants.

The side draws for heavy naphtha and LCO have side strippers. The column normally has three PA refluxes: the main column bottom PA (MCB PA), the HCO PA, and the LCO PA. At times a heavy naphtha PA is also provided to decrease the other PA refluxes if required. During simulation, the MCB PA is normally maximized by reducing the HCO PA and maintaining the HCO internal reflux (IR) reach the bottom.

The column bottom is recycled after cooling at the top of the heat transfer trays to cool the hot vapors as MCB PA. A very large amount of heat duty is removed by the MCB PA and is mostly used for preheating fresh feed and balance for steam generation.

In simulation, 14–16 stages (around 30 trays) are normally considered in the main fractionator. The desuperheating section is provided with two to three stages and uses baffle trays, disc and doughnut trays, or grids.

Simulation guidelines for fractionator and gas-con section

For the purposes of simulation, the reactor effluent is the starting point with its temperature and the simulation of the fractionator is performed like a normal petroleum distillation column with PA refluxes and product draws with strippers. Stripping steam is used in strippers. The differential temperatures (DTs) assumed for the PAs are given below:

DT MCB PA: around 150°C

DT HCO PA: around 90°C–100°C

DT LCO PA: around 55°C–60°C

Thermodynamic model GS can be used for the K value. Sour package is used ahead of the interstage condenser, reflux drum, and HP receiver to capture ammonium hydrosulfide (NH_4HS) in water and obtain the sour water composition from the compressor interstage drum, reflux drum, and HP receiver.

The gases from the fractionator are picked up by a two-stage compressor with an intercooler and vapor liquid recontactors. The gas-con section is configured as described in Chapter 7. The thermodynamic model normally used for the gas-con section for K values is SRK or PR.

Simulation guidelines for CO incineration and flue gas cooling

The flue gas generated from a single-stage regenerator or from the first stage regenerator of a two-stage configuration is simulated configuring the CO incinerator as a reactor. The combustion airflow is estimated based on oxygen kg-moles theoretically required (in stoichiometric proportions) for the full conversion of CO in the RG1 flue gas to CO_2. The actual flow of oxygen is considered to be around 15% higher than the theoretical and the actual air rate is evaluated from the actual oxygen flow. The reactions in the CO incinerator are defined as reactions of oxygen with light hydrocarbons (components of fuel gas used as support fuel) producing CO_2 and H_2O during combustion and the conversion of the CO of the regenerator flue gas to CO_2. Thus, the quantity, composition, and temperature of flue gas at the outlet of the CO incinerator are calculated. In the case of two-stage regeneration, the flue gas from the second regenerator joins the flue gas from the CO incinerator and the combined flue gas is routed to the flue gas cooler. The flue gas is cooled in the flue gas cooler to around 200°C by generating steam. The calculated temperature of the flue gas outlet from the CO incinerator is 1,017°C, and after the flue gas from RG2 is combined, the temperature at the inlet of the flue gas cooler is 941°C. The simulation of the flue gas cooling section is shown in Fig. 11.6.2. The simulation (Fig. 11.6.2) indicated that some augmentation of the flue gas cooling section would be required in an actual capacity revamp study. The results of the simulation are presented in Table 11.6.1.

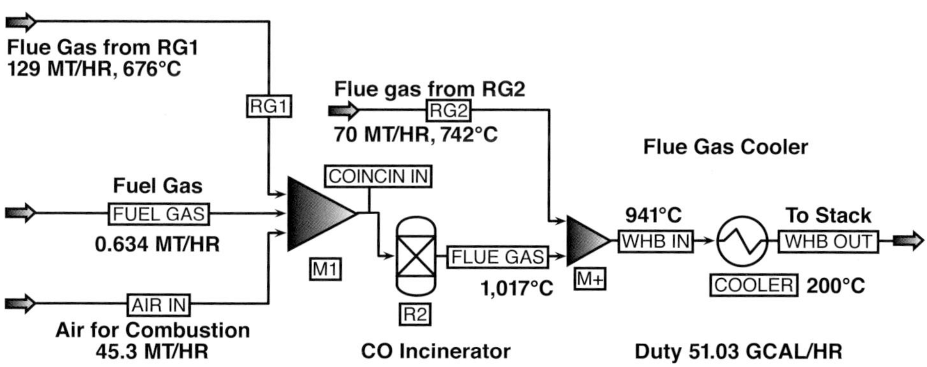

Fig. 11.6.2 Simulation of a CO incinerator and flue gas cooling of an actual plant with two-stage regeneration

Table 11.6.1 Results of the simulation of a CO incinerator and flue gas cooling

Stream name	RG-1	Fuel Gas	Air In	Flue Gas	RG2	WHB Out
Stream phase	Vapor	Vapor	Vapor	Vapor	Vapor	Vapor
Total mass rate, kg/hr	129,055.89	634.00	45,363.00	175,052.88	70,340.40	245,393.27
Total molar rate, kmol/hr	4,408.96	37.16	1,572.57	5,878.04	2,320.39	8,198.44
Temperature, °C	676.00	40.00	40.00	1,017.96	742.00	200.00
Pressure, kg/cm^2g	0.35	0.35	0.35	0.09	0.09	0.04
Total molar comp. percents						
Fuel Gas	0.00	100.00	0.00	0.18	0.00	0.13
Air	0.00	0.00	0.00	0.00	0.00	0.00
CO	6.54	0.00	0.00	0.00	0.00	0.00
CO_2	12.20	0.00	0.00	14.63	15.95	15.00
SO_2	0.16	0.00	0.00	0.12	0.13	0.12
O_2	0.04	0.00	20.90	2.08	2.44	2.18
N_2	73.52	0.00	79.10	76.31	77.55	76.66
H_2O	7.54	0.00	0.00	6.68	3.94	5.90

11.7 Concepts of FCCU Revamps

Revamping an FCCU involves revamping the following:

1. The R–R section (with a CO boiler and a flue gas cooling section)
2. Fractionators and GCU section
3. Feed preheat section

For marginal revamps, the R–R section is often not revamped with the cushion available in the riser, catalyst standpipes, terminating devices, cyclones, air blower, etc.

Often the regenerator pressure is increased marginally to keep the superficial velocity in the regenerator within 0.8–1.0 m/sec if the superficial velocity is the limiting factor. The fractionator pressure also needs to be increased by the same amount to maintain the differential pressure between the reactor and regenerator. The increase in reactor pressure also reduces the specific volume of the cracked vapors at the reactor and reduces the actual volumetric flow through the reactor. Thus, an increase in R–R pressure provides a quick solution for debottlenecking the R–R at a higher throughput.

Often, only the slides valves are changed, retaining the old catalyst standpipes to allow higher catalyst flow. Often, riser and reactor termination devices and cyclones are modified to handle a higher flow of vapor + catalyst. Typical loading (velocity) of cyclones is maintained at around 20 m/sec at the cyclone inlet. Often, the regenerator cyclones are also revamped along with the air distributors of the regenerator. To reduce the ingress of oil along with spent catalyst from the reactor to the regenerator, stripper trays are increased, or more efficient versions of stripper internals are used. This reduces the formation of coke, which in turn helps in lowering regenerator superficial velocity.

Feed nozzles are an effective and popular device for improving the atomization of feed and reducing coke lay-down on the catalyst. Often feed nozzles are changed to improved versions for better atomization and lower coke lay down. Often lift gas is used instead of steam or in conjunction with lift steam. The gas, having a higher molecular weight than steam, acts as a means for reducing the specific volume of vapor in the reactor.

In older units, HCO was recycled along with the fresh feed or at the riser to decrease the production of DO or CLO. These days most refineries try to reduce or eliminate the HCO recycle quantity entirely, as this would provide a cushion for increasing fresh feed in the FCCU. This is achieved by increasing the circulation rate, reducing the CRC on the equilibrium catalyst, or increasing the catalyst MAT. Some refineries are even routing part of the HCO to hydrocrackers to be saturated and converted to distillates. Use of CO promoters to increase the dense bed temperature and reduce the CRC is another common method used by refiners.

Coke formation can be decreased to debottleneck the regenerator by using the following:

1. Better air distributors and hole or nozzle staggering in the air grid or distributor
2. Anti-jump baffles to evenly distribute the spent catalyst in the regenerator for effective combustion and regeneration
3. Hydrotreated feed stock
4. Efficient reactor strippers to reduce the oil content in the spent catalyst
5. Low delta coke catalysts (occasional)

Methodologies used for a capacity revamp

A distillate FCCU was revamped by the licensor from 1.0 to 1.5 MMTPA using the actions listed below. No major revamp in the R–R section was done.

1. Increased the R–R pressure
2. Introduced a feed nozzle and slurry recycle nozzle

3. Changed the catalyst slide valves and DDSV
4. Added one parallel WGC and a new main air blower (MAB)
5. Changed the internals for the main fractionator and GCU
6. Augment the steam generators in the MCB PA and HCO PA

If the HCO recycle can be eliminated and equivalent fresh feed can be increased, the fresh feed capacity of one unit could be increased by around 10%–12% by reduction of HCO recycle. However, there was also a need to increase the MCB PA duty that was achieved by only installing MCB PA vs. feed heat exchanger parallel to HCOPA vs. feed heat exchanger with a change from a series configuration where MCB PA vs. feed heat exchanger was preceded by HCOPA vs. feed heat exchanger.

The limitation in fractionator pump-around section can be resolved by introducing an additional pump-around. In one unit, a heavy naphtha pump-around was introduced to lower the duties of the LCO PA and HCO PA. The debottlenecking of the gas-con section is discussed in Chapter 7.

11.8 Case Studies

Case Study 1

One very strange phenomenon observed in an FCCU fractionator column was that when the reactor temperature was increased to increase conversion, the column bottom level would drop so low that the MCB pump would nearly lose suction. Then, when the reactor temperature was decreased again, the bottom would become so full that it required an additional pump to pump out the DO.

The column had disc and doughnut trays in the desuperheating section instead of grids. It was suspected that the liquid was accumulating in that section and in the wash zone below the HCO withdrawal, so vapor–liquid load profiles for the zones were developed and sent to a reputed column internal designer. It was agreed to sacrifice stages in the wash section below the HCO tray and to pack it more loosely so that wash liquid could drop without accumulating in the zone at high vapor velocities.

The desuperheating zone was found limiting in the disc and doughnut tray section, which was causing the accumulation of liquid in the zone and loss of bottom levels. The trays were changed to grids and the problem disappeared. The unit is operating at around 15% over the design capacity. At the operating capacity, the regenerator cyclone velocity is at its limit and the fractionator has almost reached its flooding limit.

Case Study 2

One RFCCU was operating with RCO as a feed (CCR 5%–6%). This refinery was facing frequent high pressure drop problems across the main fractionator. The desalter was not doing a satisfactory job in the crude unit, as the salt content in parts per thousand barrel (ptb) was reported to be around 4–5 ptb. It was reported that after water-washing during the shutdown and in the subsequent start up, the pressure drop problem would disappear, but always reappeared after continued operation.

It was suspected that some water soluble salts were being deposited in the column dispersers at the colder upper section of the column and increasing the pressure drop. The likely culprits were NH_4Cl and NH_4HS. At a fixed temperature, NH_4Cl deposition is a function of partial pressure NH_3 and partial pressure HCl in the vapor phase (Fig. 4.10.1 of Chapter 4). Similarly, NH_4HS deposits are a product of partial pressure NH_3 and H_2S (Fig. 4.10.2 of Chapter 4).

To lower the partial pressures of NH_3, HCl, and H_2S at the top section of the column, the PA heat removal was reduced to increase the mole fraction of hydrocarbons in the upper section. This was likely to reduce the vapor phase mole fractions of NH_3, HCl, and H_2S, thus reducing their partial pressures (at constant total pressure) and the likelihood of deposition.

The top reflux and reflux ratio would increase to achieve the total heat removal requirement. This could have increased the reflux drum temperature, but did not due to the cushion in the condenser duty. The problem disappeared when the LCO PA duty was decreased. The pressure drop became normal. The top reflux returned to the column has much lower NH_3, H_2S, and HCl contents, as water is injected upstream of the top condenser to dissolve NH_4Cl and NH_4HS, reject them into the water phase, and send them out of the unit.

Case Study 3

In one RFCCU taken for capacity revamp, the MCB PA was going to the feed preheat exchangers (3 parallel), then to high-pressure steam generators (two parallel) before going to the BFW preheat heat exchanger and returning to column (Fig. 11.8.1). The differential temperature in design was around 122°C. The heat removal requirement through the MCB PA was 15 Gcal/hr. The desired flow for the MCB pump had not been achieved since the unit was commissioned.

A pressure survey was conducted across all the heat exchangers in the MCB PA circuit. A high pressure drop was observed across the feed and MCB PA heat exchanger and the MCB PA and BFW heat exchangers. It was found that the MCB PA circuit was thermally adequate but hydraulically limiting. The total actual pressure drop in the MCB PA circuit was more than the head at the design operating point in the pump characteristics curve. This was not allowing the desired flow. As a result, the refinery had been venting the steam into the atmosphere to maintain a higher differential temperature across the MCB PA to remove the heat duty.

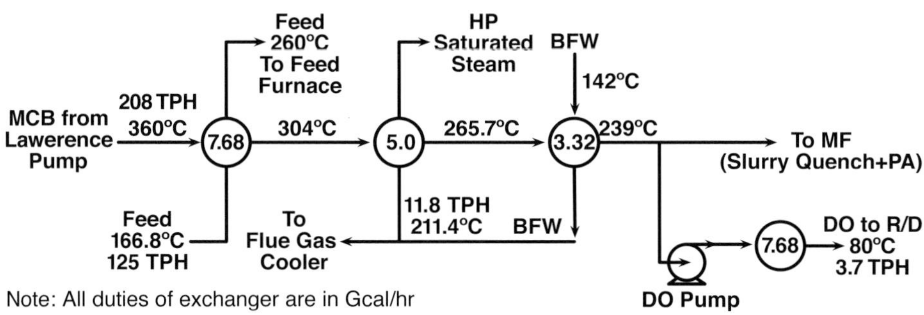

Fig. 11.8.1 MCB PA circuit

The thermal rating of the exchangers was done. It was found that the four-pass arrangement provided would lead to a high pressure drop across the feed and MCB PA and across the MCB PA and BFW heat exchangers, a high pressure drop in the entire circuit, and could lower the MCB PA flow. This corroborated the findings of the pressure survey. The rating was repeated with a two-pass arrangement in the MCB PA side and the estimated pressure drop was acceptable, without falling below the prescribed velocity (around 1.2 m/sec) limit in the MCB side in the heat exchangers. Accordingly, the tube passes in the MCB PA side were reduced from four to two in both heat exchangers. The desired flow was achieved and the steam from the steam generators was routed to steam headers. The unit was subsequently revamped for a capacity increase of around 15% with the reactor effluent data provided through the pilot plant study by the in-house research and development group.

Metallurgy

FCCUs operate at high temperatures and involve contact with the solid catalyst, so special metallurgy is used for equipment, as can be seen below.

The reactor shell is made of 1¼ Cr–0.5 Mo or 2¼ Cr–1 Mo.

The riser is a cold wall made of carbon steel with insulating refractory lining inside.

The spent catalyst line is made of carbon steel with refractory lining inside (cold wall).

The regenerator is made of carbon steel with a corrosion allowance. The inside has insulating refractory lining.

The RCSP is made of carbon steel with refractory lining inside (cold wall).

The regenerator flue gas line is made of carbon steel with insulated refractory lining inside.

The regenerator cyclones, air grid, and drip legs are made of SS 304.

The regenerator air grid is made of SS 304 with a refractory lining on the outside.

The MCB pump casing is made of SS 410 and impellor is made of abrasion resistant cast iron (with 25% Cr, 2.5% Ni, and 3% Mo).

Abbreviations

APS: average particle size

BFW: boiler feed water

BS&W: basic sediment and water

CLO: clarified oil

CRC: coke on regenerated catalyst

DDSV: double disc slide valve

DO: decanted oil

DP: differential pressure

FGDU: flue gas desulfurization unit

FST: fluid bed simulation test

HCO: heavy cycle oil

MAT: microactivity test

MCB: main column bottom

MCB PA: main column bottom pump-around

PRT: pressure recovery turbine

PSRI: Particulate Solid Research, Inc.

ptb: parts per thousand barrel

R–R: reactor–regenerator (section)
RCSP: regenerated catalyst standpipe
RCSV: regenerated catalyst slide valve
RFCCU: residue fluid catalytic cracking unit
SCSP: spent catalyst standpipe
SCSV: spent catalyst slide valve
V_2O_5: vanadium pentoxide
VSS: vortex separation system

Catalytic Reformer Units 12

12.1 Introduction

A catalytic reformer unit (CRU) is traditionally used to enhance the Research Octane Number (RON) of naphtha streams for blending in motor spirit (MS). In recent years, MS has been manufactured by blending CRU reformate, FCCU gasoline, isomerates, alkylates, and oxygenates to meet the quality requirement for MS. The typical key quality requirements are listed below.

1. Benzene: 0.5%–1% vol max
2. RON: 91
3. Motor Octane Number (MON): 81
4. Vapor Lock Index (VLI): 750
5. Oxygenates: 3% vol max
6. Olefins: 21% vol max
7. Aromatics: 35% vol max
8. Sulfurs: 50 ppm wt max

CRUs are also used as the starting process for manufacturing aromatics like benzene, toluene, and xylene (*para*-xylene).

12.2 Description

The feed stock is naphtha obtained from the distillation of crude oil or gas condensates. The full-range stabilized naphtha fraction (C5 to 140°C) is fractionated to produce typically 90°C–140°C cut as feed to the CRU. The fraction captures the C6 to C9 hydrocarbons comprised of naphthenes (N), aromatics (A), and paraffin (P) molecules. In recent years, the C6 molecules have been selectively eliminated in the feed to reduce the benzene content in reformate. Thus, the reformer feed initial boiling point has increased, typically to around 95°C and the feed stock cut range has become 95°C–140°C.

An index is often used to characterize the feed as N + 2A, (N + A), or (0.85N + A), where N and A are the volume % of naphthenes and aromatics in the feed. A feed stock rich with A and N is considered a good feed stock and will have a high index. A lower index indicates a higher paraffins content, which makes the feed stock difficult to handle in reforming, as the dehydrocyclization of paraffin becomes a concern. A feed with an index value (N + 2A) above 55 is often considered a good feed stock. Normally, the chemical composition of the feed is determined by gas chromatography analysis.

In the CRU, naphthenes in the feed are almost fully dehydrogenated to aromatics. Paraffin molecules are first cyclized and then dehydrogenated to aromatics. Some amount of cracking takes place. The following main reactions take place in a catalytic reformer.

Main reactions

1. Naphthenes → Aromatics + H_2; strongly endothermic reaction, equilibrium controlling
2. Paraffin → Cycloparaffins → Aromatic + H_2; endothermic reaction. The reaction of paraffin dehydrocyclization to naphthene is rate controlling and naphthene conversion to aromatic is equilibrium controlling.
3. Paraffin + H_2 → Lighter paraffin (hydrocracking); exothermic reaction, rate controlling
4. Paraffin → branched paraffin (isomerization reaction); almost thermo-neutral, equilibrium controlling

Some amount of disproportionation reaction also takes place and this inadvertently produces some amount of benzene even if benzene precursors in the feed are minimized.

Catalysts

The catalyst used in a catalytic reformer contains noble metals like platinum, or a combination of platinum and rhenium or platinum and tin (bimetallic catalyst) with acid sites. The bimetallic catalysts are known to produce less coke during reforming. The acid site is contributed by chlorides. The platinum/metal is dispersed on the alumina base. Typical platinum content in the catalyst is 0.2 to

0.5 wt%. The typical range of chlorine in the catalyst is 0.8 to 1.1 wt%. Previously, CCl_4 was used to enhance and maintain the chloride levels in catalysts. Currently, ethylene dichloride/ethylene perchloride is used in most of the units.

The noble metals are temporarily poisoned (deactivated) by sulfur compounds in the feed. So the CRU is preceded by a desulfurization unit, typically using a cobalt and molybdenum catalyst. This decreases the sulfur content of the feed to less than 1 ppm wt, which makes the stock acceptable as feed to the CRU.

Water often gets in the naphtha feed due to improper operation of the naphtha hydrotreater (NHT) stripper. Water leaches the chlorides from the catalyst, affecting the acidic function of the catalyst and decreasing the dehydrocyclization of paraffin and hydrocracking. Less than 50 ppm vol of moisture (15–25 ppm vol) is typical in recycle gas. If there is more than 50 ppm of moisture in the recycle gas, the reactor inlet temperature must be lowered to avoid chloride leaching and catalyst sintering. A moisture content of greater than 50 ppm in the recycle gas often calls for restriction of the reactor inlet temperature (below 480°C) and moisture content above 100 ppm may require further reduction of the reactor inlet temperature (around 460°C). The temperature limits are accepted as a general guideline. The catalyst supplier can give the exact figures based on the catalyst supplied to avoid much leaching of chlorides from the catalyst. In a CCRU, additional moisture can get into the feed through the circulating catalyst. This moisture content of RG may also increase as a result of inadequate water removal, particularly during calcination.

Catalyst temporary poisons

The temporary catalyst poisons are listed below.

1. **Sulfur** poisoning is caused by either H_2S or organic sulfur compounds present in the feed. H_2S reacts with platinum (Pt) to produce PtS and reduces the activity of the catalyst by decreasing the metallic contact area. It manifests in a decrease in hydrogen yield, a decrease in recycle gas purity, an increase in LPG yield (increased hydrocracking), and a reduction in temperature drop across reactors. Poor operation of a naphtha pre-treater normally results in sulfur slipping to the reformer. The low activity of a pre-treater catalyst or low inlet temperature can also cause slippage of organic sulfur compounds. The poor operation of a stripper can also lead to H_2S in the reformer feed and sulfur poisoning.
2. **Nitrogen** poisoning can be caused by an injection of amine-based corrosion inhibitors in the crude and condensate units. The maximum allowable concentration in feed is 0.5 ppm wt, expressed as organic nitrogen. This can react with chlorine, decrease acid function, and

produce NH_4Cl, which can deposit in the air cooler, separator, stabilizer, and even in the recycle gas compressor (RGC) casing/impeller, as experienced by the author.

3. **Carbon** deposits on the catalyst after continued operation of the unit and reduces the catalyst activity. It is burned during regeneration to reactivate the catalyst, followed by oxychlorination, calcination, and reduction to bring the catalyst back to its original state. However, some permanent loss of activity cannot be averted.
4. **Water** is not a poison, but it needs to be eliminated from the feed to avoid the reduction of catalyst activity due to leaching of chloride and increased likelihood of the catalyst sintering if subjected to high temperatures. Excess moisture leaches chloride from the catalyst. However, some water is required to enhance the activity of the acid function often referred as the chlorine–water balance.

Estimation of RG moisture

Moisture in feed assumed 10 ppm wt, feed rate assumed 100 MT/hr of molecular weight of 100, recycle ratio assumed 2 moles H_2/mole of feed, RG H_2 purity assumed 90%. Separator pressure is 4 kg/cm^2 (a).

Feed rate in moles = 100 × 1,000/100 = 1,000 kg-moles/hr. Water rate = $(10/10^6) \times 100 \times 1{,}000 = 1$kg/hr = 1/18 = 0.055 kg-mole/hr. H_2 recycle = 2 × 1,000 kg-moles. RG recycle = 2,000/0.9 = 2,222 kg-moles/hr.

Mole fraction of water in RG = 0.055/(2,222 + 0.055) = 0.000025 = 25 ppm by volume in RG (for fixed bed units or assuming no moisture ingress through circulatiing catalyst). Partial pressure of water in RG = 0.000025 × 4 = 0.0001kg/cm^2 (a), which is lower than the saturation pressure of water of 0.075 kg/cm2 (a) at a separator temperature of 40°C. Thus, the water will not condense in the separator and will remain in RG.

Catalyst permanent poisons

Some permanent poisons for the catalyst are arsenic, lead, mercury, silicon, etc. The source of arsenic may be the straight-run naphtha, cracked naphtha, or naphtha coming from the gas condensate unit as feed. Lead (Pb) may get into the feed naphtha from slop processing in a crude unit. Mercury gets into the naphtha from gas condensates. Silicon can come from antifoaming additives. Other metals like copper and nickel may get into the feed as corrosion products. Arsenic and lead should not be more than 1 ppb and 5 ppb by weight respectively. Most of the metals in the feed poison the metal function of the catalyst. Metal poisoning can be experienced initially in the first reactor and after breakthrough may affect the subsequent reactors.

12.3 Fixed-Bed CRU

A catalytic reforming unit can be fixed-bed CRU or a continuous catalytic reformer (CCR) type.

In fixed-bed reforming, the high-pressure loop operating pressure ranges from 16–22 kg/cm^2. Lower pressure and higher temperature favors the equilibrium conversion to aromatic. Coke is deposited on the catalyst during operation. The regeneration of catalyst to burn the coke deposit is performed periodically.

Typically three to four reactors are used in a series. The catalyst volume in the reactors increases progressively from the first to the last reactor as the reactors in the beginning handle fast reactions (equilibrium controlling) and the reactors at the end handle slow reactions (rate controlling: hydrocracking and paraffin dehydrocyclization). Thus the catalyst volume and consequently the size of the reactors increase from first to last reactor. A fixed-bed (semi-regenerative) configuration is shown in Fig. 12.3.1.

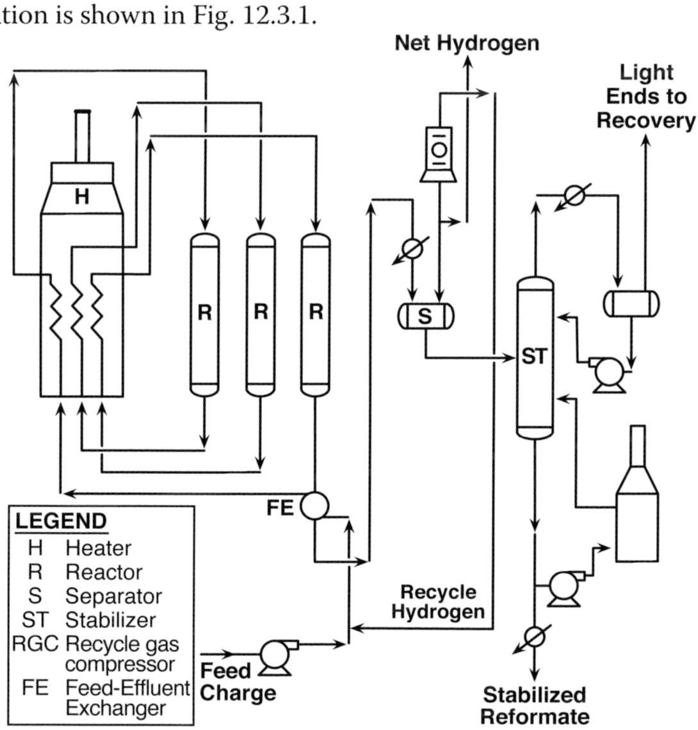

Fig. 12.3.1 Fixed-bed (semi-regenerative) scheme

The feed is typically mixed with recycled hydrogen (typically 1.6–4 moles of H_2/mole of feed, a higher ratio for fixed bed units) and heated in heat exchangers with hot effluent from the last reactor, then heated again in the furnace before being fed to the first reactor. The reactors are preceded by furnaces to increase the feed temperature to the reaction temperatures. Often Packinox-type heat exchangers are used to obtain a high heat recovery from the effluent and a high temperature at the inlet of first reactor furnace.

The outlet of the first reactor cools due to the predominantly endothermic reaction occurring in the first reactor. The effluent of the first reactor is heated in the second reactor furnace to the desired inlet temperature and fed to the next reactor. The flow scheme for other reactors is similar, with a furnace preceding each.

The CRU reactions occurring in the reactors are mostly endothermic. The endothermicity (temperature drop) reduces from first reactor to last reactor. The last reactor temperature drop often becomes positive at the end of the run due to a pronounced hydrocracking reaction and decreased aromatization reaction.

The outlet from the last reactor exchanges heat with the incoming feed and recycle gas (RG) mixture, then is further cooled in air and water coolers. The cooled product is then sent to the separator. In the separator, the hydrogen-rich gas is separated from the liquid and recirculated with the help of an RGC along with the feed. Hydrogen is produced in a CRU through dehydrogenation reactions and is typically bled from separator top under the pressure control of the separator.

A recycle gas hydrogen purity of around 85%–95% is commonly encountered. The purity of the recycle gas reduces when it nears the end of run (EOR). At the start of run (SOR), reactor inlet temperatures are usually lower and progressively increase (with deactivation of catalyst due to formation of coke on catalyst) to maintain the catalyst activity and the RON. The increase in reactor inlet temperature increases the cracking reactions and leads to the production of a higher proportion of lighter hydrocarbons (lower C_5+ reformate yield) and coke. This lowers the hydrogen purity of recycle gas as the EOR is approached. Often a higher recycle ratio is used near the EOR to decrease coke formation.

The temperature drops across the reactors are largely dependent on the naphthene content of the feed and the extent of paraffin dehydrocyclization reaction. The inherent naphthenes are almost fully converted to aromatics.

Catalyst regeneration

During the catalytic reforming operation, coke is formed on the catalyst and reduces the activity. The decreased activity of the catalyst is typically compensated for by increasing the reactor inlet temperature in a fixed-bed reactor. But after a certain amount of coke is on the catalyst, the catalyst is regenerated.

The regeneration involves the following essential steps:

1. **Controlled burning** of the coke with nitrogen circulation with 0.2%–0.5% mole of oxygen in the circulating gas. This is typically done at a temperature of 460°C–480°C.
2. **Full burning** of the coke with higher oxygen content in the circulating gases (N_2, CO_2, O_2, H_2O) to ensure complete burning of all the coke.
3. **Oxychlorination**, in which chlorine is attached to a platinum catalyst and metal is redispersed.
4. **Calcination**, in which water is removed and metal is transformed into its oxide form.
5. **Reduction**, in which metal is reduced to metallic form before the catalyst is again deployed for reforming.

12.4 Continuous Catalytic Reforming Unit (CCRU)

A CCRU operates at around 3–5 kg/cm^2 pressure in the reaction loop. The lower pressure favors the aromatization and dehydrocyclization reactions. The CCRU also operates at higher temperatures to further favor the endothermic reactions and their equilibrium conversion. Greater amounts of coke are produced and deposited on the catalyst. The catalyst in the CCRU moves continuously from reactor to reactor, from the last reactor to the regenerator, and, after regeneration, back to the first reactor.

The conversion to aromatic is high in the CCRU due to the low pressure, high reactor temperature, and higher catalyst activity. The dehydrocyclization and hydrocracking reactions are also high due to higher reactor inlet temperature. Typically, the C_5+ reformate yield is higher with low pressure in the reaction system of a CCRU.

The catalyst is housed in wire screens in the reactors. The flows of feed are radial (from the circumference toward the center) and the reformate leaves through the center (center pipe) of the reactors. Often scallops are used to contain the catalysts in the reactors. Multiple scallops are used circumferentially. The use of scallops is often preferred over screens as damaged scallops can be quickly replaced, and maintenance of the reactor internals is easier. Lower pressure drop on the reaction section is always desired, and can be achieved by improved design of reactor internals and reactor furnaces, and through use of Packinox-type heat exchangers for feed preheat by the effluent.

A CCRU typically contains several sections similar to a fixed-bed CRU:

1. Feed preheat section
2. Reactor section
3. Effluent cooling and stabilization section
4. Regenerator section
5. Hydrogen-rich gas recontacting section

In a CCRU, the addition of a regenerator section is an obvious requirement, as it ensures that the catalyst is regenerated continuously or in batches during the operation. The hydrogen-rich gas recontacting system is also very typical in a CCRU. The recontacting section is also occasionally used in fixed-bed CRUs, normally without refrigeration after compression. The CCRU uses two popular configurations:

1. Series reactor
2. Stacked reactor

In the series reactor configuration, the reactors are placed side-by-side in a series with furnaces at the upstream to heat the reactor effluents and the regenerator at one side (Fig. 12.4.1). The catalyst is circulated from the regenerator to the first reactor and then to the subsequent reactors and returned from the last reactor to the top of the regenerator.

In the stacked reactor configuration, the reactors are placed one above the other with furnaces at the upstream and the regenerator at one side. The catalyst is circulated from regenerator to first reactor and subsequent reactors from top to bottom and from the last reactor to the top of regenerator.

Catalyst circulation (series reactor)

Catalyst is introduced to the regenerator through the upper surge drum and lock hopper. Bottom hoppers are used to collect catalysts from reactors and regenerators and also to establish a seal. Circulation pipes, often called spider legs (8–12 of legs) are used to transfer the catalyst from the reactor or regenerator to the bottom hopper. The catalyst is typically transferred from the bottom hoppers to lift pots, then through lift lines to transfer the catalysts. For the series reactor configuration, the longest lift line is normally from the last reactor to the regenerator. The lift pot below the regenerator and last reactor operates with nitrogen as the lift gas, while the others use H_2 as the lift gas.

The principle of catalyst transfer using lift lines is that the lift gas velocity is kept higher than the settling velocity of the catalyst particles. Once the catalyst reaches the upper surge drum, the lift gas velocity reduces and the catalyst drops out into the upper hopper. The catalyst velocity is typically maintained at 2–3 m/sec in the lift pipes to avoid attrition loss.

The coke deposited on the catalyst is transferred continuously or in batches to the regenerator for coke burning, oxychlorination, and calcination before being returned to the first reactor bed after reduction. The coke burning, calcinations, and oxychlorination are done in the regenerator. The first and second burn are done in the upper section, then the oxychlorination and calcination are performed in the lower section.

The reduction chamber is placed at the top of the first reactor below the top hopper in some licensed units.

A N_2 blower (elutriation blower) collects the nitrogen lift gas from the top of the regenerator and the first reactor and filters and circulates it in a closed loop. The gas is filtered to eliminate fine particles produced by attrition.

Calculation of catalyst circulation rate in a CCRU

The first step for the calculation is to determine the coke burning capacity of the regenerator. The coke production in a CCRU is typically 0.03–0.05 wt% of the feed rate to the unit. This may be a function of the catalyst and feed stock quality. For a unit with a 50 MT/hr feed capacity, the coke deposit would be 50 × 1,000 × (0.05/100) = 25 kg/hr. Thus, the coke burning capacity of the regenerator would be 25 kg/hr. For a unit with a higher feed capacity, the regenerator coke-burning capacity will also be higher. Typically, the coke-burning capacity of the regenerator determines the capacity of the nitrogen compressor for regeneration, the air blower, the size of the neutralizing section, and other associated equipment. The coke-burning capacity of the regenerator and the amount of coke on the catalyst govern the catalyst circulation rate as explained below. Normally, 6% coke on catalyst is considered the maximum, so if the coke burning capacity of the regenerator is 25kg/hr, then the catalyst circulation rate will be 25/0.06 = 416.6 kg/hr. The catalyst circulation rate would then in turn determine the size of the lift lines and other associated equipment like the top surge drum, bottom surge drum, lift pots, lift gas flow, etc. The series reactor configuration is shown in Fig. 12.4.1.

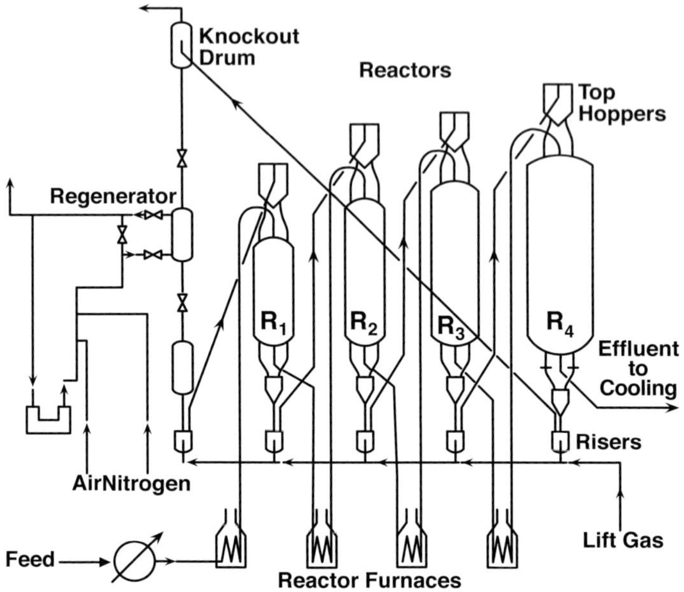

Fig. 12.4.1 Catalyst circulation in series reactors

Stacked reactor configuration (3 reactors) is shown in Fig. 12.4.2. It has two lift lines, one from the bottom of the last reactor to the regenerator top for routing spent catalyst to the regenerator and one for routing regenerated catalyst from the regenerator to the first reactor top (reduction chamber). Devices similar to a series reactor like a reactor bottom catalyst collector, L valve, regenerator top disengaging hopper, and bottom lock hopper are used for catalyst intermediate receipt and delivery, and subsequent transportation through lift lines are used in this configuration by technology suppliers.

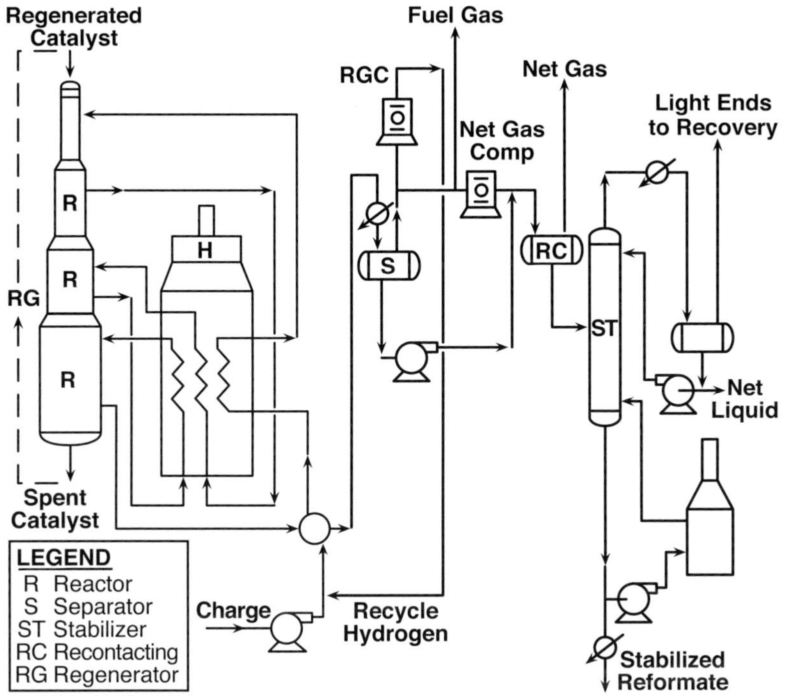

Fig. 12.4.2 Flow scheme of stacked reactors

12.5 Rating the Unit

The conversion in reactors and consequent temperature drop across the reactors are best estimated using a pilot plant or a tailored model developed by experts. These are based on particular feeds, operating conditions, and the type of catalyst used. However, for a tentative rating of the unit at higher capacity, the other parameters such as furnace duty, feed and effluent exchanger duty, stabilizer column adequacies, etc., can be estimated by a process engineer with a simulation model of the plant.

In the simulation, the reactors can be configured as conversion reactors and main reactions like naphthene conversion to aromatic, paraffin conversion to aromatic, isomerization of paraffins, hydrocracking, etc., can be defined in each reactor. A simulation model of a CCRU used for *para*-xylene production has been developed to rate the unit at a higher capacity.

The feed considered in the model below is P:N:A = 47:36:17 by weight with a composition that can be seen below (Table 12.5.1).

Table 12.5.1 Feed composition considered for simulation

Stream Name		Feed
Temperature	°C	350
Total mass rate	kg/hr	100,000.1
Total molecular weight		109.8828
Total std. liq. sp. gravity		0.750532
Total weight comp. percents		wt%
1T4MCH		30
2MHEPT		47
MXYLENE		10
MCYHXN		6
TOLUENE		7

The following components representing paraffin, naphthene, and aromatic are considered:

1. Naphthene components that are undergoing a reaction are assumed to be dimethylcyclohexane (1T4MCH, C_8 naphthene).
2. Paraffin components undergoing reaction are assumed to be methylheptane (2MHEPT, C_8 paraffin).

Reactions assumed are as follows:

Reactions

1. Naphthenes to aromatics reaction
   $C_8H_{16} = C_8H_{10} + 3H_2$
2. Paraffin to aromatic reaction
   $C_8H_{18} = C_8H_{10} + 4H_2$
3. Paraffin hydrocracking reaction
   $15C_8H_{18} + 25H_2 = 8CH_4 + 8C_2H_6 + 8C_3H_8 + 8C_4H_{10} + 8C_5H_{12}$
4. Naphthene hydrocracking reaction
   $15C_8H_{18} + 40H_2 = 8CH_4 + 8C_2H_6 + 8C_3H_8 + 8C_4H_{10} + 8C_5H_{12}$

The heat for the reactions is supplied by furnaces placed ahead of the reactors to bring the reactant temperature to an adequate level for the reaction to occur in the succeeding reactor. For simplicity, the naphthene hydrocracking reaction can be ignored. The simulation is given in Fig. 12.5.1.

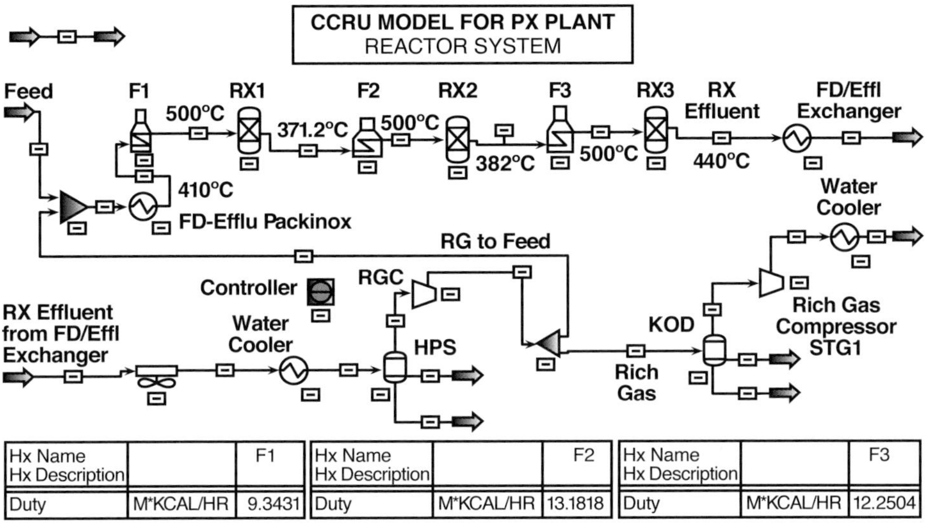

Fig. 12.5.1 Simulation of a CRU for tentative rating of the unit

A model has been developed using the following:

K value or vapor-liquid equilibrium (VLE) by GS

Liquid–vapor enthalpy by Carl Pitzer

Liquid density by API

Liquid–vapor entropy by Carl Pitzer

A controller is used to blend the recycle gas in the stipulated proportion with the fresh feed. The results of the furnace duties of simulation and rating (Fig. 12.5.1) of the unit are given in Table 12.5.2. Typically, the furnace downstream of the first reactor has a higher duty requirement than subsequent furnaces.

Table 12.5.2 Furnace heat duties from the simulation

Duty of reactor furnaces	Gcal/hr
F1	9.3
F2	13.2
F3	12.2

This is to show that a process engineer can also carry out a tentative rating of a CRU by assuming the representative components of paraffins, naphthenes, and aromatics in the feed and their reactions and the extent of the reactions. The reactor models used are conversion reactors where conversion and the extent of the reaction are defined.

Alternatively, the extent of the reaction can be assumed and reactor outlet temperature can be calculated from a conversion reactor model and validated with plant or pilot plant data for reactor outlet temperatures. After few trials, the conversion matching the actual plant reactor outlet temperature can be considered for rating the unit at a higher capacity (with the same feed constituents and composition). The extent of each reaction in each reactor is defined below.

The reaction of C8 naphthene conversion to aromatic is considered to be 98% in all three reactors. The same is distributed in the three reactors. In the first reactor, a conversion of around of 55% of the feed naphthene has been assumed. In the second reactor, 50% of the remaining feed naphthene is considered and the last 90% of the naphthene conversion is assumed. This makes the total naphthene conversion to aromatics around 98%. The reaction of naphthene to paraffin is not considered, as the conversion is insignificant.

1. Paraffin conversion to aromatic is typically around 70%. The balance is cracked to produce lighter paraffins. The converted paraffin is distributed in the three reactors. In the first reactor, 40% of the paraffin is converted to aromatic. Hydrocracking of 5% of the feed moles is considered in the first reactor.

2. Conversion from paraffin to aromatic in the second reactor is 35% of the balance. Hydrocracking of 10% of the feed moles is considered in the second reactor.

3. Conversion from paraffin to aromatic in the third reactor is 40% of the balance. Hydrocracking of 15% of the feed moles is considered in the third reactor.

These conversions were taken based on pilot plant inputs for the conversion of feed and validated with the actual temperature difference across reactors. Interreactor furnace duties were calculated to deliver the inlet temperature (500°C–510°C) to reactors. This calculation was only performed to estimate the tentative modifications required in the unit for capacity increase to the desired level. The entire reactor loop of the unit can be rated at the increased capacity using the simulation (Fig. 12.5.1). The adequacies of all the equipment, such as the RGC, rich gas compressor, feed vs. effluent heat exchanger, and the effluent cooler can be also checked.

Any modifications in the reactor section should be validated by the licensor of the unit, a catalyst vendor, or research and development (R&D) experts, as the above procedure only allows process engineers to estimate the extent of modification.

Recontacting section

Typically, a CCRU produces higher hydrogen (3.2–3.5 wt%) than a fixed-bed CRU (1.8–2.2 wt%) because the CCRU operates at a lower pressure and higher reactor inlet temperature and uses regenerated catalyst. The hydrogen-rich gas produced by a CCRU normally contains a considerable amount of lighter hydrocarbons. Thus, it becomes necessary to improve the purity of the hydrogen-rich gases and recover the large amount of hydrogen. This is typically achieved by contacting the gas and reformate at a higher pressure and absorbing part of the lighter hydrocarbons at a higher pressure in the reformate stream (Fig. 12.5.2).

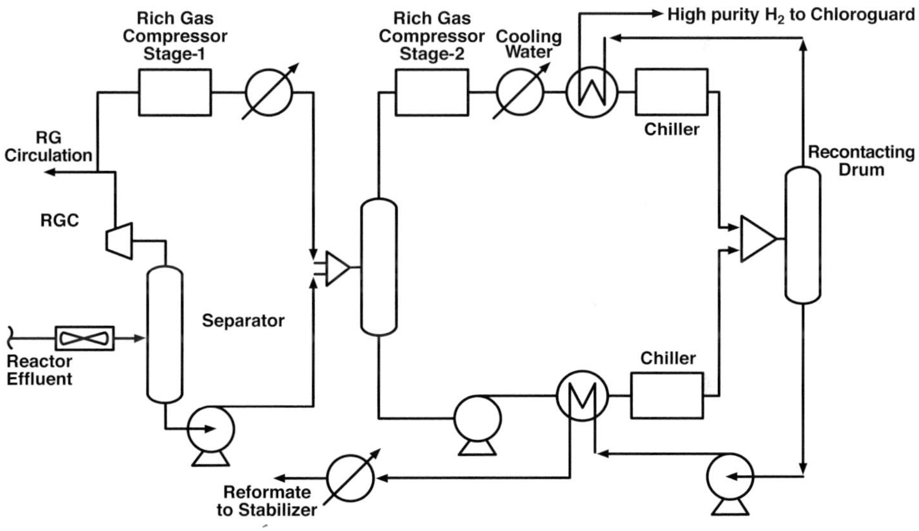

Fig. 12.5.2 Recontacting system with chillers

The rich gas compression and recontacting system may be a little different from plant to plant. In the simulation described in Fig. 12.5.3, a two-stage rich gas compressor is considered. This increases the pressure to 25 kg/cm^2 before it enters the hot–cold exchanger and refrigeration cooler, where the product is typically cooled to −2°C. If the pressure is increased to only around 10 kg/cm^2 before chilling, the chilling requirement is typically lower (around −15°C) to bring the H_2 purity of the lean gas to around 95–96 mole %. In such a case, the lean gas needs to be recompressed to a pressure of 22–25 kg/cm^2 before routing it to H_2 recovery systems such as a pressure swing adsorption (PSA) or a membrane system. Often, instead of contacting the chilled vapor and chilled reformate in the single-stage contact (recontacting drum), recontacting can be done using three to four stages in a column. The chilled vapor is introduced in the column (cold-insulated) at the bottom and the chilled reformate is introduced at the top. This marginally increases the lean gas or net gas hydrogen purity. The chilled reformate stream from the recontacting drum or the column bottom will gain heat while cooling the liquid from the first drum. It will then be further heated before being fed to the stabilizer. Similarly, chilled lean gas from the recontacting drum or column will exchange heat with incoming gas to cool the incoming gas and gain heat.

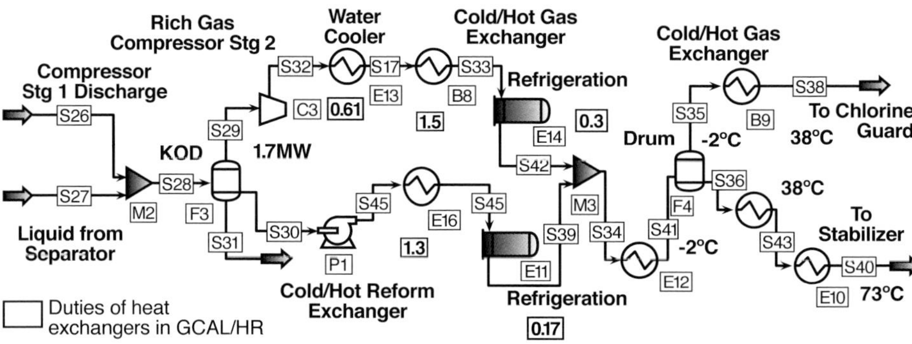

Fig. 12.5.3 Simulation of recontacting system with chillers

There are also recontacting systems without refrigeration. The rich gas from the RGC discharge is compressed, reformate from the separator is pumped, and both are contacted at a higher pressure and separated in the drum. The net gas purity increases due to the absorption of lighter components in the reformate at a higher pressure. This is typically done with two stages to increase the gas pressure from around 3.5 kg/cm^2 (g) to around 23–25 kg/cm^2 (g) in order for the lean gas to be processed in a chlorine guard and to feed to the PSA, or to perform direct coupling of the hydrogen in consumer units. The rich reformate is fed to the stabilizer after heating.

If the lean gas must be routed to a consumer unit, it has to be fed through a bed of chloroguard with a lead–lag facility. Two chloroguard vessels are used in the series. The fresh one is kept as a guard and normally succeeds the first one. This is to ensure that no chloride is passed to the hydrotreaters to cause NH_4Cl deposition in the effluent heat exchangers.

In the stabilizer section, reformate is stabilized. It produces gas and drag streams containing LPG components (Fig. 12.5.4). The drag stream is often routed to the stabilizers of other units to recover the LPG.

From the simulations of a recontacting system (Fig. 12.5.3) and a stabilization section (Fig. 12.5.4), a process engineer can fully rate the unit. The simulation of the reaction section (Fig. 12.5.1) gives a fairly good idea of the furnace, heat exchanger, and cooler duties in the reaction loop.

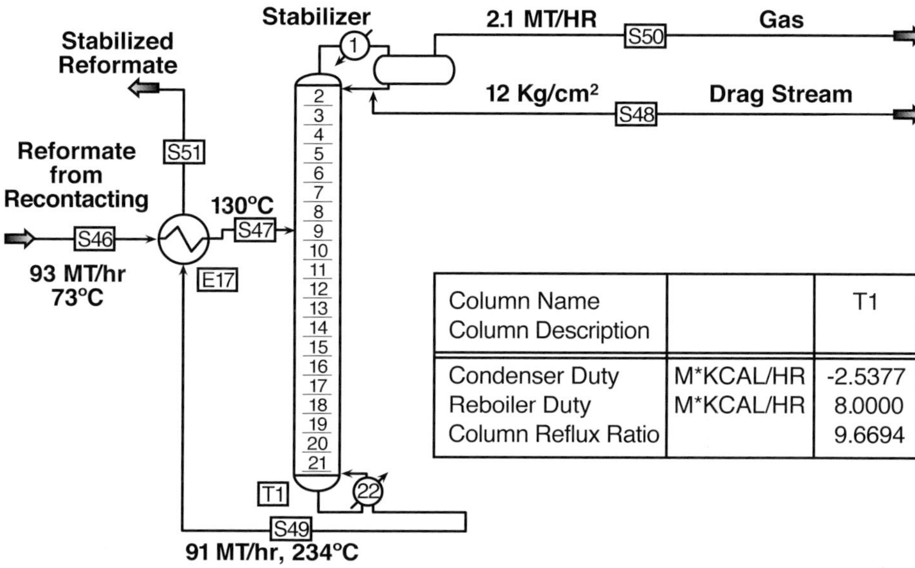

Fig. 12.5.4 Simulation of reformate stabilizer section

Treatment of reformate

Further treatment of reformate has become unavoidable with the imposition of a stricter benzene limit in MS. Typically, reformate after stabilization is again fractionated to remove the benzene-rich fraction to decrease the benzene content. Some benzene is produced from aromatization and dehydrocyclization of C_6 precursors present in the feed and some is produced through the hydrodealkylation and alkyl disproportionation reactions of toluene. The benzene-rich fraction from the top of the reformate splitter is typically fed to an isomerization unit to saturate the benzene along with the normal isomerization unit feed.

Process objectives and concerns

The primary objective of the unit is the desired RON of reformate and C5+ reformate yield. The typical reformate RON ranges from 96 to 102, and typical C5+ reformate yield is about 90% wt. Often the RON declines and is mostly attributed to lower catalyst activity contributed by lower platinum dispersion on the catalyst and chloride content. Desired platinum dispersion is around 85%–90% with chloride content of 0.9%–1.1% wt, is evaluated by catalyst experts and is normally restored with the advice of experts, usually by injecting a higher amount of chloriding agent and maintaining a higher temperature during oxychlorination (around 505°C). Often, damaged catalyst retainer screens result in catalyst getting out of the retainer screen, and the feed flows (short circuits) through the open area to the center pipe without coming in contact with the catalyst. This results in a big drop in RON. Sometimes corrosion products block the lower part of the retainer screens and feed flows preferentially through unblocked area leading to uneven/lower contact of feed with catalyst. Carburization and metal dusting are forms of high-temperature corrosion (metallic coke formation) that occur in the hottest part reactor/inter-reactor furnace tubes. The metal disintegrates to a dust of metal particles and graphite carbon that gets into the reactor screens and also corrodes the outer surface (circumference) of the screen or scallop, thereby possibly causing damage/plugging. The metallurgy of furnace tubes are upgraded from the usual 5Cr–0.5 Mo to 9 Cr–1 Mo, and DMDS is injected in the feed to maintain around 1 ppm of H_2S in RG to reduce the problem, but it cannot fully suppress carburization. The C5+ reformate yield declines as end of run (EOR) is approached, or in the case of excessive hydrocracking in the reactor as result of higher reactor inlet temperature, localized heating, higher chloride level, lower recycle ratio, and heavier feed processing (higher FBP). Higher coke yield may also result when the feed with a higher FBP is processed. Hydrogen purity can reduce due to changes in feed composition (P, N, A content), heavier feed processing, higher hydrocracking, or abnormalities in the recontacting and chilling section.

12.6 Case Studies

Case Study 1

The stabilizer off-gas in one unit was connected to the inlet of the second stage of a rich gas compressor, presumably so that it could recover additional LPG components and H_2 from the stabilizer off-gas. This caused acute NH_4Cl deposition in the valves of the reciprocating compressor, resulting in frequent passing of the suction valves and a frequent high discharge temperature and consequent trip. The stabilizer off-gas was given a caustic wash and routed to the fuel gas system. This solved the problem. Alternatively, a chloroguard could have been installed to treat the stabilizer off-gas and route it back to the rich gas compressor suction. Accordingly, the refinery was advised to install a chloroguard to remove chlorides and resort to the original routing of stabilizer off-gas.

Case Study 2 (A CCRU pressure drop across the stabilizer used to increase in continued operation)

This unit was designed to produce LPG from the stabilizer after stripping. Thus, the stabilizer operated at a lower top temperature, and pressure drop across the stabilizer used to increase. It was suspected that this might be causing NH_4Cl to deposit at a lower temperature region of the stabilizer, resulting in a higher pressure drop. The top temperature was increased during the trial and the heavy drag liquid was drawn from the drum instead of producing LPG meeting its specification of pentane content. The drag stream (heavy LPG) was routed to the crude unit stabilizer to recover the LPG from there. This resolved the problem. Alternatively, a chloroguard could have been installed at the stabilizer feed to trap the HCl and the unit could have continued production of LPG as before, but the option would have been expensive.

Case Study 3

An NHT unit produced desulfurized naphtha (DSN) that was normally fed to CRU and was also stored in the DSN tank for startup after an emergency shutdown of the CRU. Typically, the DSN may absorb moisture during prolonged storage in tank. Normally, a CRU is started with stored DSN to produce hydrogen for the startup of NHT, if the refinery does not have another source of hydrogen. This may admit a considerable amount of moisture from the DSN feed into the CRU system, which can cause the moisture in the recycle gas to increase and adversely affect the operation of the CRU. As a solution to the problem, the DSN was routed to the NHT stripper after preheating in a heat exchanger with the LP steam, which was then routed to the CRU as feed. This reduced the problem of increased moisture in the recycle gas of the CRU and improved operation while processing the DSN from storage.

Case Study 4

This case study describes the failure of a feed and effluent Packinox exchanger. In this CCRU, reformate octane was dropping by four to five units. Some reduction of hydrogen production was also observed. Many trials were carried out, such as increasing the reactor inlet temperatures and the CCl_4 injection, but no improvement was observed. It was found that the reformate color was less yellow than normal and this led to suspicion of a leak in the Packinox. There was a reformate sample point upstream of the Packinox, so a sample was drawn through that point and analyzed. It showed a higher RON than the sample obtained at the downstream, confirming that there was a leak in the Packinox. A density calculation of effluent samples across the Packinox was carried out and showed a leak of around 10%–15%. The unit operated with the small leak for quite some time and loss of octane was compensated by increasing the reactor inlet temperature.

One day, the leak began to get larger, causing the RON to drop considerably. The unit was brought down. One leaky Packinox available from another unit was brought in. Part ports were blocked by Packinox experts. The repaired Packinox was pressure-tested and installed. It was apparent that the flow area reduction in the Packinox would cause higher pressure drop in the feed and effluent circuits. The likely pressure drop was estimated based on reduced flow area and the allowable minimum feed flow. It was understood that the RGC anti-surge control valve might open during the operation due to the higher discharge pressure necessary to overcome the circuit pressure drop (caused by the partly plugged Packinox). Recycle gas, along with the feed (recycle ratio), was also somewhat reduced in order to provide some additional relief in the system pressure drop. The unit operated in this condition at 50%–60% capacity for almost six months, with the RGC anti-surge controller kept on auto mode and around 5%–10% open for the entire period of time. The new Packinox arrived and was installed, and the refinery survived.

Reformate is also required to liquidate the FCCU gasoline by blending with it to produce MS. At times reformate from other refineries was brought by rail wagons to compensate for the lower reformate production and liquidate the FCCU gasoline and produce MS.

Abbreviations

DSN: desulfurized naphtha

EOR: end of run

MON: Motor Octane Number

MS: motor spirit

NHT: naphtha hydrotreater

PSA: pressure swing adsorption

R&D: research and development

RON: Research Octane Number

SOR: start of run

VLI: Vapor Lock Index

General Process Calculations 13

In this chapter, some of the general process calculations often required by process engineers are presented. This covers some live applications of refrigeration, furnace and transfer lines, furnace flue gas side calculations, and solvent recovery fundamentals for heat optimization with examples.

13.1 Refrigeration

Refrigeration is applied in different process units such as

1. Solvent dewaxing units (SDUs) in refineries using ammonia as the refrigerant
2. Recontacting section of a CRU or CCRU
3. Methane, ethylene, or propylene refrigeration in olefin production units and propane refrigeration in natural gas condensation units.
4. Air liquefaction units for the production of oxygen and nitrogen

13.2 Refrigeration in an SDU

In an SDU, the feed is typically cooled to around –15°C to –20°C to precipitate wax from the lube oil base stocks, which are subsequently filtered through the rotary drum filters to remove the wax and improve the pour point of the lube oil fractions.

Ammonia is typically used as a refrigerant because its saturation temperature is around –32°C at 1.2 kg/cm^2 (a), making it able to cool the feed and solvent to around –15°C to –20°C, assuming a temperature approach of 10°C–12°C necessary for heat transfer. In the actual process, the ammonia at 40°C is first cooled to 22°C through self-refrigeration by reducing the pressure of part of the ammonia stream from 21 kg/cm^2 to 6.4 kg/cm^2 (a) for cooling the main ammonia stream in a heat exchanger using a controller.

In the simulation depicted in Fig. 13.2.1, a controller is again used to determine how much of the ammonia stream must be vaporized to produce the required cooling in the main stream. The heat exchanger is configured in simulation to convert part of the ammonia stream to full vapor at 2.3 kg/cm^2 (a). Again, a controller is used to control the flow of the vaporizing ammonia stream. The balance of liquid ammonia is cooled from 1°C to −16°C, and then to −32°C using self-refrigeration and heat exchange as usual. This chilled ammonia stream is used to cool the feed and the solvent, a mixture of methyl ethyl ketone (MEK) and toluene, to around −15°C.

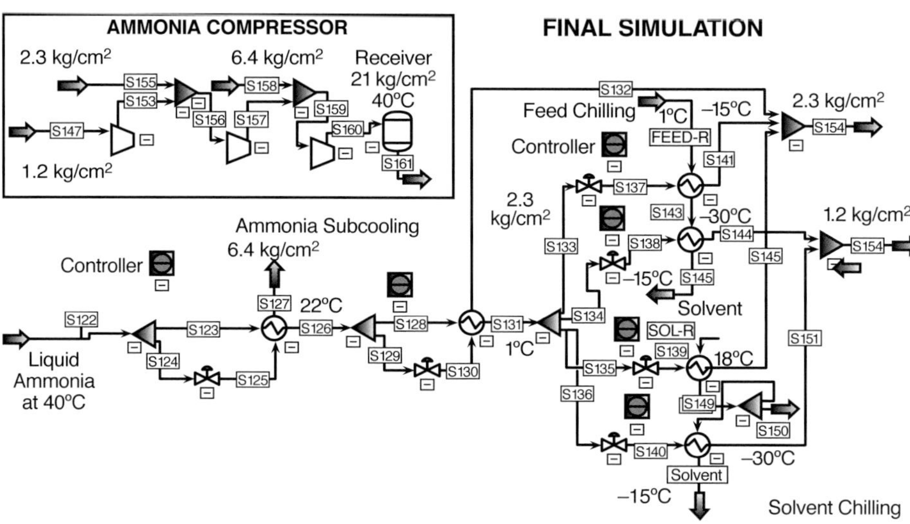

Fig. 13.2.1 Simulation of chilling with ammonia in SDU

The vaporized ammonia is fed to a three-stage ammonia compressor with three different stage pressures: 6.4 kg/cm^2, 2.3 kg/cm^2, and 1.2 kg/cm^2 (a). The final discharge pressure is around 21 kg/cm^2 (a). At this pressure, the saturation temperature of ammonia is around 40°C, so it can be condensed with cooling water. The condensed ammonia is sent to a holder at 21 kg/cm^2 and 40°C and recirculated for cooling.

After the ammonia is chilled, it is available in its vapor state at three pressure levels: 1.2 kg/cm^2 (a), where it is routed to the first stage of the ammonia compressor); 2.3 kg/cm^2 (a), where it is routed to the second stage of the ammonia compressor; and 6.4 kg/cm^2 (a), where it is routed to the third stage of the ammonia compressor. The final discharge pressure is 21 kg/cm^2 (a). Ammonia circulation rate is calculated and all the parameters of the ammonia compressor can be evaluated and can be specified along with other relevant details. The feed and solvent is cooled by heat exchange with cold streams like raffinate, cold rotary drum solvent, etc., before being cooled by the ammonia chilling system. Chilling of the feed in an SDU using ammonia as the refrigerant is shown in Fig. 13.2.1.

Ammonia compressor power requirements are given in Table 13.2.1.

Table 13.2.1 Ammonia compressor power requirements

Compressor name		C9	C11	C12
Compressor description		1st stage	2nd stage	3rd stage
Discharge pressure	kg/cm^2	2.3	6.40	22
Polytropic efficiency	%	72	73	73.5
Actual work	kW	124	714	1,061

13.3 Refrigeration with Propylene

An example simulation of refrigeration with propylene is shown in (Fig. 13.3.1). Liquid propylene is available at 18 kg/cm^2 and 40°C. The propylene is divided in two streams to remove 1 Gcal/hr of heat from the heat exchanger E1. The controller (Controller 1) controls the flow split in the splitter delivering the propylene flow through E1 to remove 1 Gcal/hr of heat (assumed for sample calculation) in E1. The heat exchanger E1 is set to have zero propylene liquid fraction at its outlet in the cold side, implying that the propylene flowing through the exchanger for chilling, should fully vaporize.

The outlet pressure of the pressure-reducing control valve is specified as 9.1 kg/cm^2 to produce an adequate chill and provide the temperature for cooling the main liquid stream in E1. The temperature at the outlet of the valve is 15°C. The vapor from the drum is released and routed to the compressor suction at the appropriate (fourth) stage. The liquid from the drum at 15°C is again split through the splitter with the controller (Controller 2) controlling the total flow of propylene to achieve a heat removal of 2 Gcal/hr (assumed) from the second heat exchanger (E2) through the full vaporization of propylene. The outlet pressure downstream to the second control valve is maintained at 1.3 kg/cm^2 (a) to cool the liquid to around –43°C and achieve the desired cooling of the fluid in E2 to around –35°C (with a differential of around 8°C).

The vapors from both drums are to be routed to the compressor suction at two levels of pressures and enter in appropriate stages to minimize the use of compressor power. The maximum chill that can be attained with the lowest pressure of 1.3 kg/cm^2 (a) is around –43°C, while a stream can be cooled to around –35°C to –36°C with temperature differential of around 7°C–8°C.

A propylene compressor would need four stages for increasing from a pressure of 1.3 kg/cm^2 (a) to 18 kg/cm^2 (a). The pressure increase would be 1.3 kg/cm^2 (a) to 2.3 kg/cm^2 (a) in the first stage, 2.3 kg/cm^2 (a) to 4.8 kg/cm^2 (a) in the second stage, 4.8 kg/cm^2 (a) to 9.1 kg/cm^2 (a) in the third stage, and 9.1 kg/cm^2 (a) to 18.0 kg/cm^2 (a) in the fourth stage of the refrigeration compressor. The vapor produced during chilling can be routed to the appropriate stage of the four-stage compressor. The vapor would finally be compressed to 18 kg/cm^2 before being cooled and condensed by a water cooler to 40°C and recirculated in the cooling process.

In the simulation, the Controller 2 controls the total flow in the propylene loop. A simulation of propylene refrigeration is shown in Fig. 13.3.1. In the simulation, stage one is shown as a combination of three stages, increasing the pressure from 1.3 kg/cm^2(a) to 9.1 kg/cm^2 (a) for simplicity, although this is not actually feasible due to a high temperature rise during compression . In actual practice, three stages with intercoolers need to configured to contain the temperature rise and minimize the power requirement. In the three stages, the compression ratio per stage would be an average of 1.9, which is typical. The propylene circulation rate is calculated and propylene compressor can be specified by a process engineer.

The propylene compressor power requirement is given in Table 13.3.1.

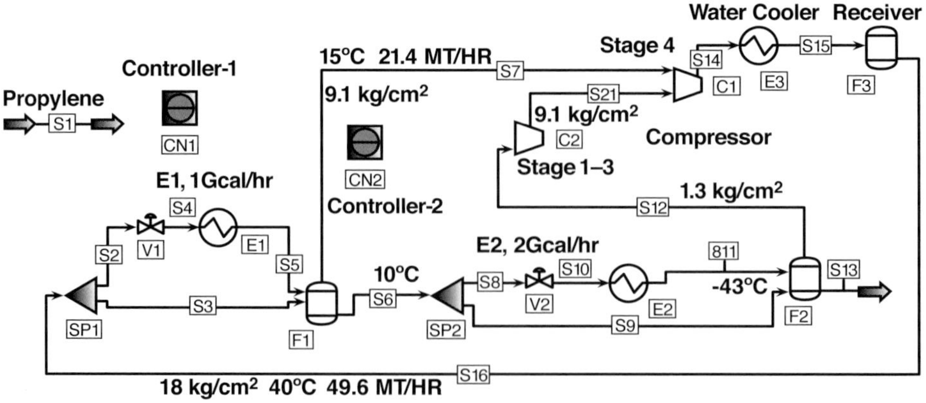

Fig. 13.3.1 Simulation of propylene refrigeration

Table 13.3.1 Propylene compressor power requirement

Compressor name	Propylene	C2	C1
Compressor description		1st–3rd stage	4th stage
Discharge pressure	kg/cm^2	9.1	18
Polytropic efficiency	%	74	71
Actual work	kW	1,068	741

13.4 Furnace Pressure Drop Calculation (Vacuum Furnace)

The pressure drop in a furnace can be estimated using tailored commercial software or developed programs based on a steady state simulator. A simulation program developed for a vacuum furnace is described below (Fig. 13.4.1).

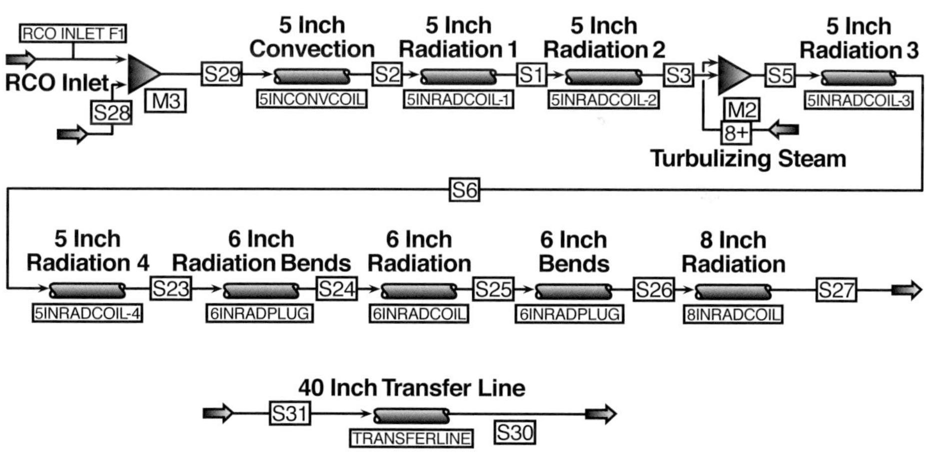

Fig. 13.4.1 Furnace pressure drop calculation of a vacuum furnace

In the program, a line section (heated pipe) (Fig. 13.4.1) represents part of the length of a furnace tube pass. The furnace tube is divided into sections and the total of all the sectional lengths will be equal to the actual length of a furnace tube pass. The sections are connected by streams.

In a furnace tube, hydrocarbon vaporization increases as a result of temperature rise with heat input and pressure loss as the tube outlet is approached. Heat input considered in a section = (heat flux per unit area considered in the design) × (external surface area π × D × actual length of the section of the furnace tube with 180 degree bends), where D is the diameter of the pipe section. Total heat input in all the sections is equal to the absorbed duty of one pass and, when multiplied by the number of passes, will deliver the absorbed duty of the furnace. Different tube diameters are considered as existing in different sections. The heat input is to be calculated using the surface area of the actual length of the section.

RCO with a small amount of water dissolved in it is entered as input to a pipe section at the convection. Introduction of turbulizing steam is also considered at a specific location, as in the actual furnace. An equivalent length is considered for the pressure drop calculation. An inlet pressure is assumed at the inlet of furnace tube and the coil outlet pressure is calculated by the program (forward calculation).

If the coil outlet pressure is higher, the inlet pressure is lowered to match the actual outlet pressure. Conversely, if outlet pressure is lower, then the coil inlet pressure is increased to get the required outlet pressure. The trial is continued until outlet pressure matches.

This method is the same as the two-phase pressure drop calculation, but with the additional feature of heat input to a section, just like in a real furnace tube. This feature is supported by any standard simulator. The method for pressure drop rating can be based on the correlation of Mukherjee and Brill. For a new design, the correlation of Beggs–Brill–Moody–Palmer can be considered.

The calculations can be performed backward also. The calculation provides velocity, acceleration, flow regimes (slug, annular, dispersed, etc.), and others as in two-phase calculations. The heat input to the 180° bend associated with a tube needs to be calculated separately if the bends are outside the fire box (in a header box), as the fire box heat flux will not impinge on the bends. Rather, some heat loss can be considered in the bend. If the bends are inside the fire box, then heat input should be considered as (heat flux) × π × D × (actual length of the bends) and can be considered added with the tube's straight length. Thus, any furnace pressure drop can be calculated by this method using a steady state simulator. This method is shown in Example Calculation 13.4.1 below.

Example Problem 13.4.1

An example problem is worked out for an actual vacuum furnace with an RCO feed, eight passes, and a total absorbed duty of 28 Gcal/hr.

The total flow of RCO through the furnace tubes = 508,420 kg/hr.

Average heat flux design = 17,940 kcal/hr m^2

Convection section detail

Convection tube length = 20.2 meters (each)

No tubes per pass in convection = 10 numbers

Total convection tube length = 201.6 meters with 5 inch diameter

Total convection tube length including U bends = 264 meters

Convection tube calculation

In the first section of the tube, the actual length considered is 264 meters (5-inch diameter, convection) as the total convection length, including the U bends at the end of the header. The heat input considered is 25% of the absorbed duty considered in convection. The furnace has eight passes.

Total furnace heat duty is 28 Gcal/hr

Convection duty = 0.25 × 28 = 7 Gcal/hr

Convection duty/pass = 7/8 = 0.875 Gcal/hr

This heat is used as input to the tube for calculation and entered as input in the section of pipe (Fig. 13.4.1) with the equivalent length of 284 meters in the convection pipe section.

The flow of RCO per pass of the furnace = 508,420/8 = 63,552.5 kg/hr.

For pressure drop calculation, the equivalent length considered for the section = 264 meters based on the number of bends per pass in convection (9).

For a more accurate estimation, the convection length can be divided into even more sections. The radiation section, which contributes more to pressure drop, can also be divided into more tube sections.

Radiation section detail

Number of 5-inch (NB) tubes = 13, Length= 20.32 meters (each)

Number of 6-inch (NB) tubes = 1, Length = 22.3 meters

Number of 8-inch (NB) tubes = 1, Length = 23.5 meters

Radiation section calculation

5-inch tube

Average heat flux design = 17,940 kcal/hr m^2

Surface area of each tube = $\pi \times D \times L = \pi \times 0.1413 \times 20.32 = 9.02$ m^2

Surface area of 13 tubes = 9.02 × 13 = 117.2 m^2

Heat input to the section = 17,940 × 117.2 = 2.1 Gcal/hr

6-inch tube

Surface area of a tube = $\pi \times D \times L = \pi \times 0.168 \times 22.3 = 11.7\ m^2$

Heat input to the tube = $17{,}940 \times 11.7 \times 1 = 0.21$ Gcal/hr

8-inch tube

In the outlet tube in radiation: 8-inch nominal

Surface area of a tube = $\pi \times D \times L = \pi \times 0.219 \times 23.5 \times 1 = 16.1\ m^2$

Heat input to the tube = $17{,}940 \times 16.1 \times 1 = 0.29$ Gcal/hr

Total radiation duty

Total radiation duty per pass = 2.1 + 0.21 + 0.29 = 2.6 Gcal/hr

Total radiation duty (8-pass) = 2.6 × 8 = 20.8 = 21 Gcal/hr

Total convection duty = 7 Gcal/hr

Total furnace duty = 28 Gcal/hr

The total furnace absorbed duty is compared with the total of the absorbed duties in each pipe section of the tubes and must be found to match exactly. The total heat absorbed in all the sections matches with the absorbed duty of the furnace. This step is a prerequisite for proceeding with the simulation program for estimation of pressure drop across the furnace (Fig. 13.4.1). Now the heat duties to the individual pipe section need to be entered in the simulation program.

Calculation of pressure drop

In the simulation, the heat input and equivalent length of each section of pipe are entered as input and the simulation program (Fig. 13.4.1) is run after assuming an inlet pressure. The inlet pressure assumed is matched with the outlet pressure of 0.41kg/cm^2 (a) or 302 mm Hg (a) in trial and error. The inlet pressure at which outlet pressure matches is the calculated inlet pressure of the furnace (6 kg/cm^2). Pressure drop is calculated as inlet pressure – outlet pressure = 6 – 0.4 = 5.6kg/cm^2. Any section pressure drop is available in the program. The vaporization, fluid property, and flow regime at the outlet of any tube or section is given by the model (by the stream interconnecting the section with the subsequent section). The calculated total pressure drop is 5.6 kg/cm^2. The transfer line is also calculated in a similar way without a heat input or with some amount of heat loss from the pipe. Typically transfer line size (diameter) is kept equal to the column nozzle.

13.5 Furnace Flue Gas Side Calculation

A furnace is fired with fuel to deliver the absorbed duty. In the process of heating the process fluid, the flue gas cools. The heat extracted from the flue gas to heat the process fluid can be evaluated using the chart below. For the purpose of explanation, a problem is considered.

Example Problem 13.5.1

A furnace is delivering an absorbed duty of 10 Gcal/hr and is fired with liquid fuel with a 15° API. The flue gas in convection cools up to 300°C, operating with 25% excess air. Calculate the efficiency of the furnace and heat that can be extracted by cooling the flue gas to 200°C.

Solution

The flue gas temperature of 300°C is equal to 572°F. Heat available at 572°F with 25% excess air is 15,600 BTU/lb of fuel (Fig. 13.5.1). Efficiency is equal to heat available at 572°F/calorific value of the fuel or efficiency is equal to (15,600/17,500) × 100 = 89.1% in this indirect method. Efficiency obtained after discounting the typical loss of efficiency of 2.5% (due to heat loss to the surroundings) is 86.6%. This method is often called the indirect method.

The fired duty will be equal to 10/0.866 = 11.55 Gcal/hr = 11.55 × 10^6 kcal/hr. The calorific value of fuel, or its lower heating value (LHV), is 17,500 BTU/lb or 9,722 kcal/kg. Thus, fuel fired = 11.55 × 10^6/9,722 = 1.188 MT/hr.

Now, if the flue gas is cooled from 300°C to 200°C (392°F), the heat available at 392°F with 25% excess air is read again from the Fig. 13.5.1 as around 16,000 BTU/lb fuel. The heat that can be extracted from the flue gas by cooling it from 300°C to 200°C is (16,000 – 15,600) = 400 BTU/lb fuel, which is equal to (400/1.8) × 1.188/10^3, which is equal to around 0.264 Gcal/hr. This heat can be utilized for heating other streams based on requirement and the temperature of the flue gas after ensuring a temperature difference of around 70°C with the stream (which is typical in the case of convective heat transfer).

The revised efficiency of a furnace with a flue gas temperature of 200°C (392°F) and 25% excess air will be 16,000/17,500 = 91.4% from the chart using the indirect method. Revised efficiency after discounting typical 2.5% as heat loss to the surroundings is 88.9%.

Now the calculation is performed by the direct method. The absorbed duty of the furnace is defined as 10 Gcal/hr. So % efficiency by direct method is (heat absorbed)/(heat released by fuel) × 100. So efficiency is (10/11.55) × 100 = 86.6 % with a flue gas exit temperature of 300°C. The total absorbed heat duty with the flue gas exit temperature of 200°C becomes 10 + 0.264 Gcal/hr, which is equal to 10.264 Gcal/hr. The revised efficiency of the furnace through this direct method is (absorbed duty)/(fired duty), or 10.264/11.55 = 89%.

The dew point of SO_2 in the flue gas is the limit to which it can be cooled. The SO_2 dew point is a function of the sulfur content in the fuel. It can be estimated by charts given in API 560.[1] The typical minimum flue gas temperature is the SO_2 dew point temperature + 20°C. A furnace is normally designed on the minimum flue gas temperature thus obtained.

Typically, in a furnace, around 65% -70% of the absorbed duty is picked up in the radiation section. Thus for the example problem, 7 Gcal/hr (10,684 BTU/lb fuel) will be absorbed in the radiation section. The radiation outlet flue gas temperature is determined considering heat absorbed in the radiation section to be 7 Gcal/hr or 10,684 BTU/lb. Fuel is 1,450°F or 787.7°C (see Fig. 13.5.1). The surface area required in the radiation section can be estimated using the heat transfer in the radiation and average heat flux for the service. Typical average heat fluxes for various services in the radiation section are given in Table 13.5.1.

Table 13.5.1 Typical heat transfer rates (heat flux) for some services

Service	Average heat flux in radiation in Btu/hr/ft² based on tube OD
Crude oil heaters	9,000 to 12,000
Vacuum heaters	7,000 to 9,000
Delayed coker	9,000 to 10,000
Hot oil circulation heaters	8,000 to 10,000

Assuming the service is a crude oil heater and the average radiant heat flux is 10,000 BTU/hr/ft^2 (Table 13.5.1), the area required in the radiation section will be $(7 \times 10^9/252)/10{,}000 = 2{,}778$ ft^2 or 266 m^2. A very approximate convection area requirement can be estimated in similar way as a typical convective heat transfer calculation using the average convective heat transfer coefficient U_D (around 7–9 BTU/hr ft^2°F) and the actual log mean temperature difference (LMTD) between the fluid in the tube and flue gas. The minimum differential temperature between the flue gas and fluid in the tube is always maintained at around 70°C minimum to compensate for the low convective heat transfer coefficient. However, the actual convective area requirement necessitates rigorous calculations and heat transfer in the shock tubes to be done separately considering heat transfer both by convection and radiation. This calculation will also need a heat transfer area including fins in the convection tubes.

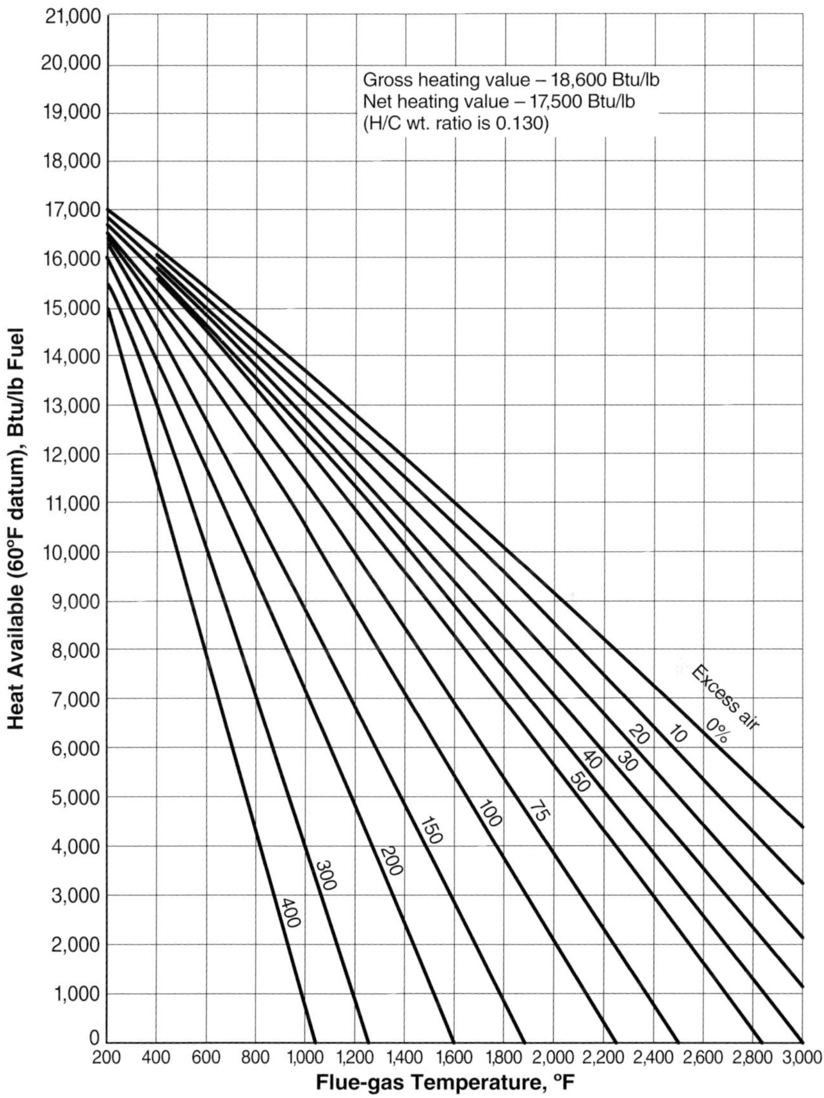

Heat available from the combustion of a 15° API fuel oil

Fig. 13.5.1 Heat available chart with flue gas temperature[2]

13.6 Solvent Recovery System in Solvent Extraction Units

13.6.1 Introduction

Solvent extraction units or extractive distillation units use solvents or entrainers. Some applications are mentioned below.

1. Extraction of benzene–toluene with sulfolane, N-Methyl-2-pyrrolidone (NMP) or other proprietary solvents in refineries and petrochemicals
2. Extraction of 1,3-butadiene with NMP, acetonitrile (ACN), or another proprietary solvent in naphtha and gas crackers
3. Extraction of styrene with proprietary solvent in naphtha and gas crackers
4. Extraction of aromatics from lube-based oils using furfural or NMP in refineries

In an extractive distillation unit for benzene–toluene separation with sulfolane or NMP as the entrainer, the entrainer phase is fed to the recovery column and the sulfolane or NMP is recovered from the bottom of the recovery column. The process is similar for ethylene glycol (EG) recovery: after the water dries from the natural gas (NG), EG is recovered from the bottom of the flash column used for recovery (sections 1.11 and 1.12 of Chapter 1). Therefore, the vaporization of large quantities of solvent that is used to modify the relative volatilities in ED columns is not required. This reduces the heat requirement in ED while recovering the solvent.

In solvent extraction units, the solvent is distributed in two phases: the extract phase and the raffinate phase. The extract phase is rich in solvent and the raffinate phase is lean in solvent. The solvent needs to be recovered from the extract and raffinate phases and recirculated again for extraction. In a lube oil complex, the solvents used are low boiling and recovered from the top of flash vessels or the columns of the recovery section. This process requires substantial heat for vaporizing the solvent. The heat content of the solvent vapors needs to be reused for energy conservation. For optimal energy, multi-effect flash vessels or columns are normally used. Increasing the number of stages (or effects) may save more energy, but at the cost of higher investment. Typically, two to three stages are optimal and accordingly used for industrial applications. The heat of solvent vapors is used to heat the feed or provide heat duty to the preheaters of preceding stages.

A furfural extraction unit (FEU) for lube oil production is shown with two stages of flash vaporization for the extract phase, followed by recovery under a vacuum and steam stripping under a vacuum (Fig. 13.6.1). The overhead heat of the first flash vessel, operating at 1.4 kg/cm^2 (a), is coupled with the feed to deliver the heat to the feed. The second stage overhead, operating at a pressure of 2.5 kg/cm^2 (a) and a temperature around 200°C, is also coupled with the feed and has the required hot end differential temperature of 10°C–12°C. Thus, all the heat of the solvent vapor is used to heat the feed and a part of the heat of the solvent vapor from the last stage is used to generate steam and is completely utilized.

A third additional stage between the two stages could have utilized the heat fully for feed heating. In such a case, the last stage overhead would have preheated the new second-stage feed (at around 2 kg/cm^2 (a)) for the recovery of NMP (see Fig. 13.6.3).

The pressure of the last stage is normally governed by the degradation temperature of the solvent. The degradation temperature of furfural is around 230°C. A gap of 10°C–20°C below the degradation temperature (230°C) is to be the saturation temperature of furfural in the last flash vessel. This fixes the temperature and consequently, the pressure of the last flasher at 2.5 kg/cm^2 (a). The first stage's pressure is fixed to have a hot end differential temperature of around 10°C–12°C across the overhead and feed heat exchanger (the first exchanger). The temperature is fixed at 173°C and the corresponding saturation pressure is the stage pressure (1.4 kg/cm^2 (a)).

The boiling point and pressure curve of the solvent, often called the boiling point rise (BPR) curve, is required for calculation and design. As the degradation temperature of furfural is low, the pressure range may also be low. The recovery of the solvent furfural from the extract phase for the extraction of aromatics from the lube oil fraction is given in Fig. 13.6.1.

The BPR curve of furfural is shown in Fig. 13.6.2.

An NMP aromatic extraction unit for lube oil production is shown with three stages of flash vaporization for the extract phase followed by recovery under vacuum and steam stripping under vacuum (Fig. 13.6.3).

The three flash stages can be used, as the higher degradation temperature of NMP (around 260°C) can provide larger temperature span. The overhead heat of the first flash vessel, operating at 1.6 kg/cm^2 (g) is coupled with feed to heat it to a temperature of 212°C. The second stage overhead, operating at 2.1 kg/cm^2(g) and 235°C, is also coupled with the feed. The third stage overhead, operating at 2.84 kg/cm^2 and 250°C, is used to heat the feed to the second stage. Thus all the heat from the recovered solvent vapor is used to heat the feed and is completely utilized. The number of stages will be decided based on the total recovery and reuse of the heat from the solvent vapor with nothing injected into the cooler/condenser.

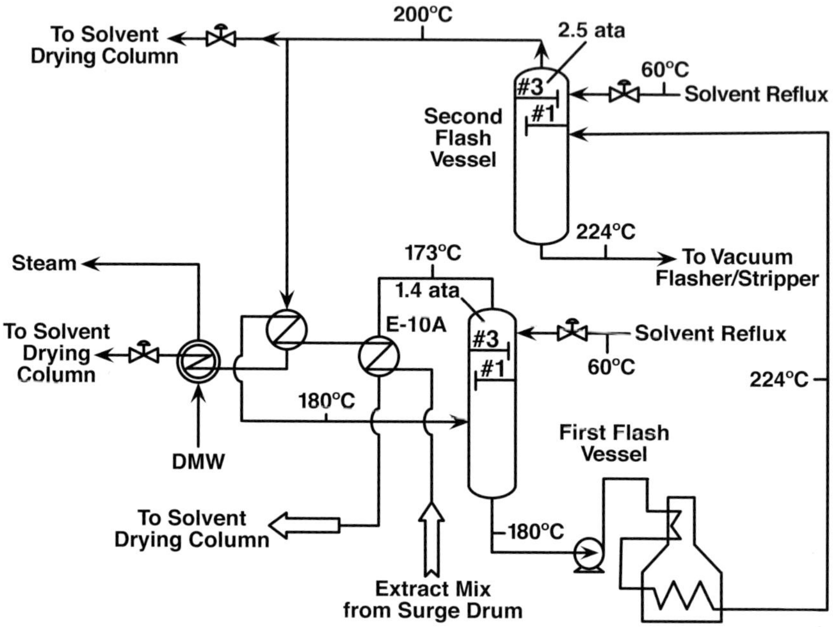

Fig. 13.6.1 Solvent (furfural) recovery from extract phase for aromatic extraction from lube oil fraction

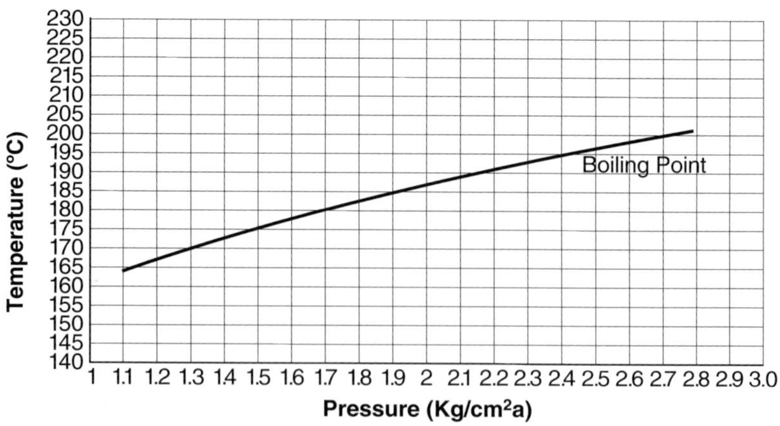

Fig. 13.6.2 Boiling point of furfural with pressure (BPR curve)

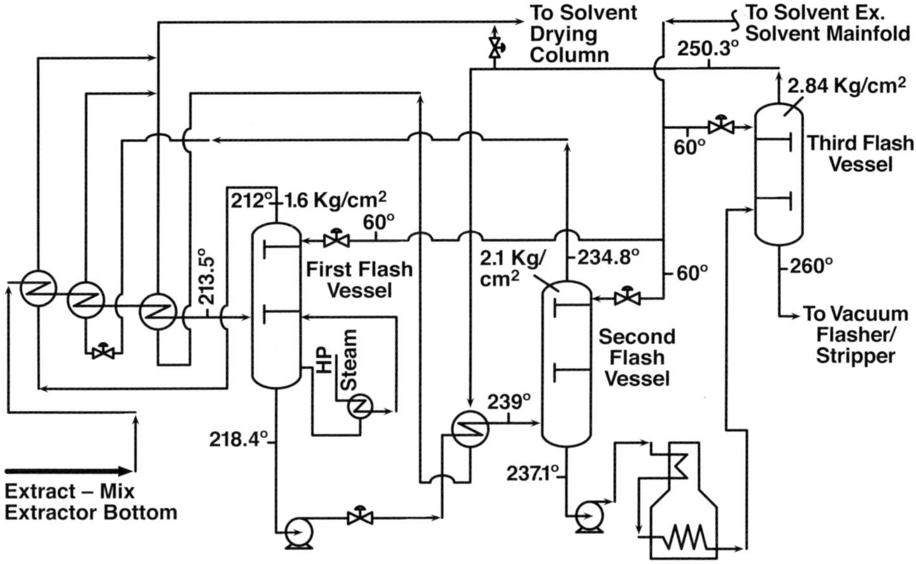

Fig. 13.6.3 Solvent (NMP) recovery from extract phase for aromatic extraction from lube oil fraction

The last stage's pressure will be governed by the degradation temperature of the solvent. The degradation temperature of NMP is 270°C. A gap of 10°C below the degradation temperature (250°C) is to be the saturation temperature of NMP in the last flash vessel. This fixes the pressure of the last flasher at 2.84 kg/cm^2(g). The pressure of the second stage is fixed to have a hot end differential temperature of around 10–12°C across the overhead and feed heat exchanger. The temperature of second stage is fixed at 235°C and the corresponding saturation pressure is the stage pressure (2.1 kg/cm^2 (g)). Thus the second stage's feed can be heated by the last stage's vapor with a temperature of 250°C. The first stage's pressure in turn is used to provide the hot end temperature difference of around 10°C with incoming feed.

Abbreviations

ACN: acetonitrile

BPR: boiling point rise

BTU: British thermal unit

EG: ethylene glycol
FEU: furfural extraction unit
LHV: lower heating value
MEK: Methyl ethyl ketone
NMP: N-Methyl-2-pyrrolidone
SDU: solvent dewaxing unit

Notes

1. American Petroleum Institute, *API 560: Fired Heaters for General Refinery Services*, 3rd ed., API standard (n.p.: American Petroleum Institute, 2001).
2. Reproduced from the article by Herbert L. Berman, "Fired Heaters III, How Combustion Conditions Influence Design and Operation," *Chemical Engineering* 85, no. 18 (1978): 133.

Isomerization 14

14.1 Introduction

Traditionally, straight-run light naphtha containing a pentane–hexane fraction was blended in motor spirit (MS). In recent years, MS has been manufactured by blending CRU reformate, FCCU gasoline, isomerate, and often alkylate and oxygenates to meet the revised quality requirement of MS (given in Chapter 12). The pentane–hexane components of the naphtha fraction is isomerized to enhance the RON and blended in the MS. Isomerate blended with MS helps to contain the aromatic content, olefin content, etc., within the specified levels, while at the same time meeting the RON of MS. Typical feedstock to the unit is C5-90°C cut straight-run naphtha. In this process of isomerization, the RON of the feedstock is typically increased to 87–92 from a level of 60–70 so that when it is blended with MS, the desired RON is met as well as the other typical requirements of the MS blend. In an isomerization unit, C5 and C6 paraffins in the feedstock are essentially converted to isoparaffins to boost the octane number of the feed stock. RONs for common components are given in Table 14.1.1 along with their normal boiling points (NBP).

The main reactions that occur in isomerization reactors are as follows:

1. N-C5 $\rightarrow$ I-C5
2. N-C6 $\rightarrow$ I-C6: 2-methylpentane (2-MP), 3-methylpentane (3-MP)
3. N-C6 $\rightarrow$ 2,2-dimethylbutane
4. N-C6 $\rightarrow$ 2,3-dimethylbutane
5. I-C6 (2-MP, 3-MP) $\rightarrow$ 2,2-dimethylbutane, 2,3-dimethylbutane

Table 14.1.1 RONs for common components

	RON	NBP, °C
Isopentane	93.5	27.84
N-pentane	61.7	36.07
Cyclopentane (CP)	102.3	49.25
N-hexane	31.0	68.73
N-heptane	0	98.43
2,2-dimethylbutane (2,2-DMB)	91.8	49.73
2,3-dimethylbutane (2,3-DMB)	101.7	57.98
2-methylpentane (2-MP)	74.4	60.26
3-methylpentane (3-MP)	75.5	63.27
Methylcyclopentane (MCP)	91.3	71.81
Cyclohexane (CH)	84.0	80.72
Benzene	114.0	80.09

I-C6 components (2-MP, 3-MP) have a low octane number, so in the deisohexanizer (DIH), they are drawn from the column and recycled along with fresh feed to convert the isohexane components to 2,2-dimethylbutane (2,2-DMB) and 2,3-dimethylbutane (2,3-DMB) to enhance the RON of the isomerate. The feed stock for the isomerization (ISOM) unit is produced through the fractionation of full range (C5-140°C) naphtha producing ISOM feed as the top cut and CCRU feed as the bottom cut. Typical C7 content in ISOM feed is around 10% wt maximum. A higher C7 content is likely to result in deposition coke in the reactors. Often full range naphtha (C5-140°C) is hydrotreated then fractionated into ISOM feed and CCRU feed. This avoids the requirement of separate feed naphtha hydrotreatment (NHT) units.

14.2 Description

Typical feed to the ISOM unit is constituted with C5-90°C cut straight-run naphtha, the top fraction of a CCRU reformate splitter unit (RSU) column, and often a side cut from the FCCU gasoline desulfurization unit (FCC GDU) column. In the RSU, the C_5+ reformate is split to remove the light reformate to lower the benzene content in the heavy reformate that goes for blending to MS. This benzene-rich light reformate is fed to the ISOM unit primarily to saturate the benzene. A side cut containing benzene components is often produced from the FCC GDU and this stream is also fed to the ISOM unit to saturate the benzene. The C5-90°C cut, along with the side cut from the FCC GDU, is first desulfurized in a naphtha hydrotreater unit to meet the sulfur specification for the isomerization catalyst before being fed to the ISOM unit. The feed is dried in a dryer to lower its moisture content to below 0.1 ppmw. Similarly, the makeup hydrogen used in the unit is also dried, almost to the same level. The hydrotreated feed and light reformate from the RSU are mixed with the hydrogen and preheated to around 90°C by heat exchange with the bottom of the isomerization reactors. The feed is then heated by around 115°C–120°C by a steam heater before being fed to the first reactor for benzene saturation and isomerization. Typically, two reactors are deployed. The first reactor saturates benzene and performs part of the isomerization, and the second reactor completes isomerization with a very low approach to equilibrium. Chlorine compounds are injected before the entry to the first reactor. The benzene saturation reaction is exothermic and the outlet temperature of reactor 1 (RX1) increases by around 40°C. The effluent from the reactor is cooled to around 110°C and fed to the second (equilibrium) reactor. The isomerization reaction that takes place in the second reactor is almost thermoneutral. The effluent from the reactor is cooled and fed to the stabilizer for stabilization. The stabilized isomerate is either fed to the DIH or to a molecular sieve to separate the isomerate and recycle the low-RON stream back to the reactors to further isomerize and improve the RON of the isomerate. A unit with a DIH is common (Fig. 14.2.1). The gases from the stabilizer reflux drum containing HCl are fed to the column with a caustic scrubbing section and a water-wash facility in the upper section of the same column. The liquid from the stabilizer reflux drum containing LPG components is fed to the stripper to remove lighter components and reduce the Reid vapor pressure (RVP) of the LPG. The recovered LPG is sent to storage through chloroguard.

The majority of the ISOM units typically consist of a reaction section followed by a stabilization section and a column (DIH) to fractionate the light isomerate and heavy isomerate and draw a cut from a location below feed to capture the components with a low RON (primarily 2-MP, 3-MP, n-Hexane). The low-octane components are recirculated to the reaction section to further isomerize the stock to high-RON components (2,2-DMB and 2,3-DMB) (Fig. 14.2.1). A typical configuration includes a benzene saturator reactor (with isomerization) followed by an equilibrium isomerization reactor (Fig. 14.2.2) for almost attaining equilibrium conversion. Typically, the reactors operate at a pressure of about 30–35 kg/cm^2. Often a deisopentanizer (DIP) is installed ahead of the isomerization unit to remove the isopentanes if the isopentanes are high (more than 8%–9% of the feed). This increases the equilibrium conversion of n-pentane to isopentane and reduces the hydraulic load to the ISOM unit. Thus, the installation of a DIP can also be used for capacity augmentation of the unit.

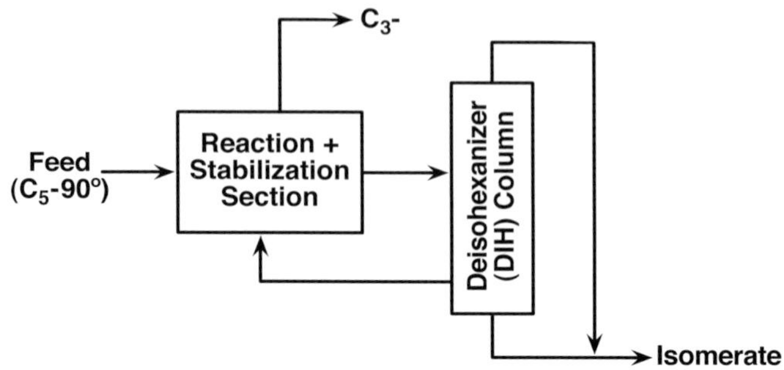

Fig. 14.2.1 Isomerization with DIH

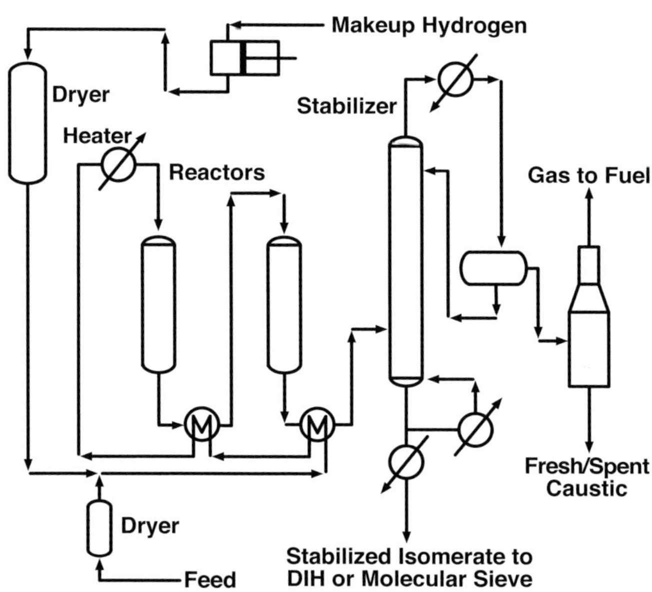

Fig. 14.2.2 Scheme of an isomerization reactor and stabilization section

The recycle draw from the column is typically taken from a location below the feed tray. Some heavy fraction builds up in the recycle stream and needs to be purged to the bottom. The unit does not need a recycle gas compressor. The catalyst used is a noble metal catalyst (platinum), which contributes to the isomerization capability of the catalyst. As a noble metal catalyst is used, the conventional unit has an extensive drying facility for the feed and makeup hydrogen to avoid moisture ingress along with the feed and makeup hydrogen and a feed hydrotreater upstream. Some configurations use molecular sieves to separate low-octane components for recycling in the reactors instead of having the DIH column.

Some isomerization units are known to use a zeolite type of catalyst to promote isomerization and do not need an acid (halide) activator. However, some amount of noble metal is also used in the zeolite catalyst for stabilization and normally operates with a higher reactor temperature. These types of units do not need the facility for the elaborate drying of the feed and makeup hydrogen and feed hydrotreatment, but the typical RON of the isomerate is marginally lower in these units (about 2 to 3 units).

Tentative guide line for simulation

A simulation of the isomerization unit can be carried out by a process engineer with the component analysis of the hydrotreated feed. All the components are defined in the feed along with the flow rate. The flow scheme in simulation can be developed where the feed is mixed with the specified hydrogen quantity and heated with RX1 effluent. It is further heated by reactor 2 (RX2) effluent and then heated with steam to enter RX1. The RX1 can be configured as a conversion reactor and all the benzene can be saturated by hydrogen to cyclohexane ($C_6H_6 + 3H_2 = C_6H_{12}$) and achieve around 80% completion of the isomerization reactions described in section 14.1. The effluent will be automatically generated and can be cooled to RX2 inlet temperature. The RX2 can be configured preferably as an equilibrium reactor (or at times a conversion reactor with defined conversions). The equilibrium constants for the reactions can be obtained from standard publications and pilot plant data. These conversions are functions of feed compositions and operating conditions of the RX2 and catalyst activity.

When the ISOM reactor is configured as an equilibrium reactor, a temperature approach of around 10°C over equilibrium can be considered. The RX2 outlet composition and temperature will be generated by the simulation. This will be cooled and fed to the stabilizer, and the bottom of the stabilizer will be fed to the DIH. The recycle will be generated from the DIH and will be recycled to RX1 inlet. The RON of the isomerate increases with an increased recycle of low-RON components.

The equilibrium conversion of N-C5 to I-C5 is around 64% at 260°C and increases to around 82% at 120°C in a mixture of N-C5 and I-C5. Similarly, the equilibrium conversion of N-C6 to I-C6 (2-MP, 3-MP, 2,2-DMB, 2,3-DMB) in a mixture of N-C6 and I-C6 also increases at a lower temperature, particularly the conversion of N-C6 to 2,2-DMB. The equilibrium conversion N-C6 to 2,2-DMB at 260°C is around 23%, and increases to around 40% at 120°C. The equilibrium conversion of N-C6 to 2,3-DMB increases marginally and the equilibrium conversion to (2-MP+3-MP) remains almost at the same level over the temperature range from 260°C to 120°C. Thus, to deliver the equilibrium conversion with a small temperature approach, a catalyst that has high activity at lower temperatures is preferred.

14.3 Case Study

An ISOM unit was not able to meet the specified isomerate RON of 87.5 after commissioning. The main reason was attributable to the change in actual feed compositions, mainly due to a very high content of naphthenes (e.g., methylcyclopentane [MCP] and cyclohexane [CH]), as compared to the compositions considered in the design. The guaranteed isomerate RON of the unit was revised and reduced to 85.0 from 87.5 by the licensors due to the change in feed composition to the unit. The original configuration of the unit is shown in Fig. 14.3.1. In the original configuration, a part of the recycle stream drawn is recycled to the first reactor and a part is recycled to bottom of the column (at tray 17 to contain the buildup of heavier components in recycle) after heat exchange with the feed (Fig. 14.3.1).

The isomerization unit's fresh feed analysis is given in Table 14.3.1.

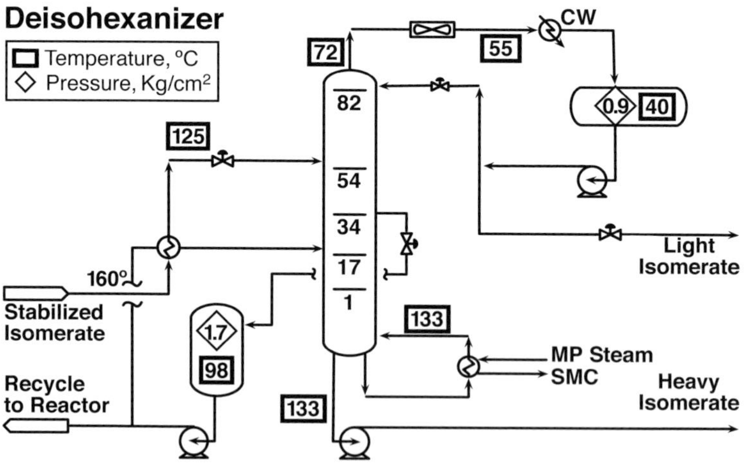

Fig. 14.3.1 Original scheme

Table 14.3.1 Fresh feed analysis compared to design

HC, %wt	Design as in process book	PGTR feed	Test run
n-paraffin	26.85	24.67	23.163
i-paraffin	52.74	39.27	36.065
Napthenes	12.63	30.37	34.880
Aromatics	7.78	5.71	5.892
Olefins	0.00	0.00	0.000
X factor	32.00	40.24	44.768
N6+A6	13.43	31.32	36.700
C5/C6 paraffin ratio	0.42	0.95	0.900

The fresh feed processed during the test run and the actual run was significantly different from the design feed. Very high percentages of naphthene components (MCP, CH, etc.) were found in the fresh feed (approximately 35 %wt as compared to around 12 %wt in the design). A higher C5:C6 paraffin ratio was found in the fresh feed (approximately 0.90 compared to approximately 0.42 in the design) as well as increased isopentane content (13.0–14.0 %wt compared to approximately 8.0 %wt in the design).

Higher naphthene content in the fresh feed inhibits the isomerization reaction of C5 and C6 n-paraffins and will result in low conversion of n-paraffins to isoparaffins, thus adversely affecting the isomerate RON. The higher the C5:C6 n-paraffin ratio, the lower the delta isomerate RON that can be achieved compared to the design, as C6 conversion to iso-C6 (2,2-DMB, 2,3-DMB) will allow a higher delta RON than C5 to iso-C5 conversion.

An increase in isopentanes in the fresh feed will reduce the equilibrium conversion of n-pentane to isopentane, as isomerization reactions to convert n-C5 to iso-C5 and NC6 to IC6 paraffins are equilibrium-controlled reactions.

A high quantity of naphthenes in feed, particularly MCP, which has a boiling point very close to that of n-hexane (desirable in the DIH recycle stream), will not be separated in the DIH effectively and may require higher reboiler and condenser duties to provide the desired separation. Thus, higher naphthenes in the feed may contribute to high naphthenic content in the DIH recycle stream. This will increase the reaction inhibitor load and further reduce the conversions.

The C6 content of the DIH recycle stream in the design and the test run is given in Table 14.3.2.

Table 14.3.2 DIH recycle stream's C6 content in design and test run

Components	Design (%wt)	Test run (%wt)
2-MP	33.91	18.43
3-MP	21.17	13.54
Hexane	14.54	11.05
Total	**69.62**	**43.02***

The purity of desirable components (2-MP + 3-MP + n-hexane) in the DIH recycle has been observed at around 43 %wt, which is considerably low compared to the 69 %wt in the design case. All of the above factors are decreasing the isomerate RON in the unit. Various alternatives were studied through a simulation of the entire unit along with the reactors to optimize the operation of the existing DIH column.

Alternatives

The following alternatives for optimizing the operation of the existing DIH column were studied:

Option 1. Enhancing reboiler and condenser duties to achieve higher 2-MP, 3-MP, and NC6 content in the recycle stream to reactor.

Option 2. Changing the location of the feed tray and DIH recycle drawoff tray to enrich the recycle with 2-MP, 3-MP, and NC6.

Option 3. Install a DIP to enhance the isomerate RON.

Option 1 and Option 3 required more intense modification so Option 2 was examined thoroughly. The tray compositions of the components were plotted. With the feed relocated to tray 34 and the recycle draw relocated to 53, higher concentrations of 2-MP, 3-MP, and the n-hexane were observed in tray 53 compared to the actual composition of the recycle. There was also a reduction of MCP and CH proportions in the recycle stream. Accordingly, it became imperative to draw the recycle from tray 54 (above feed tray).

The DIH recycle stream component analysis is shown in Table 14.3.3.

Table 14.3.3 DIH recycle stream component composition

Sl. no	Components	Design case purity (wt%)	Test run case purity (wt%)	Revised configuration purity (wt%)
1	23-DMB and minus	15.46	6.06	15.30
2	2-MP + 3-MP + N – C6	69.62	43.01	58.00
3	A6 + N6 + C7	14.92	50.93	26.70
	Total	100.00	100.00	100.00

The DIH top stream (light isomerate) analysis is shown in Table 14.3.4.

Table 14.3.4 DIH light isomerate composition

Sl. no	Components	Design case purity (wt%)	Test run case purity (wt%)	Revised configuration purity (wt%)
1	2,3-DMB and lighter	87.01	85.98	90.20
2	2-MP and heavier	12.99	14.02	9.80
	Total	100.00	100.00	100.00

The DIH bottom stream (heavy isomerate) analysis is shown in Table 14.3.5.

Table 14.3.5 DIH heavy isomerate composition

Sl. no	Components	Design case purity (wt%)	Test run case purity (wt%)	Revised configuration purity (wt%)
1	N-C6	3.01	3.21	1.70

The DIH tray relocation plan is given in Table 14.3.6.

Table 14.3.6 DIH tray relocation plan

Sl. no	Stream	Existing location (tray no.)	Revised/new location (tray no.)
1	Feed to DIH	54	33
2	Recycle draw	34	53
3	Recycle PA to column	17	Not required
4	Bottom recycle stream return		Below tray 1 reboiler side

Feed tray location was lowered from 54 to 33 (Fig. 14.3.2). The recycle draw tray was raised from the original location at 34 to 53 (Fig. 14.3.2). A condenser was added in parallel to the existing one in order to meet the higher condenser duty requirement, as the existing duty was found to be less than desired. Bottom pump capacity was augmented by the installation of an additional pump. The revised configuration of the DIH section of an isomerization unit is given in Fig. 14.3.2. In the revised configuration, a part of the DIH bottom is recycled to preheat the feed in the feed vs. recycle stream heat exchanger as in the original configuration. This required a higher flow of the column bottom liquid, necessitating an additional bottom pump. The modifications were small and had the flexibility to revert to the original configuration in case the original feed became available.

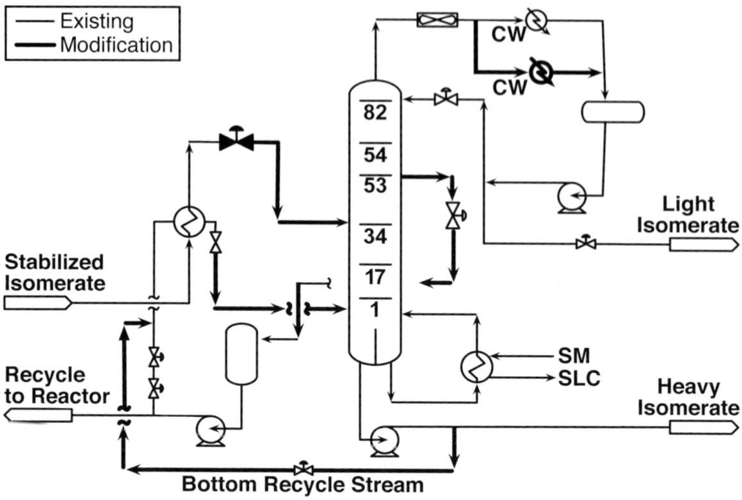

Fig. 14.3.2 DIH modified scheme

Abbreviations

CH: cyclohexane

DIH: deisohexanizer

DIP: deisopentanizer

IC5: isopentane

MCP: methylcyclopentane

NBP: normal boiling point

NC5: normal pentane

NC6: normal hexane

2-MP: 2-methylpentane

2,2-DMB: 2,2-dimethylbutane

2,3-DMB: 2,3-dimethylbutane

3-MP: 3-methylpentane

15 Troubleshooting

In this chapter, some selected troubleshooting case studies carried out by the author are described, along with some simple tools used for the troubleshooting. More case studies are described in different chapters of the book.

15.1 Troubleshooting Tools

The following calculations and procedures can be used as tools for general troubleshooting.

Pressure survey

Pressure drop ΔP is proportional to V^2, Q^2, and $(1/A^2)$.

V = linear velocity, Q = volume metric flow rate, A = cross-sectional area for flow

Heat transfer rate calculation:

$$Q = U_D \times A \times \Delta T_m = M \times C_p \times \Delta T$$

U_D = Overall heat transfer coefficient

A = surface area for heat transfer, ΔT_m = Mean temperature difference, C_p = specific heat, ΔT = temperature differential, M = the mass flow rate of the fluid

Synchronous speed calculation

$$n = (120\ f)/P$$

Where f = frequency in cycles/second, n is synchronous speed (rpm) and P = number of poles in the motor

Use of affinity laws (centrifugal device)

Flow rate Q proportional to n, d

Head (H) proportional to n^2, d^2

Power proportional to n^3, d^3

n is speed in RPM and d is the impeller diameter.

Net positive suction head (NPSH) available at the suction of a pump

$$= (P_s/\rho) - (P_v/\rho)$$

P_s = Pump suction pressure

P_v = Vapor pressure at pumping temperature

ρ = Density at pumping temperature.

For liquid at its bubble point, the elevation of the vessel or column containing the liquid only provides the NPSH available as the suction pressure and vapor pressure cancel out.

15.2 Good Design Practices

General guidelines

1. Velocity of fluids
   - Coke drum vapor space velocity: 0.3–0.6 ft/s
   - Column overhead line velocity: 15–20 m/s
   - Rich amine line velocity: <1.0 m/s
   - Lean amine line velocity: <2.0 m/s
2. Column draw-off nozzles and lines are required to meet self-venting criteria. This is required so that the dissolved gases do not get into the draw-off pumps and result in pump cavitations. This is discussed in detail in Section 2.10 of Chapter 2.
3. Residence time can be maintained in the column or vessel bottom spout when the liquid is picked up by a pump and fed downstream. A vortex breaker may also be advisable. Adequate residence time is required to eliminate the likelihood of unreleased vapor going to pumps and causing pump cavitation. The residence time should be estimated based on the volume of liquid between the low-level alarm (LLA) and high-level alarm (HLA). Typical residence time ranges from 2–10 minutes. Residence time is discussed in detail in Section 2.11 of Chapter 2.

4. Compressor suction lines are to be made self-draining toward the suction knock-out drum. Suction lines are to be designed in accordance with API 617 or API 618.
5. Equipment downstream of any pump is to have the minimum design pressure equal to the shutoff pressure of the pump + suction equipment maximum pressure (safety setting) + suction height.
6. Highly oversized safety valves may lead to column internal damage due to higher superficial velocities. It may be better to set the design pressure of the equipment and safety valve higher (minimum at 3.5 kg/cm^2 (g)). This will reduce the specific volume of the vapor at safety discharge pressure conditions and would reduce drag force (proportional to ρV^2) on internals. ρ is the vapor density at the pressure and temperature and V is the vapor or gas velocity.

Ways to avoid water in reflux to columns

Water in reflux causes many problems in distillation towers. The author has experienced many of the problems caused by water in reflux. This can be avoided by using the correct design for the reflux drum, which includes the points mentioned below.

1. The outlet line needs to be raised by 6 inches inside the vessel (with vortex breaker) (see Fig. 15.2.1).
2. The vessel should be large enough, with some room beyond that required by the design.
3. The gas line can be taken out with an impingement separator, if the flow is large.
4. The boot should be designed with adequate margin (over-designed).
5. Oil–water interface controller tapping is to be taken from the vessel (never from the line). Tapping the interface bottom from water outlet lines instead of the boot will result in low static pressure at the lower tapping due to the higher velocity in the water line and will show an incorrect interface reading. In one unit, this resulted in continuous oil ingress in the boot water going to sour water stripper.

A hat may be provided on the water outlet if the vessel is lined inside above the 6-inch protrusion.

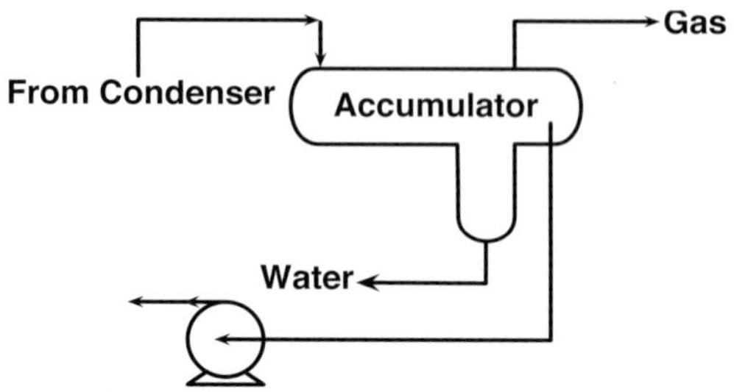

Fig. 15.2.1 Accumulator outlet pipe extension inside vessel

15.3 Case Studies of Pump Failure in Different Process Units

Case Study 1 (Frequent loss of suction of vacuum residue pumps at vacuum column bottom)

A set of two pumps located in the vacuum column bottom in the fuel-type vacuum unit of a refinery were losing suction frequently. They were vertical-split, multistage pumps with four stages in each. The suction line to each pump entered the pump vertically. A new set of pumps with a newer mechanical seal design was purchased to solve the problem and installed, but the loss of suction of the pump in the vacuum unit still continued. The supplier of the pump, along with the supplier of the mechanical seal and other pump experts, inspected the unit but could not come to an agreement on what was causing the loss of suction, so the problem continued.

To identify the root cause, a pressure survey was carried out with a compound gauge during the operation of the pump (see Fig. 15.3.1). An equation can be written as $P_2 = P_1 + h - DP_F$. The pressure is noted at P_1 and P_2 and the height (h) between points P_1 and P_2 is measured. The pressure drop due to flow (DP_F value) was calculated using the above equation and found to be almost negligible, indicating that there was no choking in the suction line or the column bottom coke trap. The pressure at point 4 (P_4, at the pump suction strainer downstream) in Fig. 15.3.1 was found to be 2 meters of liquid column (mLC), which is lower than the NPSH requirement (NPSHR) of 3.5 mLC for the pump. Thus, it was concluded that lower NPSH available was causing the pump to lose suction.

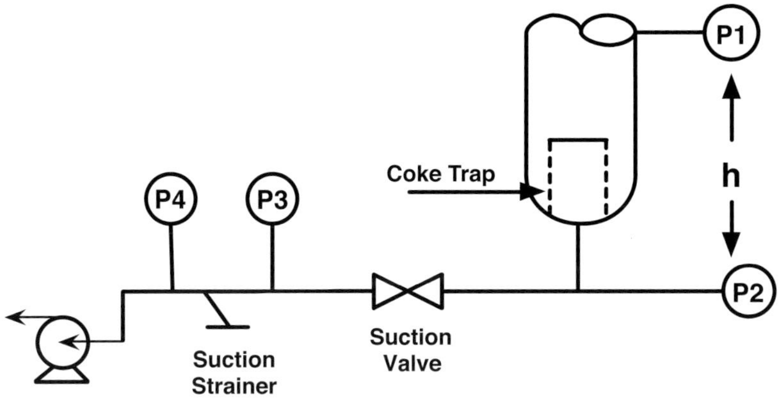

Fig. 15.3.1 Pressure survey of the pump suction line

The lower NPSH available was attributed to higher pressure drop across the strainer. But no muck was found in the strainer to cause the pressure drop. The inlet and outlet pressures were measured again with a calibrated compound gauge. The openings of the strainer were also checked and found to be around five times the cross-sectional area of the suction pipe, which is acceptable. The reason for the higher pressure drop across the suction strainer continued to be a mystery. The suction line oil-soaked insulation was removed by the maintenance group, possibly for a change of insulation, and the bare suction pipe section became visible. It was discovered that the strainer was too close to the reducer in the suction line. It was suspected that the convex part of the strainer may be leaving very low clearance for the flow due to its proximity to the reducer (Fig. 15.3.2). The suction spool was taken out and a modification was carried out to move the reducer away from the suction strainer and toward the suction of the pump (a distance of more than five times the diameter of the pipe). The new spool was fabricated and fitted and the pump tested. The pressure drop disappeared across the strainer and the strainer downstream pressure was found almost equal to the strainer upstream pressure and higher than the NPSH required by the pump. The problem totally disappeared and performance of pump became normal.

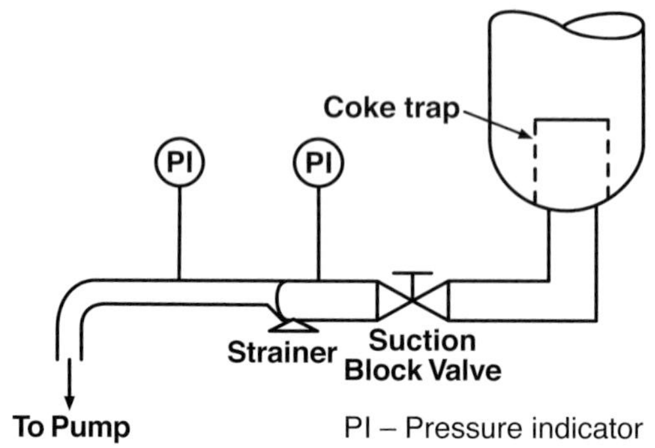

Fig. 15.3.2 Pressure survey of the pump suction line across strainer

Case Study 2 (Failure of pumps in a hydrocracker)

In a newly commissioned hydrocracker, the bottom pumps of the stripper and the main fractionator as well as some other pumps were repeatedly losing suction, making the unit unstable. The pumps had a suction line arrangement as shown in Fig. 15.3.3. As shown in Fig. 15.3.3, the strainers were installed from the side and the line to the pump suction was taken through a tee pipe below the strainer. It was decided to replace the pumps with new pumps with a lower suction NPSHR. The problem was revisited by the author and to diagnose the problem, a pressure survey was carried out in the suction line as shown by the points P_1 and P_2. The height between P_1 and P_2 is indicated by h. The calculation is as follows: $P_2 = P_1 + h \times \rho - DP_f$. The pressure drop due to flow (DP_f) was calculated from the measured data and was found to be negligible, confirming no choking in the lines or column outlet nozzles.

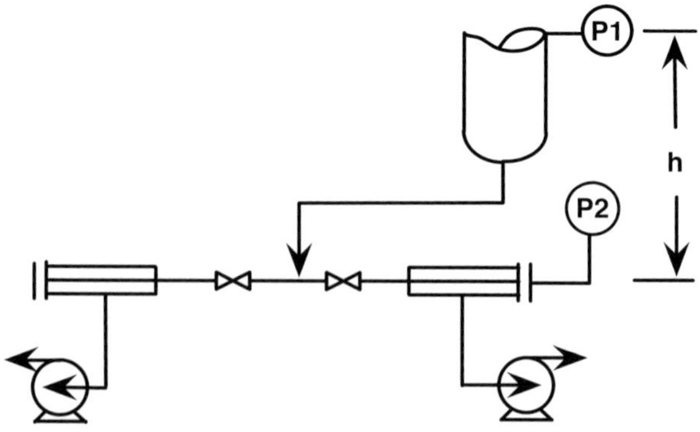

Fig. 15.3.3 Column outlet piping scheme and pressure survey

The strainer could not be inspected as the pump isolation could not be achieved because the suction valves were not holding properly. A drawing of the strainer was found and is reproduced in Fig. 15.3.4. The strainer was similar to a straight pipe with perforations, showing that the strainer design was not correct, as the perforations on the straight portion of the strainer would provide less flow area through the perforations (lower than the cross-sectional area of the suction pipe). The strainer design was modified with a curved surface facing the pump suction to provide more holes in the curved surface and consequently greater flow area (see Fig. 15.3.4). The unit was brought down to replace all the pump strainers with the modified strainer arrangement and the problem was resolved. Performance of all the pumps became normal.

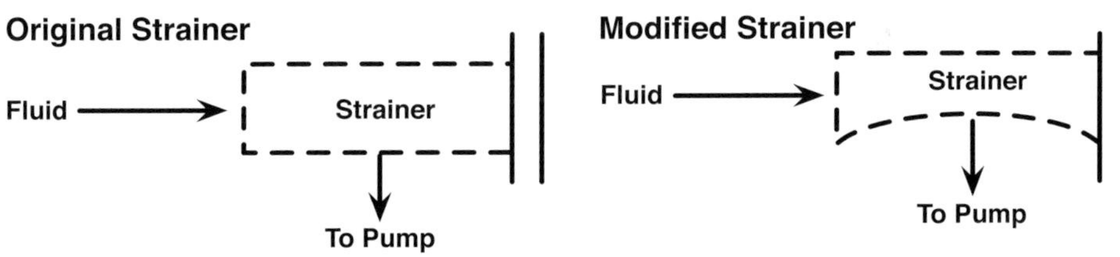

Fig. 15.3.4 Original and modified strainer

Case Study 3 (Frequent low flow in the hydrocracker unit high-pressure feed pump)

The problem was experienced in a two-stage hydrocracker unit (HCU). The feed flow to the unit would drop frequently in both stages, and the feed flow control valves had to be fully open to maintain the set flow of feed to the unit. The feed to the reactor section (high pressure) was supplied by the feed charge pump (high-pressure multistage). The first- and the second-stage feed pumps were both driven by motors through step-up gears, which are used to increase the speed of the pumps more than the motor speed based on the gear ratio. The first pump had a gear ratio of 1:1.6, and the second-stage feed pump had a gear ratio of 1:1.57.

It was found that the decrease in flow coincided with a frequency drop in the power supply by around 6%. The lower frequency of power supply resulted in a drop in the speed of the motor, which in turn lowered the speed of the pump considerably due to the high gear ratio. This reduced the discharge flow and pressure significantly. The actual speed of the motor was estimated from the synchronous speed at the low frequency and then actual speed was calculated from the specified slip and synchronous speed. The actual motor speed at lower frequency was found to be almost the same as that calculated. The speed of the pumps was evaluated using the gear ratio of individual pumps. The actual speed of the pumps was also found to be low and almost matching the estimated speed. The change in the head and capacity characteristics at a lower speed were estimated using affinity laws. The head and capacity estimated at a lower speed almost matched the actual performance of the pump at a lower speed. Thus, it was concluded that the reduction of motor speed and consequent lower pump speed was contributing to the decreased flow and discharge pressure. A pressure survey of the feed circuits was carried out as an additional check.

A higher pressure drop (3–4 kg/cm^2) was observed across one stop-check valve at the pump discharge line of both the first and second stages. The captive power plant supply was fully restored for the drives, delinking the power supply from the local area grid to ensure a nearly steady supply of frequency of 50 cycles per second. This reduced the problem of the first-stage feed flow considerably. The second-stage feed flow still had a problem with the frequency dropping to 48 cycles per second. The gear ratio was increased from 1:1.57 to 1:1.65 for the second-stage pump and the dead weight of the stop-check valve plugs was reduced, ensuring a lower pressure drop across the valves in both the stages. The problem was totally resolved.

15.4 Case Study of Repeated Rich Gas Compressor Failure in a CCRU

The rich gas compressors in a CCRU were having frequent problems. The compressors were the reciprocating type, with two stages driven by the same motor. Typically, the rich gas is produced from the separator of a CCRU and compressed, cooled, and refrigerated to condense the hydrocarbons and improve the H_2 purity of the gas. A CCRU produces typically around 3% weight of H_2 and very pure H_2 (95% mole) is recovered as a by-product from the CCRU through recontacting and chilling.

The rich gas compressors in this CCRU had many problems. One compressor broke and other two needed frequent repairs. The rich gas compressor piston rings were frequently getting worn, sticky material was found in the suction valves leading to passing of the compressor valves and high discharge temperature, and the compressors experienced high vibrations and frequent tripping. The unit was visited by the compressor vendor and an engineering contractor, and it was felt that the compressors needed to be replaced. The problem was revisited and the observations of the author are described below.

The compressor suction line arrangement is shown in Fig. 15.4.1.

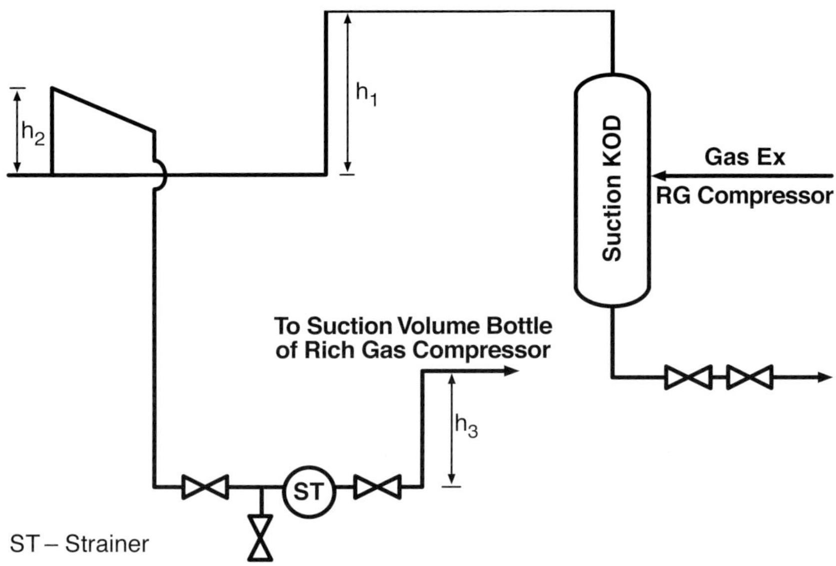

Fig. 15.4.1 Compressor suction line arrangement

The elevation of the main suction line was lower than the suction drum top, thus the suction line was not self-draining toward the separator and no loop was allowed in the suction line, which is the primary requirement as per API 618. Individual suction lines had loops before entering the suction volume bottle in the existing arrangement. Outlet from the suction volume bottle to the casing was drawn from the bottom of the volume bottle (no protrusion) in one of the compressors that broke. A continuous draining provision from the volume bottle is normally provided, but as the suction pipe from volume bottle was not extended (extension of 2 inches is suggested to avoid condensate entry into casing), the liquid ingress to the volume bottle had direct entry to the compressor casing, which could damage the compressor (Fig. 15.4.2). In the other two compressors, the protrusion was present as required. It was further found that the casing coolant water temperature was lower than the suction vapor temperature, which can condense the liquid as the vapor is saturated. API recommends that the water temperature be 6°C higher than the vapor temperature to avoid condensation of the vapor due to the low temperature of the cooling water. The compressor suction valve deposits were analyzed and tests showed that the content was mostly NH_4Cl. It was also found that the stabilizer off-gas was connected to the second-stage suction, presumably for the recovery of H_2 and higher hydrocarbons from stabilizer off-gas. It was suspected that chloride from the stabilizer off-gas was getting into the system and that NH_4Cl was getting deposited in the compressor valves.

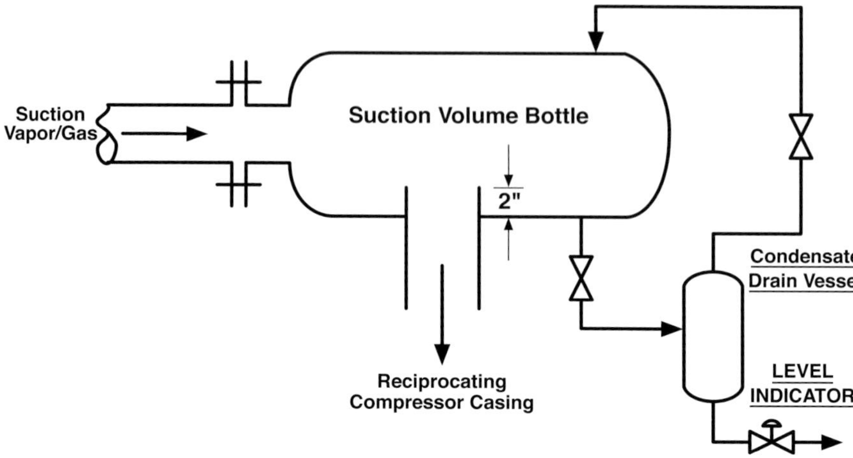

Fig. 15.4.2 Correct suction volume bottle arrangement of a compressor

The suction line needed to be self-draining toward the suction knockout drum. If this could not be done, as an improvised solution, a continuous drain was provided in the common suction line to drain any condensate formed or carried over from the suction knockout drum (Fig. 15.4.3). The steam tracer can be isolated in the common suction line to facilitate draining the condensate from the common suction header through a continuous drain. Loops in the individual suction lines needed to be eliminated but the steam tracing and insulation had to be kept in service so that no liquid could condense and get into the suction volume bottle. The casing cooling water temperature had to be increased by 6°C above the suction temperature to avoid the condensation of gas in the compressor casing. The volume bottle needed to be modified to raise the suction pipe inside the volume bottle by 2 inches in order to avoid any liquid entering the compressor casing. This would also ensure draining (integral draining) of the liquid through the volume bottle drain if there is any liquid ingress.

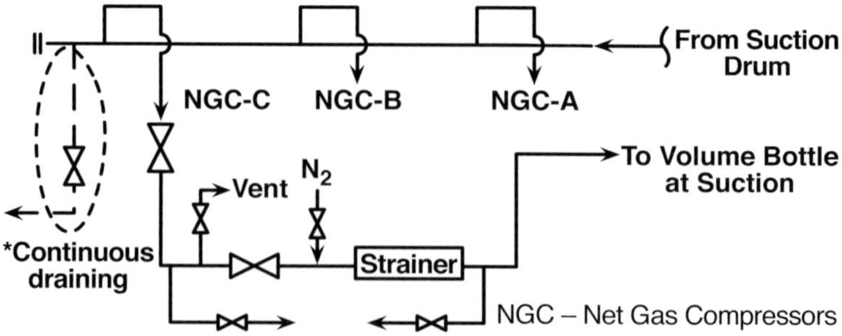

Fig. 15.4.3 Existing suction piping of the compressor

The stabilizer off-gas line was routed to fuel gas after a caustic wash to remove chlorides. Finally, it was suggested to install a chloroguard in the stabilizer off-gas line so that the gas could be rerouted back to the suction of the second stage compressor. Alternatively, a chloroguard could have been installed at the feed inlet of the stabilizer to remove all the chlorides from the stabilizer feed, but that would have been a costlier option. All the modifications were implemented, and the compressor is operating without any problem.

15.5 Case Study of the Failure of the Top Dome of a Crude Fractionator

A crude unit was operating for quite some time and had a topper column without a pre-topper upstream. Subsequently, a pre-topper was added ahead of the topper during a capacity revamp of the unit. The top dome of the topper thinned and developed a leak while in operation. It was learned that the topper was being operated at the same top temperature as before the revamp. The problem was analyzed and it was found that the operating top temperature of the topper was lower than the water dew point. The water dew point is likely to increase in the crude column (topper) with the installation of the pre-topper as a result of lower hydrocarbon partial pressure at the topper top. The lower hydrocarbon partial pressure was due to the decreased naphtha production from the topper (as part of the naphtha was recovered from the pre-topper column overhead). Thus, the topper top temperature was lower than the dew point of water. This caused severe corrosion in the column overhead as a result of water condensation followed by acid corrosion. The top dome was replaced and the Monel lining extended to cover more height from the top. An online dew point calculation block was installed to evaluate water dew point and the top temperature was continuously maintained at around 10°C above the calculated water dew point. The top pump-around reflux duty reduced, increasing the reflux ratio and the hydrocarbon partial pressure in the overhead of the topper. No thinning of the column top dome was observed thereafter.

15.6 Case Study where Vacuum Was not Achievable in a Vacuum Column

The required vacuum was not achieved when starting the vacuum distillation unit, even though the column was tested for vacuum before the feed was cut in and the test was satisfactory. Initially, higher pressure drop was suspected in the intercondensers and after-condensers of the ejector system. A pressure survey was conducted across the pre-condenser, intercondenser, and after-condenser. The pressure drop was found only marginally higher in the last condenser. It was cleaned and then taken online, but still, the vacuum was not achieved.

The reduced crude oil (RCO) feed was analyzed to find the flash point, which was discovered to be low. The RCO streams from all the crude units that were directly feeding the vacuum unit were then analyzed, and one showed a low value. The crude and RCO heat exchangers in the unit supplying the RCO were checked for leaks by analyzing RCO samples for flash point. Thus, the leaking exchanger was identified and isolated and the problem was solved.

15.7 Case Study of Heavy Corrosion in a Crude Column Overhead Circuit

This occurred in an atmospheric vacuum unit (AVU) where the RCO from the crude unit was directly fed to the vacuum distillation unit.

In a crude column, HCl corrosion was noticed in the overhead air coolers and heavy FeS deposits were found in the air cooler tubes and header box. The corrosion led to the thinning of the overhead air cooler tubes, which caused hydrocarbons to frequently leak outside. The column had a top pump-around reflux with a considerable duty. The problem was analyzed and it was found that the column was operated with a very low reflux ratio and no bleed of gas stream was maintained, as the gas would go to the flare and cause hydrocarbon loss. Low reflux ratio and lack of a bleed gas stream can cause high H_2S partial pressure in the overhead vapor. The high H_2S partial pressure may lead to H_2S accelerated acid corrosion and the same was corroborated by the heavy FeS deposits in the aircooler tubes and header box.

The chemical reactions that represent H_2S-accelerated HCl corrosion are as below: The H_2S in vapor phase can react with the $FeCl_2$ in aqueous phase and can regenerate HCl to freshly attack iron.

$$HCl + H_2O + Fe \longrightarrow FeCl_2 + Fe(OH)_2 + H_2$$

$$FeCl_2 + H_2S \longrightarrow 2HCl + FeS$$

The top pump-around heat duty was reduced to increase the reflux ratio and decrease the H_2S partial pressure in the overhead vapor. The column pressure was lowered and as a result, the gas bleed started from the reflux drum. The gas was routed to the suction of the wet gas compressor (WGC) of the fluid catalytic cracker unit (FCCU) to conserve the gas and recover LPG components from the gas. The corrosion problem reduced, and the system became trouble-free.

15.8 Case Study of FCCU Fractionator Flooding

An FCCU's capacity was revamped from 1.0 to 1.5 MMTPA. After the revamp, it was found that the fractionator bottom level became low with the increase in reactor temperature when operating at 100% capacity. The bottom level was reaching such a low value that it could cause a loss of suction in the main column bottom pump (Lawrence pump) and disrupt plant operation. To restore the bottom level, the plant operators would lower the reactor temperature, but then the bottom level would become very high, necessitating an additional pump-out of decant oil (DO). The problem and its solution is described in case study 1 in section 11.8 of Chapter 11.

15.9 Case Study of the Perils of Decreasing the Stripping Steam in a Crude Column

There was no problem apparent in the unit, but the liquefied petroleum gas (LPG) yield was low. Usually the stabilized naphtha's Reid vapor pressure (RVP) and the gas bleeds from the stabilizer, crude column, and pre-topper are monitored to maintain the LPG yield. In the unit, the stabilized naphtha RVP was found low and there were no unusual gas bleeds observed from the columns, but still the LPG yield remained low. Then attention was focused on the quantities of stripping steam to the main crude column and side strippers, as they were found to be quite low. It was learned that stripping steam quantities had been reduced below the design level to conserve energy. The unit had a kerosene splitter column to produce mineral turpentine oil (MTO) as top product and superior kerosene (SK) as bottom product. The problem was further analyzed and a continuous gas bleed was noted from the reflux drum of the kerosene splitter column. It was suspected that LPG components were escaping along with the heavy naphtha draw (110°C–140°C cut) and with the straight-run kerosene with fewer stages in the side stripper and less stripping steam in the main column and side strippers. It was suspected that the LPG components slipping into the straight-run kerosene were bled through the reflux drum when the kerosene was further fractionated in MTO and SK. The fractionation was helping to maintain the flash point of finished SK. The stripping steam to the side strippers and main column was increased to resolve the problem. The gas bleed from the kerosene splitter reflux drum stopped and LPG production from the stabilizer went up by 10–12 m^3/hr. The author has seen this phenomenon in two crude distillation units at two different locations where stripping steam was reduced considerably.

15.10 Case Study of a VGO Hydrotreater Unit Stripper that Experienced Frequent Choking

This was a VGO hydrotreater unit with around 20% conversion operating at around 80 kg/cm^2 pressure in the separator with a stripper column preceding the main fractionator. The unit had four drums in the effluent circuit: a cold high-pressure separator (CHPS), a hot high-pressure separator (HHPS), a cold low-pressure separator (CLPS), and a hot low-pressure separator (HLPS) (Fig. 4.5.1 of Chapter 4). A high pressure drop in the stripper was experienced after operating for 2–3 months, necessitating a shutdown. During shutdown, the stripper column was washed with water and the pressure drop was reported to improve after restart.

The problem and its solution is described in case study 1 in Section 4.5 of Chapter 4.

15.11 Case Study of Troubleshooting a High Pressure Drop in a Residue Fluid Catalytic Cracking Unit's Main Fractionator

One residue fluid catalytic cracking unit (RFCCU) in a refinery was operating with RCO as a feed with a Conradson carbon residue (CCR) of around 5–6 . This refinery was facing a frequent high pressure drop (DP) problem across the main fractionator. After water-washing during shutdown and in the subsequent start up, the pressure drop would disappear, then reappear after continued operation for a short duration. The refinery tried many things to resolve the issues, but the problem continued. The problem and its solution are described in case study 2 in Section 11.8 of Chapter 11.

15.12 Case Study of Troubleshooting a Partial Conversion Hydrocracker Effluent Circuit

In a partial conversion hydrocracker unit (one with 60% conversion), a new catalyst was loaded to increase the conversion to 65%. The surface area in the effluent cooling section had reached its limit. The CHPS temperature had increased from around 55°C to around 80°C. The unit had a high-pressure recycle gas amine scrubber. The gas inlet temperature to the amine scrubber increased, often causing heavy foaming. The recycle gas hydrogen purity dropped considerably, making it difficult to maintain the specified recycle ratio. The problem and its solution are included in case study 2 in Section 4.5 of Chapter 4.

15.13 Case Study of an HCGO Quality Problem in a Coker Unit

A coker unit was in operation producing gas, coker naphtha, light cycle gas oil (LCGO), and heavy cycle gas oil (HCGO). The feed to the unit was VR. Around 10% of the heavy stocks were recycled along with the secondary feed.

The LCGO was routed to a diesel hydrotreater (DHDT). The HCGO was routed to a partial conversion hydrocracker along with the VGO feed. The hydrocracker had around 30% HCGO in the feed stock. Then the capacity of the coker unit was increased by around 20% and the quality of HCGO became inferior and resulted in an increase of the frequency of filter backwash in the hydrocracker.

Operators tried increasing the column pressure to increase coke drum pressure to 1.4 kg/cm^2 (g) from 1.05 kg/cm^2 (g) to contain the drum overhead vapor velocity and avoid the likelihood of carryover from coke drum spoiling the quality of HCGO.

To analyze the problem, the coker unit drum and fractionator were simulated with the blend of actual liquid and gas products, the recycle, and the drum overhead quench to constitute the feed to the fractionator. The following points were checked:

1. The drum overhead velocity, which was found to be around 0.55 ft/sec below the typical maximum allowable velocity. Thus, it was concluded that the drum overhead velocity was not likely to cause any carryover from the coke drum.
2. Loading in the fractionator wash zone was configured with three stages (six bubble cap trays and five shed decks), with a system factor of 0.85. The diameter in the wash zone was found to be just adequate. Thus, any deposition of coke in the cross-flow trays (bubble cap trays) may induce flooding in the wash zone due to inadequate diameter.

The vapor–liquid profile was generated from simulation and sent to the internal expert to check the adequacy of the column and the wash zone in particular.

The additional field checks carried out to eliminate other likely reasons for the problem are listed below.

1. The feed surge drum had a breathing connection to the column. The breathing line was connected near the column wash zone (at the same elevation as the shed decks), so the level of the drum was checked to obviate the likelihood of surge drum overflow and the ingress of VR fresh feed to the column shed deck trays.
2. Pressure of the HCGO pump-around side and VR fresh feed in the heat exchangers was checked for any likelihood of VR leak to the HCGO pump-around circuit and consequent ingress of VR into the HCGO through pump-around return. It was found that the HCGO pump-around side pressure was higher than the VR by around 3 kg/cm^2(g) and HCGO product side pressure was still higher. Thus, VR ingress in the HCGO or HCGO PA was not likely even in the case of leakage from the HCGO vs. VR heat exchangers.

3. The HCGO sample was found to contain coke fines.

The wash section of the column had five shade trays just above the flash zone, followed by six bubble cap trays above. It was felt that the deposition of coke particles in the downcomers would lower downcomer clearances, partly choke the riser ports of the bubble caps, and cause flooding and the carryover of coke fines in the HCGO. Thus, it was suggested to change the wash zone to a grid and packing assembly (counterflow device) to reduce the accumulation of coke particles in the wash zone. Accordingly, the same was implemented, the HCGO quality improved, and the frequency of filter backwash in the hydrocracker reduced to normal.

15.14 Case Study of a Corrosion Problem in a Diesel Hydrotreater

A diesel hydrotreater was designed to process a mixture of straight-run gas oil (SRGO) and coker unit LCGO. It had two parallel trains of reactors and effluent heat exchanger trains terminating in a common HHPS. The HHPS outlet vapor was routed directly to the high-pressure air cooler (HPAC) and then to the CHPS, and the HHPS bottom liquid was routed to the HLPS and was fed to the stripper after preheating to around 250°C. The two parallel effluent trains had two Packinox exchangers heating the feed and recycle gas mixture before it was heated in the furnace prior to its entry to the reactors. The common HHPS is located downstream of the effluent vs. feed Packinox heat exchangers. The effluent temperature at the outlet of the Packinox was around 180°C; accordingly, the HHPS temperature was also 180°C. There was a gradual increase in the pressure drop across the effluent side of the Packinox. To resolve the problem, the refinery introduced water injection at the inlet of the Packinox in both trains. The unit already had the normal provision of a water injection at the inlet of the HPAC.

Severe corrosion was noticed in the stripper upper section. The corrosion product was analyzed and it was found to have been caused primarily by NH_4Cl. The suction strainer of the pressure recovery turbine (PRT) located at the downstream of the HHPS going to the HLPS had frequent deposits of solids, making the operation of the PRT almost impossible.

It was found that NH_4Cl was likely to deposit in the effluent side of the Packinox and might increase the pressure drop across the Packinox and eventually get into the HHPS liquid and choke the strainers of the PRT downstream of the HHPS. It was found from the NH_4Cl deposition charts

(Chapter 4, Figs. 4.10.1 and 4.10.2) that a small amount of chlorides (1 ppm) in the hydrotreater feed can deposit as NH_4Cl in the Packinox at the outlet temperature of the Packinox (180°C) and get into the HHPS liquid. From there it can easily get into the HLPS, which, when introduced in the stripper, can cause corrosion. The only solution to the problem would be to increase the Packinox outlet temperature and consequently the temperature of HHPS. A minimum temperature of around 235°C–240°C would be necessary to avoid the problem.

A higher HHPS temperature could be achieved by sending part of the effluent through the bypass across the feed and effluent Packinox (facility already provided in the existing scheme), but this would reject additional heat to the HPAC and reduce the feed preheat at the inlet of the reactor furnaces. The furnace heat duty requirement would increase and quickly reach its limit. Thus, as a solution, it was suggested that new heat exchangers be installed at the following locations (Fig. 15.14.1):

1. At the hot vapor line from the HHPS. The vapor from the HHPS would heat the feed and the recycle gas mixture, and cool the effluent to around 180°C (original level) before it is routed to the HPAC with a water injection. The preheated feed would be routed to the feed and effluent Packinox in the one train for further heating.
2. At the downstream (in the HLPS liquid line) to heat the feed and recycle the gas mixture before it is routed to the feed and effluent Packinox in the other train for further heating.

The HPAC duty requirement increases as the HHPS vapor quantity increases with the increase in HHPS temperature. This increases the CHPS temperature by 5°C–10°C, so a water cooler was installed in the CHPS vapor line to cool the recycle gas back to the original temperature level of 55°C.

Thus, two sets of new feed and recycle gas mixture vs. effluent heat exchangers were proposed: one at the HHPS vapor and the other at the HLPS liquid to the stripper. The installation of the new heat exchangers would also make it feasible to impart almost the same amount of heat to the feed and recycle gas mixture to maintain the reactor furnace inlet temperature. The stripper feed inlet temperature would likely reduce to some extent, as the HLPS liquid quantity would reduce as a result of the HHPS operating at a higher temperature. The stripper feed vs. stripper bottom heat exchanger have a margin in the surface area to keep the stripper feed temperature at the same level by reducing the steam generated by the stripper bottom in the downstream circuit.

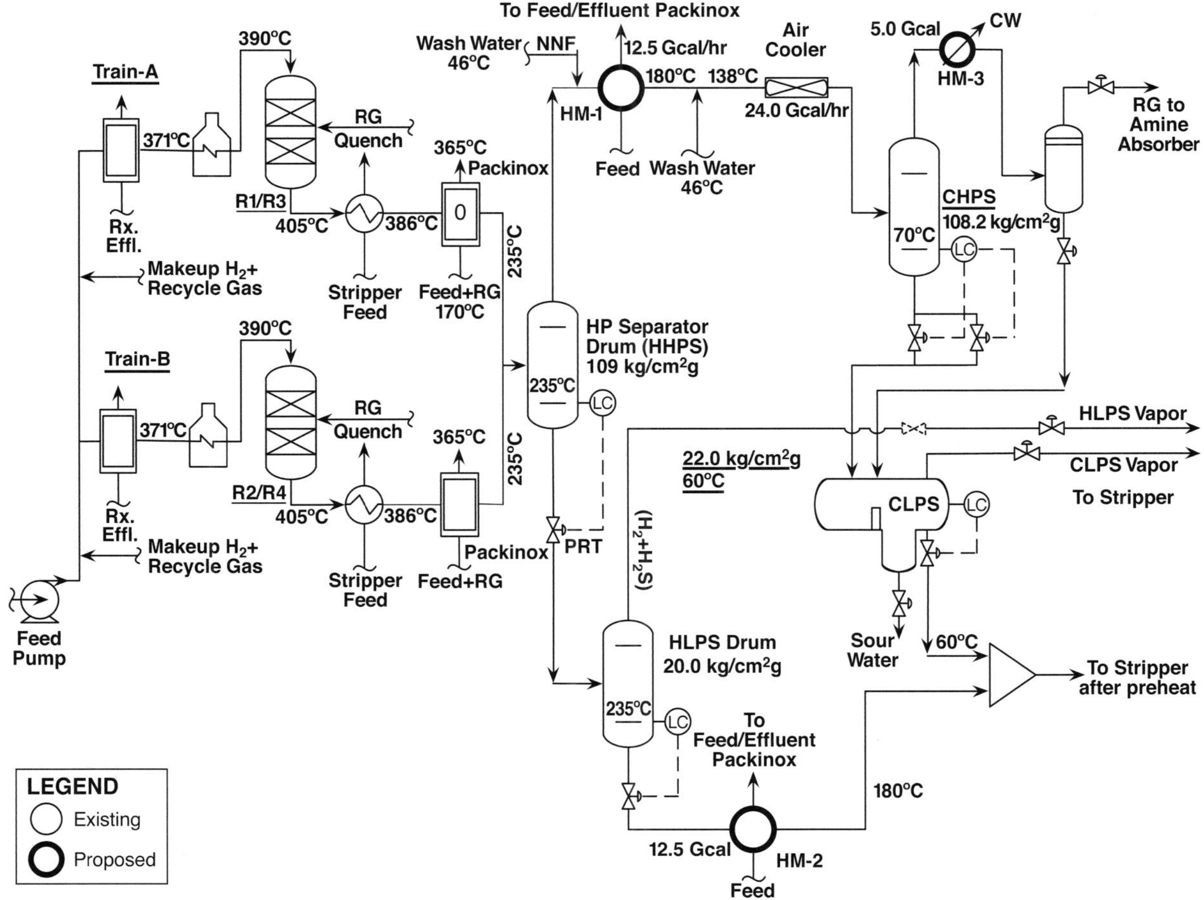

Fig. 15.14.1 Hydrotreater scheme showing modifications

In a hydrotreater, if an HHPS is provided, it can only be installed at around 240°C or higher to avoid the deposition of NH_4Cl at the upstream. The same may also apply to low-conversion hydrocrackers or VGO hydrotreaters, as the vapor quantities and consequently, the hydrocarbon partial pressure will be less and, in turn, increase the partial pressures of NH_3 and HCl in the vapor, which favors the deposition of NH_4Cl at a higher temperature.

15.15 Case Study of Deactivating Catalyst in the Second Stage of a Two-Stage VGO Hydrocracker

The hydrocracker was designed for 100% conversion of VGO feed (with 800 ppm wt of total nitrogen). It had a first-stage reactor with three beds of catalyst to attain around 40% conversion in the first stage. After cooling in feed/effluent heat exchangers and a HPAC, the effluent from the first stage was routed to the CHPS and then to the CLPS. After heating, the CLPS product was fed to the charge heater and fractionator along with liquid from the HLPS from the second-stage.

The fractionator bottom was fed to the vacuum column for further recovery of diesel. The vacuum column bottom was pumped by a low-discharge pressure pump (around 12 kg/cm^2) and fed to the second-stage high-pressure pump suction to feed the unconverted oil to the second stage reactor after preheat.

Initially, the first stage was commissioned and stabilized, and later the unconverted feed from the vacuum column bottom was fed to the second stage. The second-stage catalyst average temperature (CAT) and reactor inlet temperature were lower than the first-stage inlet to facilitate increased aromatic saturation. The second-stage catalyst was to handle treated feed only (without nitrogen), but after the feed was introduced and the specified inlet temperature attained, no hydrocracking was observed, as indicated by the absence of temperature increase and hydrogen demand in the second stage. The reactor inlet temperature increased further but no rise in temperature across any bed was observed. Quenches were not introduced as there was no temperature increase.

The second-stage feed was analyzed for nitrogen and a nitrogen content of around 35–40 ppm was found. Potential sources of nitrogen ingress in the feed to the second stage were studied, but no source was found. Finally, it was noted that the vacuum bottom low-pressure pump used raw feed as the sealing fluid in the mechanical seal. It was suspected by the team that the raw feed, (with a nitrogen content of around 800 ppm wt) could get into the second-stage feed through the mechanical seal and increase the nitrogen content of the feed to second stage. A test for nitrogen at the inlet and outlet of the pump was carried out and an increase of nitrogen at the outlet was found and confirmed the suspicion. Accordingly, the raw feed to the mechanical seal was disconnected and the sealing liquid was changed to unconverted liquid from the first stage (with no nitrogen). The problem disappeared. It may be pertinent to mention that the sealing liquid for the mechanical seal finds an entry to the pump suction through the throat bush of the pump or equalizing holes of the impeller (for a single seal).

15.16 Case Study of Low Propylene Recovery in a Propylene Recovery Unit

The propylene recovery unit (PRU) was originally designed to recover propylene from the LPG of an FCCU. The original scheme had a fractionator to recover propylene from the top and butane and propane from the bottom. The propylene was routed to a carbonyl sulfide (COS) hydrolyzer and then to a stripper to strip off the lighter fractions. The bottom of the stripper was the propylene stream with a purity of around 70% by wt. The 70% propylene-rich stream was fed to the deethanizer column of the naphtha cracker complex along with the original olefin-rich feed stream of the stream cracker unit to increase propylene production. The mixture would then go to the depropanizer column (see Fig. 1.10.3 in Chapter 1).

Later, the refinery tried to recover additional propylene from the LPG of the coker unit, as the columns in the olefin plant had enough cushion to process marginally higher. Accordingly, the LPG from the coker unit was introduced along with the FCCU LPG on the same tray (tray 54) in the propylene fractionator with the same feed inlet temperature of 90°C–100°C after preheat (as in the original scheme). The propylene recovery dropped with the introduction of the coker LPG together with the LPG from the FCCU.

The unit was simulated with the same feed point for the LPG mixture and it was found that more propylene could be recovered if the reboiler duty was increased; however, the reboiler did not have the adequate surface area to supply the additional duty.

The propylene content in the LPG from the coker unit was much lower (around 14 wt%) than the propylene content in the LPG from the FCCU (around

30 wt%). Following the fundamentals of distillation, a simulation was run where the LPG from the coker was introduced at a lower feed point (tray 36) through a separate feeding circuit with a separate preheater at about 110°C. It was discovered that with the existing reboiler duty of 3.75 Gcal/hr, it was possible to recover almost all of the propylene from both LPG feed streams using different feed points. This method was executed and a much higher recovery of propylene with almost the same purity was obtained.

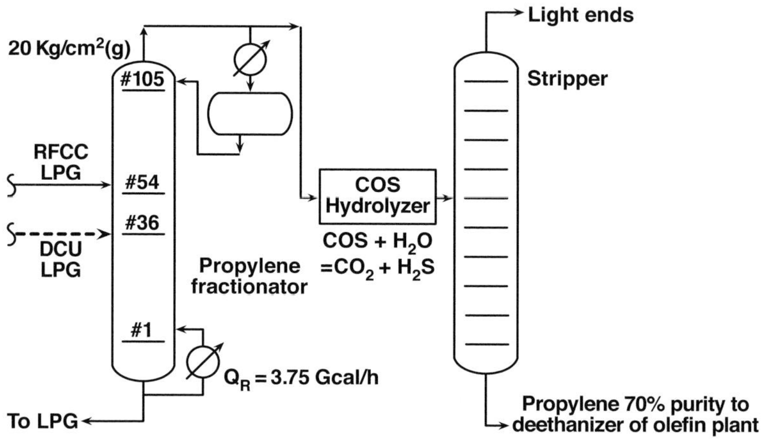

Fig. 15.16.1 Propylene recovery unit scheme with modification

Abbreviations

COS: carbonyl sulfide

HC: hydrocarbon

MTO: mineral turpentine oil

NPSH: net positive suction head

NPSHR: net positive suction head required

RVP: Reid vapor pressure

SK: superior kerosene

Index

A

B

C

D

E

F

G

H

I

K

N

O

P

Q

R

S

T

U

V

W

X